The Chemistry and Application of Dyes

Edited by
David R. Waring
Kodak Limited
Liverpool, England
and
Geoffrey Hallas
The University
Leeds, England

Plenum Press • New York and London

Library of Congress Cataloging-in-Publication Data

The Chemistry and application of dyes / edited by David R. Waring and Geoffrey Hallas.
p. cm. -- (Topics in applied chemistry)
Includes bibliographical references.
ISBN-13: 978-1-4684-7717-7 e-ISBN-13: 978-1-4684-7715-3
DOI: 10.1007/978-1-4684-7715-3
1. Dyes and dyeing. I. Waring, David R. II. Hallas, G. (Geoffrey) III. Series.
TP897.C523 1990
667'.2--dc20 89-27837
CIP

The majority of the synthetic examples described are taken from the published literature and it should be noted that claims, including those regarding yields and reaction conditions, are those of the published authors.

Before undertaking any of the described syntheses, the experimenter should consult the literature regarding the safe handling of the chemicals used. Where possible all syntheses should be carried out in a hood (fume cupboard) and personal protection should include gloves, safety glasses, and laboratory coat.

Softcover reprint of the hardcover 1st edition 1990

A Division of Plenum Publishing Corporation
233 Spring Street, New York, N.Y. 10013

The Chemistry and Application of Dyes

TOPICS IN APPLIED CHEMISTRY

Series Editors: Alan R. Katritzky, FRS
Kenan Professor of Chemistry
University of Florida, Gainesville, Florida

Gebran J. Sabongi
Laboratory Manager, Encapsulation Technology Center
3M, St. Paul, Minnesota

CHEMICAL TRIGGERING
Reactions of Potential Utility in Industrial Processes
Gebran J. Sabongi

THE CHEMISTRY AND APPLICATION OF DYES
Edited by David R. Waring and Geoffrey Hallas

STRUCTURAL ADHESIVES
Edited by S. R. Hartshorn

A Continuation Order Plan is available for this series. A continuation order will bring delivery of each new volume immediately upon publication. Volumes are billed only upon actual shipment. For further information please contact the publisher.

Contributors

S. M. Burkinshaw, Department of Colour Chemistry and Dyeing, The University of Leeds, Leeds, West Yorkshire LS2 9JT, England

P. F. Gordon, Fine Chemicals Research Centre, ICI Organics Division, Hexagon House, Blackley, Manchester M9 3DA, England

P. Gregory, Fine Chemicals Research Centre, ICI Colours and Fine Chemicals, Hexagon House, Blackley, Manchester M9 3DA, England

J. Griffiths, Department of Colour Chemistry and Dyeing, The University of Leeds, Leeds, West Yorkshire LS2 9JT, England

L. Shuttleworth, Research Laboratories, Photographic Products Group, Eastman Kodak Company, Rochester, NY 14650, U.S.A.

D. R. Waring, Kodak Limited, Acornfield Road, Kirkby, Liverpool L33 7UF, England

F. Walker, 43 Child Lane, Roberttown, Liversedge, West Yorkshire WF15 7QN, England. Retired, formerly of L. B. Holliday & Co. Ltd., Leeds Road, Huddersfield HD2 1UH, England

M. A. Weaver, Retired from Research Laboratories, Eastman Chemicals Division, Eastman Kodak Company, Kingsport, TN 37662, U.S.A.

Preface

It is particularly appropriate that a volume concerned with dye chemistry should be included in the series Topics in Applied Chemistry. The development of the dye industry has been inexorably linked not only with the development of the chemical industry but also with organic chemistry itself since the middle of the last century. The position of dye chemistry at the forefront of chemical advance has declined somewhat since 1945 and more markedly so during the last 15 years, with pharmaceutical and medicinal chemistry assuming an increasingly prominent position. Nevertheless, dye production still accounts for a significant portion of the business of most major chemical companies.

The field of dye chemistry has stimulated the publication of many books over the years but surprisingly few have concentrated on or even included the practical aspects of dye synthesis and application. Thus, the present volume is designed to fulfill that need and provide the reader with an account of advances in dye chemistry, concentrating on more recent work and giving, in a single volume, synthetic detail and methods of application of the most important classes, information which will be invaluable to both student and research chemist alike.

The volume is divided into eight chapters. The introductory chapter briefly chronicles the evolution of dye technology and the theory of color. Although the book is essentially practical, it is intended to direct the reader to literature concerned with the theory of color, a topic which has attracted considerable attention during recent years, in parallel with the increased availability of computing power. The following chapter discusses the classification or division of dyes by structure. Various structural classes of dye, notably azo and anthraquinone types, have been adapted, by careful selection of substituents, for application to a variety of fibers. Such fibers, by virtue of their own innate chemistry, place many demands on the dyes with which they can effectively be colored. The chapter discusses some of the chemistry associated with particular structural types.

Chapters 3–6 describe the dyes used for application to the four most important fiber types, namely cellulosic (essentially cotton), polyester, polyamide (nylon and wool), and polyacrylonitrile. Each chapter includes an account of the dyes used for the particular fiber and synthetic details for a selection of the structures under discussion.

Chapter 7 describes the methods used in the application of dyes to the major fiber types and their more important blends and concludes with appropriate practical details.

In some respects the final chapter points to the future of dye innovation in its discussion of nontextile applications of dyes. This important growth area for the dye industry has stimulated vigorous research activity in recent years, not least because it is in this arena that there is a greater potential for proprietary products and improved profit margins.

D. R. Waring
G. Hallas

Liverpool and Leeds

Contents

3. Dyes for Cellulosic Fibers

David R. Waring

4. Dyes for Polyester Fibers

L. Shuttleworth and M. A. Weaver

5. Dyes for Polyacrylonitrile

P. Gregory

6. Dyes for Polyamide Fibers

F. Walker

7. Application of Dyes

S. M. Burkinshaw

8. Nontextile Applications of Dyes

P. F. Gordon

1

Introduction: The Evolution of Present-Day Dye Technology

J. GRIFFITHS

I. BACKGROUND

The breadth and complexity of dye chemistry poses many difficulties for the non-expert. One way of achieving a better understanding of the subject is to examine how the chemistry and technology of the commercial synthetic dyes evolved, and this process of evolution can be considered from several different viewpoints. For example, there are those developments which were largely concerned with the application of the colorants to various substrates, as exemplified by the introduction of azoic dyeing or prereduced vat dyes. An alternative approach is to consider the discovery and development of dye chromophores, and this in turn is closely related to the historical evolution of color and constitution theory. The history of the synthetic dyes in fact consists of a complex interweaving of these separate considerations, linked closely with the contemporary development of theoretical organic chemistry, and controlled by social and economic factors in the developing industrial nations of the late nineteenth century. For the purposes of the present brief review, the development of new dyes and dyeing methods will be examined in a chronological sense, and the development of present-day understanding of the molecular origins of color will be discussed separately but with cross reference to the practical developments being made in the industrial arena.

One thing emerges clearly from such a study, namely, the great part played by dye chemistry in the development of modern structural organic chemistry.

J. GRIFFITHS • Department of Colour Chemistry and Dyeing, The University of Leeds, Leeds, West Yorkshire LS2 9JT, England.

Perhaps the efforts of dye chemists were not always motivated by purely altruistic considerations, but the energy and drive they gave to practical organic chemistry encouraged the work of academics and, after the first burst of new dye introductions in the 1850–1880 period, when structural organic chemistry lagged far behind practical developments, theory made rapid advances and was in a position to control the direction of dye research by the early 1900s. Thus, the history of dye chemistry is not only an interesting introduction to modern dye chemistry, but also has much to teach us about the structural organic chemistry that we now take for granted.

II. THE HISTORICAL DEVELOPMENT OF SYNTHETIC DYES

A. The Early History up to 1865

Prior to the mid-nineteenth century, virtually all colorants originated from animal, vegetable, or mineral sources and frequently had to be shipped over considerable distances and were subject to the vagaries of nature. The economic limitations of these colorants became increasingly apparent in the late eighteenth century as the industrial revolution took hold in Europe and, by the 1850s, the textile industry was ripe for the introduction of cheaper synthetic dyes with a greater reliability of supply. Two exceptionally valuable natural dyes were Madder and Indigo, and these presented a major challenge to dye chemists. The eventual replacement of these dyes with their synthetic equivalents marked two important milestones in the history of synthetic dyes.

Madder was obtained from the root of the *Rubia tinctorum* plant, the principal coloring component being alizarin **(1)** (1,2-dihydroxyanthraquinone). Cellulosic fabrics were dyed with this material by a mordant dyeing process, commonly employing alum, thereby forming an insoluble red metal complex of alizarin in the fiber. The popular Turkey Red process of this type was both complicated and tedious, and could take up to one month to produce the finished, dyed cloth. Although it gave bright red dyeings of excellent fastness properties, even by present-day standards, the process economics were hardly ideal. Madder was produced in Turkey, but much had to be shipped from further afield, notably the British East Indies, and this also was an economic disadvantage.

Indigo **(2)** was by far the most important natural blue colorant and, like Madder, was of vegetable origin. The major source of this dye was the Indian plant *Indigofera tinctoria,* the leaves of which are rich in indoxyl **(3)**, as a glycoside derivative. Fermentation of the leaves liberates free indoxyl, and this is rapidly oxidized by air to the desired, water-insoluble indigo. In Europe the woad plant had traditionally provided indigo for dyeing, but the Indian product began to be imported by the twelfth century, and by the 1850s all cotton was dyed with the Indian dye. Wool, however, continued to be dyed by the woad vatting

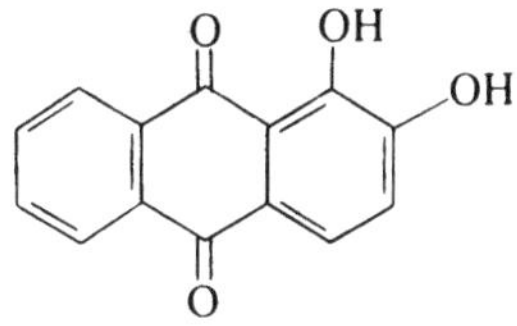

1

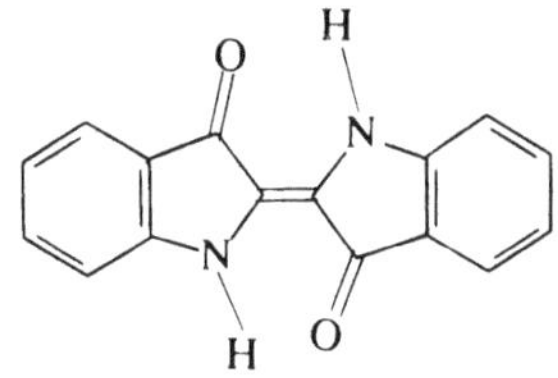

2

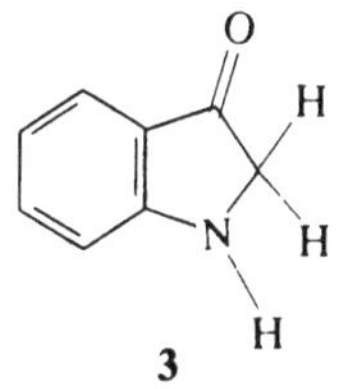

3

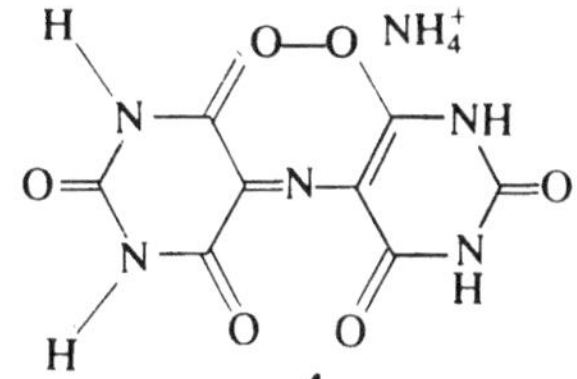

4

technique, as this used milder conditions which were less harmful to the fiber. The superior wash- and light-fastness properties of indigo meant that none of the early synthetic blue dyes posed a threat to the natural dye. However, chemists were eventually to develop an economical synthesis of indigo which had a disastrous effect on the natural indigo industry.

Before these major attacks on the natural dyes could take place, many earlier advances in dye chemistry were to occur, starting as early as 1771, when Woulfe began his investigations on indigo and discovered picric acid, arguably the first "unnatural" colorant, by reacting the natural dye with nitric acid. Although it was found that picric acid dyed silk a greenish yellow, it was not until 1845 that an economical route to picric acid from phenol was found and the "dye" could be commercialized. It was not a success however because of its poor lightfastness, but it did find limited use for shading indigo dyeings to give bright greens. Another dye that was discovered too soon for commercial use was Murexide **(4)**. This was first made by Prout in 1818 by reaction of uric acid with concentrated nitric acid followed by ammonia, but its commercialization as a mordant dye had to wait until 1851 when an economic route was found. The fastness properties were very poor, however, and it was not a success.

Indigo attracted the attentions of several chemists and, as a consequence, was responsible for the discovery of that most important dye intermediate, aniline. Thus, in 1826 Unverdorben found that "distillation" of indigo gave aniline and it is not surprising that the early dye chemists used aniline as a starting material in order to recreate colored materials. However, aniline was to remain a rare chemical curiosity until an important sequence of events had taken place. Thus, in 1825 Faraday discovered benzene, and by 1845 Zinin and Hofmann had found how to nitrate this chemical and how to reduce the product to aniline. Finally, the extensive work of Mansfield in the late 1840s established that coal tar provided an abundant source of the requisite benzene. The result of these separate investigations was to prove one of the major factors in the success story of Perkin's Aniline Purple.

The background to Perkin's discovery has been documented on numerous occasions and need only be summarized briefly here. At the age of eighteen, while on vacation from the Royal College of Chemistry during the Easter of 1856, William Henry Perkin was carrying out experiments on the dichromate oxidation of aniline in his home laboratory. From the unprepossessing black solid so formed he was able to isolate a small amount of a purple dye which evidently caught his imagination. The dye gave bright purple shades on silk, and he sought the opinion of a commercial dyer called Pullar, who gave a favorable response. The process was rapidly patented (by August of the same year) and Perkin relinquished his position at the Royal College in order to set up the firm of Perkin and Sons with his father, G. F. Perkin, and brother, T. D. Perkin, for manufacturing the dye on a commercial scale. It should be emphasized that the risks for such a venture were particularly high in those days before the Limited Liability Act of 1862 so that personal financial disaster was always a possibility. It is remarkable that Perkin senior risked the greater part of his life savings on the venture.

Work began on the plant at Greenford Green in June 1857, but this was not a straightforward operation as many technical problems had to be overcome before the dye could be made economically. Thus, large-scale nitration of benzene and the reduction of nitrobenzene had to be pioneered. The oxidation step had to be optimized, as did the dye isolation procedures, because the Aniline Purple was formed in extremely poor yields. By 1864 the process had been improved to the stage that the trade could be offered Aniline Purple as a crystalline solid. Perkin remained an active experimental chemist in spite of the wealth that his successful dye brought him. He went on to show that Aniline Purple actually consisted of two dyes, **5** and **6**, in varying amounts, the former being the main component. The structures were not deduced until many years later however, and it is interesting to note that **5** arose from toluidines present in the aniline available at that time.

If Aniline Purple proved a success in Britain shortly after its introduction, its acceptance in France was even more dramatic, but unfortunately for Perkin this brought him little financial reward. Perkin's French patent was invalidated, and consequently French manufacturers were able to step in remarkably quickly and produce the dye competitively. So successful were the French that Britain began importing the dye and the alternative name Mauveine, or Mauve, proved more

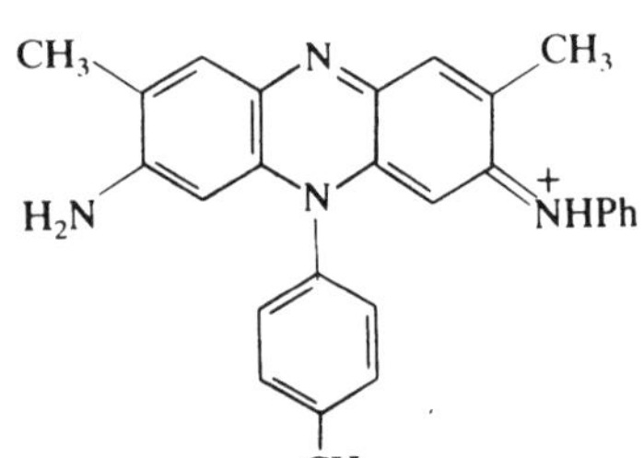

5

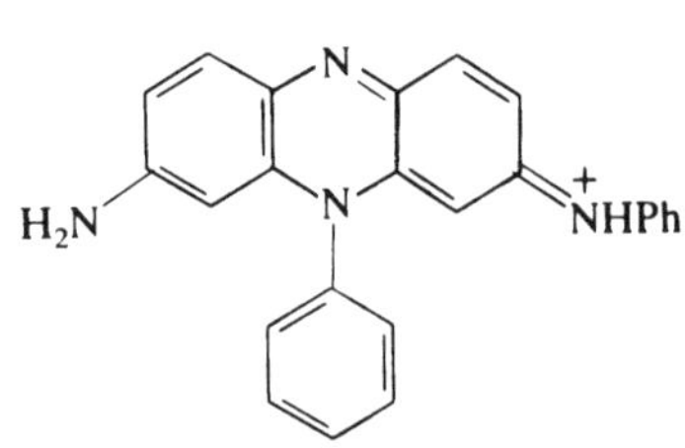

6

popular for the dye than Aniline Purple. Perkin's success with Aniline Purple had a great stimulating effect on the search for new synthetic dyes, both in Britain and France. Chemists concentrated on aniline as starting material and, adopting a purely empirical approach, since structural chemistry was in its infancy and little was known about the molecular composition of most organic compounds, they were to rapidly discover many new basic dyes with characteristically brilliant colors. For example, the first triphenylmethane dye, Magenta, was discovered in 1856 by Natanson, and by 1859 this was produced commmercially after Verguin had shown that it was formed economically by oxidizing crude aniline with stannic chloride. One of the principal components of the dye was homorosaniline **(7)**, together with rosaniline. It was Hofmann who showed that toluidine was an essential ingredient in the reaction. Magenta actually proved a greater commercial success than Aniline Purple, and it was soon manufactured in Britain by, for example, Perkin and Sons and the firm of Simpson, Maule and Nicholson.

Most other discoveries of the late 1850s and the 1860s were of the triphenylmethane type, such as Hofmann's Violet **(8)** (1863) and Girard's Violet Imperial **(9)** (1860), but the first glimmerings of a systematic approach to dye synthesis came independently from Nicholson and Gilbee in 1862 when they deliberately sulfonated *N*-phenyl derivatives of Magenta in order to obtain dyes with better water solubility. These products proved to be the first acid dyes for wool, and marked the beginning of sulfonation as a standard technique for preparing dyes with good water solubility and protein fiber affinity.

Not all discoveries in this period were of the triphenylmethane basic dye type, and a major breakthrough was made with the synthesis of the azo dye Bismarck Brown by Martius in 1863. The structure of this dye **(10)** was not really

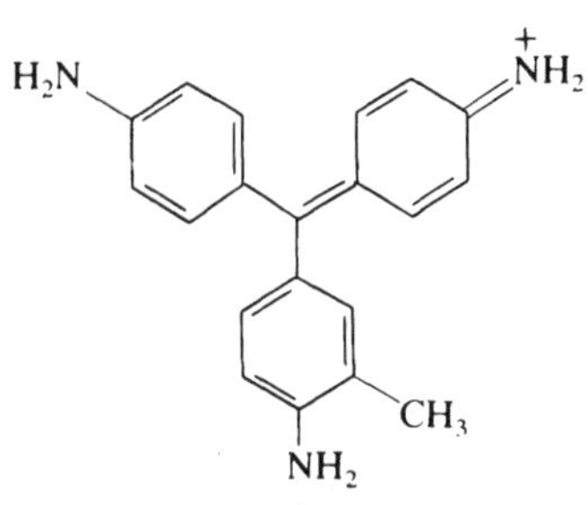

7

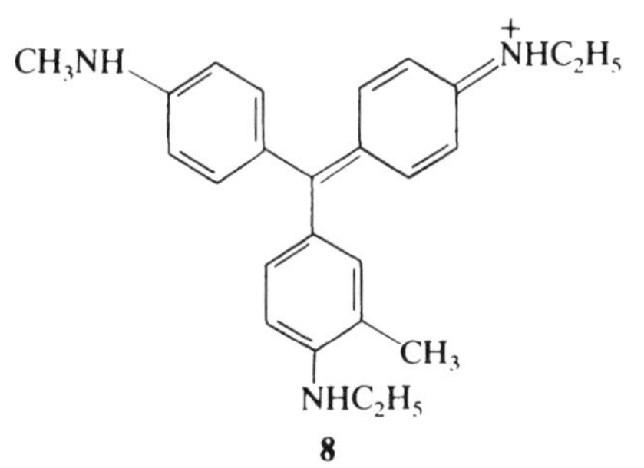

8

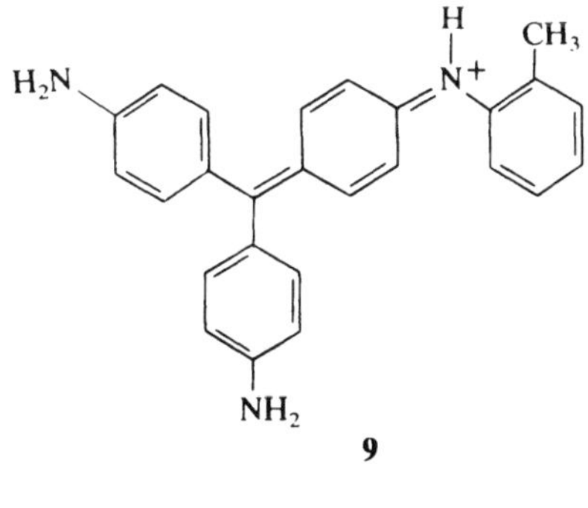

9

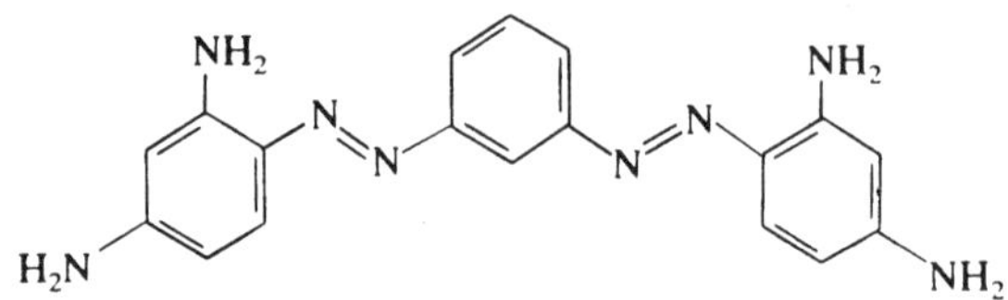

10

understood until Griess demonstrated the azo coupling reaction between diazonium ions and amines shortly after. Griess had in fact discovered the diazotization reaction in 1858, and it was in following this work that Martius stumbled across his dye. By reacting *m*-phenylenediamine with nitrous acid he established the conditions for a self-coupling reaction, so leading to **10**. These early experiments were to lead to the most important single chemical class of dyes in use today, i.e., the azo dyes.

By the mid-1860s chemists were directing more attention to the outstanding structural problems of organic chemistry and a more scientific approach to dye chemistry developed. No longer was the "try it and see" philosophy a commercial economic proposition. The need for systematic research was appreciated nowhere more keenly than in Germany. During the Aniline Purple and Magenta period Germany had kept a very low profile. The absence of any meaningful patent law in that country meant that German companies could make full use of British and French inventions, which was good for profits but did nothing to encourage industrial research. On the other hand, scientific education in Germany at that time was better than anywhere in Europe, as evidenced by the number of outstanding German chemists employed in foreign dye companies. When by 1865 the German dye industry began to assert itself, many of these chemists returned home, and the new scientific approach resulted in the rapid advance of the industry in that country, with a corresponding decline of the industries in Britain and France.

B. Developments in Dye Chemistry from 1865 to Modern Times

Although there is no clear demarcation between the era of empirical dye research and the era of a more deductive approach, the year 1865 perhaps deserves more than any to act as marker for the change, for it was in that year that Kekulé propounded the "cyclohexatriene" formulation of benzene,[1] and later went on to explain the aromatic character of benzene in terms of oscillating bonds. This major advance in structural organic chemistry was particularly pertinent for the dye chemist, as all known dyes at that time contained benzenoid residues and so paved the way for a greater understanding of the structures of dyes and the reactions which gave rise to them.

A particularly significant development in the post-1865 period was the laboratory synthesis of alizarin, which as we have seen is the principal ingredient of Madder. This advance was made by Graebe and Liebermann in 1868, some 37 years after pure alizarin had been isolated from Madder by sublimation. These workers were able to establish the basic anthraquinonoid structure of the dye and so embark on a designed synthesis from anthraquinone, involving bromination and alkali fusion. The fact that they were successful was fortuitous as the position of the two hydroxy groups in **1** was not known, but nevertheless this success did show the value of the scientific approach. The high cost of bromine prevented commercialization of this method, but others were stimulated to develop alternative syntheses that could compete economically with natural Madder. The breakthrough came in the following year and was made independently by Perkin and Caro. This new procedure involved sulfonation of anthraquinone in the 2-position followed by alkali fusion. An amicable agreement was reached between the two companies involved (Perkin and Sons in England and Badische Anilin und Soda-Fabrik in Germany) whereby production of synthetic alizarin was shared.

The new synthetic material proved to be extremely profitable, and eventually the Madder industry disappeared. In the period 1869 to 1873 Perkin and Sons had no real competition in Britain and production of the dye rose dramatically, to a high of 435 tons in 1873. However, this was to prove a turning point and, largely because of severe German competition, the company was sold in 1874 and production at Greenford ceased in 1876. Madder was the first important natural dye to become a casualty of the synthetic dye competition.

Although the diazotization and diazo coupling reactions were established by 1860, and Bismarck Brown had reached the market place, commercial azo dyes were generally slow to develop at first. Further work of a more academic nature had to follow before the potential of this group of colorants was properly appreciated. In 1866 Martius and Griess showed that the previously known dye Aniline Yellow, of uncertain structure, was 4-aminoazobenzene and was formed by coupling diazotized aniline to aniline itself, followed by acid-catalyzed rearrangement of the diazoamino compound. The coupling reaction between diazonium ions and phenols was discovered by Kekulé in 1870, and he established the structure of the hydroxyazo dyes by chemical methods. By the mid-1870s the versatility of the coupling reaction began to be appreciated by dye chemists and in 1875 Caro made the basic azo dye Chrysoidine; in the following year Roussin made the first acid azo dyes for wool, namely, Orange I and Orange II. From then on, azo dyes were introduced with increasing frequency and today they are the most important and widely used dye class.

In spite of the success of the scientific approach to dye research, there was still scope for the experimental chemist to make valuable chance discoveries, and this was well examplified by the discovery of the sulfur dyes by Croissant in 1873. He experimented with sawdust and other cellulosic materials, heating them with sodium polysulfide to give deeply colored products. These products, in their reduced form, had affinity for cotton and after dyeing they could be oxidized to the final color, the resultant dyeings showing good fastness properties. The class of chromophores was new, but the dyeing method closely resembled the vat dyeing technique and was readily accepted by conservative dyers. The first completely new dyeing method came in 1880 with the introduction of azoic dyeing.

This process was patented by the Read Holliday and Sons company of

Huddersfield, England, and involved saturating the fabric with an alkaline solution of a phenol or naphthol, and then treating the fabric with a solution of a diazotized aromatic amine. As the dyer had to carry out the diazotization process himself, the resultant dyes were called popularly "ice colors" and were very successful in calico printing. In the early 1880s cotton was still the most important textile material and its dyeing could be achieved by various methods, of which vat dyeing with indigo and mordant dyeing with alizarin gave the most satisfactory results as far as fastness properties were concerned. However, these processes were troublesome and gave very restricted shade ranges. The most convenient methods involving basic dyes on tannin-mordanted cotton gave inferior fastness results. Thus, a major step forward was made in 1883 when Griess developed disazo sulfonated dyes which had a natural affinity for cellulose without the need for a mordant. These then were the first *direct dyes* for cotton; in the following year, greatly improved dyes based on benzidine as tetrazo component were introduced by Böttiger, of which Congo Red **(11)** is perhaps the best known.

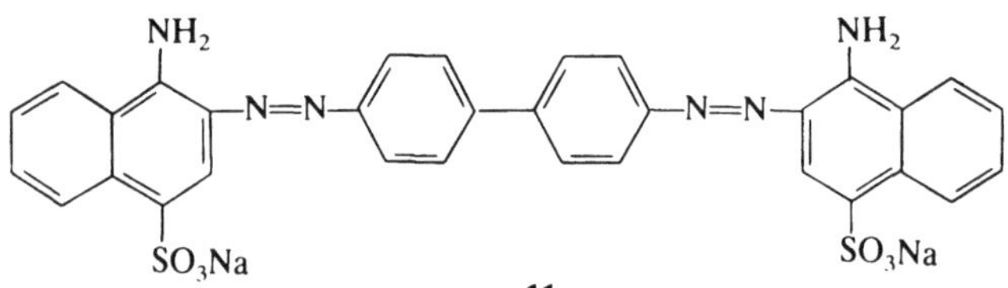

11

The mordanting process, such as the Turkey Red dyeing method, had been in use for thousands of years, but the first extension of this principle to azo dyes came in 1885. Thus, Meldola and Nietzki had prepared dyes from the coupling reaction between diazonium ions and salicyclic acid and had found that these would dye chrome-treated wool to give dyeings with good wash-fastness properties. This was the beginning of metal-complex azo dye chemistry and was to lead to many important future developments.

By the 1890s the dyer had a wide range of synthetic dyes to choose from but, even so, indigo still reigned supreme as the blue dye with outstanding fastness and was thus the most important natural dye still surviving. The European dye manufacturers were aware of the great rewards that would accrue from the discovery of a viable synthesis of indigo, and research toward this end intensified in the last decade of the nineteenth century. By 1890 the structure of indigo had long been established; the dye had been synthesized in the laboratory by Baeyer in 1878, using *o*-nitrophenylacetic acid as starting material. Neither this route nor a subsequently discovered alternative from *o*-nitrocinnamic acid were economical enough for commercialization, however, so that industrial research on indigo synthesis was initiated. Most prominent in this area was the German BASF company and, after an unprecedented investment of both capital and manpower, the company was able to place synthetic indigo on the market in 1897 as a serious alternative to the natural product. The manufacturing process was based on the conversion of anthranilic acid **(12)** into *o*-carboxyphenylglycine **(13)**, and cyclization of the latter to indoxyl by alkali fusion (Scheme 1). The economics were not

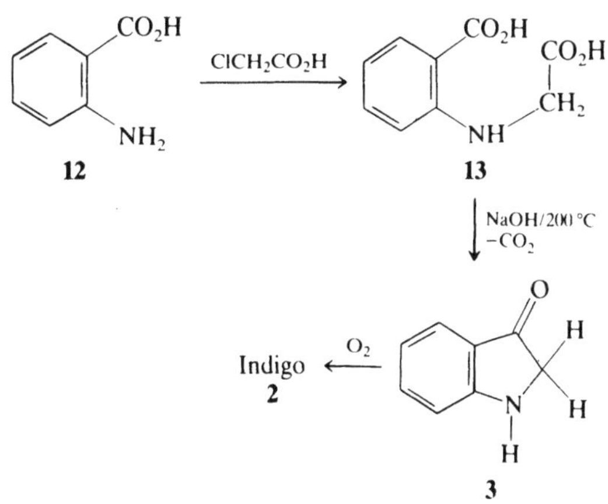

Scheme 1

ideal, however, and the indigo produced was sold at a somewhat higher price than the natural material. The higher purity and the more reproducible dyeing behavior of the former was an advantage that some dyers were prepared to pay for. A more economical route was established by Roessler in 1902 when he found that phenylglycine **(14)**, a more economical intermediate than **13**, could be cyclized to indoxyl at moderate temperatures by using sodamide as the base (Scheme 2). This proved to be a dramatic improvement over the former method and, within a short time, sales of natural indigo were badly hit; the indigo plantations went into a rapid and irreversible decline. Thus, the last surviving representative of the natural dyes had been defeated.

At the turn of the century, chemistry was a well-developed science and dye chemistry became increasingly sophisticated, with new introductions reaching the market place at an unprecedented rate. Major synthetic effort, no doubt spurred on by the great success of synthetic indigo, centered on vat dyes in the early years of this century. A discovery of major importance was made by Bohn at BASF in 1901. In an attempt to make an indigo structure containing anthraquinone residues, he adopted the principle of Scheme 2 and reacted 2-aminoanthraquinone with chloroacetic acid and subjected the product to alkali fusion. Although he obtained a blue dye, this product was not the expected indigo derivative and was later shown by Scholl to be the quinonoid structure **(15)**, now referred to as *indanthrone*. This new blue chromophore could be vatted and applied to cotton in the same way as indigo and gave outstanding fastness

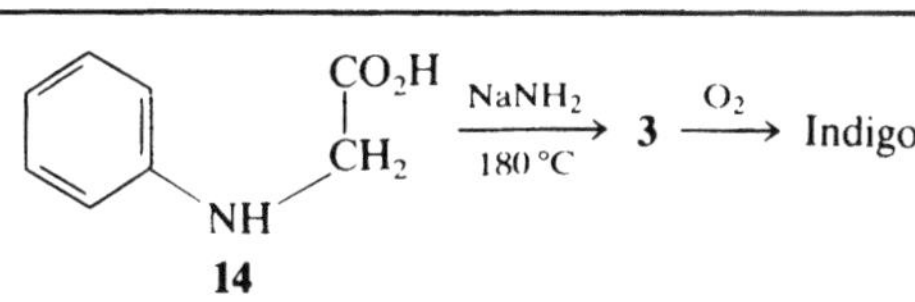

Scheme 2

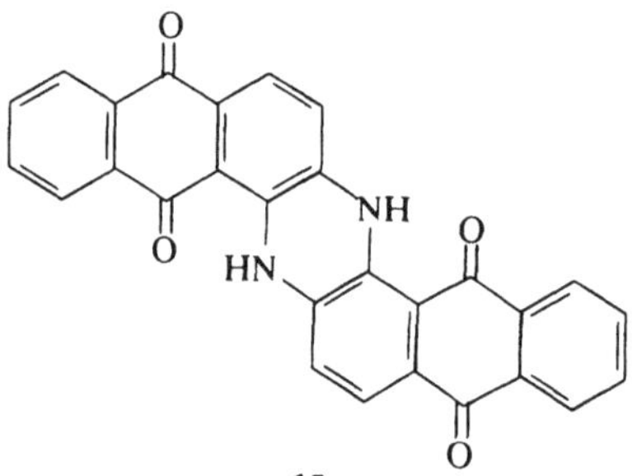

15

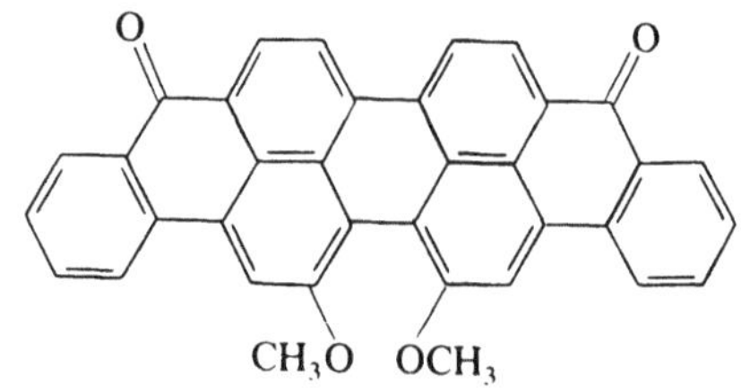

16

results, and it was introduced to the market as Indanthren Blue R. The recognition of quinones as potential vat dye systems encouraged intensive investigation of other related structures and many new vat colors were introduced in the 1900–1910 period. Examples include dyes based on indanthrone, flavanthrone, pyranthrone, violanthrone, isoviolanthrone, 1,5-dibenzamidoanthraquinone, anthraquinone-acridones, and anthraquinone-carbazoles. It should be appreciated that these dyes were commercially very important, since cellulosics were the most widely used fabrics at that time and only the vat dyes could give an outstanding performance. Even today the vats have an important and seemingly unshakable share of the total dye market.

It should not be thought, however, that dye research on vats continued to the total exclusion of other dye classes, and important developments took place in other areas. For example, the J-acid direct dyes were introduced between 1900 and 1902, and in 1912 the introduction of Naphtol AS and its analogs resulted in a significant improvement in the color range and fastness properties of azoic dyes. In the nontextile area, the first cyanine dyes to be used as photographic sensitizers were introduced. Many advances in sulfur dyes were made and, in the anthraquinone area, Herzberg in 1913 discovered the blue acid dyes formed from bromamine acid and arylamines.

With the advent of the First World War in 1914, developments in dye chemistry naturally slowed down and there were relatively few discoveries of note in the following four years as manufacturers concentrated resources on maintaining production of established dyes. However, there were some important introductions, perhaps the most significant being the Neolan dyes from the Society of Chemical Industry (Basle) in 1915. These were the preformed chromium complexes of mordant azo dyes, for dyeing wool and silk. In the war period greater interest was shown in improving the dye intermediates situation and, for example, a cheap route to phthalic anhydride was found, involving the vanadium pentoxide air-oxidation of naphthalene. This discovery had an important effect on the economics of indigo and anthraquinone synthesis and, in the latter case, led to increased interest in the development of new anthraquinone dyes.

After the war, as industry began to expand again, dye chemistry resumed its former progress but, as there were no major developments in the polymers used for textiles, this was more a period of consolidation than speculative chromophore innovation. Thus, the color ranges of the azo, anthraquinone, and vat dyes were improved and manufacturers sought new ways of applying dyes in order to make life easier for the dyer and to gain an edge over competitors. For example, the first green vat dye, Caledon Jade Green **(16)**, was introduced by ICI in 1920, so filling an important gap in the vat dye range. Developments in stabilized diazonium salts and in the sulfuric esters of pre-reduced vat dyes in the 1920–1930 period increased the versatility and convenience of the azoic and vat dyeing processes, respectively. However, one man-made fiber, cellulose acetate, had gained in importance by the 1920s and presented problems for the dyer. Thus, here was one challenge that called for a totally new approach in dye application chemistry. In 1922 the Ionamine dyes were introduced and enabled cellulose acetate to be dyed to satisfactory depths with good fastness properties.[2] These were *N*-methylsulfonic acid derivatives of amino-dyes, such as **17**. Although water soluble, in the dyebath these compounds hydrolyzed slowly to the water-insoluble dye, such as **18**, which then dyed the hydrophobic acetate fiber by what is now known as the disperse dyeing mechanism. Soon after, it was appreciated that the same effect could be achieved by using the water-insoluble dye in a very finely dispersed form in water. This simple observation meant that a much wider range of dye structures could be used than with the Ionamine approach, and thus were born the first disperse dyes, introduced as the Duranols (British Dyestuffs), and the SRA Colors (British Celanese). Interestingly, the modern classification of this group as "disperse dyes" was not adopted until 1953.[3]

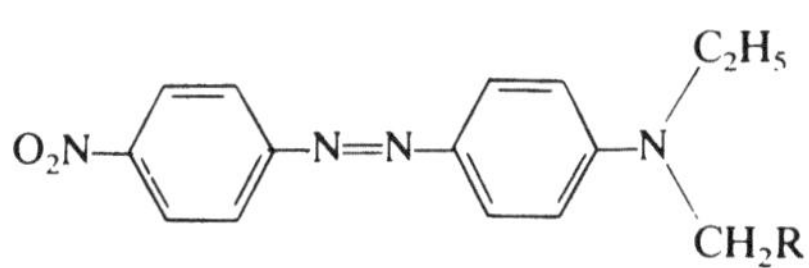

17 R = SO_3Na
18 R = H

The most important new chromophore of this century was introduced to the market in 1934. This was copper phthalocyanine, marketed by ICI as Monastral Fast Blue BS. As with many major discoveries of this type, one can often find in retrospect instances of earlier workers having observed the same or similar phenomena, but failing to capitalize on their observations. In this case, it appears that metal-free phthalocyanine had been made as long ago as 1907 by Braun and Tcherniac, when they heated *o*-cyanobenzamide above its melting point for a prolonged period, but many years were to elapse before the identity of the blue product was determined. Copper phthalocyanine itself was made by de Dierbach and Van der Weid in 1927 by heating *o*-dibromobenzene and cuprous cyanide in pyridine in a sealed tube. However, they believed the resultant blue solid to be a complex of the ingredients and the structure was not pursued further. The situation was quite different a year later when Dendridge at the Scottish Dyes Corporation observed the formation of a bright blue-green impurity in the manufacture of phthalimide from molten phthalic anhydride and ammonia when the reaction was carried out in an enamel-lined iron autoclave. Closer investigations showed that the colored material was caused by contact between the reactants and the iron surface, where cracks were present in the lining. In this case the commercial potential of the material was suspected, and some effort was put into determining its structure. In subsequent years as the structure was deduced and other metal derivatives synthesized, it transpired that copper phthalocyanine had particularly good pigment properties so that Monastral Fast Blue BS was commercialized. Copper phthalocyanine derivatives are now the most widely used blue pigments, and suitably modified phthalocyanines have found wide use as textile dyes, for example, in the acid, direct, ingrain, and reactive dye classes.

In the post-1930s perhaps the most important influence on the direction of dye research was the introduction and commercial expansion of new synthetic fibers. Thus, in 1939, nylon 66 (DuPont) and nylon 6 (IG) were simultaneously introduced to the market and production received a valuable impetus during the war years because of the military potential of these fibers. Poly(ethylene terephthalate) on the other hand was discovered in 1941 by Whinfield and Dickson, but wartime restrictions prevented commercialization of this fiber until the 1950s. By 1954 "Terylene" (ICI) and "Dacron" (Du Pont) were on the market. Poly(acrylonitrile) fibers developed from the late 1940s, commencing with the introduction of "Orlon" by Du Pont. Today the synthetic fibers hold a very significant share of the world market. The essentially hydrophobic character of the synthetics posed problems for the dyer, and the disperse dyes developed for cellulose acetate were used with some success. However, there were shortcomings, particularly in the case of polyester, and manufacturers concentrated on modifying the dyeing behavior and fastness properties of existing chromophoric systems for these new polymers. In the 1950s and 1960s, more new dyes were introduced annually than ever before, but the majority of these were essentially azo compounds and anthraquinones with modified side chains and substitution patterns as far as the disperse types were concerned. More innovative cationic structures appeared for dyeing acrylic fibers.

Although man-made fibers grew rapidly in importance during this period, cotton was still by far the most important textile fiber and continues to be so today. Thus, when the cotton reactive dyes were introduced in 1956, this was to be one of the most significant advances in dye application chemistry of the century. Work initiated at ICI in 1953 on the feasibility of obtaining covalent bonding between dye and fiber under practical dyehouse conditions led to the investigation of dichlorotriazinyl dyes by Stephen and Rattee, and culminated in the development of the Procion reactive dyes and the establishing of viable, efficient application procedures. The enormous potential of these dyes stimulated research by other companies and many other reactive dye systems were patented and even commercialized successfully. Some systems that have withstood the rigors of the marketplace include chloropyrimidines, chloroquinoxalines, fluoropyrimidines, and vinylsulfones. Reactive dyes for polyamide and wool fibers have also received much attention, though with less striking success.

Partly because the dye manufacturer had satisfied most of the needs of the dyer, and largely because of the industrial recession of the 1970s, dye research entered a period of decline after 1970. Manufacturers began to place greater emphasis on the rationalization of existing dye ranges and on improving the economics of dye production, and less on dye innovation. This trend continues today, and most new developments in dye chemistry are to be found in the high-technology areas, such as medicine, diagnostics, lasers, and electrooptics, at least as far as the patent literature is concerned. It is interesting to note that many of the chromophores used in such applications were first made more than 50 years ago, and were discarded as textile dyes as dye chemistry advanced. Thus, many of the old dye systems, long regarded as of historical interest only, are finding a new lease of life in today's technology.

III. THE DEVELOPMENT OF COLOR AND CONSTITUTION THEORY

As we have seen, dye chemistry and organic chemistry evolved hand in hand during the nineteenth century. Initially, dye developments proceeded at a faster pace than theory and such developments did much to encourage the advance of structural organic chemistry. Another major stimulus to structural theory came not from the chemistry of dyes, but from the attempts made by scientists to understand the origins of color in dyes. This was particularly true of the late nineteenth–early twentieth century period, as the electronic theory of molecular structure began to be developed. Thus it is of interest to review briefly the development of color and constitution theory from the time of the first synthetic dyes, and to see how this has led to a situation where modern computers can now be used to predict the color of a dye before it has even been synthesized.

At the time of Perkin's discovery, organic chemistry as a science was very much in its infancy. Thus, for example, it was not until Kekulé published his theory in 1865 that the structure of benzenoid compounds began to be properly understood. It should also be appreciated that the electron was not discovered until 1897, and it was only as recently as 1916 that the concept of the electron-pair bond was formulated by Lewis in his famous article "The Atom and the Molecule."[4] By 1868 sufficient data had been accumulated about synthetic dyes for Graebe and Liebermann to make the point that the unsaturation in dye molecules was in some way associated with color. This deduction was based on the observation that reduction of dyes destroyed their color. The first real attempt to interpret the color of dyes in terms of their molecular composition was made by Witt in 1876.[5] By this time, structural chemistry had progressed to the stage where functional groups were well understood, and Witt considered that a dye molecule could be dissected into three components. Thus, there were benzene, or fused benzene, rings, and to these were attached simple unsaturated groups, e.g., —NO_2, —N=N—, —C=O, which he called *chromophores*. Finally, there were basic groups, e.g., —NH_2, —OH, designated *auxochromes*, and the presence of all three units produced intense color. This attractively simple picture still finds use today, and within Witt's theory we can see the first glimmerings of the modern concepts of electron-donating (+M) and electron-withdrawing (−M) groups, and of conjugation.

Witt's theory was readily applicable to the azo and anthraquinone dyes, but could not easily be reconciled with the triphenylmethane cationic dyes. Thus, an alternative interpretation of the origin of color was proposed quite independently by Armstrong in 1888 and Nietzki in 1889. In this case, color was considered to be due to the presence of quinonoid structures (i.e., benzenoid rings modified to a cyclohexadiene residue with two exocyclic double bonds). For example, these units had to be present in the formulated structures of the cationic dyes. Generally more than one quinonoid structure could be drawn, and it was a logical step to assume that some kind of oscillation of the structure between these possibilities was responsible for light absorption. In fact, the electromagnetic wave theory of light had been accepted for a long time, and the interaction of a vibrating wave with an oscillating bond system provided an attractive mechanism for the absorption process. However, simple theoretical considerations showed that the frequency of atomic vibrations were far too low to be matched with the frequency of light. After the discovery of the electron, it was pointed out that vibrations of such a particle would involve the correct sort of frequency for light absorption, but this suggestion was not treated seriously by chemists at the turn of the century. Theoretical chemistry made significant progress in the early 1900s and, after Lewis's establishing of the electronic theory of bonding in molecules, the association of light absorption with electronic excitation was widely accepted.

One important theoretical concept to benefit greatly from color and constitution studies was that of *mesomerism*. In 1899 Thiele had published his "conjugation and partial valence hypothesis" in order to explain the anomalous chemical properties of conjugated polyene systems, and it was thereby suggested that in such systems bond equalization occurred, and there were partially unsatisfied valencies at chain termini. By 1911 Kauffmann had extended Thiele's ideas and put forward his concept of the "decentralization of chemical functions," which in effect was a clear recognition of the resonance transmission of substituent effects in conjugated molecules. In a further extension of his theory, Kauffmann introduced the concept of limiting structures for depicting the true structure of a molecule, whereby the latter corresponded to some average of these. Such ideas laid the foundations for mesomerism and Ismailsky described this "average structure" as the *chromostate* in recognition of its apparent direct relevance to color. For example, a triphenylmethane dye can be regarded as an average of three limiting structures, and the consequent chromostate is highly colored. Although such ideas were well accepted, the nature of the process connecting the various limiting structures was obscure. It was Weitz and König who in 1922 made a clear distinction between structural tautomerism (i.e., true equilibria) and valence tautomerism (mesomerism).[6]

The most important advance in theoretical chemistry was the advent of quantum mechanics, which first found mathematical formulation in 1925 with the independent contributions of Schrödinger and Heisenberg. By the early 1930s quantum mechanics had been applied to the problem of bonding in molecules and, for example, the π-bond was mathematically formulated in 1930. At this stage the concepts of mesomerism, long used by the organic chemist, and a new concept called *resonance* were to be shown as equivalent on the basis of quantum mechanical calculations. The approximation procedure that gave rise to this concept was *valence bond* theory, which was subsequently used to calculate the absorption spectra of dyes. Also developed in the early 1930s was an alternative mathematical approximation for using quantum mechanics to calculate the properties of molecules, namely, molecular orbital theory. Thus, using this method, Hückel was able to quantify many of the properties of complex conjugated organic molecules, and his approach, popularly referred to as Hückel molecular orbital (HMO) theory, was to prove of lasting importance and to lay the groundwork for modern MO methods.

With the introduction of the valence bond and HMO methods, many of the refinements and applications of these theories in subsequent years owed a great deal to dyes and attempts to predict color. For example, in 1939 Mulliken applied the HMO method with considerable success to the absorption spectra of polyenes and cyanine dyes[7] and, by the same year, Pauling had used valence bond theory to account for the color of cationic and anionic dyes.[8] In 1940 Förster calculated the spectroscopic shifts in polyene dyes caused by increasing chain length, and two years later Herzfeld and Sklar were able to predict the well-known convergent and nonconvergent vinylene shifts in unsymmetrical and symmetrical cyanine dyes, respectively, again using valence bond theory. Thus, a satisfactory physical interpretation of the structure of organic molecules, and the means for calculating the light absorption properties of dyes, were well established by 1940.

A new quantum mechanical approach to calculating the properties of conjugated organic molecules was introduced independently by three separate workers in 1942. This was the "electron gas" or "free electron" method, formulated by Bayliss,[9] Kuhn,[10] and Simpson.[11] While this method was very successful for calculating the spectroscopic properties of simple dyes, and received many refinements in later years, it failed to receive wide acceptance, largely because of the greater popularity of the HMO method. Early spectrosopic calculations suffered from the neglect of electronic interaction effects so that, before further progress in color prediction could be made, further refinements of the theories were required, and most of the developments in molecular orbital theory after 1940 were concerned with the complex problem of allowing for these effects. Nevertheless, in 1950 Dewar was able to use a suitably parameterized form of HMO theory to calculate the color of a range of cyanine-type dyes with notable success.[12] By this time, valence bond theory was no longer prominent because of its inconvenience and mathematical complexity when dealing with large molecules.

An important step forward was made in 1953 when the theoretical approaches of two groups of workers, based formally on HMO theory but taking into account electronic interaction, were introduced.[13] The new computational procedure, popularly referred to as the Pariser–Parr–Pople (PPP) method, was very successful for calculating the electronic spectra of π-electron systems and, following the development of suitable molecular parameters, has proved to be one of the most successful color prediction techniques, for routine use, at the present time.[14] One disadvantage of the PPP method is that it ignores σ-electrons and considers the π-electrons only. Various MO procedures were developed into the 1970s and include all valence electrons, e.g., the CNDO/S method of del Bene and Jaffé,[15] and even *ab initio* methods, all of which would have presented formidable computational problems 25 years ago but which, with the rapid advances made in computer technology, are within the reach of most chemists. These methods should, with suitable elaboration, become as easily applicable to complex dye structures as the PPP method in the near future, with perhaps greater predictive success.

Although enormous strides have been made in the area of mathematical color prediciton, there is still a place within dye chemistry for qualitative color and constitution theory, since this provides broad guidelines for the synthetic chemist when attempting to modify the color of existing dye types. Such theories also help to illustrate structural relationships in dyes which can be obscured by complex computational procedures. Space limitations preclude a detailed survey of qualitative or descriptive approaches to color and constitution, and readers are referred to recent general surveys for further details.[16–19] These qualitative theories range from the purely descriptive, such as the original theory of Witt, the classification favored by the present author,[16] and the Triad theory of Dähne,[19] to methods such as Dewar's perturbational MO approach,[12] and the configuration analysis theory of Fabian,[17] which are firmly based on mathematical considerations. All have a place in the context of modern dye chemistry, and give an insight not only into the origins of color in dyes but also into general electronic structural theory of organic compounds.

REFERENCES

1. A. Kekulé, *Bull. Soc. Chim. Fr. 3,* 98 (1865).
2. A. G. Green and K. H. Saunders, *J. Soc. Dyers Colour. 39,* 10 (1923).
3. C. M. Whittaker, *J. Soc. Dyers Colour. 69,* 205 (1953).
4. G. N. Lewis, *J. Am. Chem. Soc. 38,* 762 (1916).
5. O. N. Witt, *Chem. Ber. 9,* 522 (1876).
6. E. Weitz and T. König, *Chem. Ber. 55,* 2864 (1939).
7. R. S. Mulliken, *J. Chem. Phys. 7,* 364 (1939).
8. L. Pauling, *Proc. Natl. Acad. Sci. U.S.A. 25,* 577 (1939).
9. N. S. Bayliss, *J. Chem. Phys. 16,* 287 (1948).
10. H. Kuhn, *J. Chem. Phys. 16,* 840 (1948).
11. W. T. Simpson, *J. Chem. Phys. 16,* 1124 (1948).
12. M. J. S. Dewar, *J. Chem. Soc.* 2329 (1950).
13. R. Pariser and R. G. Parr, *J. Chem. Phys. 21,* 466, 767 (1953); J. A. Pople, *Trans. Faraday Soc. 49,* 1375 (1953).
14. J. Griffiths, *Chem. Br.* 997 (1986).
15. J. del Bene and H. H. Jaffé, *J. Chem. Phys. 48,* 1807, 4050 (1968).
16. J. Griffiths, *Colour and Constitution of Organic Molecules,* Academic Press, London (1976).
17. J. Fabian and H. Hartmann, *Light Absorption of Organic Colorants,* Springer-Verlag, Berlin (1980).
18. J. Griffiths, *Rev. Prog. Color. Relat. Top. 14,* 21 (1984).
19. S. Dähne and F. Moldenhauer, *Prog. Phys. Org. Chem. 15,* 1 (1985).

2

Classification of Dyes by Chemical Structure

P. GREGORY

I. INTRODUCTION

There are several ways to classify dyes. For example, they may be classified by fiber type, such as dyes for nylon, dyes for cotton, dyes for polyester, and so on. Dyes may also be classified by their method of application to the substrate. Such a classification would include direct dyes, reactive dyes, vat dyes, disperse dyes, azoic dyes, and several more types. These classifications are more suited to a book aimed at the technology of the application of dyes to a substrate rather than one dealing primarily with the synthesis and chemistry of dyes. Griffiths[1] has classified dyes on a purely theoretical basis based upon the electronic origins of color. Only four classes are defined, namely:

1. Donor–acceptor chromogens.
2. Cyanine-type chromogens.
3. Polyene chromogens.
4. $n \rightarrow \pi^*$ chromogens.

Although such a simple system may appear attractive, it has major drawbacks as far as the practicing dye chemist is concerned. For instance, the vast majority of commercially important dye types fall into the first category, with none at all in the last category and relatively few in the other two categories. Essentially, therefore, the dyes of importance are not classified at all by this method. Furthermore, this terminology is alien not only to most dye research chemists but also to the dye users, the dyehouse chemists and technologists.

The most appropriate system for the classification of dyes is by chemical structure. This system has many advantages. First, it readily identifies dyes as

P. GREGORY • Fine Chemicals Research Centre, ICI Colours and Fine Chemicals, Hexagon House, Blackley, Manchester M9 3DA, England.

belonging to a group which have characteristic properties, for example, azo dyes (strong and cost effective) and anthraquinone dyes (weak and expensive). Second, it produces a manageable number of groups (about a dozen). Third, and most importantly, it is the classification used most widely by both the synthetic dye chemist and the dye technologist. Thus, both chemists and technologists can readily identify with phrases such as an azo yellow, an anthraquinone red, and a phthalocyanine blue. Consequently, classification of dyes by chemical structure is the system adopted throughout this book.

The *Colour Index* already classifies dyes (and pigments) by chemical structure.[2] To repeat this classification here is rather pointless for several reasons. First, the reader can already consult the *Colour Index* lassification. Second, the *Colour Index* classification includes pigments as well as dyes while this book is confined to dyes only.* Thus, important pigment classes such as quinacridones are not included. Third, the classification adopted in this chapter, while maintaining the backbone of the *Colour Index* classification, is intended to progress beyond the latter classification by simplifying it. This is done by showing the structural interrelationships of dyes which are given separate classes by the *Colour Index*.[2] Finally, the classification is chosen to highlight some of the more recent discoveries in dye chemistry.

Emphasis will be placed on both the relative commercial importance of the dyes and on the more recent developments, such as the blue reactive triphendioxazine dyes, the red disperse tricyanovinyl styryl dyes, the red disperse benzodifuranone dyes, and the blue disperse styryl dyes. Some of the larger and more important classes of dye such as azo, anthraquinone, and polymethine will be subdivided further.

The overriding trend in dyestuffs research for many years has been improved cost effectiveness rather than increased technical excellence. Usually, this means replacing tinctorially weak dyes such as anthraquinone, the second largest class after the azo dyes, with tinctorially stronger dyes such as heterocyclic azos, triphendioxazines, and benzodifuranones. This theme will be pursued throughout the chapter.

II. AZO DYES

Azo dyes are by far the most important class, accounting for over 50% of all commercial dyes. Because of this, azo dyes have been studied more than any other class and a vast volume of data now exists on them. Obviously, therefore, only the more important aspects of azo dyes can be discussed in this section.

A. Basic Structure

Azo dyes contain at least one azo group (—N═N—) but can contain two, three, or, more rarely, four azo groups. The azo group is attached to two radicals of which at least one but, more usually, both are aromatic. They exist in the *trans* form **(1)** and the bond angle is *ca.* 120°; the nitrogen atoms are sp^2 hybridized.

* Unlike dyes, pigments do not pass through a solution phase during application to the substrate; they remain insoluble.

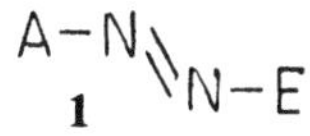

1

In monoazo dyes, which are by far the most important type, the A radical contains electron-accepting groups and the E radical contains electron-donating groups, particularly hydroxy and amino groups. If the dyes contain only aromatic radicals they are known as carbocyclic (or sometimes homocyclic) azo dyes. In contrast, if they contain one or more heterocyclic radicals, the dyes are known as heterocyclic azo dyes. Representative examples of various azo dyes are shown by structures **(2)** to **(12)**. These illustrate the enormous number of permutations possible

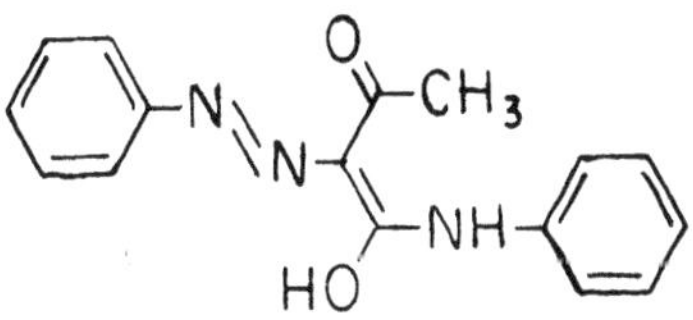

2
C.I. Disperse Yellow 16
Carbocyclic Monoazo
A→E

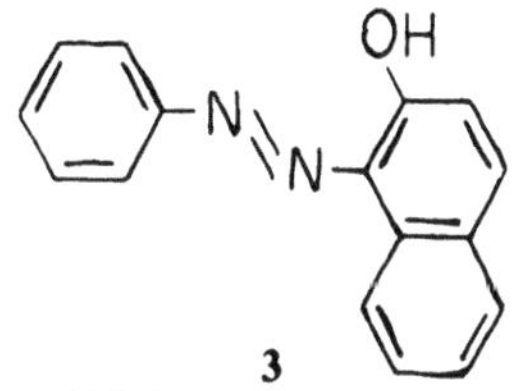

3
C.I. Solvent Yellow 14
Carbocyclic Monoazo
A→E

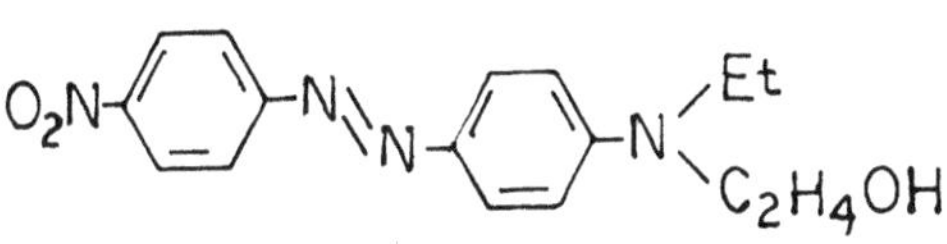

4
C.I. Disperse Red 1
Carbocyclic Monoazo
A→E

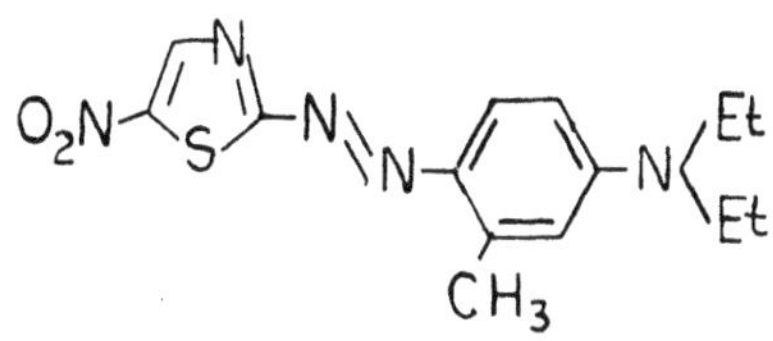

5
Disperse Blue
Heterocyclic Monoazo
A→E

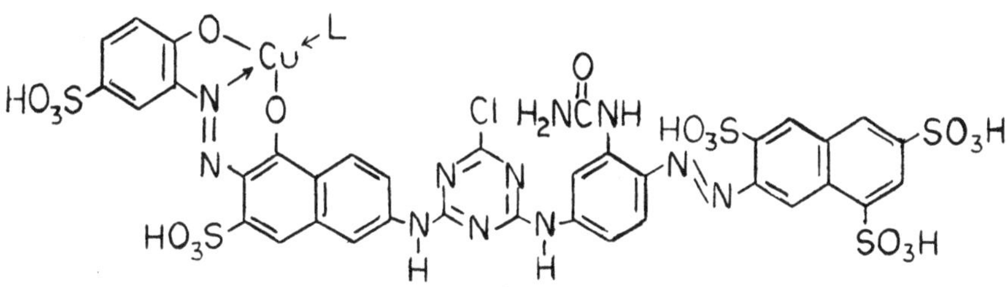

6
Reactive Brown
A → E—X—E ← A Disazo

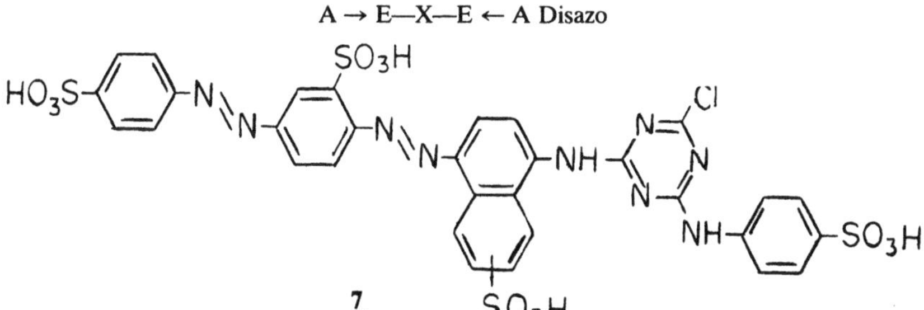

7
C.I. Reactive Brown 1
A→M→E Disazo

The definitions of A, D, E, M, X and Z can be found in ref. 2.

8

C.I. Acid Black 1

A→Z←A′ Disazo

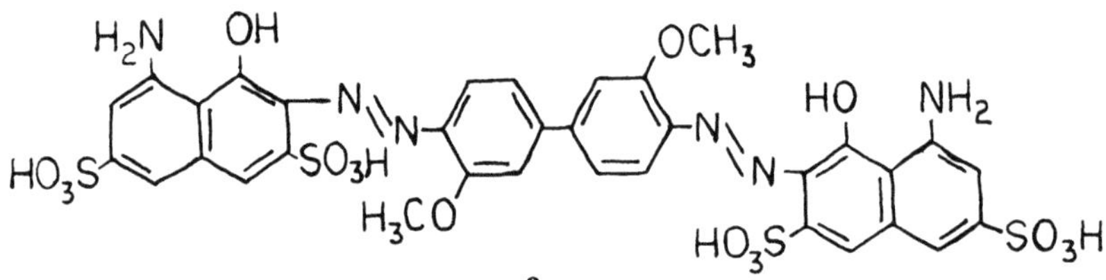

9

C.I. Direct Blue 15

E←D→E′ Disazo

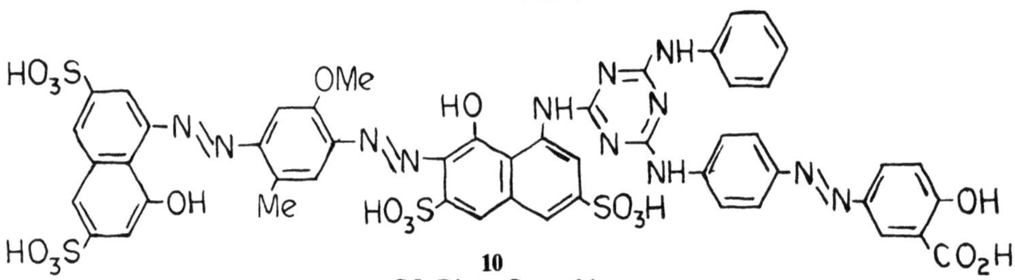

10

C.I. Direct Green 26

A→M→E—X—E←A Trisazo

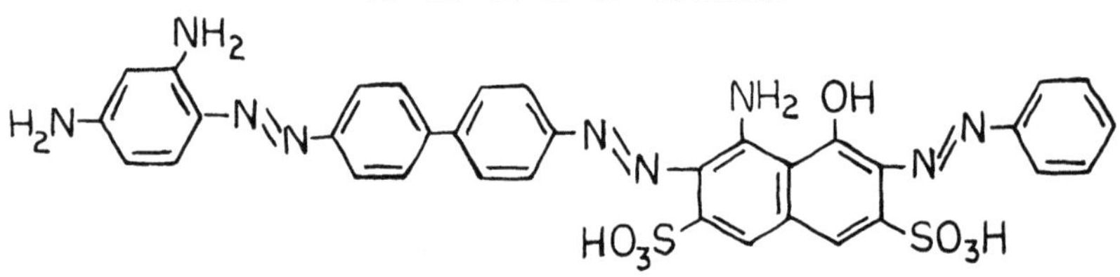

11

C.I. Direct Black 38

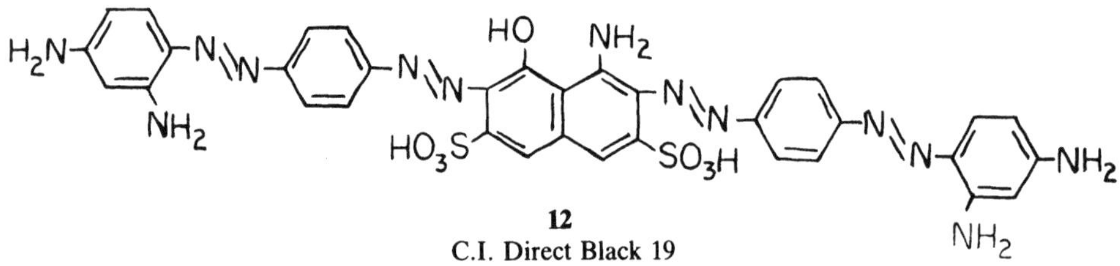

12

C.I. Direct Black 19

E←D→Z←D→E Tetrakisazo

in azo dyes, particularly with the polyazo dyes. However, it should be remembered that the vast majority of commercial azo dyes are monoazos, with relatively few trisazo and tetrakisazo dyes.

It is evident that as the number of azo groups per molecule increases, so do the possible permutations. For example, there are more types of disazo dyes than monoazo dyes, more types of trisazo dyes than disazo, and so on. Only a few types of tris- and tetrakis-azo dyes are exemplified, reflecting their minor commercial importance opposite the disazo and especially the monoazo dyes.

B. Synthesis

Almost without exception, azo dyes are made by diazotization of a primary aromatic amine followed by coupling of the resultant diazonium salt with an electron-rich nucleophile. The modern dyestuffs industry therefore owes a large debt of gratitude to Peter Greiss, the chemist who discovered the diazotization and coupling reaction in 1858.

The diazotization reaction is effected by treating the primary aromatic amine with nitrous acid, normally generated *in situ* with hydrochloric acid and sodium nitrite. The nitrous acid nitrosates the amine to generate the *N*-nitroso compound **(13)** which tautomerizes to the diazo hydroxide **(14)**. Protonation of the hydroxy group followed by the elimination of water generates the resonance-stabilized diazonium salt **(15)**; see Scheme 1. For weakly basic amines, i.e., those containing several electron-withdrawing groups, nitrosyl sulfuric acid ($NO^+HSO_4^-$) is used as the nitrosating species.

A diazonium salt, unlike for example a nitronium ion (NO_2^+), is a weak electrophile. Thus, it will only react with highly electron-rich species such as

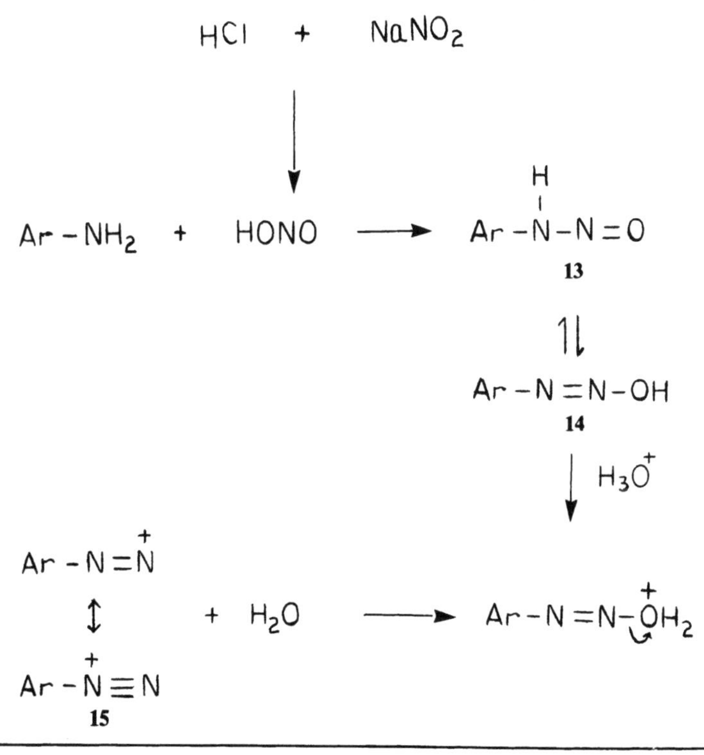

Scheme 1

amino and hydroxy compounds. Even hydroxy compounds have to be ionized for reaction to occur. Consequently, hydroxy compounds such as phenols and naphthols are coupled in an alkaline medium (pH $\geqslant pK_a$ of phenol or naphthol; see equation 1), while aromatic amines such as *NN*-dialkylanilines are coupled in a slightly acid medium (see equation 2). This provides optimum stability for the diazonium salt (stable in acid) without deactivating the nucleophile (protonation of the amine).

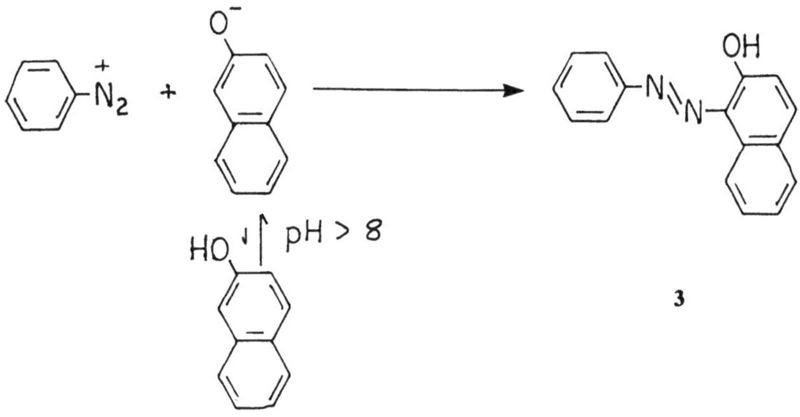

3

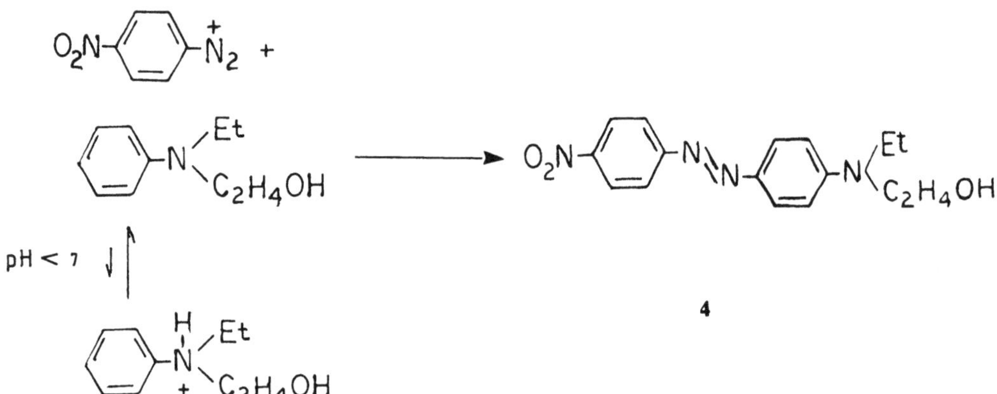

4

(2)

Coupling components containing both amino and hydroxy groups, such as H-acid **(16)**, can be coupled stepwise. Coupling is first carried out under acid conditions to effect azo formation in the amino-containing ring. The pH is then raised to ionize the hydroxy group (usually to pH $\geqslant$ 7) to effect coupling in the "naphtholate" ring, with either the same or a different diazonium salt; see Scheme 2.

The vast number of available diazo components and coupling components means that countless permutations of azo dyes is possible. This enormous synthetic versatility is one of the advantages of azo dyes; it is possible to "tailor-make" molecules to suit a particular application. The synthetic route employed also explains why most azo dyes contain hydroxy or amino groups.

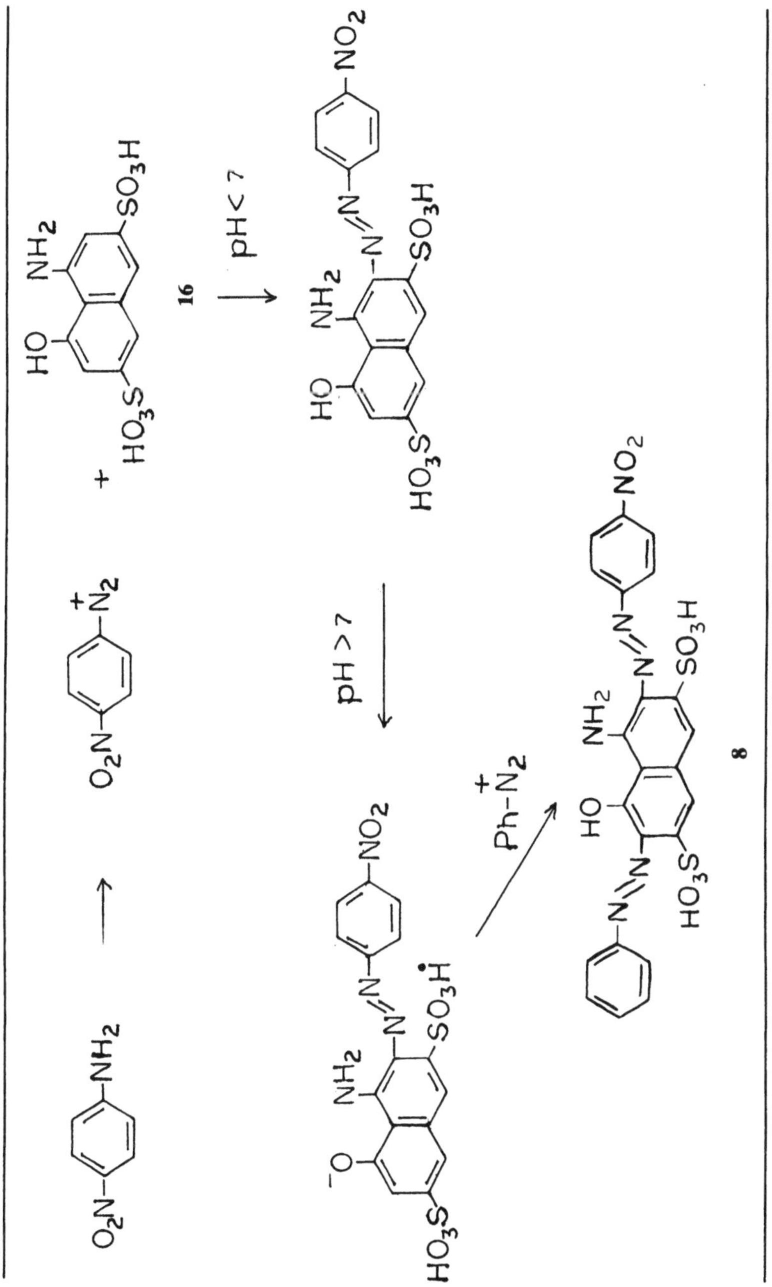

Scheme 2

The somewhat unusual conditions needed to produce an azo dye, namely, strong acid plus nitrous acid for diazotization, the low temperatures necessary for the unstable diazonium salt to exist, and the presence of electron-rich amino or hydroxy compounds to effect coupling, means that azo dyes have no natural counterparts.

C. Tautomerism

In theory, azo dyes can undergo tautomerism: azo/hydrazone for hydroxyazo dyes; azo/imino for aminoazo dyes, and azonium/ammonium for protonated azo dyes.

1. Azo/Hydrazone Tautomerism

This phenomenon was discovered by Zincke and Binderwald[3] in 1884. These workers obtained the same orange dye by coupling benzene diazonium chloride with 1-naphthol and by condensing phenylhydrazine with 1,4-naphthoquinone. The expected products were the azo dye **(17)** and the hydrazone dye **(18)**. Zincke and Binderwald correctly assumed a mobile equilibrium between the two forms, *viz.* tautomerism.

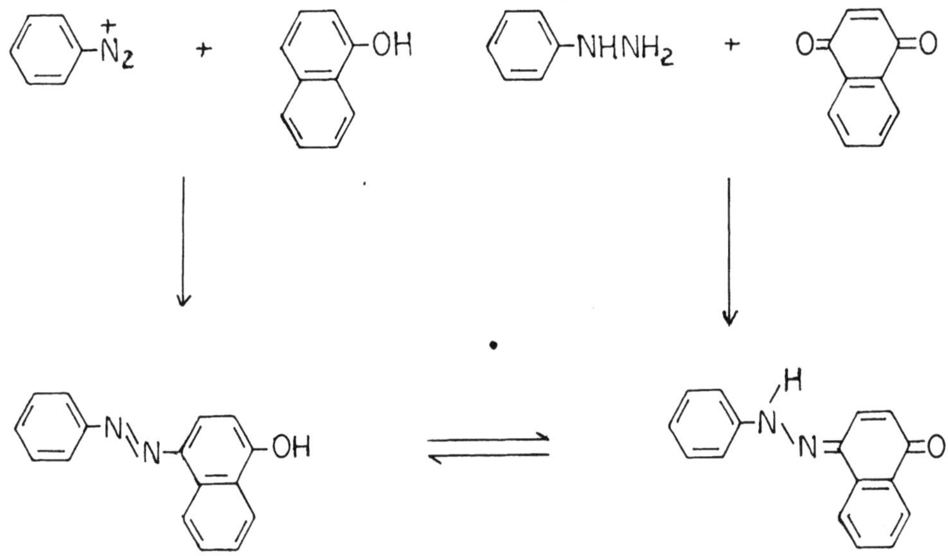

17 **18**

This discovery prompted extensive research into azo/hydrazone tautomerism, a phenomenon which is not only interesting but also extremely important as far as commercial azo dyes are concerned. This is because the tautomers have different colors, different properties, e.g., light fastness, different toxicological properties,[4] and, most importantly, different tinctorial strengths. Since the tinctorial strength of a dye primarily determines its cost effectiveness, it is desirable that commercial azo dyes should exist in the strongest tautomeric form, namely, the hydrazone form.

Hydroxyazo dyes may be grouped into three broad classes:

1. Those that exist predominantly as azo tautomers.
2. Those that can exist as a mixture of azo and hydrazone tautomers.
3. Those that exist predominantly as hydrazone tautomers.

Almost all azophenol dyes **(19)** belong to the first category. They exist totally in the azo form except for a few special cases.[5]

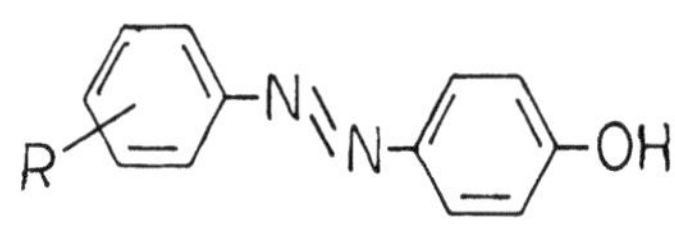

19

Azonaphthol dyes typify the second category. The energies of the azo and hydrazone forms of 4-phenylazo-1-naphthol dyes are similar and so both forms are present. The azo tautomers **(20)** are yellow ($\lambda_{max} \sim 410$ nm, $\varepsilon_{max} \sim 20{,}000$) while the hydrazone tautomers **(21)** are orange ($\lambda_{max} \sim 480$ nm; $\varepsilon_{max} \sim 40{,}000$). The relative proportions of the tautomers are influenced by both solvent (Figure 1) and substituents (Figure 2).

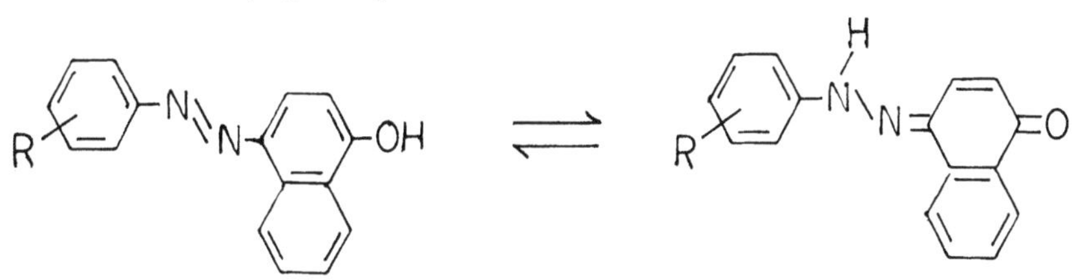

20 **21**

The isomeric 2-phenylazo-1-naphthols **(22)** and 1-phenylazo-2-naphthols **(23)** exist more in the hydrazone form than the azo form; see Figure 3.

Important classes of dyes which exist totally in the hydrazone form are azopyrazolones **(24)**, azopyridones **(25)**, and azoacetoacetanilides **(26)**.

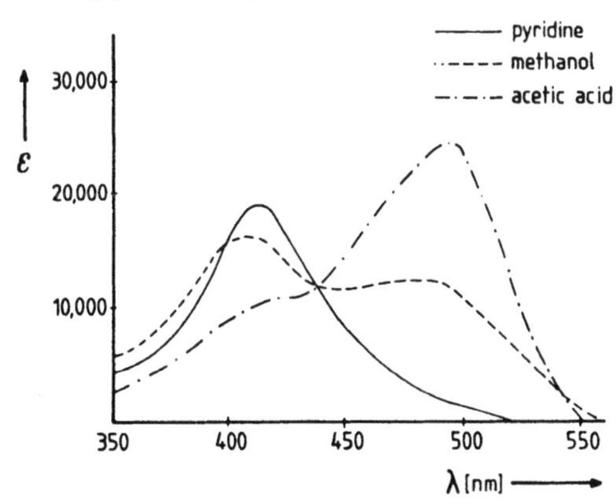

Figure 1. Effect of solvent on 4-phenylazo-1-naphthol.

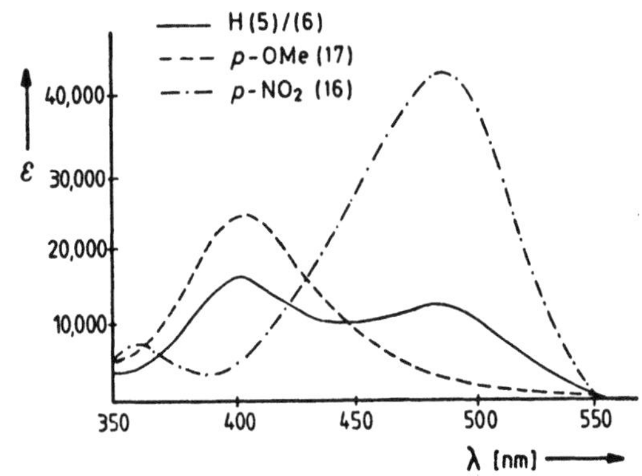

Figure 2. Electronic effect of substituents in 4-phenylazo-1-naphthol.

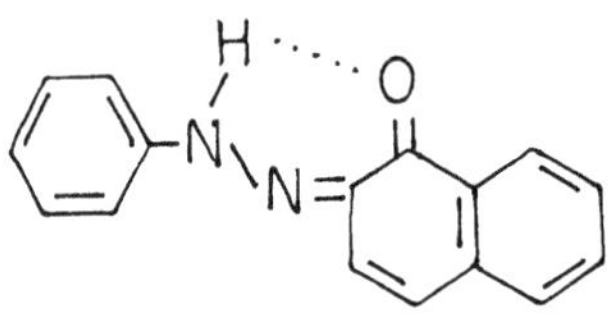

22

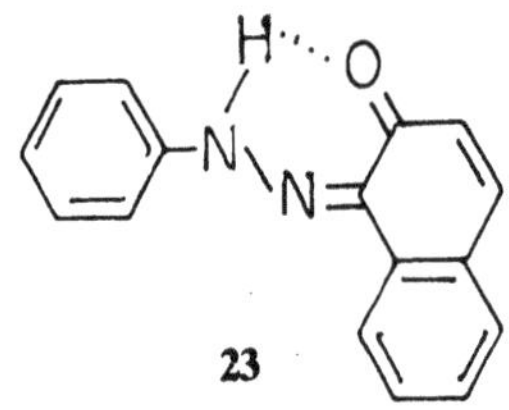

23

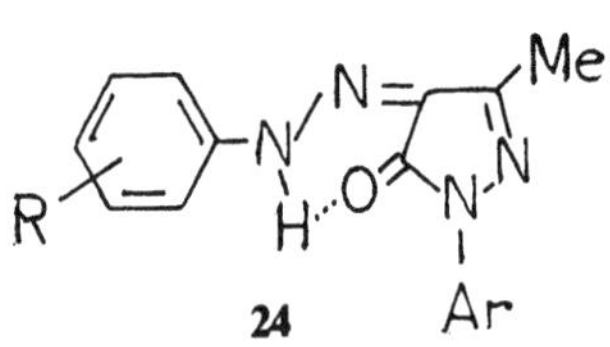

24

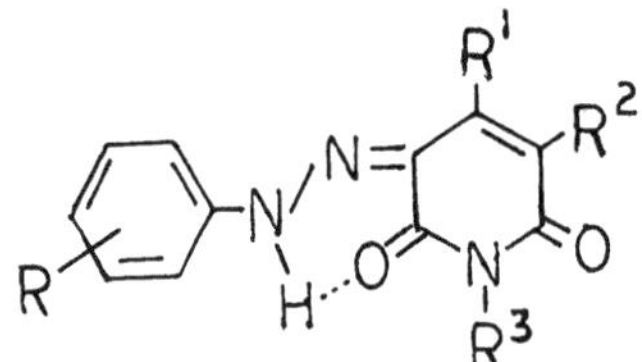

25

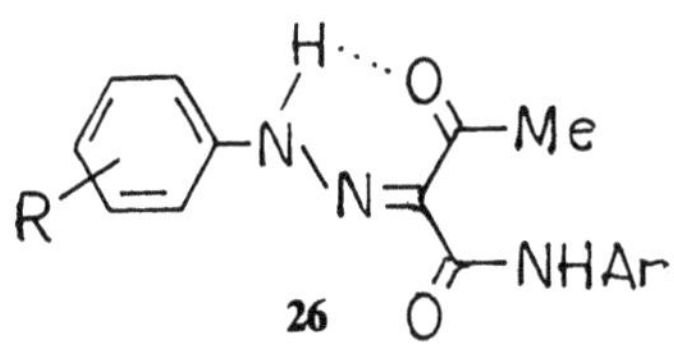

26

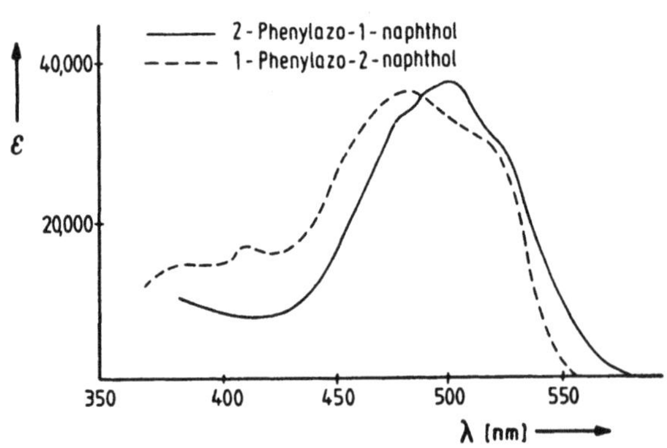

Figure 3. Spectra of 1-phenylazo-2-naphthol and 2-phenylazo-1-naphthol (in CH_3CO_2H).

2. Azo/Imino Tautomerism

All aminoazo dyes exist exclusively as the azo form **(27)**; there is no evidence for the imino form **(28)**. Presumably, a key factor is the relative instability of the imino grouping.

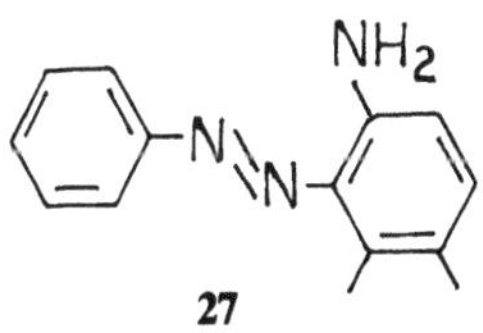

27

>C=NH

BE 614.5

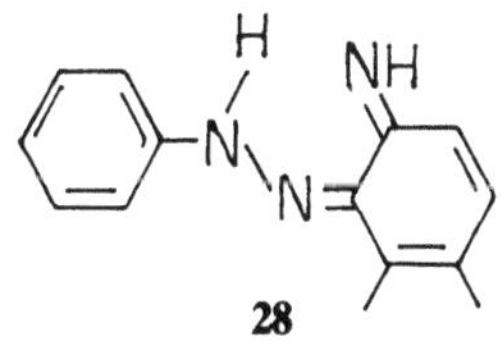

28

>C=O

744 kJ mol^{-1}

In disazo dyes from aminonaphthols, one group exists as a true azo group and one as a hydrazo group **(29)**.

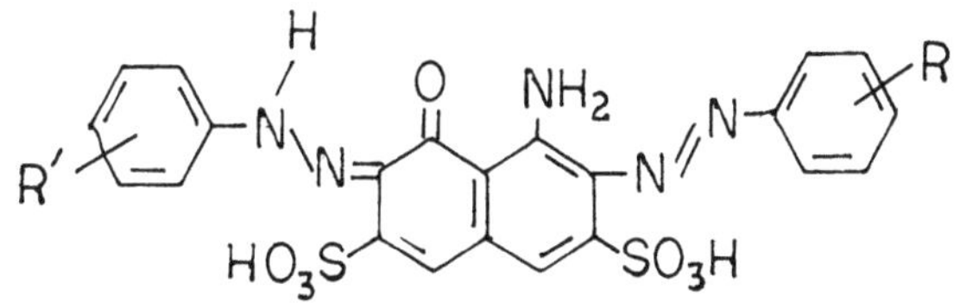

29

3. Azonium/Ammonium Tautomerism

Aminoazo dyes undergo protonation at either the terminal nitrogen atom to give the essentially colorless ammonium tautomer **(30)** ($\lambda_{max} \sim 325$ nm), or at the β-nitrogen atom of the azo group to give a resonance-stabilized azonium tautomer **(31)**. The azonium tautomer is brighter, generally more bathochromic and stronger ($\varepsilon_{max} \sim 70{,}000$) than the neutral azo dye. The azonium tautomers are related to the diazahemicyanine dyes used for coloring polyacrylonitrile (Chapter 5).

The most familiar use of the protonation of azo dyes is in indicator dyes such as Methyl Orange **(32)** and Methyl Red **(33)**.

A more detailed account of azo dye tautomerism can be found elsewhere.[6]

D. Metallized Azo Dyes

Only three metals are of any importance in this context, namely, copper, chromium, and cobalt.

The most important copper dyes are the 1:1 copper(II):azo dye complexes of formula **(34)**; they have a planar structure.

In contrast, chromium(III) and cobalt(III) form 2:1 dye:metal complexes which have nonplanar structures. Geometrical isomerism exists. The *o,o'*-dihydroxyazo dyes **(35)** form the Drew–Pfitzner or *mer* type **(36)** while *o*-hydroxy, *o'*-carboxyazo dyes **(37)** form the Pfeiffer–Schetty or *fac* type **(38)**.

Metallization of dyes originally occurred during the mordanting process to

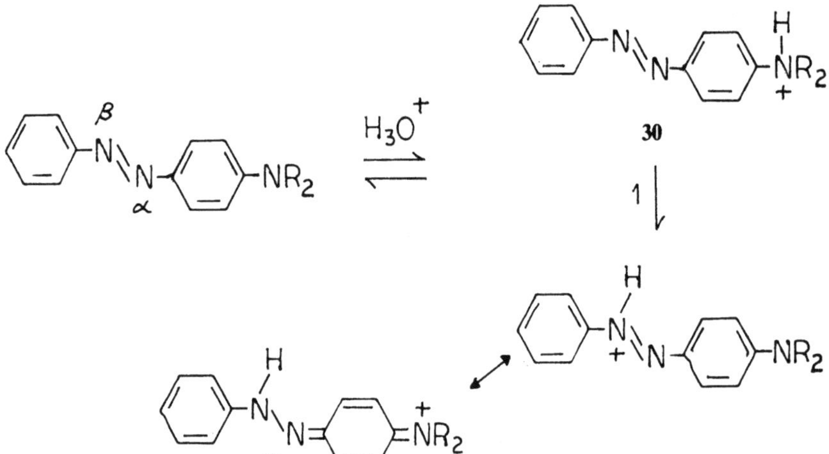

30

31

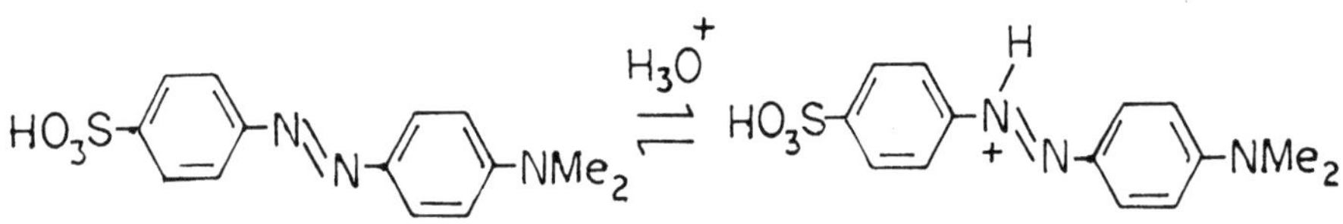

32
Yellow

Red

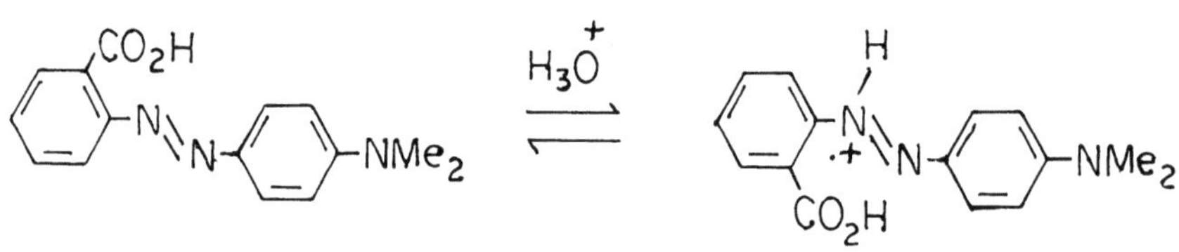

33
Orange

Red

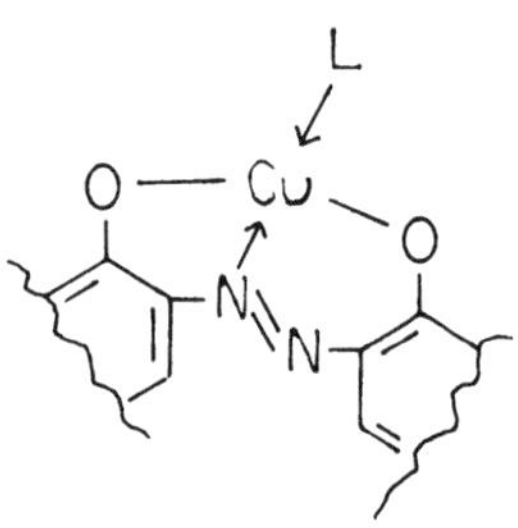

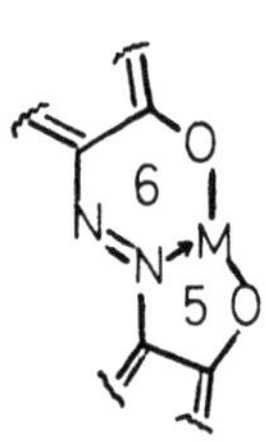

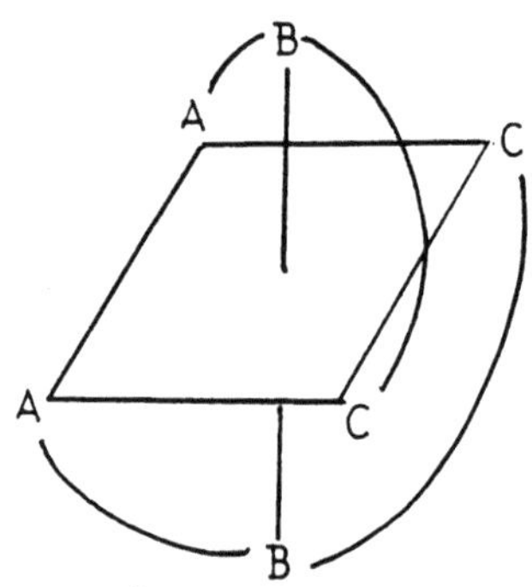

34

35

A = —O—
C = —O—

36

help fix the dye to the substrate. Premetallized dyes are now used widely in various outlets to improve the properties of the dye, particularly its light fastness. However, this is at the expense of brightness since metallized azo dyes are duller than nonmetallized dyes.

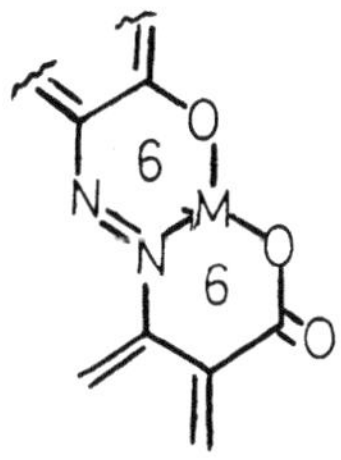

37

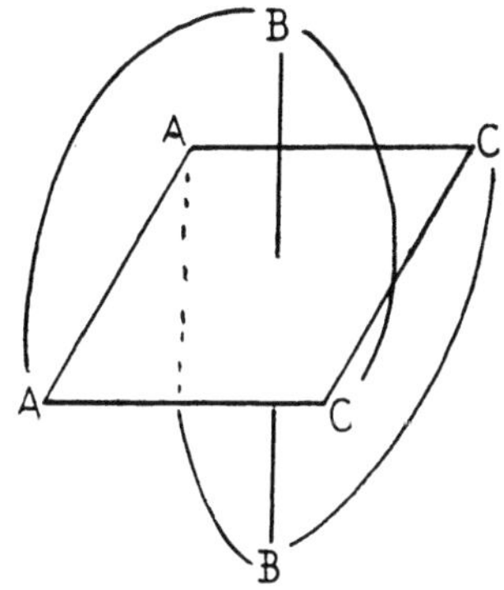

38

A = —CO_2—
C = —O—

E. Carbocyclic Azo Dyes

These bread and butter dyes are the backbone of most commercial dye ranges. Based totally on benzene and naphthalene derivatives, these dyes provide yellow, red, blue, and green colors for all the major substrates such as polyester, cellulose, nylon, polyacrylonitrile, and leather. Typical structures **39–45** are shown.

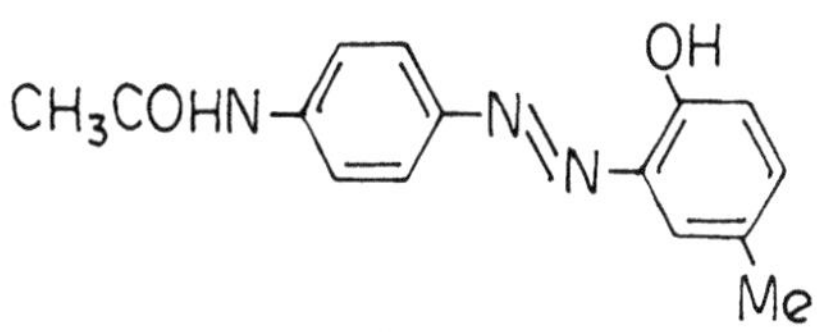

39
Disperse Yellow
(Polyester)

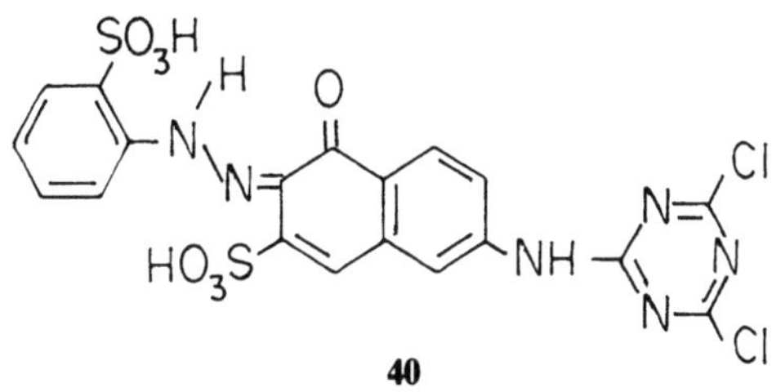

40
Reactive Orange
(Cotton)

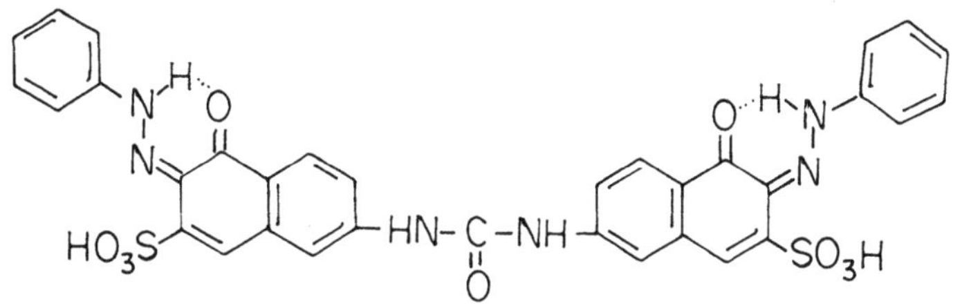

41
Direct Orange
(Paper)

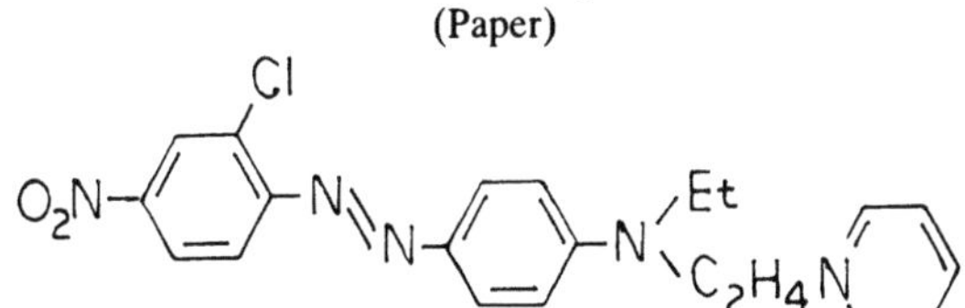

42
Basic Red
(Polyacrylonitrile)

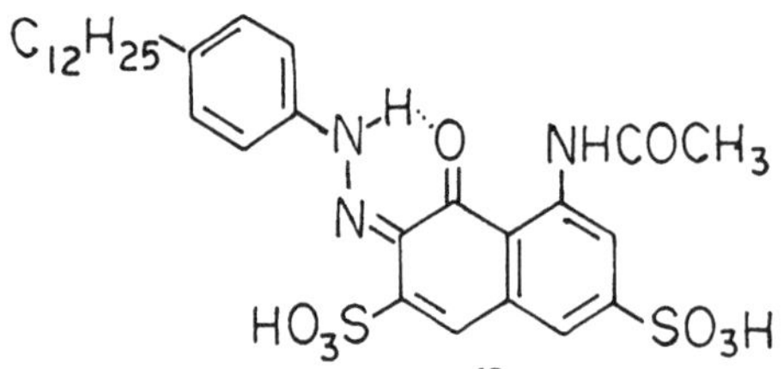

43
Acid Red
(Nylon and wool)

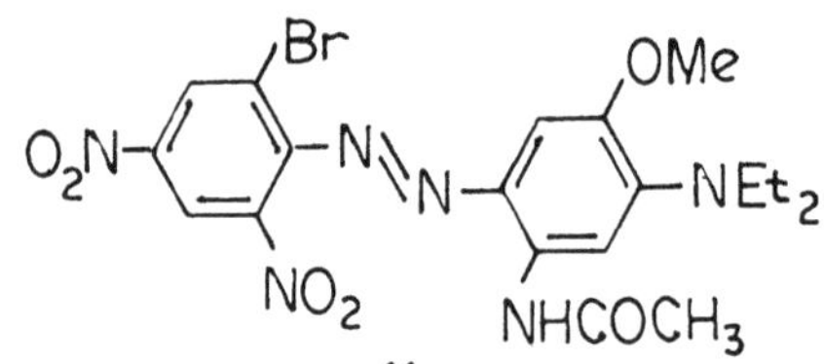

44
Disperse Blue
(Polyester)

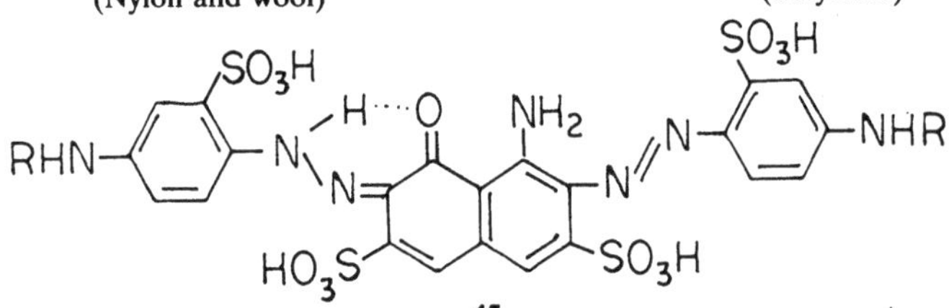

45
R = Reactive group
Reactive Green
(Cotton)

Most azoic dyes belong to the carbocyclic azo class. The only difference is that the dyes are formed in the fiber pores during the application process (see Chapter 3 and the *Color Index*[2]).

The carbocyclic azo class provides dyes having high cost effectiveness combined with good all-round fastness properties. However, one property they lack is brightness. Consequently, they cannot compete with anthraquinone dyes for brightness. This shortcoming of carbocyclic azo dyes is being overcome with the advent of the heterocyclic azo dyes.

F. Heterocyclic Azo Dyes

The two main thrusts of dyestuffs research have been improved technical excellence and greater cost effectiveness. One long-term aim had been to combine the brightness and high fastness properties of anthraquinone dyes with the strength and economy of azo dyes. This aim is now being realized with heterocyclic azo dyes.

Heterocyclic azo dyes fall rather conveniently into two main groups: those derived from heterocyclic coupling components and those derived from heterocyclic diazo components.

All the heterocyclic coupling components which provide commercially important azo dyes contain only nitrogen as the hetero atom. They are indoles **(46)**, pyrazolones **(47)**, and especially pyridones **(48)**; they provide yellow to orange dyes for various substrates.

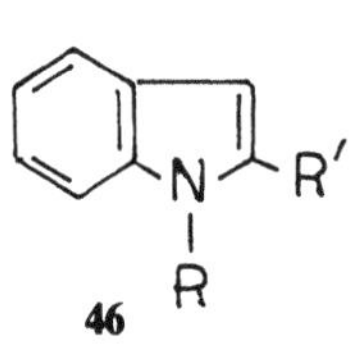

46

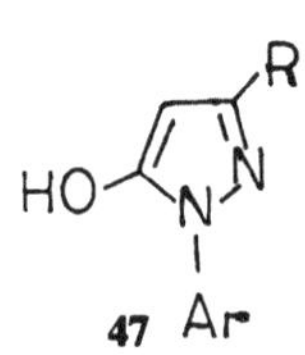

47

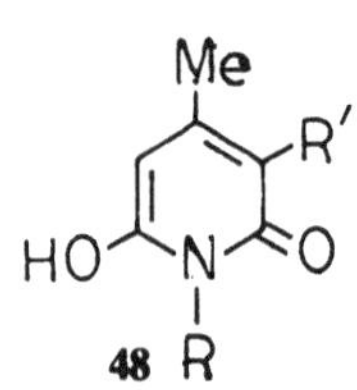

48

Until recently, many yellow dyes were of the azopyrazolone type. Nowadays, these have been largely superseded by the azopyridone dyes. Azopyridone yellow dyes are brighter, stronger, and generally have better fastness properties than azopyrazolone dyes. As discussed earlier, they exist in the hydrazone tautomeric form.

In contrast to the heterocyclic coupling components, all the heterocyclic diazo components that provide commercially important azo dyes contain sulfur, either alone, or in combination with nitrogen. These S or S/N heterocyclic azo dyes provide bright, strong shades that range from red, through blue, to green, and therefore complement the yellow/orange colors of the nitrogen heterocyclic azo dyes in providing a complete coverage of the entire shade gamut.

Representative dyes are the thiadiazole red **(49)**, the nitrothiazole reddish-blue **(50)**, and the thiophene greenish-blue **(51)**. All are disperse dyes for polyester, the second most important substrate after cellulose.

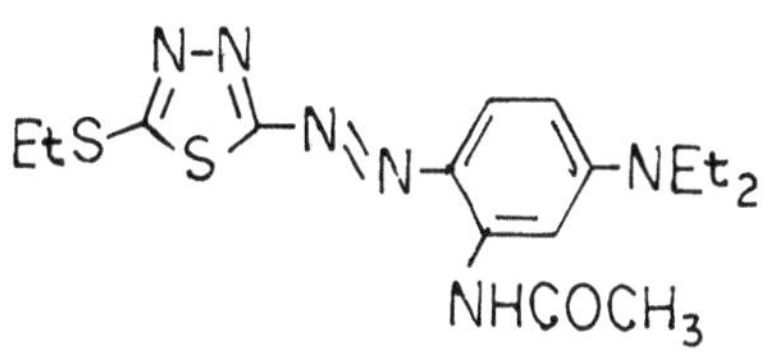

49

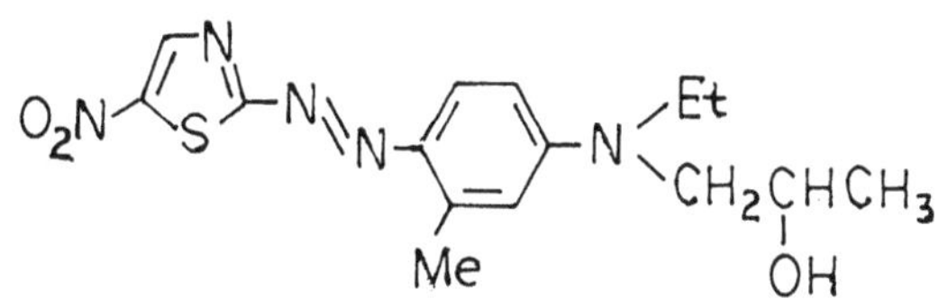

50

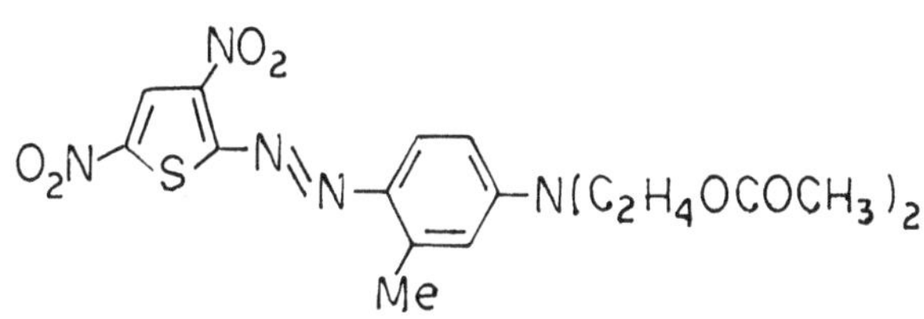

51

As a class, azo dyes are cost effective, span the whole shade gamut from yellow through red and blue to green, and possess good all-round fastness properties. Their main deficiency, which is being overcome with the heterocyclic azo dyes, is lack of brightness opposite their main rivals, the anthraquinone dyes.

III. ANTHRAQUINONE DYES

Anthraquinone dyes are the second most important class after azo dyes. They are also one of the oldest types of dyes since they have been found in the wrappings of mummies dating back over 4000 years. In contrast to the azo dyes, which have no natural counterparts, all the important natural red dyes were anthraquinones.[7]

A. Basic Structure

Anthraquinone dyes are based on 9,10-anthraquinone **(52)**.

9,10-Anthraquinone is essentially colorless. To produce commercially useful

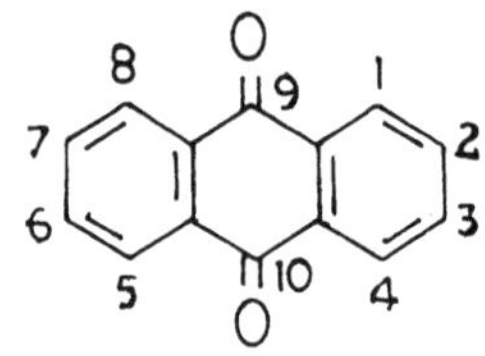

52

dyes, powerful electron-donor groups such as amino or hydroxy are introduced into one or more of the four alpha positions (1, 4, 5, and 8). The most common substitution patterns are 1,4-, 1,2,4-, and 1,4,5,8-. To maximize the properties, primary and secondary amino groups (not tertiary) and hydroxy groups are employed. These ensure the maximum degree of π-orbital overlap, enhanced by intramolecular hydrogen-bonding, with minimum steric hindrance. These features are illustrated in C.I. Disperse Red 60 **(53)**.

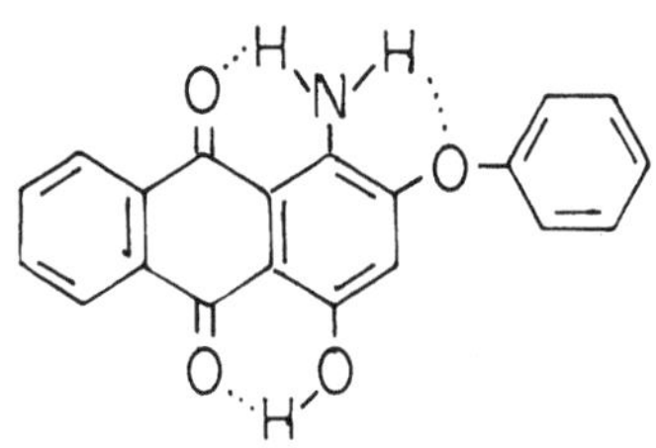

53

The strength of electron-donor groups increase in the order: OH < NH_2 < NHR < NHAr. Tetra-substituted anthraquinones (1,4,5,8-) are more bathochromic than di- (1,4-) or tri-substituted (1,2,4-) anthraquinones. Thus, by an appropriate selection of donor groups and substitution patterns, various colors can be achieved.

Typical dyes **(54)**–**(61)** are shown.

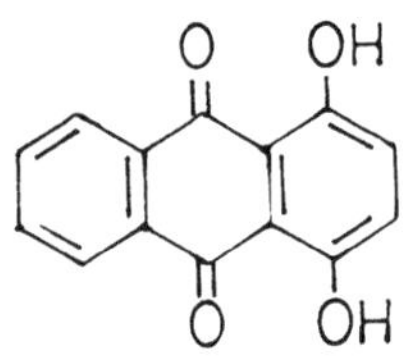

54
Quinizarin
(Orange)

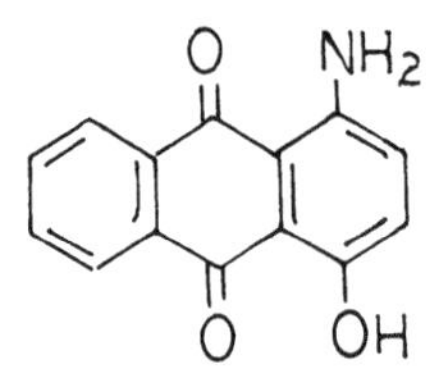

55
C.I. Disperse Red 15

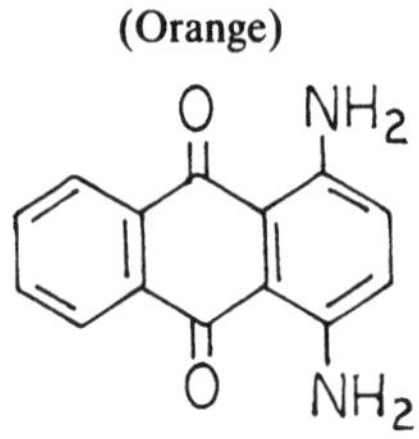

56
C.I. Disperse Violet 1

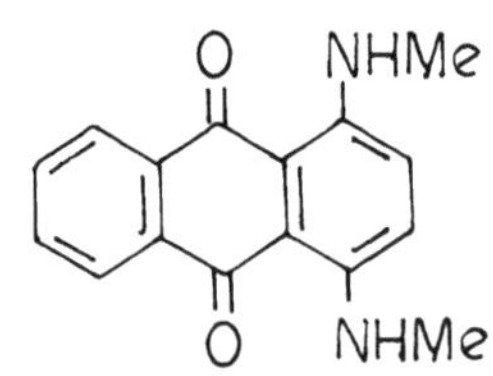

57
C.I. Disperse Blue 14

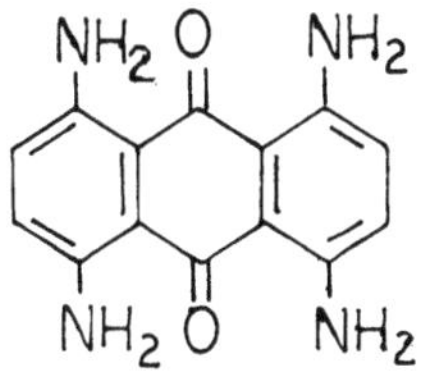

58
C.I. Disperse Blue 1

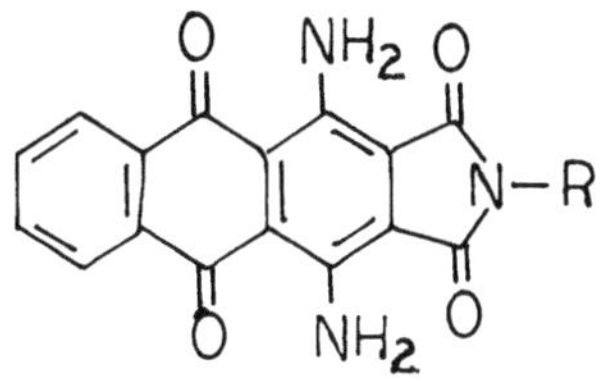

59
Disperse Turquoise

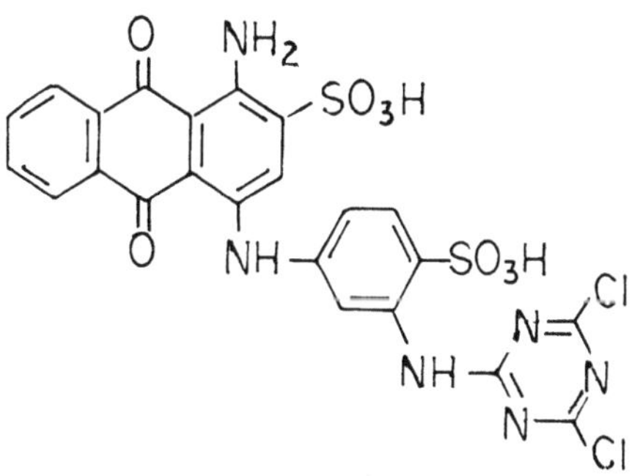

60
C.I. Reactive Blue 4

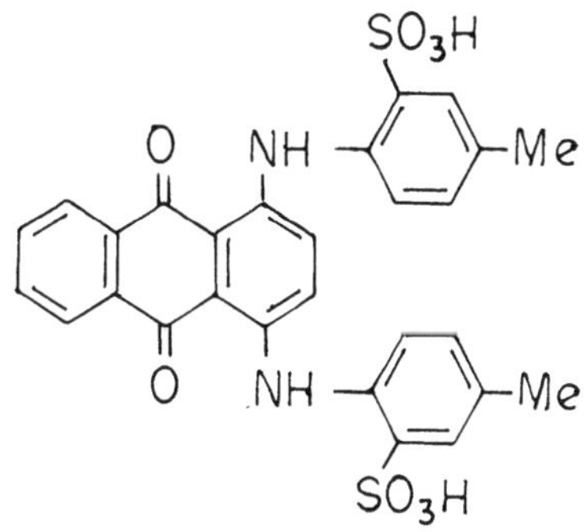

61
C.I. Acid Green 75

B. Synthesis

In complete contrast to the azo dyes, anthraquinone dyes are prepared by the stepwise introduction of substituents into the preformed anthraquinone skeleton **(52)**; see equation (3).

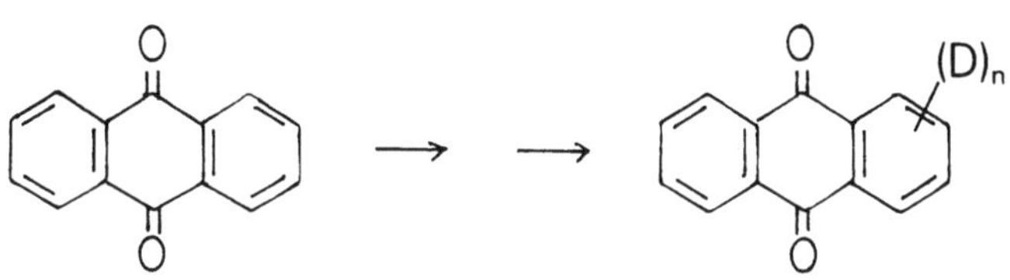

(3)

52

This highlights two very important points. First, the degree of freedom for producing a variety of different structures is restricted. Second, the availability of only eight substitution centers imposes a further restriction on synthetic flexibility. Therefore, there is nowhere near as much synthetic flexibility as there exists in azo dyes, and this is one of the drawbacks of anthraquinone dyes.

C. Tautomerism/Metal Complexes

Although tautomerism is theoretically possible in amino and hydroxy anthraquinone dyes, none has been observed. Studies by ^{13}C NMR have shown convincingly that amino and hydroxy dyes of 9,10-anthraquinone exist as such.[8]

The main attributes of anthraquinone dyes are brightness and good fastness properties, including light fastness. Metallization would detract from the former and there is no need to improve the latter. Consequently, metallized anthraquinone dyes are of no importance.

D. Properties

The major advantages of anthraquinone dyes are brightness and good fastness properties. However, these attributes are offset by their major disadvantage: they are not cost effective because they are both expensive and tinctorially weak. That they are still used extensively, particularly in the red and blue shade areas, is because other dyes could not provide the combination of properties offered by anthraquinone dyes, albeit at a price. That situation is now changing and anthraquinone dyes are being attacked increasingly by other dye types.

IV. BENZODIFURANONE DYES

Benzodifuranone (BDF) dyes are one such type that are attacking anthraquinone dyes.

The BDF chromogen is one of the very few novel chromogens to have been discovered this century. As with several major dye discoveries[9] the BDF chromogen was discovered by accident. Greenhalgh[10] and co-workers at ICI questioned the pentacenequinone structure **(62)** assigned by Junek to the intensely red-colored product obtained from the reaction of *p*-benzoquinone with cyanoacetic acid; see Scheme 3. (A compound such as **(62)** should hardly be colored owing to the lack of conjugation.) Instead, the red compound was correctly identified as the benzodifuranone **(63)**.

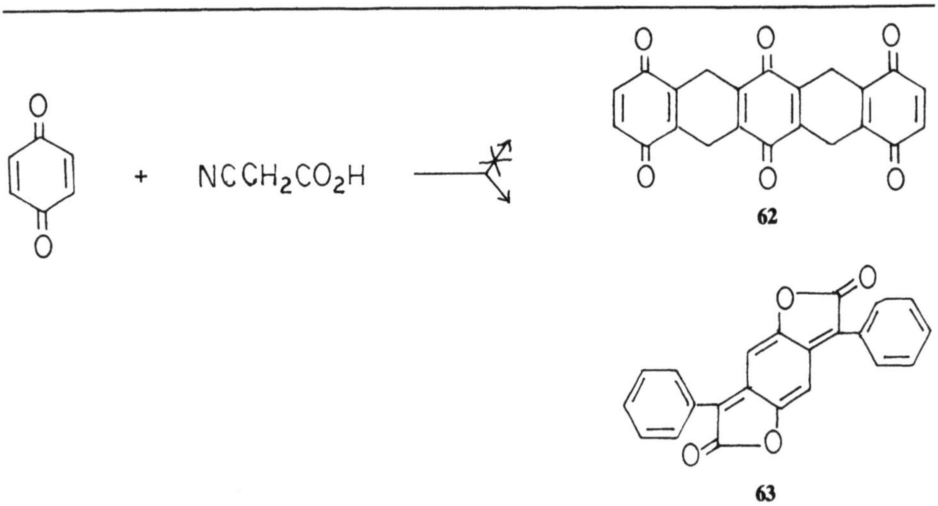

62

63

$\lambda_{max}^{CHCl_3}$ 466 nm ε_{max} 51,000

Scheme 3

BDFs are unusual in that they span the whole color spectrum from yellow through red to blue.[11–13] The first commercial BDF, a red disperse dye for polyester, is already making a tremendous impact. Its brightness even surpasses that of the anthraquinone reds, while its high tinctorial strength (*ca* 3–4 times anthraquinones) makes its cost effective.

V. POLYCYCLIC AROMATIC CARBONYL DYES

Structurally, these dyes contain one or more carbonyl groups linked by a quinonoid system. They tend to be relatively large molecules built up from smaller units, typically anthraquinones. Since they are applied to the substrate (usually cellulose) by a vatting process (see Chapter 7), the polycyclic aromatic carbonyl dyes are often called the anthraquinonoid vat dyes.

Although thc colors of the polycyclic aromatic carbonyl dyes cover the entire shade gamut, only the blue dyes and the tertiary shade dyes, namely, browns, greens, and blacks, are important commercially. Typical dyes are the blue **(64)**, indanthrone, the brown **(65)**, CI Vat Brown 3, the black **(66)**, CI Vat Black 27, and the green **(67)**, CI Vat Green 1, probably the most famous of all the polycyclic aromatic carbonyl dyes.

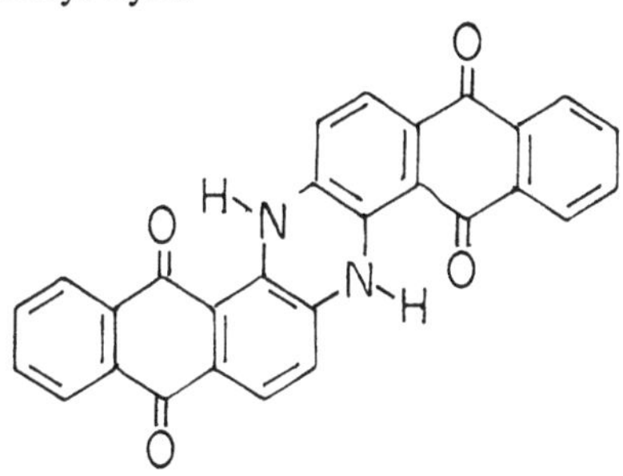

64

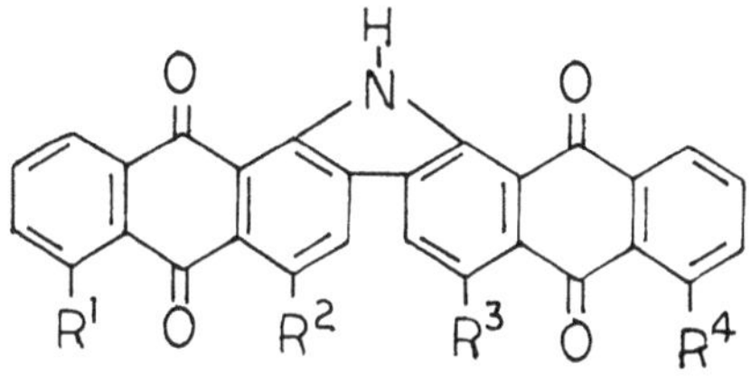

65 $R^1 = R^3 = PhCONH$; $R^2 = R^4 = H$
66 $R^1 = R^4 = H$; $R^2 = R^3 = PhCONH$

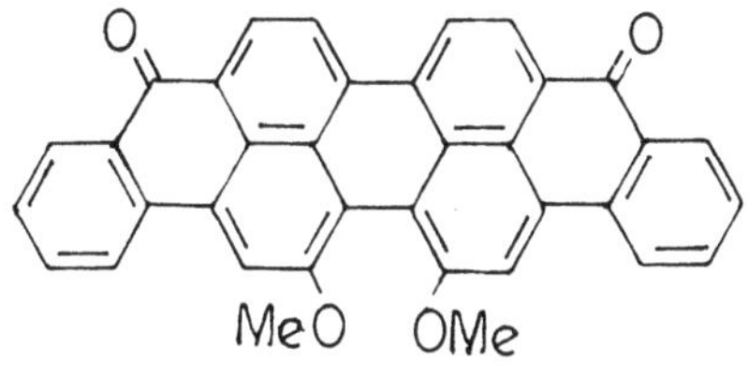

67

As a class, the polycyclic aromatic carbonyl dyes exhibit the highest order of light fastness and wet fastness. The high light fastness is undoubtedly associated with the absence of electron-donating and further electron-withdrawing groups, thus restricting the number of photochemical sites in the molecules. The high wet fastness is a direct manifestation of the application process of the dyes (see Chapter 7).

VI. INDIGOID DYES

Like the anthraquinone, benzodifuranone, and polycyclic aromatic carbonyl dyes, the indigoid dyes also contain carbonyl groups. They are also vat dyes.

Indigoid dyes represent one of the oldest known classes of dyes. For example, 6,6′-dibromoindigo **(68)** is Tyrian Purple, the dye made famous by the Romans. Tyrian Purple was so expensive that only the very wealthy were able to afford garments dyed with it. Indeed, the phrase "born to the purple" is still used today to denote wealth.

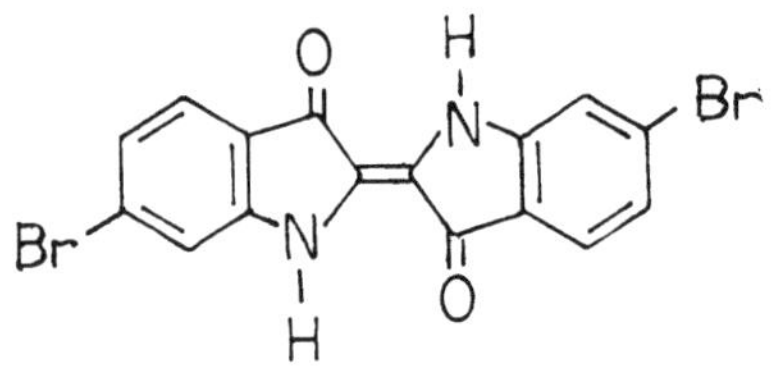

68

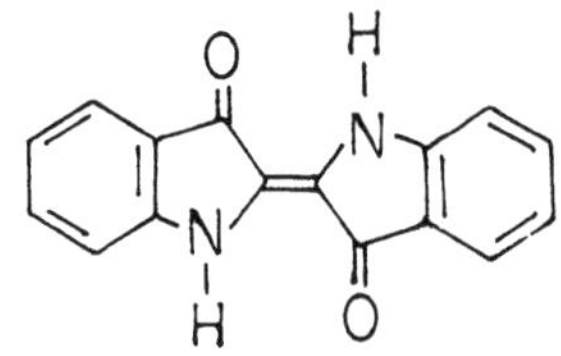

69

Although many indigoid dyes have been synthesized, only indigo itself **(69)** is of any importance today. Indigo is used almost exclusively for dyeing denim jeans and jackets. It is held in high esteem by the young who like its blue color and the fact that it fades on tone to give progressively paler blue shades.

One of the main fascinations of indigo is that such a small molecule should be blue. Normally, extensive conjugation, e.g., phthalocyanines (Section X), and/or several powerful donor and acceptor groups, e.g., azo and anthraquinone blues, are required to produce blue dyes. Several explanations[14] have been proposed to account for the blue color of indigo but the controversy was only resolved recently due to the classic work of Klessinger and Luttke. These workers proved beyond doubt that the chromogen (i.e., that molecular species responsible for the color) of indigo is the crossed conjugated system **(70)**.[15–17]

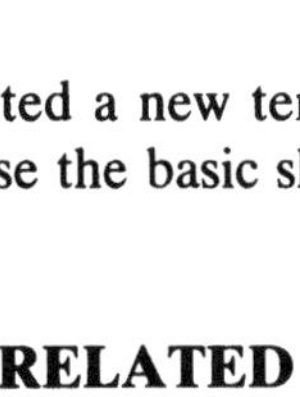

70

This surprising discovery prompted a new term for this kind of chromogen: it was called an H-chromogen because the basic shape resembled a capital H.[18]

VII. POLYMETHINE AND RELATED DYES

Cyanine dyes **(71)** are the best known polymethine dyes. Nowadays, their commercial use is limited to sensitizing dyes for silver halide photography.

However, derivatives of cyanine dyes provide important dyes for polyacrylonitrile (see Chapter 5).

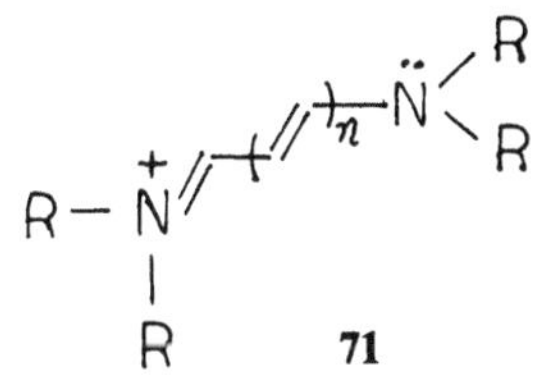

$n = 0, 1, 2, \ldots$

71

A. Azacarbocyanines

A cyanine containing three carbon atoms between the heterocyclic nuclei is called a carbocyanine **(72)**. Replacing these carbon atoms by one, two, and three nitrogen atoms produces azacarbocyanines **(73)**, diazacarbocyanines **(74)**, and triazacarbocyanines **(75)**, respectively. Dyes of these three classes are important yellow dyes for polyacrylonitrile (Chapter 5).

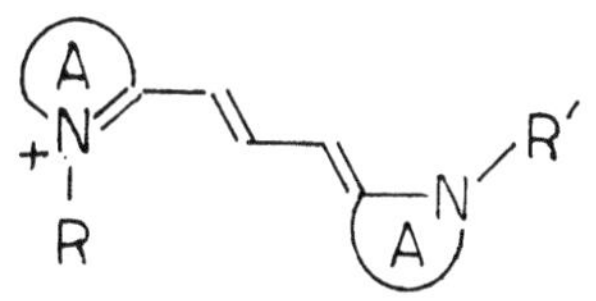

72

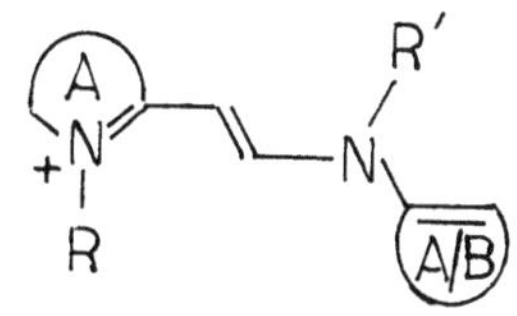

73

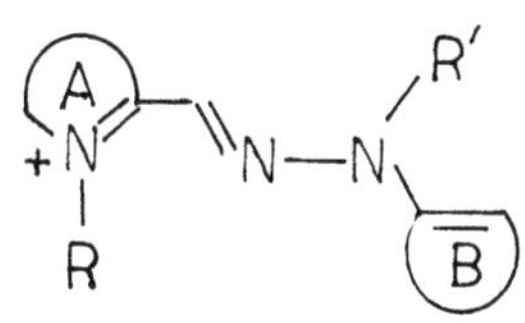

74

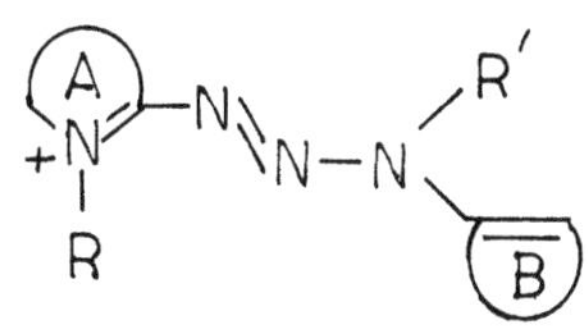

75

A = Heterocyclic ring
B = Carbocyclic ring

B. Hemicyanines

These "half-cyanine" dyes may be represented by structure **(76)**. They may be considered as cyanines in which a benzene ring has been inserted into the conjugated chain. Hemicyanines provide some bright fluorescent red dyes for polyacrylonitrile.

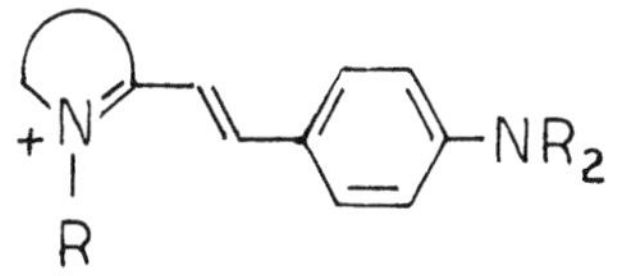

76

C. Diazahemicyanines

Diazahemicyanine dyes are arguably the most important class of polymethine dyes; they have the general structure **77**.

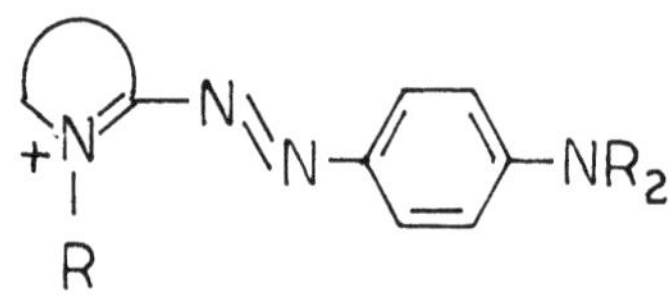

77

The heterocyclic ring is normally composed of one (e.g., pyridinium), two (e.g., pyrazolium and imidazolium), or three (e.g., triazolium) nitrogen atoms, or sulfur and nitrogen atoms [e.g., (benzo)thiazolium and thiadiazolium]. Triazolium dyes provide the market-leading red dyes for polyacrylonitrile, while a benzothiazolium dye is the market-leading blue dye (Chapter 5).

VIII. STYRYL DYES

Styryl dyes have been given a separate classification from polymethine dyes because they are not polymethine dyes. Polymethine dyes are defined by the *Color Index*[2] as dyes containing a conjugated chain of carbon atoms terminated by an ammonium group and, in addition, a nitrogen, sulfur, or oxygen atom, or an equivalent unsaturated group. The styryl dyes in this section are uncharged molecules containing a styryl group ($\rangle$C=C$\langle$) usually in conjugation with an NN-dialkylaminoaryl group.

Styryl dyes were once a fairly important group of yellow dyes for a variety of substrates. They are synthesized by condensation of an active methylene compound, especially malononitrile (**78**; X = Y = CN) with a carbonyl group, especially an aldehyde (**79**); see equation (4). As such, styryl dyes have small molecular structures and are ideal for dyeing densely packed hydrophobic substrates such as polyester. CI Disperse Yellow 31 (**80**), R = Bu, R^1 = C_2H_4Cl, R^2 = H, X = CN, Y = CO_2Et, is a typical dye.

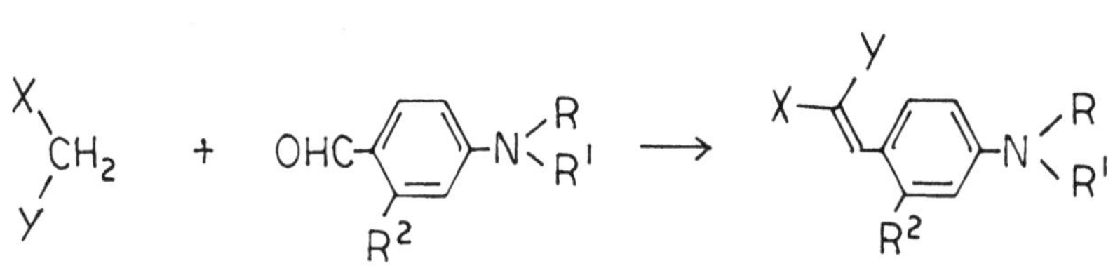

(4)

78 **79** **80**

Yellow styryl dyes have now been largely superseded by superior dyes such as azo pyridones (Section II.C), but there has been a resurgence of interest in red and blue styryl dyes.

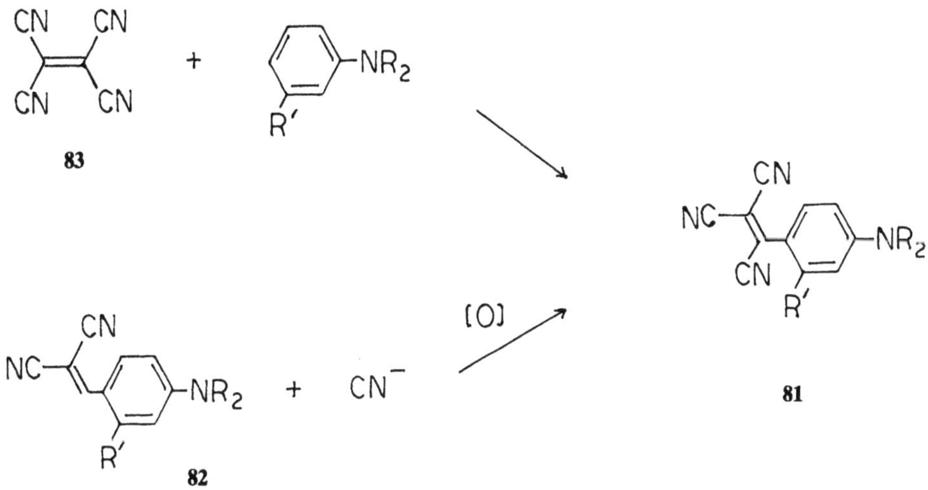

Scheme 4

The addition of a third cyano group to produce a tricyanovinyl group causes a large bathochromic shift: the resulting dyes (**81**, Scheme 4) are bright red rather than the greenish-yellow color of the dicyanovinyl dyes (**82**). These tricyanovinyl dyes have been patented by Mitsubishi[19] for the transfer printing of polyester substrates. The two synthetic routes to the dyes are shown in Scheme 4. One is by the replacement of a cyano group in tetracyanoethylene (**83**) and the second is by oxidative cyanation of a dicyanovinyl dye with cyanide. The use of such toxic reagents could hinder the commercialization of the tricyanovinyl dyes.

Blue styryl dyes are produced when an even more powerful electron-withdrawing group than tricyanovinyl is used. Thus, Sandoz discovered that the condensation of (**84**) with an aldehyde gives the bright blue dye (**85**) for polyester.[11]

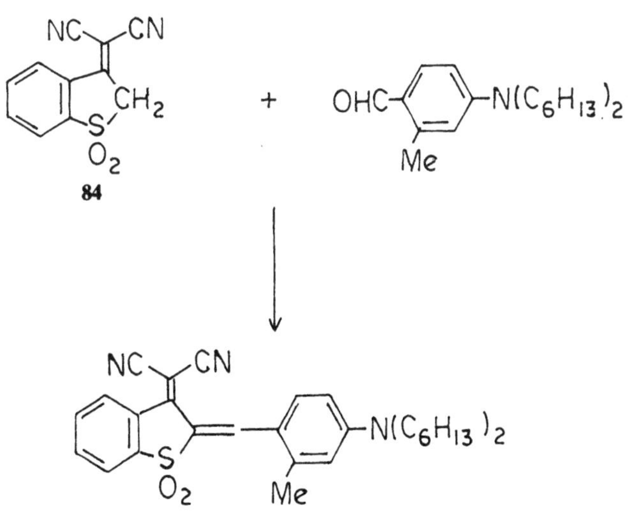

85

In addition to exceptional brightness, this dye also possesses high tinctorial strength ($\varepsilon_{max} \sim 70{,}000$). However, its light fastness is only moderate.

IX. DI- AND TRI-ARYL CARBONIUM AND RELATED DYES

The structural interrelationships of the diarylcarbonium dyes **(86)**, triarylcarbonium dyes **(87)**, and their heterocyclic derivatives are shown in

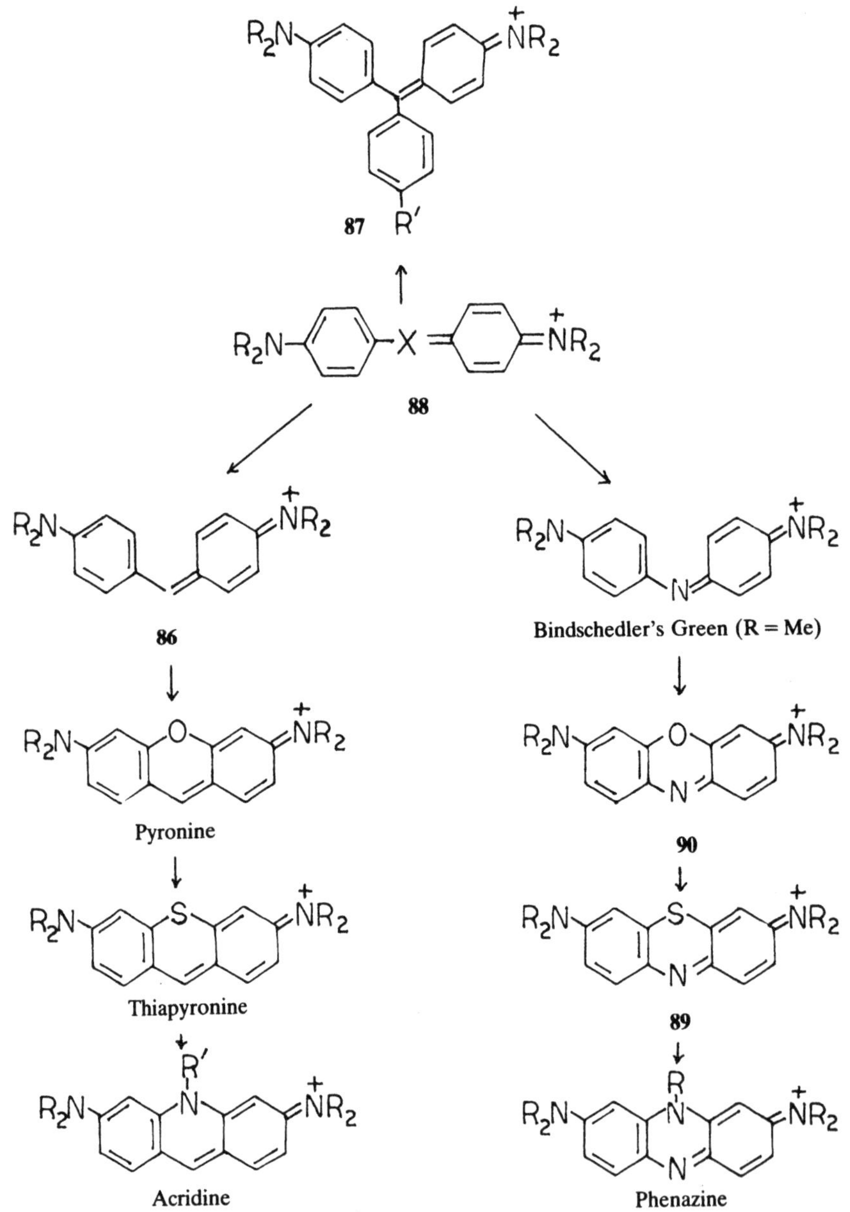

Scheme 5

Scheme 5. As a class, the dyes are bright and strong but are generally deficient in light fastness. Consequently, they are used in outlets where brightness and cost effectiveness, rather than permanence, are paramount, for example, the coloration of paper. Many dyes of this class, especially derivatives of pyronines (xanthenes), are among the most fluorescent dyes known.

Typical dyes are the diphenylmethane, Auramine O (**88**; R = Me, X = C—NH_2; CI Basic Yellow 2) the triphenylmethane, Malachite Green (**87**; R = Me; R′ = H; CI Basic Green 4), the thiazine dye Methylene Blue (**89**; R = Me; CI Basic Blue 9)—used extensively in the Gram staining test for bacteria—and the oxazine dye, CI Basic Blue 3 (**90**; R = Et). The last dye is used to some extent for dyeing acrylic fibers (see Chapter 5).

It is worth discussing the oxazine dyes in a little more detail due to the notable advances made in recent years in triphendioxazine dyes. Triphendioxazine direct dyes have been known for many years with CI Direct Blue 106 (**91**) being a typical dye.[2] However, ICI[20] and Parton created a resurgence of interest in triphendioxazine dyes by successfully modifying the intrinsically strong and bright triphendioxazine chromogen to produce blue reactive dyes for cotton. These blue reactive dyes combine the advantages of azo dyes and anthraquinone dyes. Thus they are bright, strong dyes with good fastness properties. Structure (**92**) is typical of these reactive dyes.

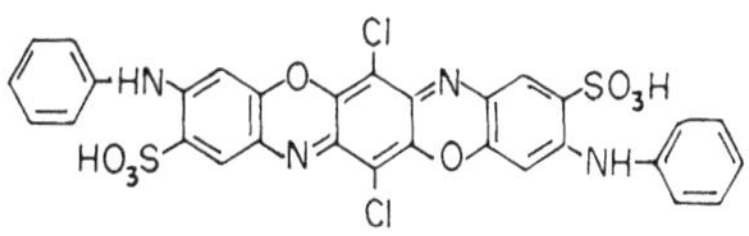

91

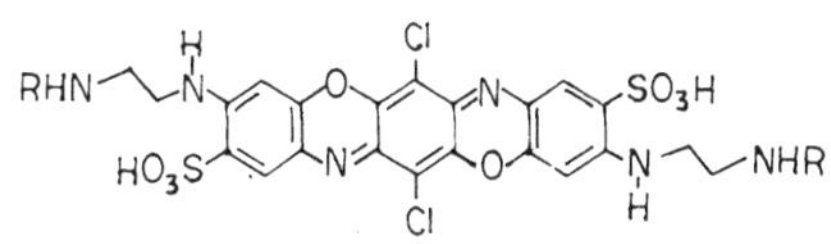

92
R = Reactive group

Like phthalocyanines (see Section X), triphendioxazines are large molecules and therefore their use will be restricted to coloring the more open-structured substrates such as paper and cotton.

X. PHTHALOCYANINES

Phthalocyanine is the only novel chromogen of major commercial importance discovered since the nineteenth century. As was the case for many of the major dye discoveries, phthalocyanines were discovered by accident. In 1928, during the routine manufacture of phthalimide from phthalic anhydride and ammonia (equation 5) it was found that the product contained a blue contaminant. Chemists of Scottish Dyes Ltd, now part of ICI, carried out an

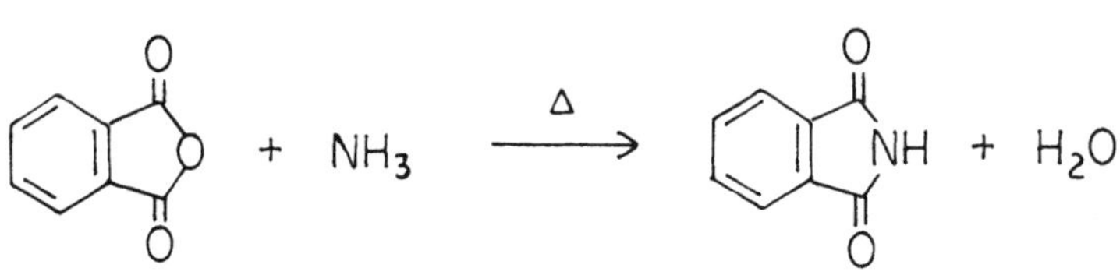

(5)

independent synthesis of the blue material by passing ammonia gas into molten phthalic anhydride containing iron filings. The importance of the colorant was realized (it was intensely colored and very stable) and a patent application was filed in the same year.

The structure of the blue material was not elucidated until 1934, when Linstead, from the Imperial College of Science and Technology, London, showed it to be the iron complex of **(93)**.[21] Linstead christened the new material phthalocyanine, reflecting both its origin (from phthalic anhydride) and its beautiful blue color (like cyanine dyes). A year later Robertson,[22] in one of the first uses of X-ray crystallography, confirmed Linstead's structure.

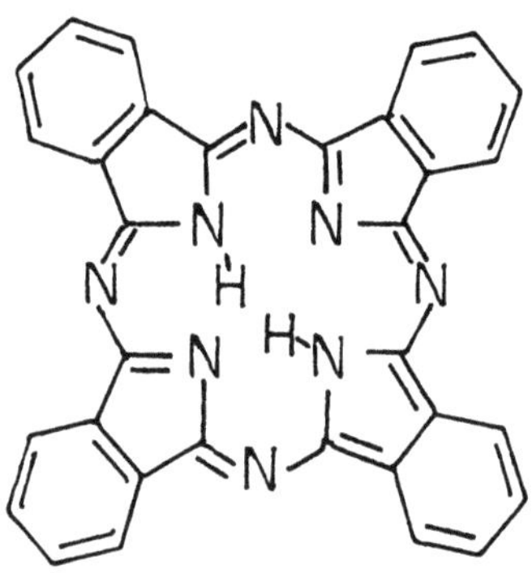

93

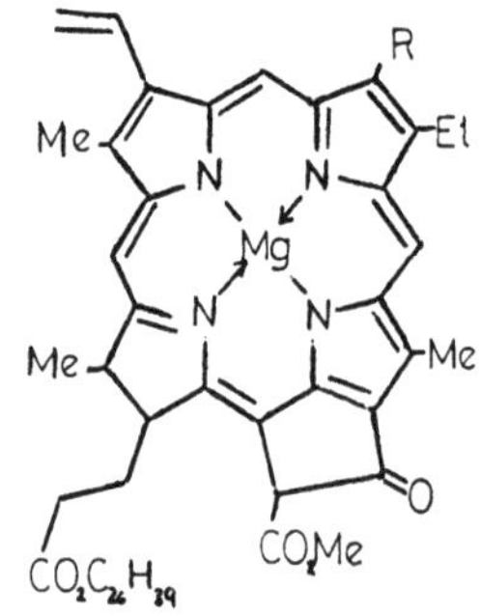

94
R = Me (Chlorophyll a)
R = CHO (Chlorophyll b)

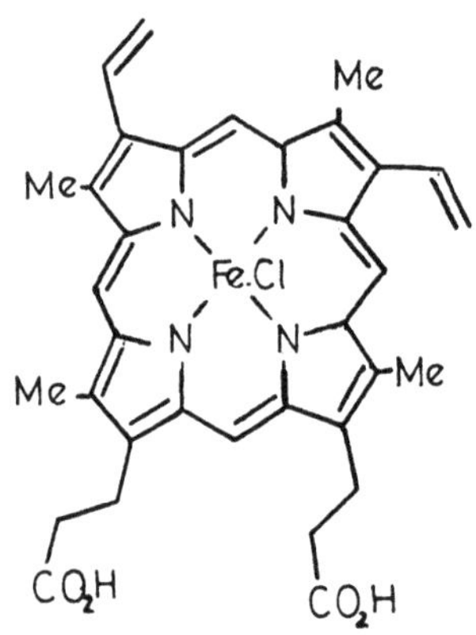

95

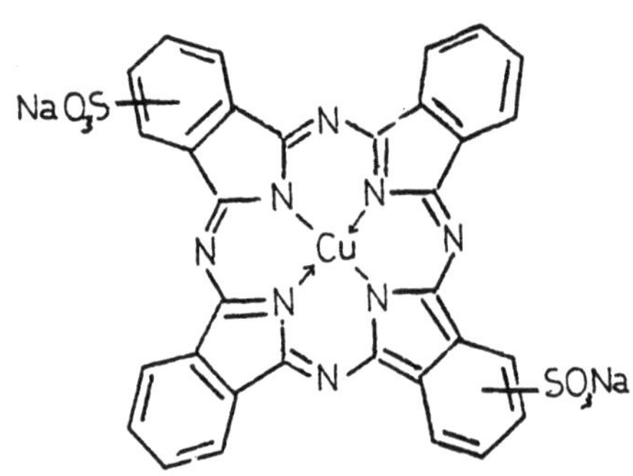

96

Phthalocyanines are analogs of the natural pigments chlorophyll **(94)** and haemin **(95)**. However, unlike these natural pigments, which have extremely poor stability, phthalocyanines have exceptional stability and are probably the most stable of all the colorants in use today. Substituents have only a minor effect on the color of phthalocyanines and so their hues are restricted to blue and green.

Of all the metal complexes evaluated, copper phthalocyanines gave the best combination of color and properties and consequently the majority of phthalocyanine dyes are based on copper phthalocyanine; CI Direct Blue 86 **(96)** is a typical dye.

As well as being extremely stable, phthalocyanines are bright and tinctorially strong ($\varepsilon_{max} \sim 100{,}000$); this renders them cost effective. Consequently, phthalocyanines are used extensively in printing inks and paints. The preponderance of blue and green labeling on products is testament to the popularity of phthalocyanine-based printing inks and the majority of blue and green cars are the product of phthalocyanine-based paints.

XI. QUINOPHTHALONES

Like the hydroxy azo dyes (Section II.C), quinophthalone dyes can, in theory, exhibit tautomerism. Since the dyes are synthesized by the condensation of quinaldine derivatives **(97)** with phthalic anhydride **(98)**, they are often depicted as structure **(99)** see Scheme 6. This is incorrect, since the two single bonds prevent any conjugation between the two halves of the molecule. The dyes exist as structure **(100)**, in which the donor pyrrole-type nitrogen atom is conjugated to the two acceptor carbonyl groups via an ethylenic bridge. In addition to the increased conjugation, structure **100** is stabilized further by the six-membered intramolecular hydrogen-bond between the imino hydrogen atom and the carbonyl oxygen atom.

Quinophthalones provide important dyes for the coloration of plastics (e.g., CI Solvent Yellow 33; **100**, R = H) and for the coloration of polyester. For example, CI Disperse Yellow 54 (**100**; R = OH) is the leading yellow dye for the transfer printing of polyester.

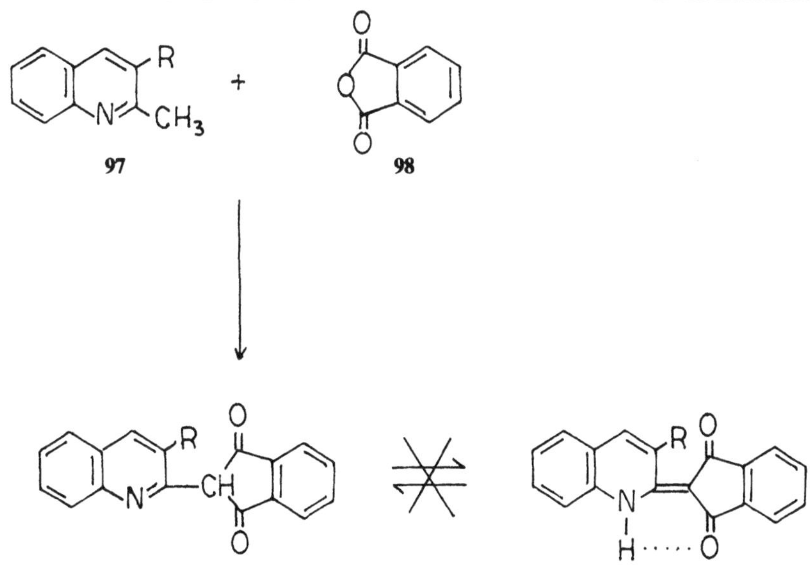

Scheme 6

XII. SULFUR DYES

Sulfur dyes are synthesized by heating aromatic amines, phenols, or nitro compounds with sulfur or, more usually, alkali polysulfides.[23] Unlike most other dye types, it is not easy to define a chromogen for the sulfur dyes. It is likely that they consist of macromolecular structures of the phenothiazonethianthrone type **(101)**, in which the sulfur is present as (sulfide) bridging links and as thiazine groups.[24]

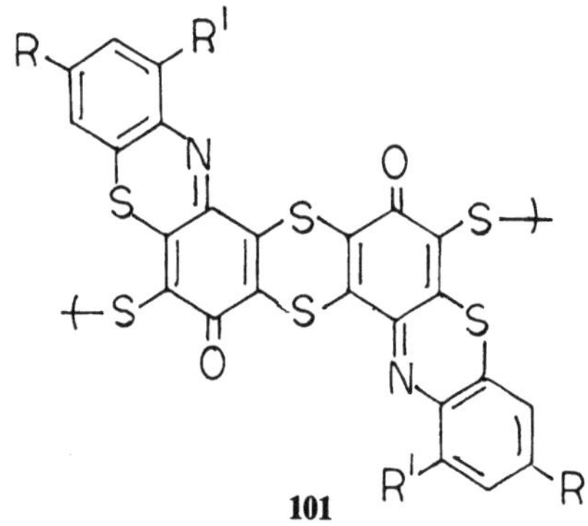

101

Sulfur dyes are used for dyeing cellulosic fibers. They are insoluble in water and are reduced to the water-soluble leuco form for application to the substrate by using sodium sulfide solution. The sulfur dye proper is then formed within the fiber pores by atmospheric oxidation.[25]

Sulfur dyes constitute an important class of dye for producing cost-effective tertiary shades, especially black, on cellulosic fibers. One of the most important dyes is CI Sulfur Black 1, prepared by heating 2,4-dinitrophenol with sodium polysulfide; see equation (6).

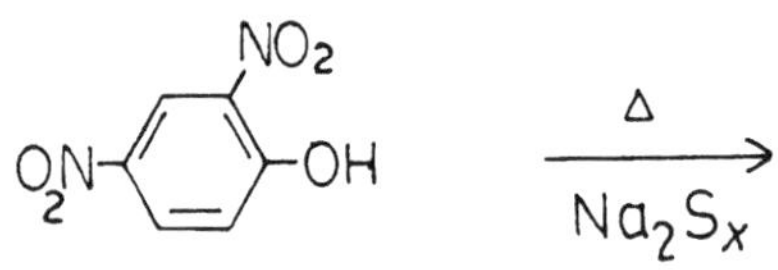

C.I. Sulfur Black 1 (6)

XIII. NITRO AND NITROSO DYES

These dyes are now of only minor commercial importance but are of interest for their small molecular structures.

The early nitro dyes were acid dyes used for dyeing the natural animal fibers such as wool and silk. They were nitro derivatives of phenols, e.g., Picric Acid, CI 10305 **(102)**, or naphthols, e.g., CI Acid Yellow 1 **(103)**.

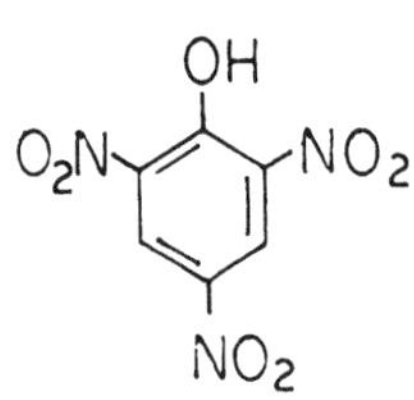

102

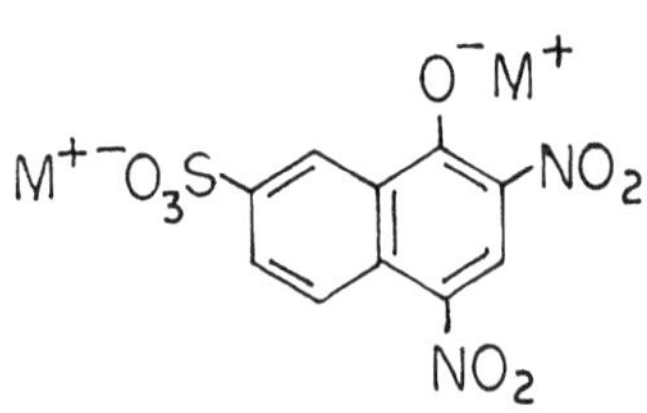

103

$M^+ = Na^+, K^+$

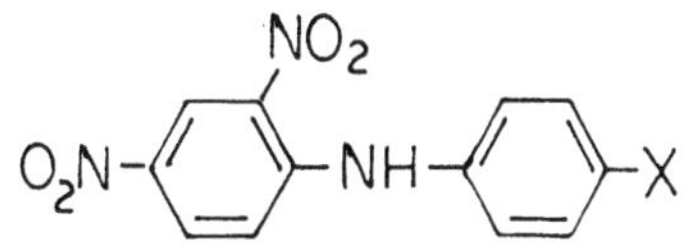

(104)

X = H C.I. Disperse Yellow 14
OH C.I. Disperse Yellow 1
NH_2 C.I. Disperse Yellow 9

The most important type of nitro dyes are the nitrodiphenylamines of general structure **104**. These small molecules are ideal for penetrating dense fibers such as polyester and are therefore used as disperse dyes for polyester. All the important dyes are yellow. Although the dyes are not terribly strong ($\varepsilon_{max} \sim 20{,}000$), they are cost effective because of their easy synthesis from inexpensive intermediates (equation 7).

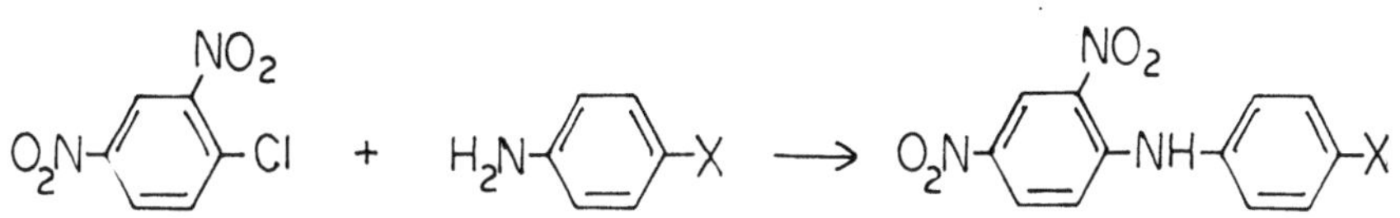

(7)

Nitroso dyes are metal-complex derivatives of *o*-nitrosophenols or naphthols. Tautomerism is possible in the metal-free precursor between the nitrosohydroxy tautomer **(105)** and the quinoneoxime tautomer **(106)**.[26]

The only nitroso dyes important commercially are the iron complexes of sulfonated 1-nitroso-2-naphthol, e.g., CI Acid Green 1 **(107)**; these inexpensive colorants are used mainly for coloring paper.

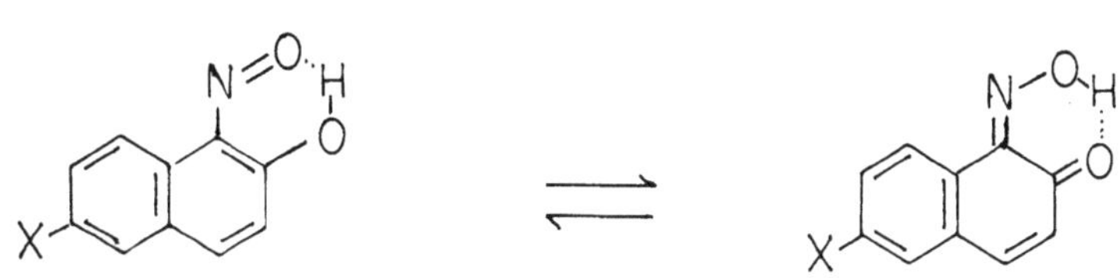

105 **106**

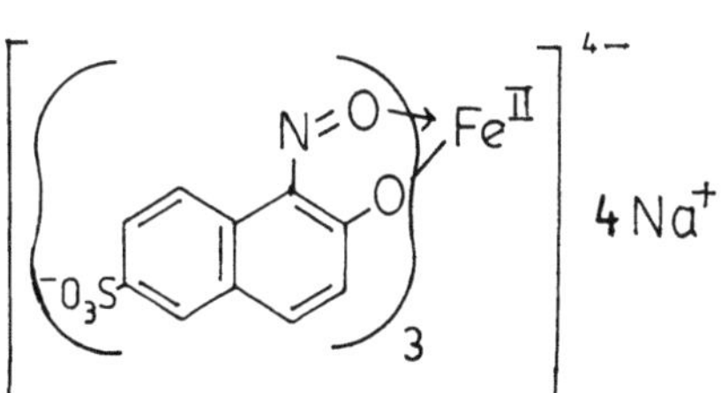

107

XIV. MISCELLANEOUS DYES

Other classes of dyes which still have some importance are the stilbene dyes and the formazan dyes.

Stilbene dyes are in most cases mixtures of dyes of indeterminate constitution which are formed from the condensation of sulfonated nitroaromatic compounds in aqueous caustic alkali either alone or with other aromatic compounds, typically arylamines.[2] The sulfonated nitrostilbene **(108)** is the most important nitroaromatic and aminoazobenzenes are the most important arylamines. CI Direct Orange 34, the condensation product(s) of **(108)** and the aminoazobenzene **(109)**, is a typical stilbene dye.

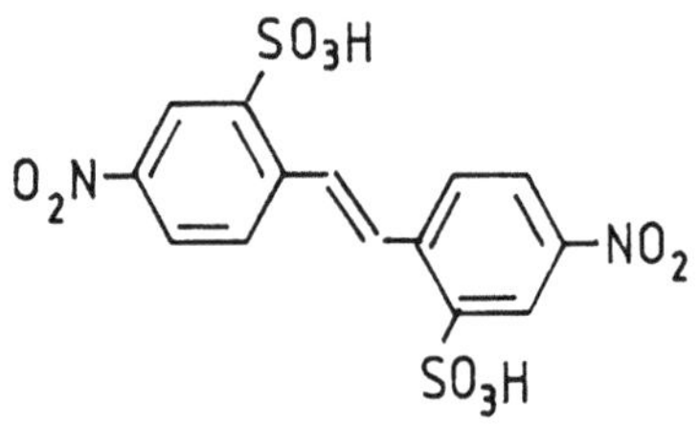

108

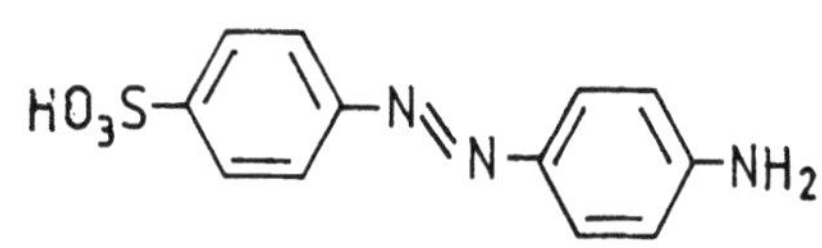

109

Formazan dyes bear a formal resemblance to azo dyes, since they contain an azo group but have sufficient structural dissimilarities to be considered as a separate class of dyes. The most important formazan dyes are the metal complexes, particularly copper complexes, of tetradentate formazans.[27,28] They are used as reactive dyes for cotton, with **(110)** being a representative example.

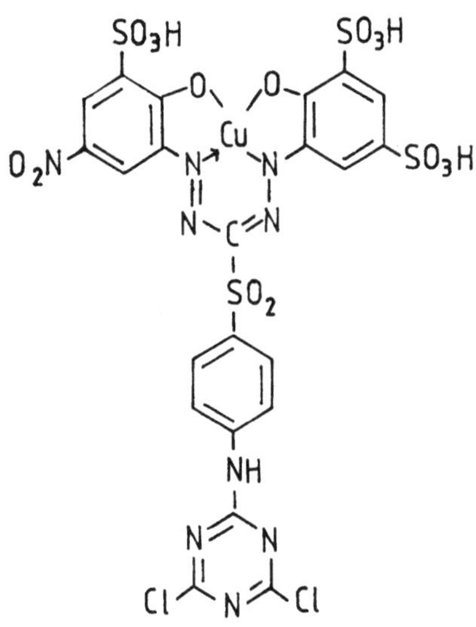

110

XV. SUMMARY

The best way to classify dyes as far as the dye chemist and end user is concerned is by chemical structure. This produces a manageable number of about one dozen classes.

Azo dyes are by far the most important class. Indeed, of the current commercial dyes over 50% are azo dyes. Anthraquinones constitute the second most important class of dye but are under attack from both heterocyclic azo dyes and from the newer chromogens such as benzodifuranones.

REFERENCES

1. J. Griffiths, *Color and Constitution of Organic Molecules,* Academic Press, London (1976).
2. *Color Index,* 3rd Edition, Vol. 4, The Society of Dyers and Colorists (1971).
3. T. Zincke and H. Binderwald, *Chem. Ber. 17,* 3026 (1884).
4. P. Gregory, *Dyes and Pigments 7,* 45 (1986).
5. P. F. Gordon and P. Gregory, *Organic Chemistry in Color,* pp. 104–108, Springer-Verlag, Berlin, Heidelberg, New York (1983).
6. Ref. 5, pp. 96–115.
7. Ref. 5, pp. 3 and 4.
8. Ref. 5, pp. 164–169.
9. Ref. 5, Chapter 1.
10. C. W. Greenhalgh, J. L. Carey, and D. F. Newton, *Dyes and Pigments 1,* 103 (1980).
11. J. Griffiths, in: *Developments in the Chemistry and Technology of Organic Dyes* (J. Griffiths, ed.), p. 16, Blackwell, Oxford (1984).
12. C. W. Greenhalgh, J. L. Carey and D. F. Newton, British Patent 1,568,231 (1976).
13. C. W. Greenhalgh and J. L. Carey, British Patent 2,055,881A (1980).
14. Ref. 5, pp. 211–213.
15. W. Luttke and M. Klessinger, *Chem. Ber. 97,* 2342 (1964).
16. W. Luttke, H. Hermann, and M. Klessinger, *Angew. Chem. Int. Ed. Engl. 5,* 598 (1966).
17. M. Klessinger, *Tetrahedron 19,* 3355 (1966).
18. M. Klessinger and W. Luttke, *Tetrahedron 19* (Suppl. 2), 315 (1963).
19. Mitsubishi, Japanese Patent 59,078,896 (1982).
20. B. Parton, British Patent 1,349,513 (1970).
21. R. P. Linstead, *J. Chem. Soc.,* 1016, 1017, 1022, 1027, 1031, and 1033 (1934).
22. J. M. Robertson, *J. Chem. Soc.,* 615 (1935).
23. Ref. 2, p. 4475.
24. Ref. 5, p. 16.
25. Ref. 2, Vol. 3.
26. Ref. 5, p. 257.
27. R. Price, The Chemistry of Metal Complex Dyestuffs, pp. 373–383, in: *The Chemistry of Synthetic Dyes* (K. Venkataraman, ed.), Vol. III, Academic Press, New York and London (1970).
28. K. H. Schundehutte, Reactive Dyes: Chromophore Systems, pp. 287–291, in: *The Chemistry of Synthetic Dyes* (K. Venkataraman, ed.), Vol. VI, Academic Press, New York and London (1972).

3

Dyes for Cellulosic Fibers

DAVID R. WARING

I. INTRODUCTION

Although the term "cellulosic fibers" is frequently used, the type is massively dominated by cotton. Other fibers which have been embodied as cellulosic are viscose rayon, linen, cuprammonium rayon, and jute and these can be dyed with dyes related to those used on cotton. Paper and leather can also be dyed similarly. Cellulose acetates, although clearly cellulosic in structure, behave very differently to cotton and are dyed using hydrophobic dyes of the disperse type (Chapter 4). In a survey conducted in 1982, cotton production accounted for 42.8% (at 15.5 Mtons) of the total world fiber production.[1] Thus, cotton production equalled the total output of man-made fibers. The importance of cotton, alone or in fiber blends, particularly with polyester, is self-evident and during the development of the dye industry several dye classes have emerged for its coloration.

It is generally considered that the most significant technical advance in the dye industry in recent years was the introduction of fiber-reactive dyes for cellulose by ICI in 1956.[2] The observations and work of Rattee[3] and Stephen[4] on the formation of a covalent bond between cellulose and 2,4-dichloro-*s*-triazinyl dyes, with subsequent improvements in wash fastness, led to a proliferation of reactive systems and a profusion of choice of hue for the colorist.

It might perhaps have been expected that such an advance would have led to a domination of cotton dyeing and printing in the thirty years or so since the introduction of reactive dyes. In fact, fiber-reactive dyes currently account for only 6% of the total world dye production and as little as 9·7% of the market for dyes applied to cellulosics.[5] Sulfur dyes and direct dyes, first discovered around 100 years ago, continue to dominate the market with 29% and 27.4%, respectively. The remainder is made up by vat dyes, 19.4%; azoic dyes, 8.1%; and pigments, 6.4%.

DAVID R. WARING • Kodak Limited, Acornfield Road, Kirkby, Liverpool L33 7UF, England.

The regional variation in dye usage and the employment of older traditional methods in less sophisticated markets was reflected in a survey carried out in Japan in 1979[2,6] where reactive dyes have steadily increased their market share mainly at the expense of sulfur dyes and in 1979 accounted for 33.3% of that market. The survey indicated that the remainder of the cellulosic market was taken up by direct dyes, 31.5%; sulfur dyes, 16.3%; azoic dyes, 9.6%; and vat dyes, 9.3%. It is not unreasonable to assume that the usage in the Western world more closely approximates the Japanese picture than that of the worldwide survey. However, sulfur and direct dyes maintain their major share worldwide mainly because of economic considerations, due to low price and ease of application.

Cellulose is an amorphous and hydrophilic fiber, being a linear polymer of glucose units. As such, the only functional groups are alcoholic hydroxyl groups. Thus, dye retention mechanisms fall into one of three types. In the case of azoic, vat, and sulfur dyes, water-soluble precursors are applied to the fiber and then, after generation of the dye, are caused to aggregate to form water-insoluble molecules which are, essentially, trapped within the fiber pores. Direct dyes rely upon van der Waals forces and the formation of a number of hydrogen bonds between a long planar dye molecule, containing a multiplicity of hydrogen bonding centers, and the minimum number of sulfonic acid groups required to give adequate water solubility. Fiber reactive dyes, which usually have excellent wash fastness, achieve fixation by the formation of covalent bonds between the reactive groups on the dye and the hydroxyl groups on the cellulose.

II. FIBER REACTIVE DYES

A. Historical Development

Reactive dyes are colored compounds capable of forming a covalent bond between the dye molecule and the fiber. Rys and Zollinger,[7] in their definition, described the covalent bond as being formed between a carbon atom of the dye molecule and an oxygen, nitrogen, or sulfur atom of a hydroxy, amino, or mercapto group of the substrate. Commercial fiber reactive dyes have been developed for both wool and polyamide but, undoubtedly, the major success has been in the application to cotton and its blends.

Attempts to produce fiber reactive systems are not confined to recent times, although until 1953 they were rather unsuccessful and spasmodic. In the 1890s Cross and Bevan[8] showed that if cellulose was treated with strong alkali it was converted into "soda cellulose," which would undergo esterification or ether formation. Practical application of this original observation was not achieved until 1953 despite the demonstration in 1924 of a process for coloring cellulose using isatoic anhydride, applied at 95 °C in the presence of sodium carbonate (equation 1).[9] Similar conditions were applied in 1926 to the reaction with 4-nitrobenzyldimethylphenylammonium hydroxide.[10] The resulting cellulose ether was reduced to the primary aromatic amine derivative, which was then diazotized

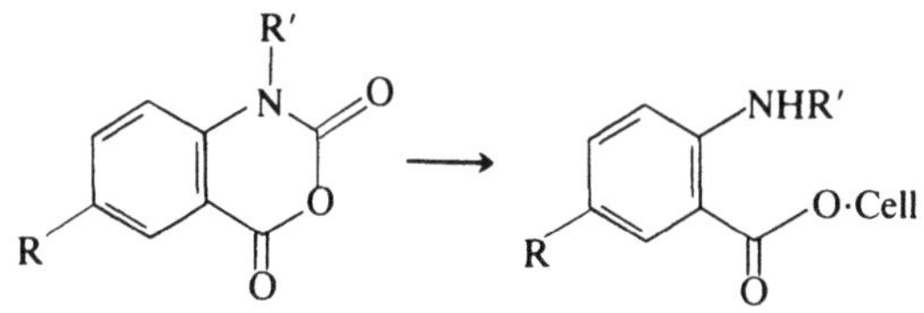

(1)

and coupled to produce an azo dye covalently bound to the fiber. In 1943, the Russians Kursanov and Solodkov[11] approached the problem in a similar manner using 3-nitrobenzyloxymethylpyridinium chloride.

A major breakthrough came in 1953, following the work of Rattee[3] and Stephen[4] at ICI, with the application of 2,4-dichloro-1,3,5-triazin-6-yl dyes to cotton. The first commercial reactive dyes were the initial members of what is now known as the Pocion MX range.[12] The dyes are formed by reaction of an amino group on the chromogen or dye-base with cyanuric chloride in weakly acidic conditions. Usually a primary aromatic amino group is involved although secondary amino groups and, less frequently, aliphatic amino groups have also been used. Cyanuric chloride, produced by trimerization of cyanogen chloride over charcoal at 250–500 °C or similarly over charcoal impregnated with an alkaline earth metal chloride, is a very reactive molecule which behaves essentially as an acylating agent. The electronegative character of the ring nitrogen atoms and of the chlorine substituents induces a relative positive charge on the ring carbon atoms, which are thus highly susceptible to nucleophilic attack by molecules containing amino (equation 2), hydroxyl, or alkoxyl groups. Since the inclusion of an amino group in the chromogen is usually straightforward, a full spectrum of dyes containing the reactive system was quickly established. Dye formation (equation 2) is achieved smoothly at 0–5 °C and the usual procedure involves mixing an aqueous solution of the dye with a suspension of finely divided cyanuric chloride. The pH is maintained as close to 6 as possible by titration of alkali into the reaction mixture, so as to neutralize the liberated hydrogen chloride.

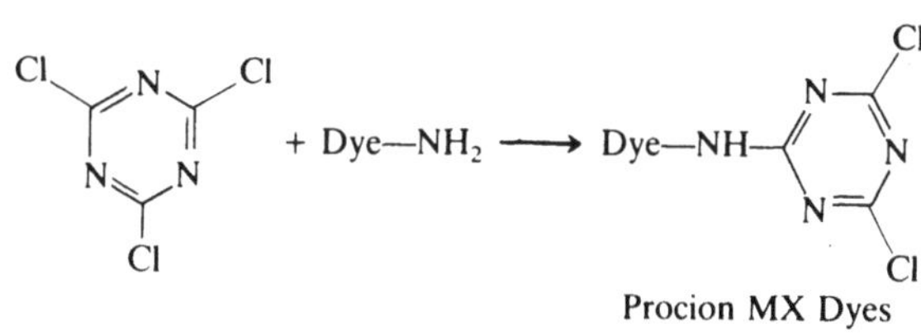

(2)

The dye structures thus produced are essentially in two parts, the chromogen and the reactive system, and it is convenient to consider them separately since many chromogens are common to several dye ranges and only the reactive systems differ.

While considering the *s*-triazinyl system, it is worth noting that it has been used for many years in the synthesis of direct dyes, notably the Chlorantine dyes of Ciba. Subsequently, Ciba were able, because of earlier patents disclosing the use of cyanuric chloride, to market the Cibacron reactive dyes suitable for hot dyeing applications.[13]

B. Reactive Systems

The great convenience of cyanuric chloride lies in the specificity of the reactivity of its chlorine atoms. Thus, one chlorine atom only can be displaced by a nucleophile at 0–5 °C, the second at 35–40°C, and the third at 80–85 °C. Much use of this relative reactivity is made in designing dyes for specific applications.

Following condensation of the chromogen with cyanuric chloride, the reactivity of the remaining chlorine atoms is diminished by the influence of the lone pair of electrons on the bridging nitrogen atom which donates electron density into the triazine ring. Nevertheless, the chlorine atoms are still highly susceptible to further reaction, notably hydrolysis, and under acidic conditions the relative stabilizing effect of the nitrogen lone pair is negated by protonation and the chlorine atoms increase in reactivity. Initially, this reactivity diminished the potential shelf life of a commercial product but the problem was solved by the incorporation of a buffer in the dye powder,[14] thereby preventing the essentially autocatalytic hydrolysis.

The dichloro-*s*-triazinyl systems (Procion MX) are thus susceptible to nucleophilic attack at the second chlorine atom at 35–40 °C and it is at this temperature that reaction takes place with cellulose. The basis of the application process is to place the fiber in an aqueous solution of the dye, which is stable in neutral solutions, and, by addition of salt, to displace the dye from solution and on to the fabric. Addition of alkali results in reaction of the dye with the fiber.[15] Replacement of one of the chlorine atoms in a dichloro-*s*-triazinyl dye with a second noncolored species such as an amine or an alkoxide (**1**, **2**; R = alkyl or

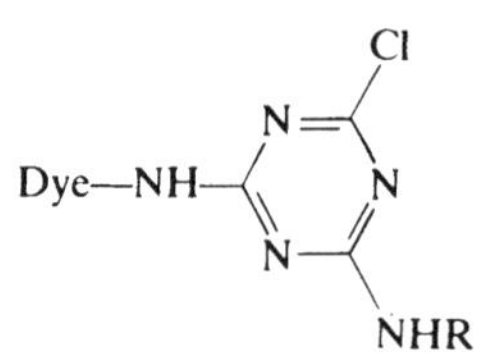

Procion H; P; Cibacron Dyes

1

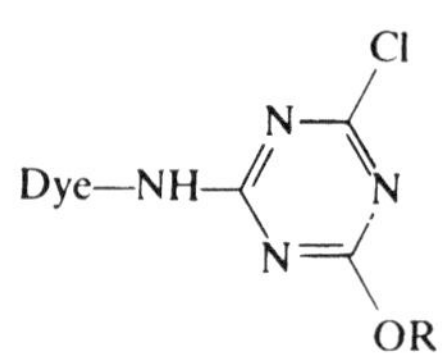

2

aryl) results in dye ranges (Procion H, Procion P, and Cibacron) which can be applied to cellulosics at 80 °C under alkaline conditions. Improved exhaustion (removal of dye from the liquor onto the fabric) is achieved in the Procion HE dye range by incorporation of two monochloro-*s*-triazinyl rings into the molecule (**3**), so affording two opportunities for reaction with the fiber. Structure **3** is usually prepared by condensation of two molecules of a dichloro-*s*-triazinyl dye with a suitable diamine at 35–40 °C. Common diamines used are the phenylenediamines and their sulfonic acids, 4,4′-diaminostilbene-2,2′-disulfonic acid or simple aliphatic diamines. It is possible[16] in structure **3**, or in **1**, where R is a dye

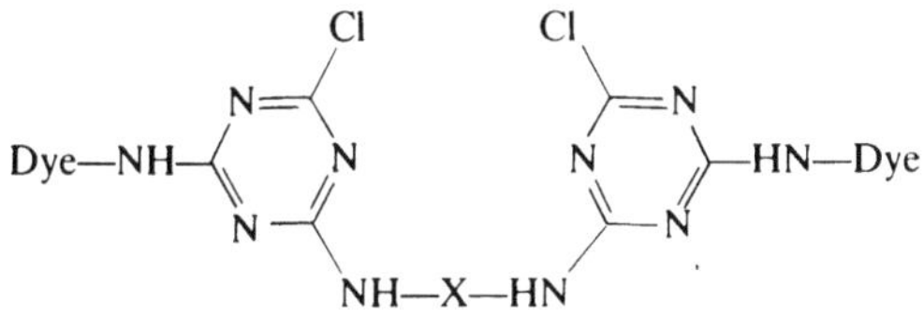

Procion HE Dyes

3

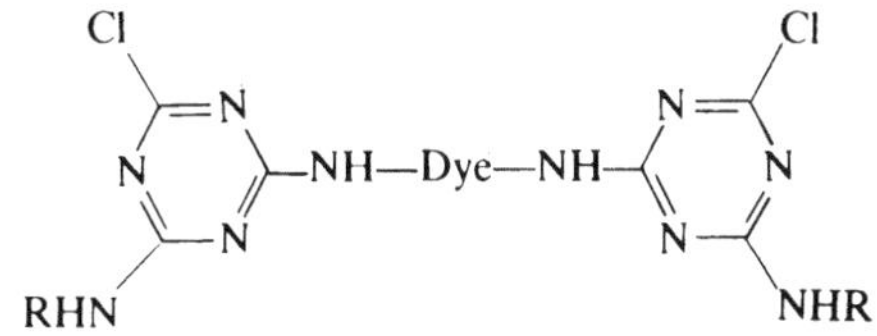

Procion Supra and HE Dyes

4

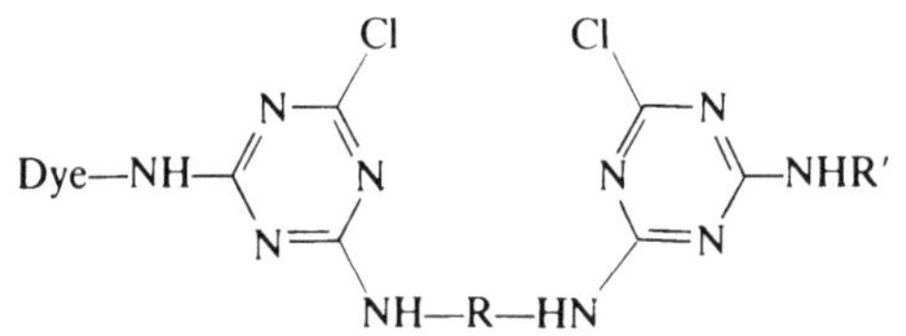

Procion Supra Dyes

5

molecule, to incorporate two different chromogens into one molecule and thereby produce hues otherwise difficult to prepare in a single chromogenic system.

The monochloro-*s*-triazinyl dyes are particularly suited to printing applications, as are members of the Procion Supra range which also incorporates two reactive systems **(4,5)**. Such dyes are usually applied as pastes which include the necessary alkali for fixation, usually achieved at 100–150 °C.

More than 50% of the current sales of reactive dyes are based on the *s*-triazinyl system (patent protection has now largely expired); however, the initial launch by ICI stimulated vigorous activity in all the major dye companies and the appearance of several other reactive systems. Hoechst had, in fact, been simultaneously approaching the problem from a different viewpoint and the Remazol dyes were introduced in 1958.

The Remazol dyes[17,18] are the sulfate esters of hydroxyethylsulfonyl dyes (**6**, equation 3), which, on treatment with mild alkali, generate the vinylsulfone group,[19] which in turn reacts with ionized cellulose to form the dye–fiber bond. The relatively low substantivity of these dyes made them especially useful for printing.

$$\text{Dye—SO}_2\text{CH}_2\text{CH}_2\text{O—SO}_3\text{Na} \longrightarrow \text{Dye—SO}_2\text{—CH=CH}_2 \quad (3)$$

Remazol Dyes

6

The major efforts, however, were concentrated on alternative chloroheterocycles and, particularly, on diazinyl types. The structural similarity between the 1,3,5-triazinyl system and the 1,3-diazinyl system of pyrimidine was turned to advantage and a number of reactive dye ranges containing the pyrimidine ring[20] have been marketed.[21] Thus, 2,4,5,6-tetrachloropyrimidine when condensed with a suitable chromogen containing a free amino group gave rise to the Drimarene[22] and Reactone dyes (**7**) of Sandoz and Geigy, respectively, which were introduced

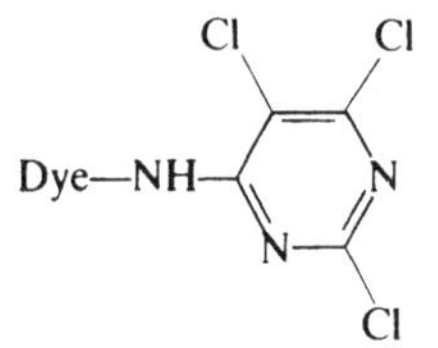

Drimarene, Reactone Dyes

7

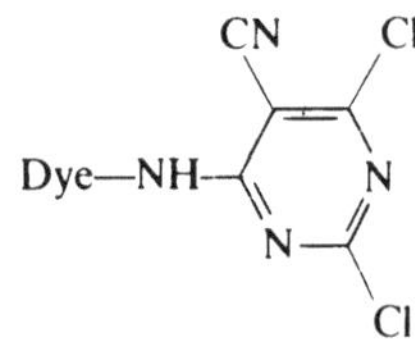

8

in 1959. The chlorine atom most reactive to nucleophilic attack is that at the 4-position of the pyrimidine ring. The closely related dyes based on 2,4,6-trichloropyrimidine were claimed in patents but were commercially unimportant. The reactivity of the pyrimidinyl dye systems is reduced, in comparison with that of the triazinyl based systems, due to reduced relative positive charge located on the ring carbon atoms. Consequently, the corresponding dyes possess similar reactivity to the monochlorotriazinyl dyes previously described. The reactivity can be enhanced by the presence of an electron-withdrawing group at the 5-position. Examples include dyes prepared from 5-cyano-2,4,6-trichloropyrimidine **(8)**[23] and the Reactofil dyes of Geigy **(9)**[24] in which the chromogen and the reactive system are linked by a carbonylamino group. The carbonyl group within the linkage exerts sufficient influence to increase the reactivity of the pyrimidine ring. Other substituted pyrimidines which have found commercial success have employed leaving groups other than chlorine. Bayer's Levafix EA dyes **(10)** and Levafix P dyes **(11)** use 2,4-difluoro-5,6-dichloropyrimidine[25] and 4,5-dichloro-6-methyl-2-methylsulfonylpyrimidine,[26] respectively, the leaving groups being fluorine and methylsulfonyl. Both of these dye systems have a relatively high level of reactivity and may be applied under mild conditions. A good level of reactivity is also manifested in the Cibacron F dyes, which use a monofluorinyltriazinyl system in which cyanuric fluoride is condensed with the chromogen and subsequently with a colorless amine **(12)**. The Levafix EA and Cibacron F dyes can both be applied to cellulose at 50 °C.

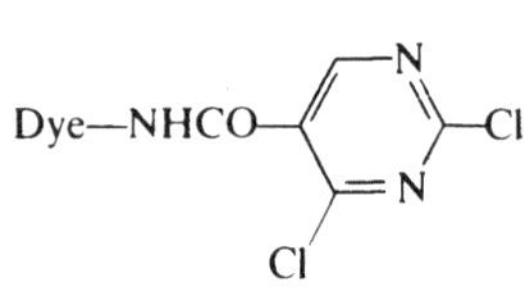

Reactofil Dyes

9

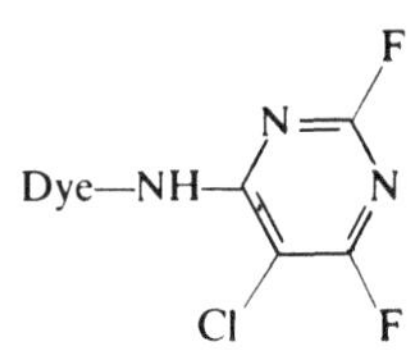

Levafix EA Dyes

10

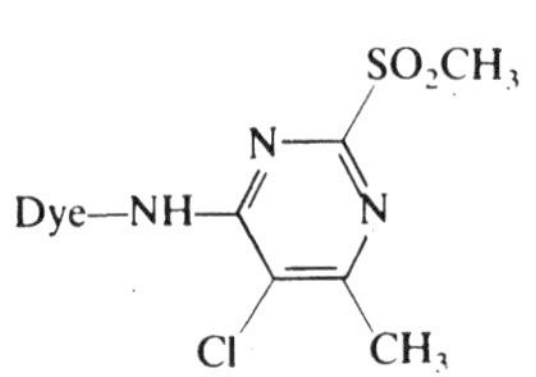

Levafix P Dyes

11

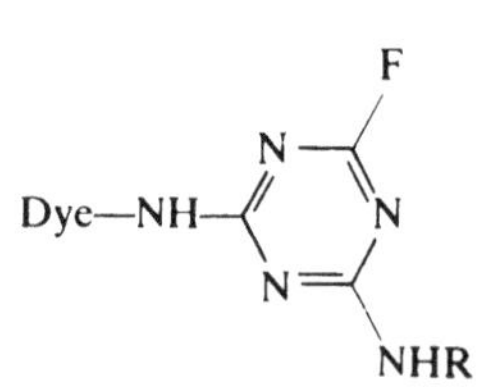

Cibacron F Dyes

12

Of the diazinyl systems theoretically available, pyrimidine has proved to be the most important commercially. Others have been used, however, including pyridazine, quinoxaline, and phthalazine. The 1,2-diazinyl ring system was used in the Primazin P dyes of BASF **(13)** and the Solidazol dyes of Cassella **(14)** although neither were particularly successful. In both instances the reactive group is attached to the chromogen via a carbonylamino group. In the 2,3-dichloroquinoxalinyl dyes used by Bayer in their Levafix E **(15)**, range, introduced in 1961, the reactivity is markedly increased and is comparable to that of the dichlorotriazinyl dyes. In this instance, as in the case of the Eliasane dyes of Francolor **(16)**, the reactive system is also attached to the chromogen via a carbonylamino group. The dyes are prepared via the acid chloride of the reactive heterocyclic system and involve condensation with a free amino group on the chromogen.

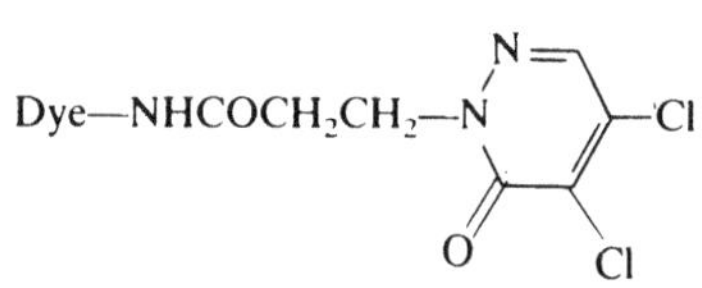

Primazin P Dyes
13

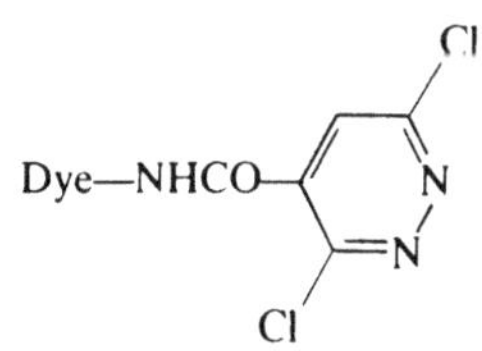

Solidazol Dyes
14

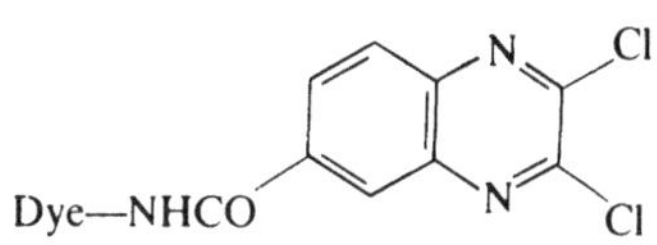

Levafix E Dyes
15

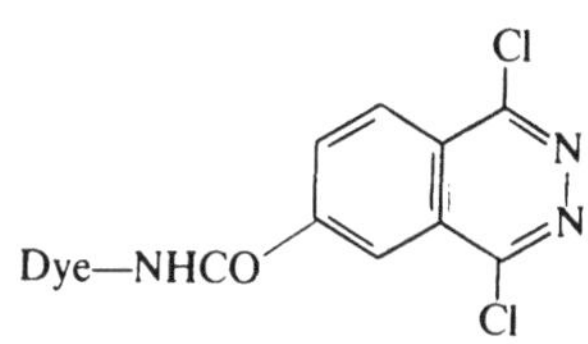

Elisiane Dyes
16

An alternative approach to increasing dye reactivity lies in the introduction of more labile leaving groups and this was achieved to some extent with the fluorine and methylsulfonyl systems previously mentioned. However, reaction of a monochlorotriazinyl dye with a suitable tertiary amine[27] such as pyridine (equation 4), nicotinic acid, trimethylamine, or 1,4-diazabicyclo[2,2,2]octane produces dyes sufficiently reactive for cold dyeing. The choice of the leaving group is important in commercial dyes because of the potential of dye-bath odor. Consequently nicotinic acid has proved to be the most successful and is used in the Kayacelon React range of Nippon Kayaku.

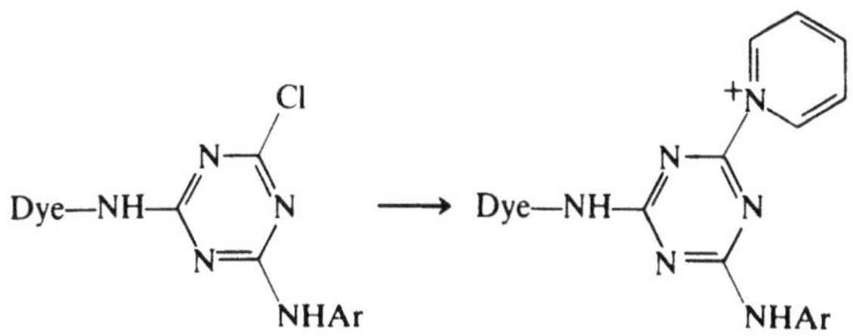 (4)

A recent novel approach by ICI[28] was the introduction of Procion T dyes particularly for the dyeing of cotton in polyester/cotton blends. The dyes contain phosphonic acid groups which readily undergo esterification reactions in the presence of dicyandiamide. The fixation process involves baking at 200 °C for 2–3 min and requires acidic conditions, so that dyeing of polyester/cotton blends in conjunction with disperse dyes is very successful. The use of conventional reactive dyes in this process is inappropriate since alkaline conditions, which adversely affect the disperse dye, are required. The quaternized triazines, which react under neutral conditions, however, can be used. The Procion T dyes are of little value for use in the dyeing of pure cotton because the acidic conditions required cause a deterioration in fiber strength. This is not evident in the polyester/cotton blends which derive their strength and wear characteristics from the polyester.

C. Chromogens in Reactive Dyes

In view of the fact that dye fixation is achieved by formation of a covalent bond, the chromogenic structures chosen in the dyes can be much simpler than the multiazo structures favored in direct dye ranges. In practice, monoazo, disazo, metallized monoazo, metallized disazo, formazan, anthraquinone, triphenodioxazine, and phthalocyanine chromophores have all been used for the preparation of reactive dyes. As pointed out previously, the chromogens in use are frequently common to the various dye ranges and only the reactive system varies.[29]

1. Azo Reactive Dyes

Yellow dyes are usually of the simple monoazo type or, particularly in the case of greenish yellows, monoazo dyes utilizing heterocyclic couplers. The simple homocyclic monoazo yellow dyes, usually mid to reddish yellows, are prepared by diazotization of a suitable primary aromatic amine, usually a substituted aniline containing at least one sulfonic acid group, a naphthylamine-sulfonic acid, or alternatives such as 4-amino-4′-nitrostilbene-2,2′-disulfonic acid, and reaction with an aniline or naphthylamine coupling component or their mono *N*-alkyl derivatives. The reactive group is then condensed with the amino group. Thus, for example, 2-naphthylamine-4,8-disulfonic acid is diazotized in aqueous hydrochloric acid at 0–5 °C by dropwise addition of sodium nitrite solution and coupled in acidic medium with *m*-toluidine to produce the monoazo chromogen. Condensation with cyanuric chloride results in the dye originally introduced as Procion Yellow M-RS, now Procion Yellow MX-R (**17**, CI Reactive Yellow 4). The dye is thoroughly mixed with disodium hydrogen phosphate and potassium dihydrogen phosphate. Condensation of the primary amine with cyanuric chloride results in a significant hypsochromic, or blue, shift equivalent to acylation. Condensation of the dichlorotrazinyl dye, as in **17**, with, for example, metanilic acid (aniline-3-sulfonic acid) produces a dye of the Procion H type. In Procion Yellow H-A (**18**, CI Reactive Yellow 3) the coupler is 3-acetamidoaniline and one chlorine atom has been displaced with ammonia.

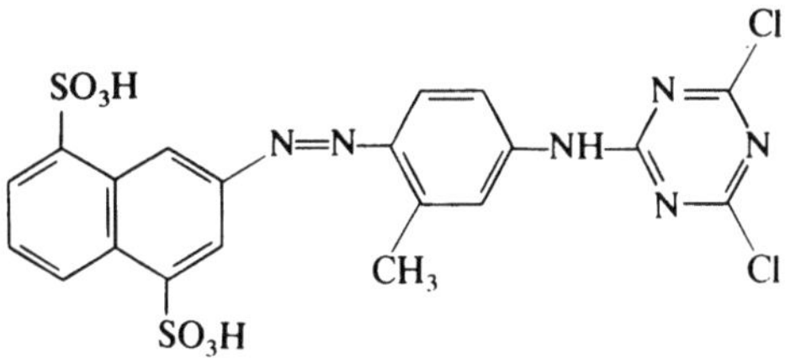

Procion Yellow MX-R; CI Reactive Yellow 4

17

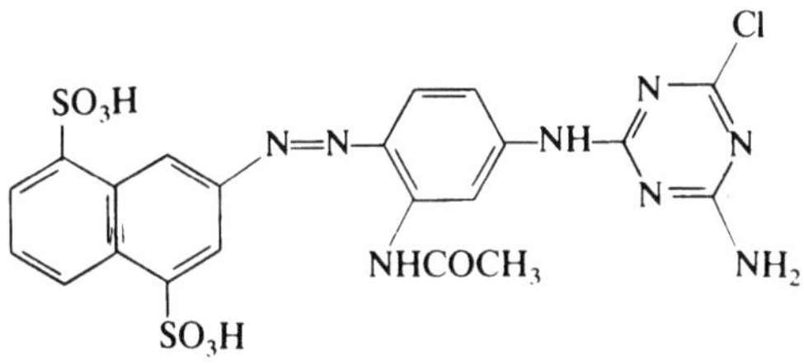

Procion Yellow H-A; CI Reactive Yellow 3

18

Good yellow shades are produced from 4-amino-4′-nitrostilbene-2,2′-disulfonic acid when diazotized and coupled with, for example, *N*-methylaniline followed by condensation with a suitable reactive system, such as cyanuric chloride.[29]

Heterocyclic coupling components have also been used in greenish-yellow dyes, particularly couplers of the active methylene type.[30] 1-Aryl-3-methyl-5-pyrazolone compounds,[31] where the aryl group usually contains a sulfonic acid substituent, are coupled with a diazonium component bearing a reactive group. Thus, for example, 1,3-phenylenediamine-4-sulfonic acid is selectively condensed with cyanuric chloride, the amino group situated *para* to the sulfonic acid group being the more reactive. One of the reactive chlorine atoms of the triazinyl ring may or may not be replaced with an amino compound before diazotization and coupling with the pyrazolone **(19)**. It is possible to prepare the same compound using the *N*-acetyl derivative of the diazonium component to couple with the

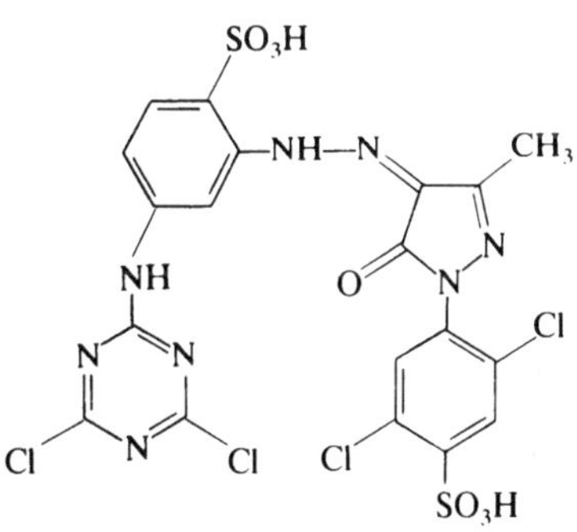

Procion Brilliant Yellow MX-6G
CI Reactive Yellow 1

19

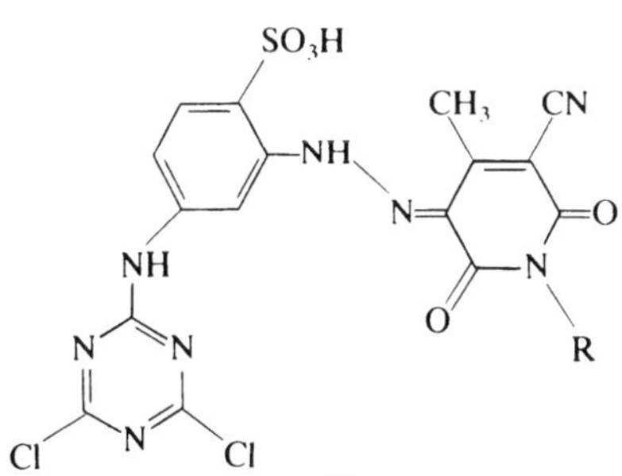

20

pyrazolone followed by hydrolysis and condensation with the reactive system. Very greenish yellow hues are also produced by the use of pyridone coupling components.[32] Dyes prepared from coupling components of the active methylenic type exist essentially in the hydrazone form **(20)**.

Pyrazolones are most conveniently synthesized via condensation of a hydrazine with a β-keto ester. The hydrazines are commonly substituted phenylhydrazines prepared by diazotization and subsequent reduction of a suitable aniline derivative. Condensation with a β-ketoester, such as ethyl acetoacetate, produces the hydrazone which is cyclized by heating in alkali. Clearly, a wide variety of aryl derivatives is possible and β-keotesters commonly used are ethyl acetoacetate and diethyl oxalacetate.

Pyridones are synthesized by condensation of a β-keto ester with an active methylenic compound such as diethyl malonate or ethyl cyanoacetate in the presence of an amine. The amine may be omitted if malononitrile or a suitable amide is used.[33]

The most generally used coupling components are the aminonaphthols, which can be used to give hues from orange through to black. J-acid (6-amino-1-naphthol-3-sulfonic acid), so commonly encountered in direct dyes, is used in the production of orange colors[34] as in dye **21**,* prepared by diazotization of aniline-2-sulfonic acid, coupling with J-acid and condensation with the reactive

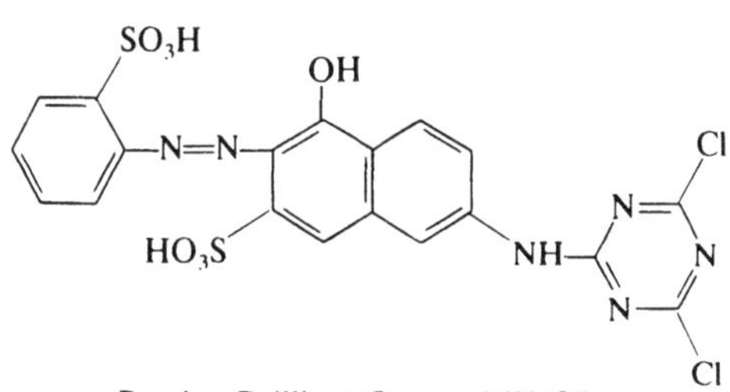

Procion Brilliant Orange MX-GS
CI Reactive Orange 1
21

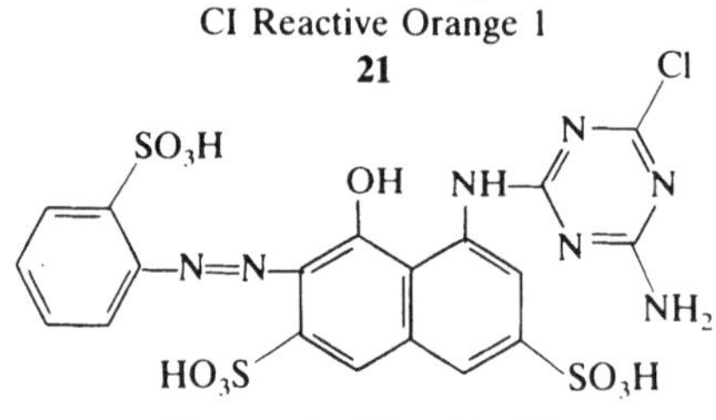

Cibacron Brilliant Red B
CI Reactive Red 12
22

* As discussed in Chapter 2, azo dyes prepared from naphthol couplers exhibit "keto-enol" tautomerism and an equilibrium exists between the hydroxy-azo form and the keto-hydrazone form. Throughout this chapter such dyes are depicted in their hydroxy-azo form despite the fact that the alternative tautomer predominates. This decision was made because the latest issue of the *Color Index* describes the dyes in this form. There is a body of opinion, however, that this situation should be changed in forthcoming issues to reflect the more favored structural forms of the dyes.

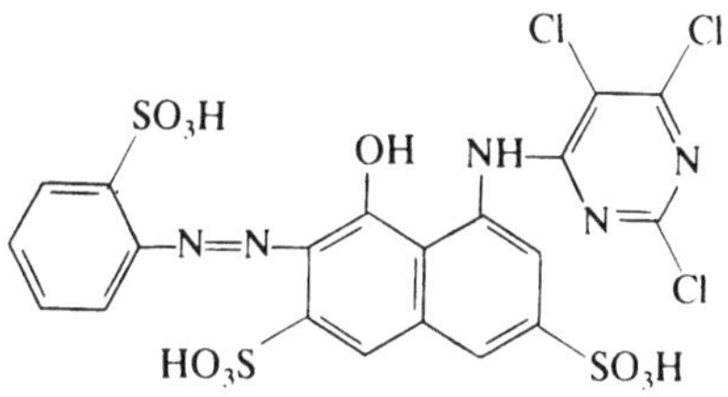

Drimarene Red Z-2B
CI Reactive Red 17
23

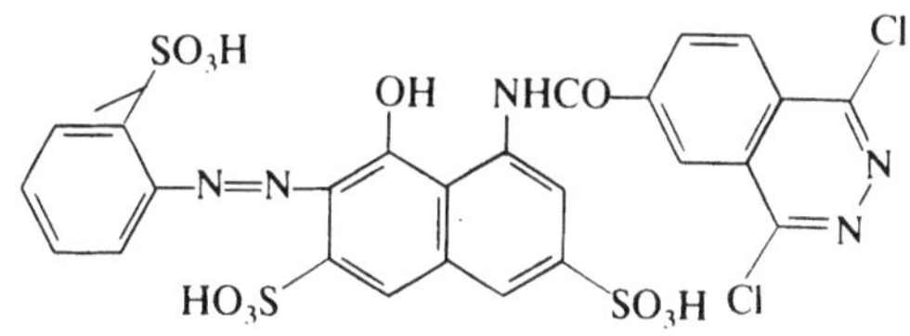

Elisiane Brilliant Red B
CI Reactive Red 96
24

group. It is possible to condense the reactive system with the coupler prior to reaction with the diazonium component. Replacement of J-acid by H-acid, as in dyes **22–24**, which illustrate the use of a single chromogen in different reactive dyes,[35,36] produces a bathochromic shift and a bright bluish-red hue. There are, of course, many possible variations on this theme using different substituents in the parent aniline and by replacement of one of the reactive chlorine atoms, as in dye **22**. Two molecules of a dichlorotriazinyl dye can be linked by an aromatic diamine and thereby produce dyes exemplified in the Procion HE range.

To produce more bathochromic monoazo dyes it is neccessary to metallize the azo bond using *o*-hydroxy ligands to link to the metal atom. Rubine, violet, and blue dyes are produced mainly by using copper complexes which are planar and, consequently, more substantive. 2-Aminophenol-4-sulfonic acid, when diazotized and coupled to sulfo-J-acid followed by reaction with copper sulfate and sodium acetate solution, produces a chromogen which, on condensation with cyanuric chloride, gives the rubine dye Procion Rubine MX-B (CI Reactive Red 6, **25**). Violet colors are obtained by the use of H-acid as coupling component with simple *o*-aminophenol diazonium components. More bathochromic diazonium components, which can be achieved by introduction of electron-withdrawing groups, such as nitro, or by use of aminonaphthols, particularly 1-amino-6-nitro-2-naphthol-4-sulfonic acid or aminoazobenzene derivatives, give blue or navy blue colors such as dye **26**. Bright blue and navy blue dyes are also produced by using Chicago acid (8-amino-1-naphthol-5,7-disulfonic acid). In this particular case it should be noted that the reactive group is attached to an amino group in

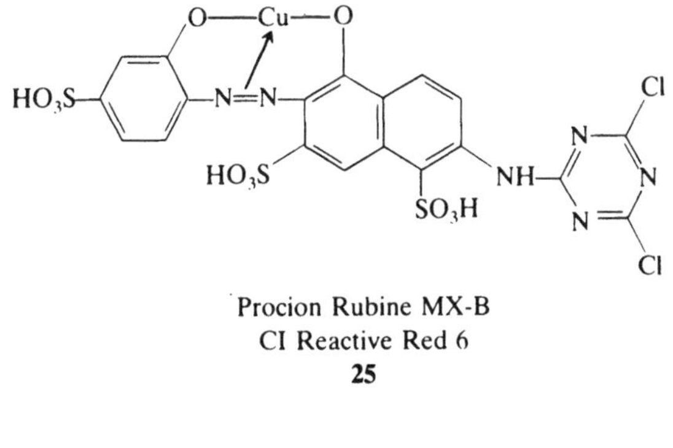

Procion Rubine MX-B
CI Reactive Red 6
25

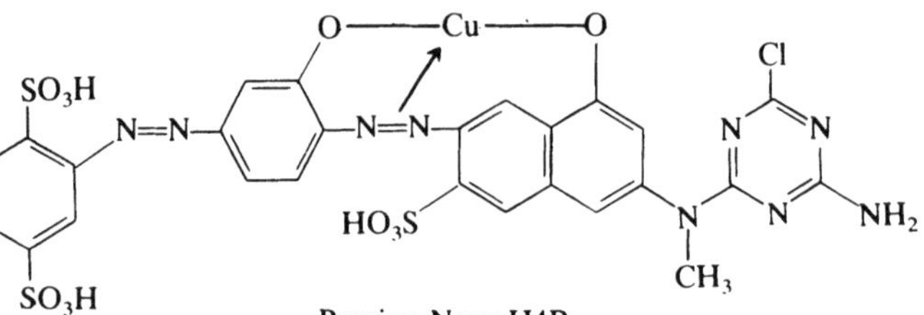

Procion Navy H4R
CI Reactive Blue 40
26

the diazonium salt so that the 8-amino group in Chicago acid, which is relatively inert, remains unreacted in the dye.

Cobalt and chromium complexes are used in the production of gray and black hues, but these dyes are large molecules since the octahedral spatial arrangement of the bonds of chromium and cobalt chelates requires that one metal atom combines with two molecules of the chromogen. The resulting nonplanar structures have low substantivity on cotton, restricting these dyes mainly to printing applications.

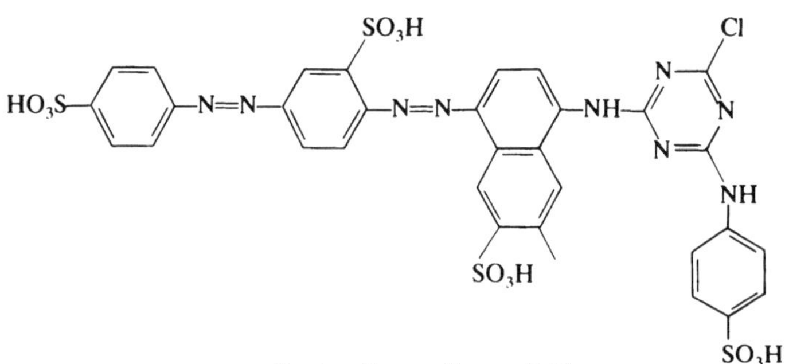

Procion Orange Brown H-G
CI Reactive Brown 1
27

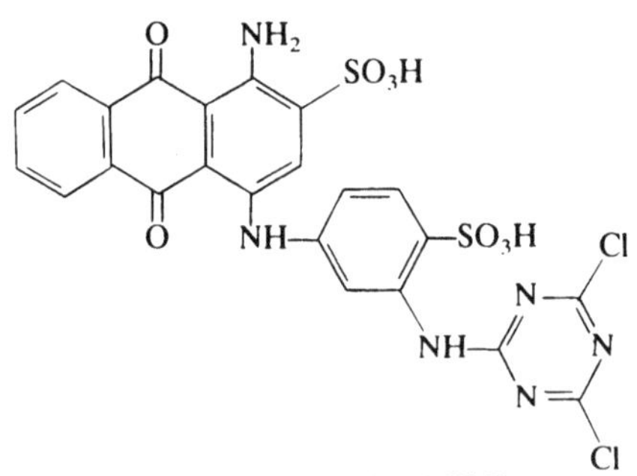

Procion Blue MX-R
CI Reactive Blue 4
28

Colors such as brown, olive, and green which are difficult to obtain in self-shade dyes can be produced by mixtures of dyes and, less successfully, in some mixed chromogen dyes. Brown hues can also be produced in disazo molecules by using one, two **(27)**, and often three naphthylamine molecules.[37] Green dyes can be prepared by combining a blue chromogen with a yellow chromogen via the reactive system which usually acts as a chromophoric blocking group.[16]

2. Anthraquinone Reactive Dyes

Blue dyes can also be prepared by using anthraquinone, triphenodioxazine, formazan, and phthalocyanine chromogens. Anthraquinone dyes are second only in overall importance to azo dyes and the chromogen is commercially important in reactive dyes. The most common structures are derived from bromaminic acid (1-amino-4-bromoanthraquinone-2-sulfonic acid) and, by variation of substituent, give rise to bluish violet to bluish green colors with the bright reddish to mid-blues being the most important. Syntheses from bromaminic acid are straightforward, involving condensation using an aromatic amine containing the reactive system or using an aromatic diamine such as 1,3-phenylenediamine-4-sulfonic acid, which affords a dye conveniently substituted with a free amino group capable of condensation with the appropriate reactive system, as in the case of Procion Blue MX-R (CI Reactive Blue 4, **28**).

Various substituted 1,3- and 1,4-phenylenediamines have been used, the 1,3-diamines producing redder blues than the 1,4-diamino compounds. Other diamines such as derivatives of benzidine, 4,4′-diaminostilbene, and 4,4′-diaminoazobenzene have also been used successfully.

3. Phthalocyanine Reactive Dyes

Solubilized phthalocyanine reactive dyes are used for bright turquoise hues, which cannot be achieved using either azo or anthraquinone dyes. Copper phthalocyanine is the most commonly used, the basic structure **(29)** being solubilized by the introduction of sulfonic acid groups via sulfonation or chlorosulfonation. Generally, the phthalocyanine molecule is treated with chlorosulfonic acid and the resulting tetrasulfonyl chloride condensed with one molar equivalent of a diamine, such as a phenylenediamine or an aliphatic diamine. The residual sulfonyl chloride groups are then hydrolyzed and the reactive system, such as cyanuric chloride **(30)**, introduced in a final condensation step. Condensation of a proportion of the sulfonyl chloride groups with ammonia results in brighter hues.[38]

4. Triphenodioxazine and Formazan Reactive Dyes

Relatively recently, the domain of blue reactive dyes has been infiltrated by the introduction of triphenodioxazines[39] and by an increased interest in formazans.[29] Triphenodioxazine structures have been used in direct dyes for

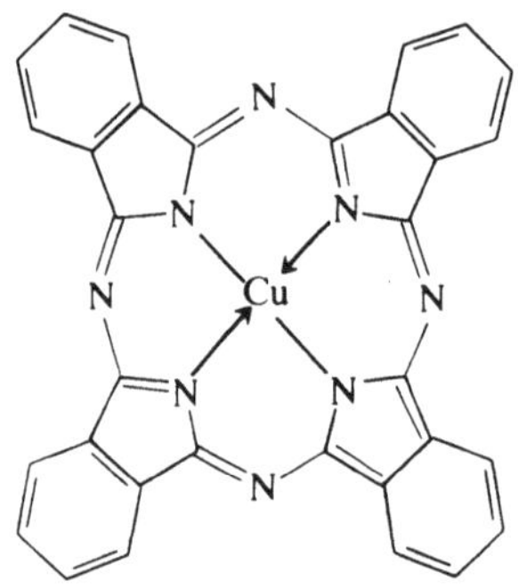

CI Pigment Blue 15
29

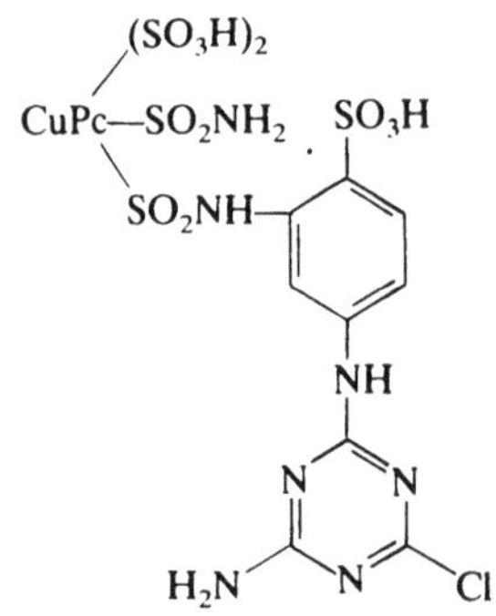

Cibacron Turquoise Blue G-E
CI Reactive Blue 7
30

many years but in recent times have won approximately 10% of the blue reactive dye market, mainly at the expense of anthraquinones, since they offer bright hues and, being tinctorially stronger, have a marked commercial advantage over anthraquinones. Currently, research effort tends to concentrate in this area, and further new products are anticipated. Synthesis of the parent molecule **31** has been significantly improved[39] so that economical products, such as CI Reactive Blue 163, have been introduced. The precise structures have not been officially disclosed but it would appear that groups A and B in structure **31** are selected from hydrogen or sulfonic acid and the reactive system is attached in the R-substituent through an arylamino group.

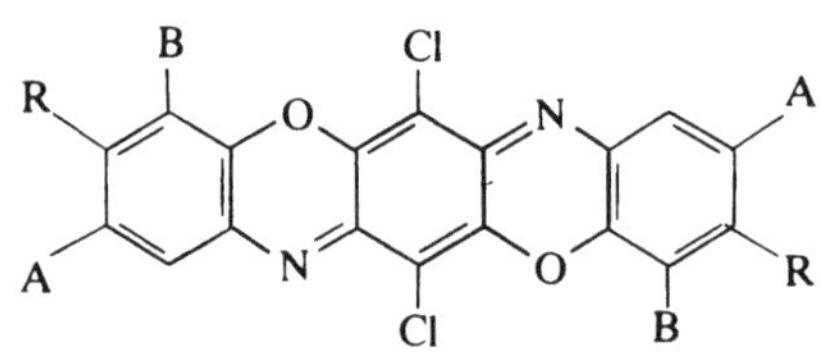

31

CI Reactive Blue 163 (ICI) is a dichlorotriazinyl dye with a greenish-blue color. Other triphenodioxazine dyes introduced by ICI include a monochlorotriazinyl derivative (CI Reactive Blue 198), a phosphonic acid derivative (CI Reactive Blue 172), and, more recently, by Ciba-Geigy, a fluorotriazine derivative (CI Reactive Blue 204).

There has also been increased research activity in formazan structures (**32**; R = reactive group). It is possible to locate the reactive group on any of the aromatic rings each of which may be benzene or naphthalene. Ring B frequently contains a hydroxy group *ortho* to the formazan nitrogen, thus acting as a ligand. Other ligands used include carboxyl groups (**33**). Formazans are capable of producing bluish red to blue copper complexes.[40]

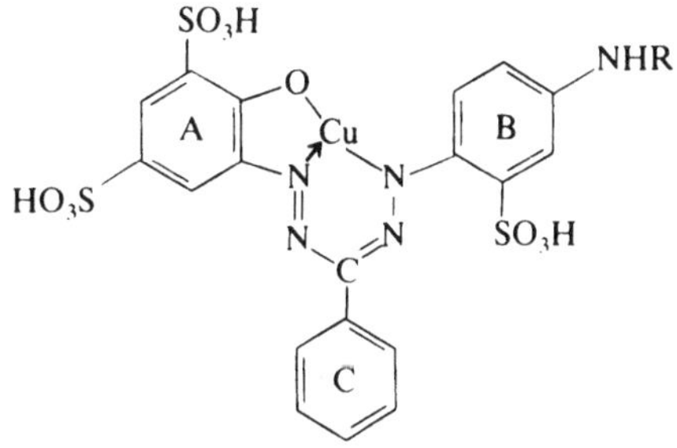

32

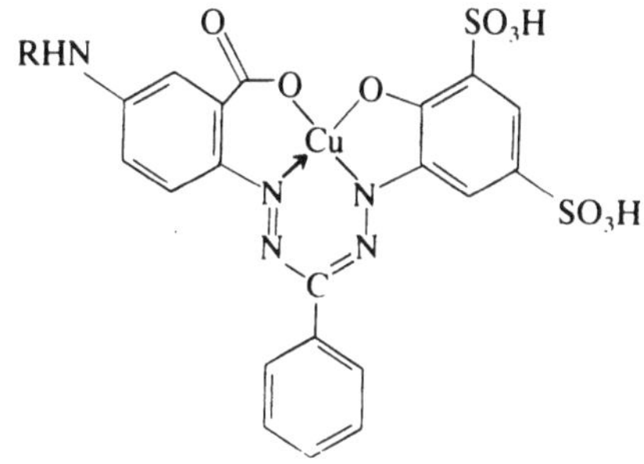

33

III. DIRECT DYES

A. Azo Direct Dyes

Despite the fact that many direct dyes were developed up to 100 years ago, these dyes still command 27.4% (figures published in 1981) of the cellulosic market, mainly because their ease of application renders them economic for the dyer. This factor is particularly valuable when sophisticated dyeing machinery is not available.

Direct dyes are planar anionic dyes, with substantivity for cellulosic fibers, which are usually applied from an aqueous dyebath containing an electrolyte, and not requiring the use of a mordant.[41] (Application of a mordant dye requires that the fiber be treated with a substance, the mordant, such as tannic acid, prior to dyeing so as to confer affinity of the dye for the substrate.) Direct dyes are also used for paper and leather.

1. Monoazo Dyes

An extensive account of the early work on direct dyes was published in 1952[42] and more recent developments have been reviewed.[43,44] This class of dyes is dominated by azo structures, usually containing more than one azo group. Relatively few monoazo dyes are still in use (approximately 30) and of those the majority are based on J-acid (6-amino-1-naphthol-3-sulfonic acid) as coupling component. Thus, for example, CI Direct Red 118 **(34)** uses a coupler prepared by condensation of J-Acid with 3-nitrobenzoyl chloride followed by reduction of the nitro group. Coupling in alkaline solution with benzenediazonium chloride completes the synthesis of the dye, which is also known as Rosanthrene O. The fastness of dyes of this type is usually improved by aftertreatment on the fiber (see Section III.B). Further reaction of *N*-(3′-aminobenzoyl)-J-acid with 3-nitrobenzoyl chloride and reduction of the nitro group results in the coupling component **35**. Reaction of **35** with diazotized aniline and 2-methoxyaniline gives Diazo Brilliant Orange GR (CI Direct Orange 75) and Diazo Brilliant Scarlet ROD (CI 17845), respectively.

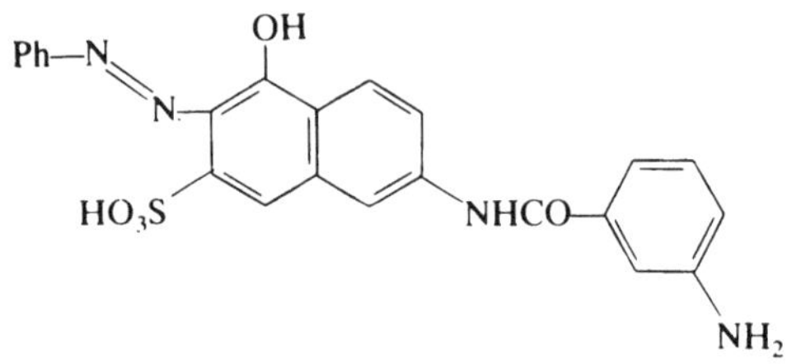

Rosanthrene O
CI Direct Red 118
34

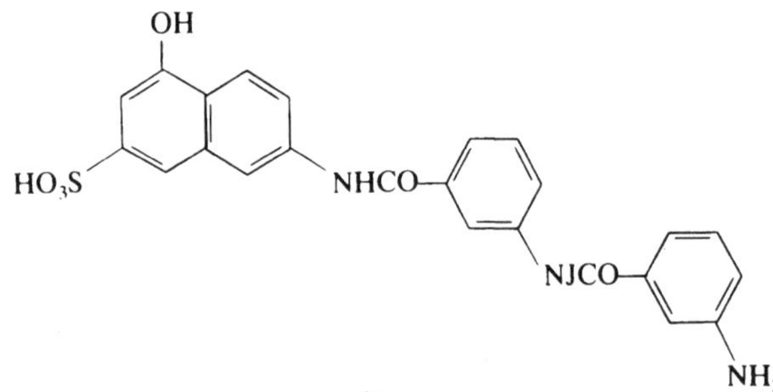

35

2. Disazo Dyes

The use of disazo dyes is prevalent because of their increased wash fastness and substantivity. There are several ways in which disazo molecules can be synthesized and these dyes are usually classified according to the nature of the diazonium component and the coupling component used. A "shorthand" classification, known as Winther symbols, has evolved. Thus, $A \rightarrow Z \leftarrow A'$ molecules, also known as primary disazo dyes, are formed by diazotization of compound A and coupling with Z. The synthesis is completed by diazotization of A' and coupling with $A \rightarrow Z$. In fact, there are but few examples of $A \rightarrow Z \leftarrow A'$ dyes applied to cellulosic fibers (some are used as acid dyes on polyamide fibers), although one important example is CI Direct Orange 18 **(36)** which is derived from resorcinol coupled with diazotized dehydrothio-*p*-toluidinesulfonic acid and benzenediazonium chloride. Disazo dyes of this type derived from carbonyl-J-acid **(37)**, prepared by phosgenation of J-acid, are more useful, as with CI Direct Orange 26, obtained by reacting the intermediate with two molecules of benzenediazonium chloride, and Benzo Fast Scarlet 4BS (CI Direct Red 23)

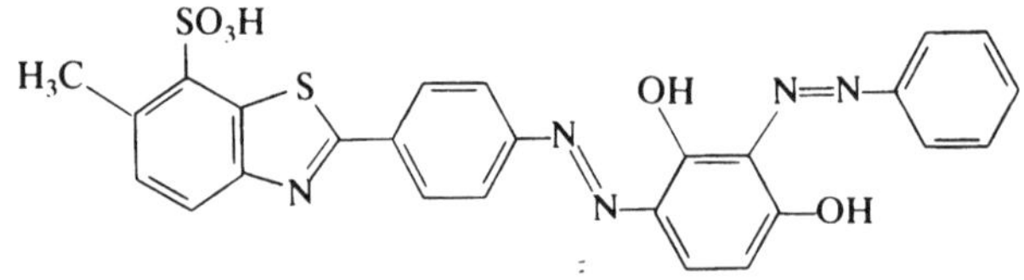

CI Direct Orange 18
36

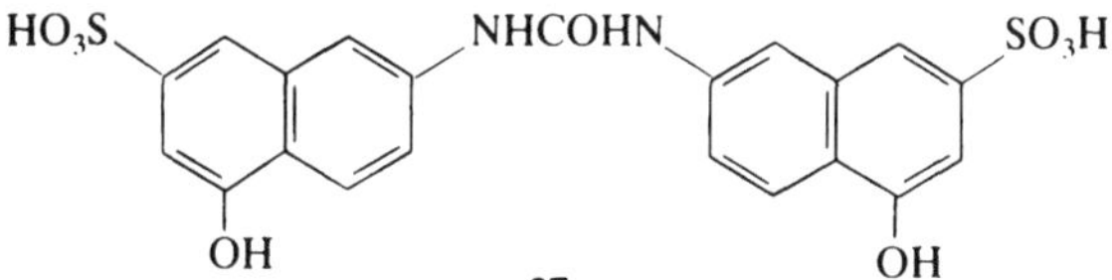

37

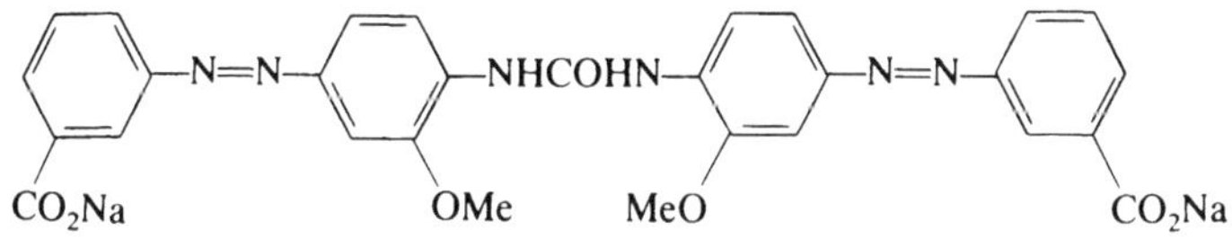

Sirius Yellow G
CI Direct Yellow 49
38

prepared by reaction with one molecule of benzenediazonium chloride and one of diazotized 4-aminoacetanilide. Dyes of this type are also made by phosgenation of aminoazo compounds. An example is Sirius Yellow G (**38,** CI Direct Yellow 49) prepared by phosgenation of the monoazo dye obtained from *m*-aminobenzoic acid and *o*-anisidine. Unsymmetrical disazo dyes can be prepared by coupling twice, once in alkaline solution and again in acid solution, to aminonaphthol couplers such as H-acid.

Secondary disazo dyes, classified as A → M → E, number some 150 disclosed direct dye structures, the possible combinations of components being almost endless and, with appropriate selection, can yield from red to green colors. A components are selected from substituted anilines or naphthylamines (including aminonaphthols), M components from aniline or α-naphthylamine derivatives, or occasionally an aminonaphthol, and E components are chosen from J-acid, H-acid, Chicago acid (8-amino-1-naphthol-5,7-disulfonic acid), γ-acid (7-amino-1-naphthol-3-sulfonic acid), and Schaeffer's acid (2-naphthol-6-sulfonic acid). The most useful dyes are those having E components based on J-acid, either alone or as an *N*-substituted derivative. The direct dyes of the Sirius range, marketed by I.G. Farbenindustrie until 1945 and subsequently by BASF, contain examples of this type which possess good fastness properties, particularly to light. The middle component is selected from aniline or α-naphthylamine derivatives since the subsequent aminoazo dye is conveniently diazotized for further coupling. When A and M components are aniline derivatives the dyes are red, as in the case of Sirius Red 4B (**39**) and Sirius Supra Red 5B (**40**). Use of J-acid as a middle component also gives red dyes, such as CI Direct Red 16 (**41**).

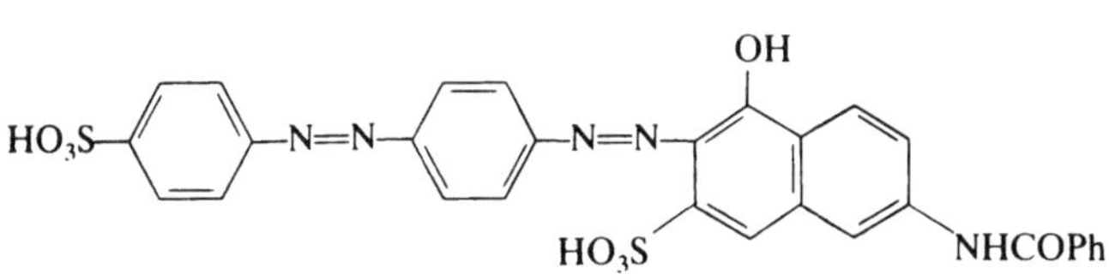

Sirius Red 4B
CI Direct Red 81
39

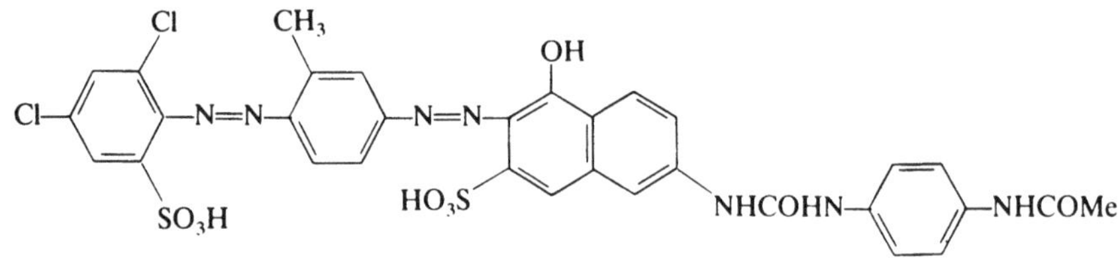

Sirius Supra Red 5B
CI Direct Red 110
40

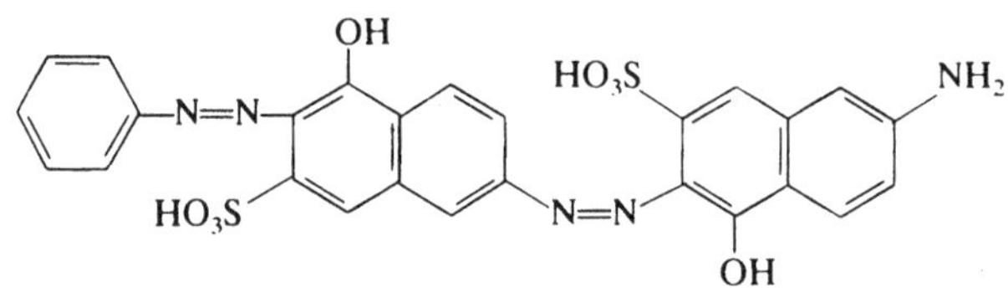

CI Direct Red 16
41

Replacing aniline derivatives with naphthylamines gives rise to bathochromic shifts such that violet and blue dyes are available. Aminonaphthols, typified by H-acid, are used as A components in violet to blue dyes, such as Sirius Supra Blue F3R (CI Direct Blue 67, **42**) which is a bright dye with good fastness properties. Green dyes are produced by further increasing the length of the conjugated chain in both the A and M components. Thus, for example, using 4-aminobiphenyl-3-sulfonic acid as the A component and 2-ethoxy-1-naphthylamine-6-sulfonic acid as the M component results in the green dye Brilliant Benzo Fast Green GL (**43**). γ-acid is often used as an E component in black and gray dyes such as Neutral Gray G (CI Direct Black 3, **44**), which is prepared by coupling aniline to α-naphthylamine and the product to γ-acid in alkaline solution.

The third type of disazo dye is classified using Winther symbols as $E \leftarrow D \rightarrow E'$ and is usually produced via tetrazotization of a suitable diamine. Many dyes have been based on benzidine, the use of which has now ceased in the Western world, since this intermediate is known to be a potent bladder carcinogen. For a number of years dyes based on benzidine have been imported

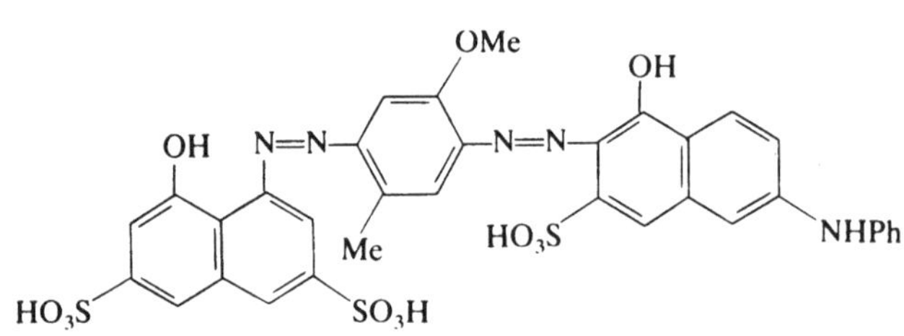

Sirius Supra Blue F3R
CI Direct Blue 67
42

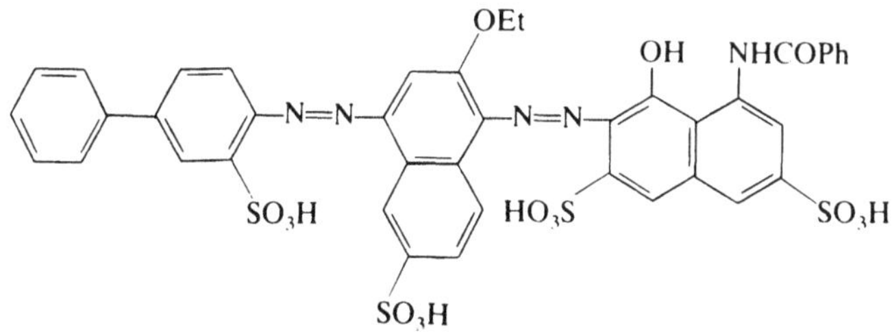

Brilliant Benzo Fast Green GL
CI Direct Green 13
43

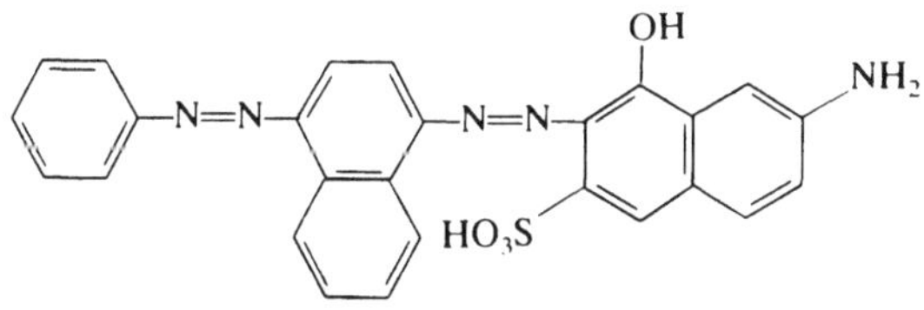

Neutral Gray G
CI Direct Black 3
44

from Asia, where they continue to be made, but it has been shown that degradation in the dyebath results in detectable amounts of benzidine, so that their use is now also to be strongly discouraged. Nevertheless, since the discovery of Congo Red by Bottiger in 1884, in which benzidine is tetrazotized and coupled twice with naphthionic acid (1-naphthylamine-4-sulfonic acid), many thousands of tons of this and similar dyes have been used and the *Color Index* describes approximately 300 known structures. Other diamines which have been employed are 2,2′-dichlorobenzidine, 2,2′-dimethoxybenzidine, 2,2′-dimethylbenzidine, 4,4′-diaminodiphenylurea, 4,4′-diaminodiphenylthiourea, 1,5-diaminonaphthalene, and 4,4′-diaminostilbene. End components have been chosen from substituted naphthols and phenols; typically H-acid, J-acid, and salicyclic acid; *m*-phenylenediamine, naphthylaminesulfonic acids, and pyrazolones. The resultant dyes are characterized by poor lightfastness, but good washfastness, and achieve commercial usage due to their low cost.

Bright yellow dyes are obtained by tetrazotization of 4,4′-diaminodiphenylurea or its 3,3′-disulfonic acid and coupling with a phenol such as salicyclic acid. Using benzidine as the diamine produces a bathochromic shift and in CI Direct Orange 1 the coupling components are salicyclic acid and 3-methyl-1-*p*-sulfophenylpyrazol-5-one. Use of one phenolic and one naphthol derivative as the E components produces red dyes, and two naphthol derivatives, as in CI Direct Violet 3, give violet dyes. Introduction of electron-donating groups, such as methoxy or methyl, exemplified by *o*-dianisidine (2,2′-dimethoxybenzidine), produces a further bathochromic shift as with CI Direct Blue 1, which is prepared by tetrazotization of *o*-dianisidine and coupling with two molecules of Chicago acid.

3. Trisazo Dyes

Clearly, the number of possibilities for disazo structures is very large and the choice is increased still further in the trisazo and polyazo series, having four or more azo groups. There are four of the possible structural combinations which have some commercial importance as designated by the Winther symbols **45** → **48**. The dyes produced have dull colors such as olive green, navy blue, and black and, of the structures of the type **45** and **46** disclosed in the Color Index, many contain benzidine and are, therefore, of no further value. Type **47** is the most useful, yielding brighter hues of blue and gray than other trisazo molecules, particularly when the E component is J-acid or phenyl-J-acid.

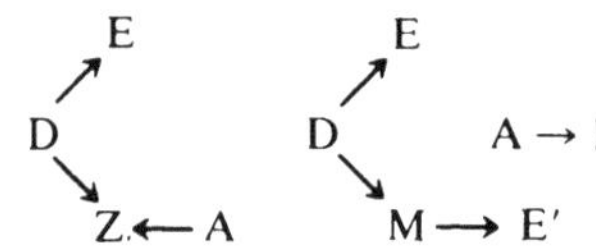

A → M → M′ → E   A → M → Z ← A′

45   **46**   **47**   **48**

4. Metal Complex Dyes

In certain cases, the light fastness of direct dyes can be improved by aftertreatment with copper or chromium salts (see Section III.B) but the need to include a further treatment is clearly undesirable and soluble metal complexes are now available for direct application. Copper forms planar complexes with *o,o*′-dihydroxyazo dyes or with *o*-hydroxy-*o*′-alkoxyazo dyes where the alkyl group is displaced during chelation. Some dyes in, for example, the Sirius and Durazol ranges are of this type, the structures of many of the earlier German-produced dyes becoming available through the intelligence reports following the Second World War (BIOS and FIAT Reports). The light fastness of certain of these dyes is greater than 7 and can only be equaled by the anthraquinonoid vat dyes. Fastness to washing is not always so good, however. Sirius Supra Blue 3RL **(49)** is prepared via tetrazotization of 3,3′-benzidinedicarboxylic acid and coupling with 3′-carboxyphenyl-J-acid followed by reaction with ammoniacal copper sulfate at 90 °C. Useful diazonium components are of the 2-aminophenolic, 2-aminocarboxylic acid, or 2-aminoalkoxy type and coupling components based on J-acid predominate.

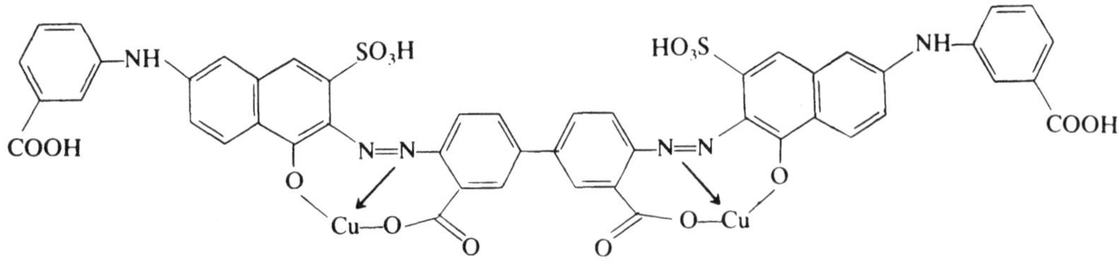

Sirius Supra Blue 3RL
CI Direct Blue 93
49

5. Mixed Chromophore Dyes

It has been mentioned previously that before the value of cyanuric chloride in reactive dyes was realized this molecule was used extensively in the synthesis of direct dyes and many patents were published by Ciba, as long ago as 1923. The dyes were studied by Fierz-David and Matter[45] and a number of the commercial products available at that time were identified. Cyanuric chloride was used because it serves as a chromogenic or chromophoric blocking group which effectively allows the discrete chromogens their unique influence within a single molecule. Consequently, blue and yellow chromogens combined in a single molecule produce a green dye. This is illustrated by Chlorantine Fast Green BLL (CI Direct Green 26, **50**) which is prepared by condensing the disazo dye, H-acid → cresidine → H-acid, with cyanuric chloride at 0–5 °C and the product with 4-aminobenzeneazosalicylate at 35–40 °C. Finally, two molecules of aniline are added and the condensation is completed at 90–95 °C. It is also possible to produce green dyes by using anthraquinone molecules as the blue chromogen[46] as in Chlorantine Fast Green 5GLL (CI Direct Green 28, **51**).

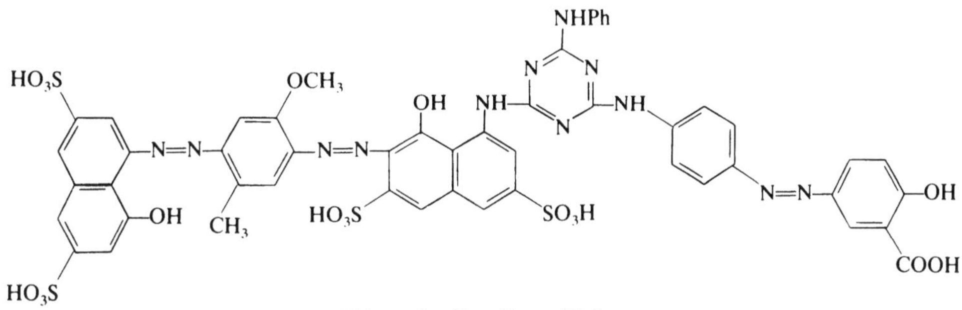

Chlorantine Fast Green BLL
CI Direct Green 26
50

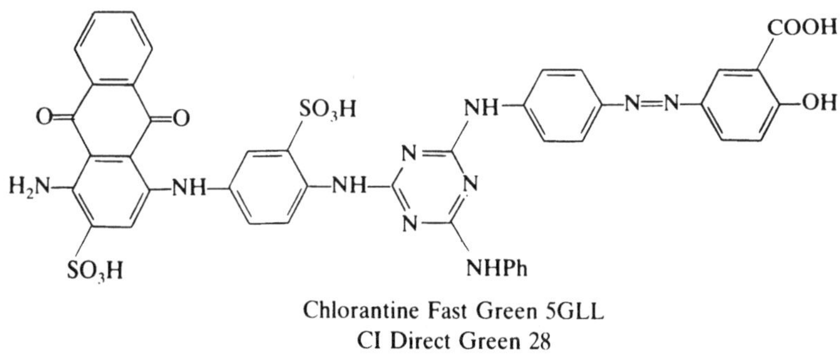

Chlorantine Fast Green 5GLL
CI Direct Green 28
51

B. Aftertreatments

A feature of direct dye chemistry is the use of a so-called aftertreatment process of the dyed fabric in order to improve the fastness properties of the adsorbed dye. One such aftertreatment process is known as developing the dye and involves the diazotization of a suitable aromatic amino group contained in the dye on the fiber and coupling with a suitable "developer" such as phenol, β-naphthol, *m*-phenylenediamine, or 3-methyl-1-phenylpyrazol-5-one. Monoazo dyes such as Rosanthrene O **(34)** are usually developed in this way. Alternatively, dyes containing a coupling component with an unoccupied coupling position may be further developed by reaction with a diazonium component.

Other aftertreatments include the use of formaldehyde,[47] after metallization, particularly with copper, by reaction of *o,o'*-dihydroxyazo or *o*-hydroxy-*o'*-carboxyazo dyes with copper sulfate solution, and the use of various cationic fixing agents.[44]

C. Triphenodioxazine Dyes

The *Colour Index*[48] lists many blue direct dyes and the most numerous structures are of the multiazo type. However, the need for bright blue direct dyes is frequently satisfied by the use of triphenodioxazines.[39,42] Used since the 1930s by I.G. Farbenindustrie, the basic structure is typically synthesized by condensation of two molecules of an aromatic amine with chloranil to produce a dianilide intermediate which is converted into the triphenodioxazine dye by oxidative cyclization, occasionally in sulfuric acid in order to effect simultaneous sulfonation. Sirius Light Blue FFRL **(52)** was prepared by condensation of two molecules of 3-amino-*N*-ethylcarbazole with one molecule of chloranil and cyclization using benzenesulfonyl chloride, followed by sulfonation. Other commercially important dyes such as CI Direct Blue 109 **(53)**, prepared from 1-aminopyrene, and CI Direct Blue 106 **(54)**, obtained from 4-aminodiphenylaminesulfonic acid, were

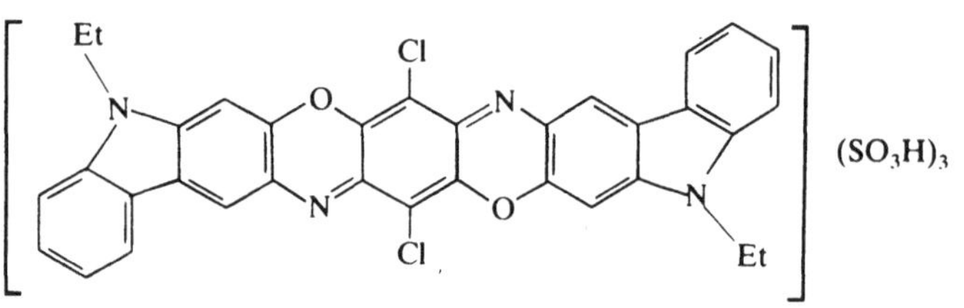

Sirius Light Blue FFRL
CI Direct Blue 108
52

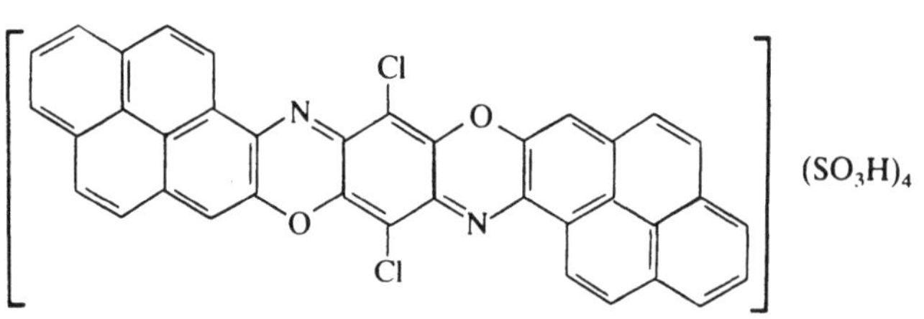

CI Direct Blue 109
53

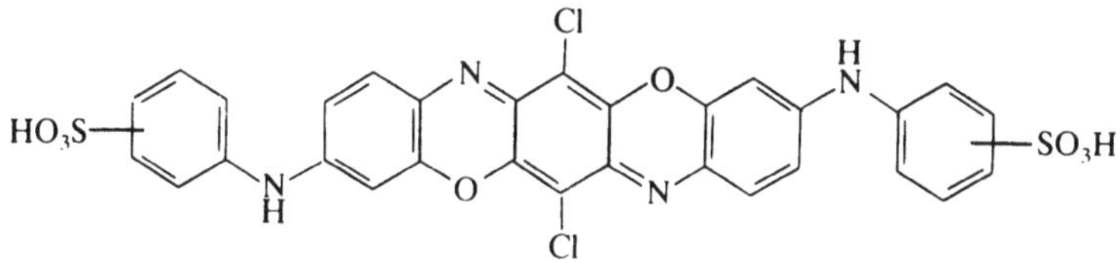

CI Direct Blue 106
54

also introduced by IG. The dyes are noted for their superior lightfastness and brightness of hue.

IV. AZOIC DYES

In 1880, work in the Huddersfield factory of dye manufacturer Read Holliday, subsequently L.B. Holliday, led to the discovery that when cotton cloth impregnated with an alkaline solution of 2-naphthol was treated with a solution of benzenediazonium chloride, a red-orange dye was formed within the lattice of the fibers. This discovery gave rise to the azoic class of dyes, essentially pigments, having little or no water solubility when formed in the fiber structure. The method of application remains essentially unaltered in that a solution of the coupling component is padded onto the fabric, which passes between two rollers and through a bath of the diazonium component. Boiling the dyed fabric leads to further dye aggregation and to improved wash fastness. Dyes of this type enjoyed an 8.1% market share throughout the world in the 1981 survey by Aeberhard.[5]

In view of the fact that the dye is actually formed during the dyeing process, the dyer is supplied with the coupler and the diazonium component either as a free base or as a stabilized diazonium salt. Consequently, the *Colour Index*[48] lists Azoic Coupling Components (around 110) and Azoic Diazo Components (around 130), not all of disclosed structure, and it would be theoretically possible to react almost any coupler with almost any diazo component. In practice, however, certain combinations are known to produce particular shades and fastness characteristics.

Azoic dyes are used mainly on cotton but can be applied to wool and silk, although such use is limited. The chemistry of these dyes has been reviewed.[41,42,44]

In 1912, a significant advance was made by Winther, Laska, and Zitscher who observed that the anilide of 2-naphtol-3-carboxylic acid had much greater affinity for cotton than 2-naphthol. The compound was subsequently marketed as Naphtol AS (Griesheim-Elektron originally marketed the Naphtol AS range and the company became part of the IG Farbenindustrie combine) and its success led to the development of several arylamides of the same acid under the Naphtol brand name. Similar products are now marketed by other manufacturers such as ICI (Brenthols) and Ciba-Geigy (Cibanaphthols). The naphtholarylamides are prepared from 2-naphthol by introduction of the 3-carboxy groups using the Kolbe–Schmidt reaction followed by condensation with an arylamine. The condensation can be effected by reaction of the arylamine with 0.5 gram mole equivalent of phosphorus trichloride, forming a phosphazine intermediate, which is not isolated, followed by reaction with 2-naphthol-3-carboxylic acid (BON acid). An alternative route involves the careful preparation of 2-hydroxy-3-naphthoyl chloride, avoiding self-condensation, and reaction with the free arylamine. About 70% of the coupling components used are of this type and give rise to, mainly, orange and red hues, although, with appropriate selection of the diazonium salt, blue dyes are available. Despite the introduction of many arylamides into commercial use Naphtol AS **(55)** remains the most important. The arylamines used were initially substituted anilines and naphthylamines.[49]

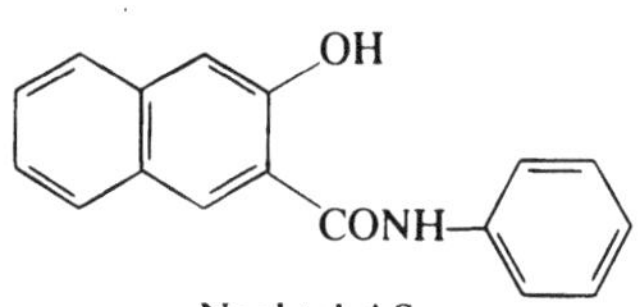

Naphtol AS
55

Yellow hues are not available from 2-naphthol-3-carboxylic acid arylamides and consequently a series of acetoacetamido couplers was introduced. The first was diacetoacetyltolidide **(56)**, which is conveniently prepared by reaction of 3,3′-dimethylbenzidine with diketene. The resulting compound is clearly not a naphthol but was marked as Naphtol AS-G and was the first of a series of arylides derived from, for example, *o*-toluidine, *o*-anisidine, and the chloroanilines to be commercialized under the Naphtol AS name. Naphtol AS-LG **(57)** is an example of such a product which gives dyes with improved light fastness.[50]

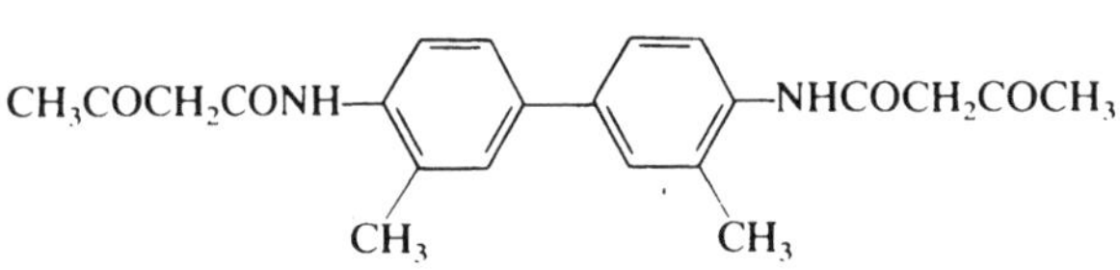

Naphtol AS-G
56

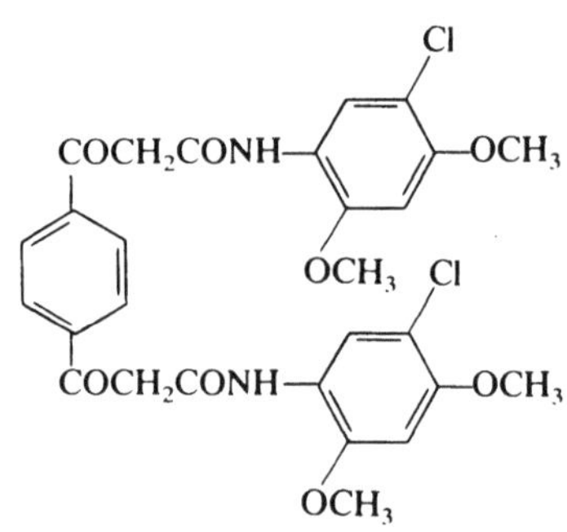

Naphtol AS-LG
57

Arylides are also prepared from other *o*-hydroxycarboxylic acids such as 3-hydroxyanthracene-2-carboxylic-2′-methylanilide (Naphtol AS-GR) for dull green dyes, 2-hydroxycarbazole-1-carboxylic-4′-chloroanilide (Naphtol AS-LB, **58**),[50] and 2-hydroxydibenzofuran-3-carboxylic-2′,5′-dimethoxyanilide (Naphtol AS-BT, **59**) for brown dyes.

The predominant diazonium components are substituted anilines devoid of solubilizing groups such as sulfonic acids. IG Farbenindustrie originally marketed

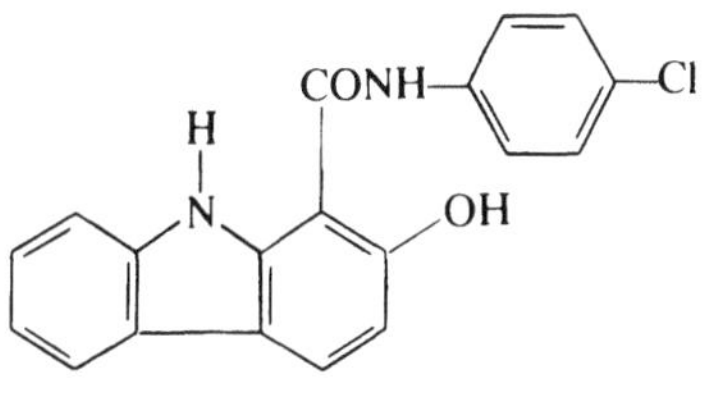

Naphtol AS-LB
58

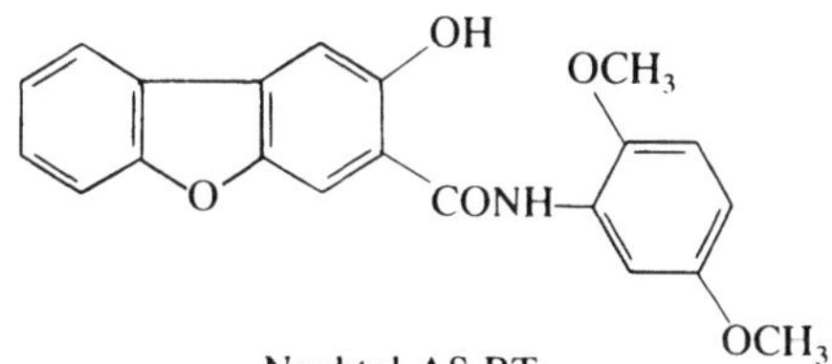

Naphtol AS-BT
59

the compounds as Fast Bases and tended to reflect the color of the dyes which they were used to produce. Thus, Fast Orange GP Base is 3-chloroaniline hydrochloride, Fast Red TR Base is 4-chloro-2-methylaniline, and Fast Bordeaux GP Base is 4-methoxy-2-nitroaniline. Blue and black azoic dyes are usually produced by using aminodiphenylamine diazonium components, and some substituted aminoazobenzenes, including Variamine Blue B Salt (4′-methoxydiphenylamine-4-diazonium chloride), Fast Black Salt G (tetrazotized 4,4′-diphenylamine coupled with one equivalent of 2-ethoxy-5-methylaniline, the amino group diazotized, and the resulting tetrazo compound isolated as the zinc chloride double salt), and Fast Black Salt K (a stabilized diazonium salt derived from the aminoazo compound 4-nitroaniline → 2,5-dimethoxyaniline).

When azoic dyeing and printing was in its infancy, it was necessary for the dyer to diazotize the amines in the dyehouse and this obviously caused some problems. However, in 1923 BASF commercialized stabilized diazonium salts in the form of their zinc chloride double salts. The compounds are easily prepared by addition of zinc chloride to a concentrated solution of the diazonium chloride and, if necessary, salting the product out of solution. The stability of zinc chloride double salts is variable and tends to increase with molecular weight. The simpler products are explosive when dry and are therefore unsafe. Other means of stabilizing diazonium salts have been discovered and naphthalene-1,5-disulfonic acid, sodium borofluoride, and cobalt chloride have been used. The salts are adjusted to a standard strength by dilution, commonly with aluminum sulfate which acts as a fire retardant, buffer, and desiccant. In specific cases, zinc sulfate and magnesium sulfate are also used. The chemical constitution of some commercial products have been listed by Allen.[44]

Although, in theory, it is possible to produce dyes by reaction of any of the diazonium components with any of the Naphtol AS couplers, experience has shown that certain specific combinations produce better fastness characteristics than others. For example, 2,5-dichloroaniline (Fast Scarlet GG base) is used to produce a greenish yellow of light fastness 5–6 in combination with Naphtol AS-G **(56)** and with Naphtol AS-OL (2-naphthol-3-carboxylic acid 4′-methoxyanilide) a scarlet of light fastness 6–7.

More recently, research in this area of dye chemistry has been directed toward single application techniques for producing azoic dyes. The Azanil salts introduced by Hoechst[51] are water-soluble diazoamino compounds, which are applied with a Naphtol AS coupler and the dye formed on the fiber by acidification of the dyebath.

Printing mixtures containing both diazonium salts and coupling components were fist produced in 1894 by BASF using alkali metal isodiazotates, the diazonium salt being regenerated in the fabric by use of steam. In 1930, IG improved the system in their range of Rapidogen colors in which the diazonium salt is in the form of a diazoamino compound produced by reaction of the diazonium salt with a secondary amine such as sarcosine. The stabilized diazonium salt is applied with the Naphtol AS coupler and the dye is formed by releasing the diazonium salt with acid steam. The use of acid steam is costly in terms of plant maintenance and further development led to the marketing of the Neutrogen dyes in which the diazonium salt is stabilized with *o*-carboxyphenylglycine and related compounds which revert to the diazonium salt at higher pH.

V. VAT DYES

Vat dyes are dyes which are converted into an alkali soluble form before application to the fiber. This process is usually achieved by using alkaline sodium dithionite. The soluble form is applied to the fabric and the insoluble dye is then regenerated by air oxidation and the whole process is known as vatting. Vat dyes have been regarded as the aristocrats of the cellulosic dyes due to their remarkable chemical stability and outstanding fastness properties, and account for approximately 19% of the worldwide usage.[5] The disadvantages of vat dyes are associated with the high price and the relative difficulty of application. Consequently, these products are used in outlets of high life expectancy. The compounds are frequently difficult to prepare, involving small batch size, multistage processes less economically viable than other dye classes.

The discovery of vat dyes is often credited with René Bohn[42] who, in 1901, attempted to synthesize an anthraquinone analog of indigo from 2-aminoanthraquinone. In fact, vat dyes have been known and used since antiquity through such natural products as indigo **(60)** and its 6,6′-dibromo derivative, Tyrian Purple. Considerable work by Baeyer, following his structure elucidation in 1883, led to the development of synthetic routes to indigo and hence Bohn, in

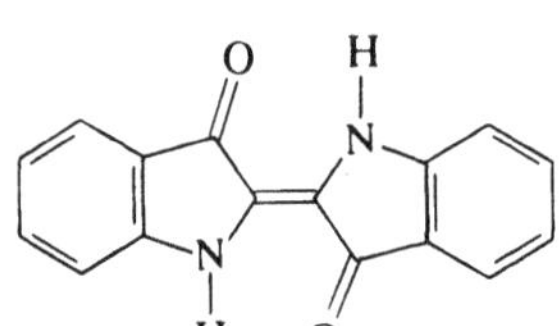

CI Vat Blue 1
60

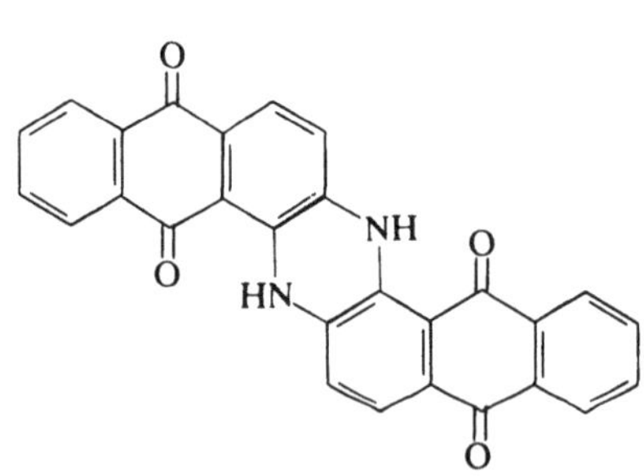

CI Vat Blue 4
61

attempting to apply this new knowledge to synthesize an anthraquinonoid analog of indigo, produced indanthrone **(61)**, sold as Indanthren Blue R, the name indicating it to be an indigo from anthraquinone. Thus, Bohn actually obtained the first known anthraquinone vat dye, of which there are now approximately 200 in commercial use providing from yellow to black dyes. The fastness properties of Indanthren Blue led to the introduction of further members of the Indanthren range by BASF (called Indathrene in the U.S. where they are marketed by GAF) and several other vat dye ranges from other manufacturers such as Caledon (ICI) and Cibanone (Ciba).

Usually the dyes are marketed as the insoluble or oxidized form, although it is also possible to obtain them as stable leuco (colorless) esters prepared (equation 5) by treatment of the anthraquinone dye with sulfur trioxide in dry pyridine in the presence of a metal such as copper.[41]

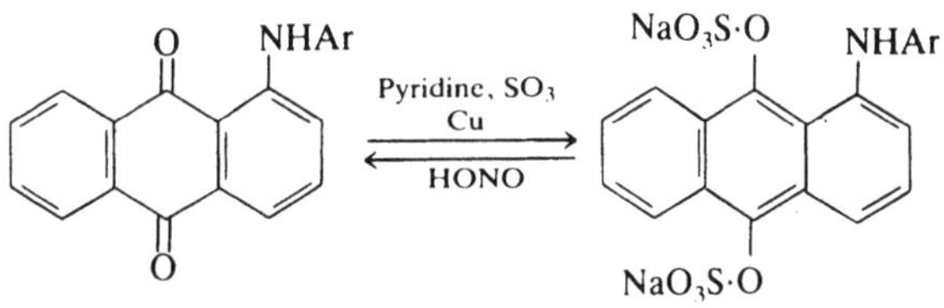

(5)

A. Anthraquinone Vat Dyes

The acylaminoanthraquinones are used as vat yellows, in the main, although it is possible to produce reds and violets. A simple example is 1,5-dibenzoylaminoanthraquinone (Indanthren Yellow GK, **62**) obtained by the dibenzoylation of 1,5-diaminoanthraquinone. Reaction of 2 moles of 1-amino-5-benzoylaminoanthraquinone with 1 mole of terephthaloyl chloride produces Sandothrene Yellow 2GW **(63)**. The introduction of electron-donating groups in the 4,8-positions of **62** produces the expected bathochromic shift and thus 1,5-dibenzoylamino-4,8-dihydroxyanthraquinone is a violet dye.

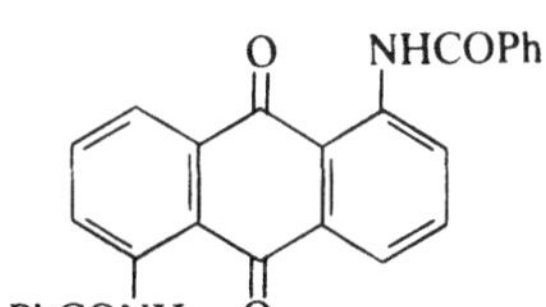

Indanthren Yellow GK
CI Vat Yellow 3
62

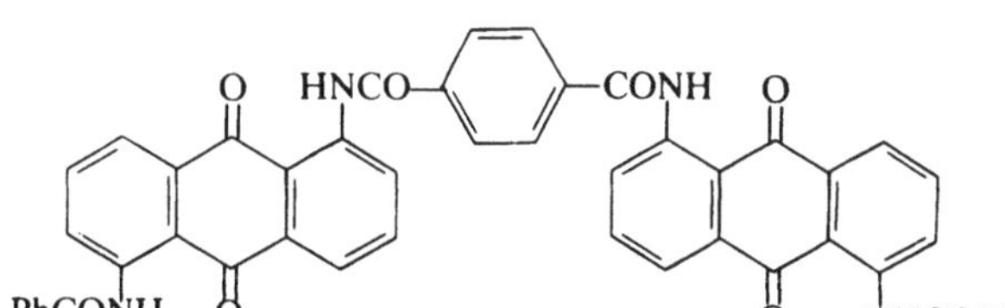

Sandothrene Yellow 2GW
CI Vat Yellow 23
63

1,4-Diaroylaminoanthraquinones are also bathchromically shifted and give red shades.[52] Thus, Cibanone Brilliant Pink 2R **(64)** is 1-benzoylamino-4-(4′-dimethylaminosulfonylbenzoylamino)anthraquinone. Indanthren Yellow 4GF is a monoacylaminoanthraquinone derived from 2 anthraquinone molecules **(65)**.

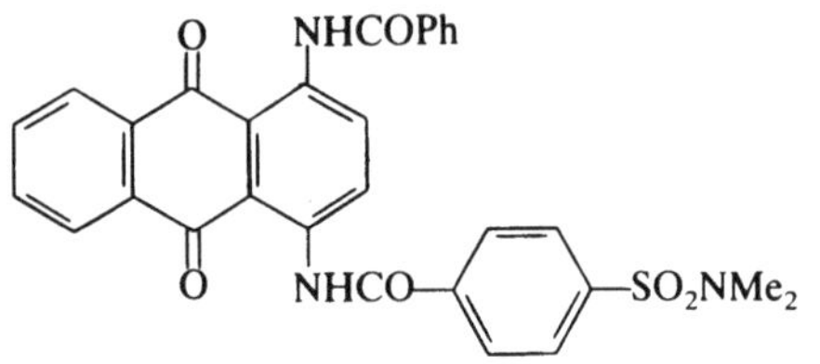

Cibanone Brilliant Pink 2R
CI Vat Red 44
64

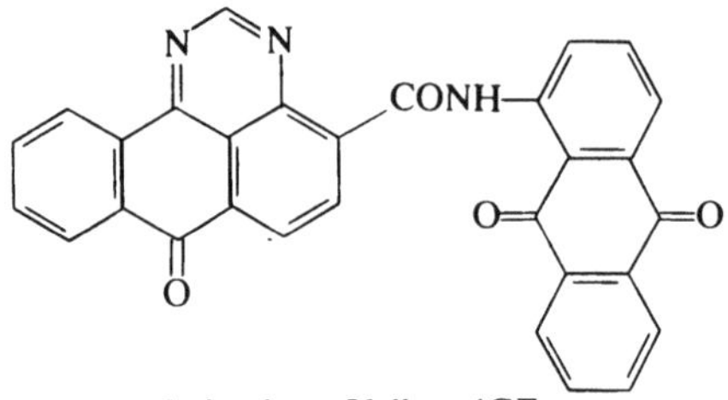

Indanthren Yellow 4GF
CI Vat Yellow 20
65

It has already been mentioned in the section on reactive dyes that cyanuric chloride is an acylating agent and Ciba have produced vat dyes of the acylaminoanthraquinone type, such as Cibanone Red 4B which is derived from three molecules of 1-amino-4-benzoylaminoanthraquinone and one molecule of cyanuric chloride, and Cibanone Red G **(66)** which is obtained from two molecules of 1-amino-4-methoxyanthraquinone and one molecule of cyanuric chloride, followed by substitution of the reactive chlorine with ammonia.[42]

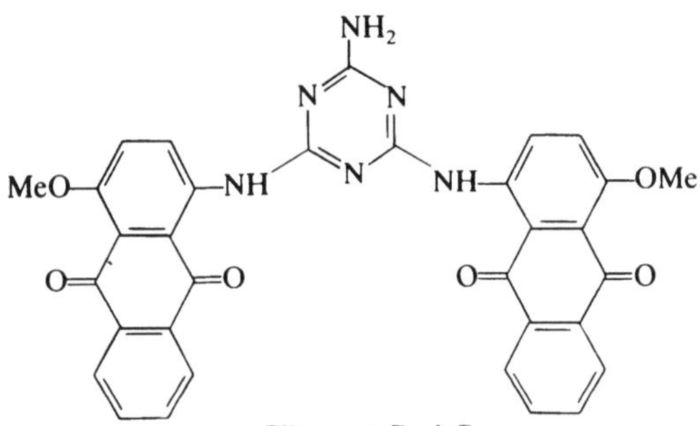

Cibanone Red G
CI Vat Red 28
66

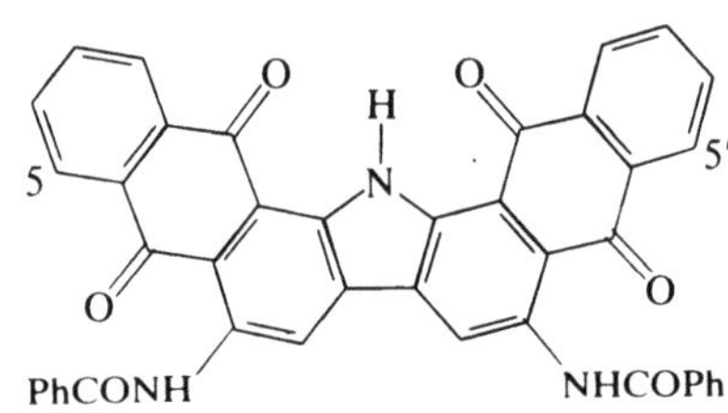

Indanthren Olive R
CI Vat Black 27
67

The "anthrimides" are anthraquinone structures where two such molecules are linked via a secondary amino group. They are commercially unimportant as such but are used as intermediates in the synthesis of anthraquinonecarbazoles, which are dyes of outstanding fastness to wet treatments. When an anthrimide is heated with, for example, aluminum chloride or sulfuric acid, cyclization to the carbazole takes place and structures of this type have been used to give yellow, orange, brown, and olive hues. For example, the anthraquinonecarbazole **67** may be prepared by condensation of 1-aminoanthraquinone with 1-chloroanthraquinone to form 1,1′-dianthrimide which, on dinitration, reduction, benzoylation, and cyclization in sulfuric acid, yields Indanthren Olive R. The 4,5′-dibenzoylamino derivative is brown and the 5,5′-derivative is golden orange. A molecule derived from three anthraquinone ring systems in a similar manner gives Indanthren Brown GR **(68)**.

The pyrazoloanthrones are also commercially important dyes. An example is Cibanone Red 6B **(69)**, which can be prepared from 1-hydrazinoanthraquinone

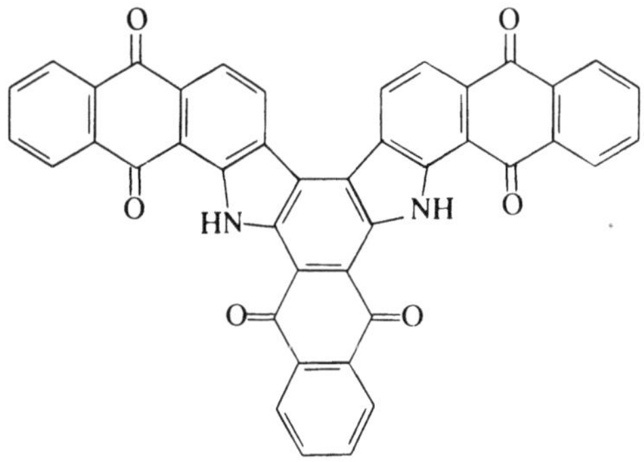

Indanthren Brown GR
CI Vat Brown 44
68

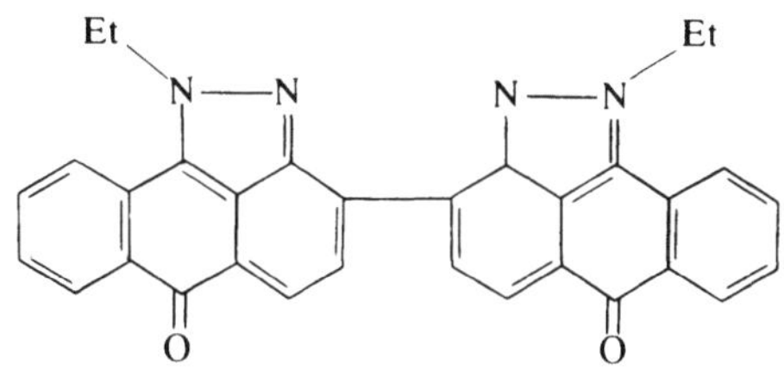

Cibanone Red 6B
CI Vat Red 13
69

by ring closure in acid followed by heating in alkali and ethylation of the resulting pyrazolodianthryl which is bluish-red. Indanthren Grey M **(70)**, contains both a pyrazole and a carbazole ring system and is prepared by condensing pyrazoloanthrone with 3,9-dibromobenzanthrone and then with 1-aminoanthraquinone, the product being cyclized using sodium hydroxide, followed by air oxidation.

Other five-membered heterocyclic rings have been successfully annellated in the anthraquinone series including thiophene, thiazole, oxazole, and imidazole. The oxazoles are the most important, dyes from imidazole systems being obsolete. The heterocycles are formed by making use of *ortho*-substituted amino, hydroxyl, or thiol groups which can, for example, be acylated and cyclized. CI Vat Red 10 (e.g., Cibanone Red 2B, **71**) is produced by several companies and, despite the presence of a free amino group, has good fastness properties. The high fastness levels have been attributed to the possible presence of zwitterionic forms and hydrogen bonding.[52]

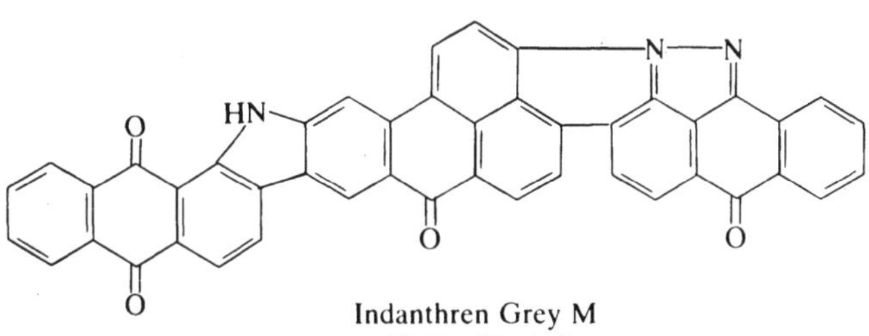

Indanthren Grey M
CI Vat Black 8
70

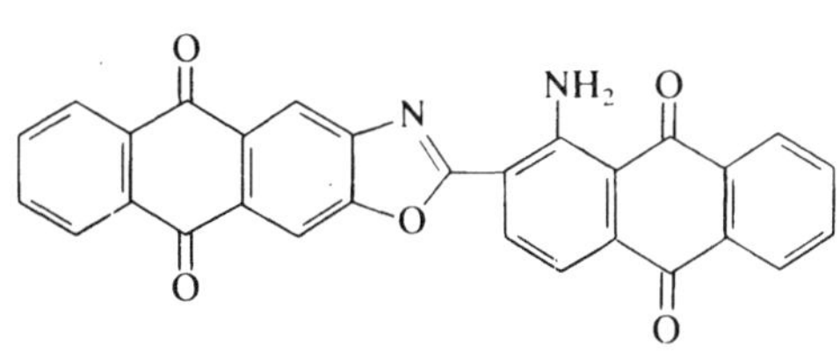

CI Vat Red 10
71

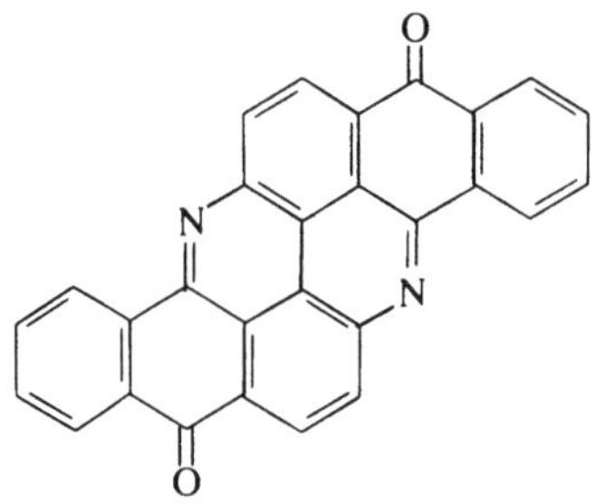

Flavanthrone
CI Vat Yellow 1
72

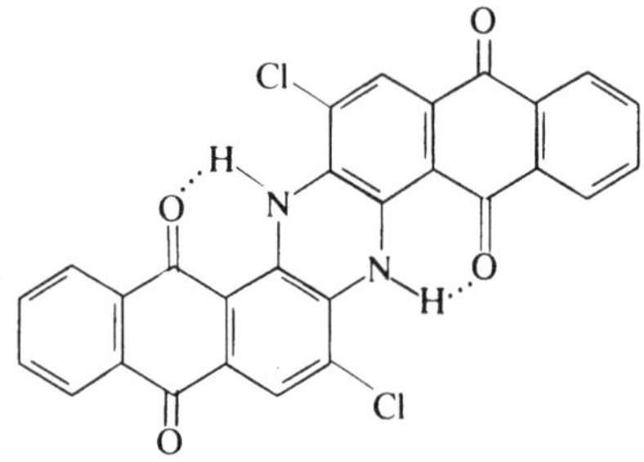

Caledon Blue XRC
CI Vat Blue 6
73

Similarly, six-membered heterocycles such as acridones, quinoneazines, pyrazines, and pyrimidines are frequently incorporated into such molecules.[44] The best known of the anthraquinone vat dyes are indanthrone **(61)** and flavanthrone **(72)**. Indanthrone is a beautiful blue anthraquinoneazine prepared by the alkali fusion of 2-aminoanthraquinone at 220 °C.[53] Indanthrone itself does not have good all-round fastness but its 3,3′-dichloro derivative (Caledon Blue XRC, **73**) is somewhat improved and is used widely for bright blues. Flavanthrone, Indanthren Yellow G, **(72)** is a byproduct in the preparation of indanthrone from 2-aminoanthraquinone and is better prepared, using the Ullmann reaction, by condensing 2 molecules of 2-amino-1-chloroanthraquinone in the presence of copper powder, followed by conversion into the diphthalimido derivative with phthalic anhydride. The dye is then produced by boiling in aqueous sodium hydroxide. A further improved synthesis[54] involves the oxidation of the Ullmann condensation product with aqueous hydrochloric acid and sodium chlorate.

The anthraquinoneacridones are capable of producing hues from orange to blue and can be prepared, for example, by reaction of 2-aminonaphthalene-1-sulfonic acid with 1-chloroanthraquinone-2-carboxylic acid[55] which produces Indanthren Red RK **(74)**. Clearly, anthranilic acid and its derivatives can replace naphthalene derivatives as in the condensation with 1,5-dichloroanthraquinone which gives Indanthrene Violet FFBN **(75)**.

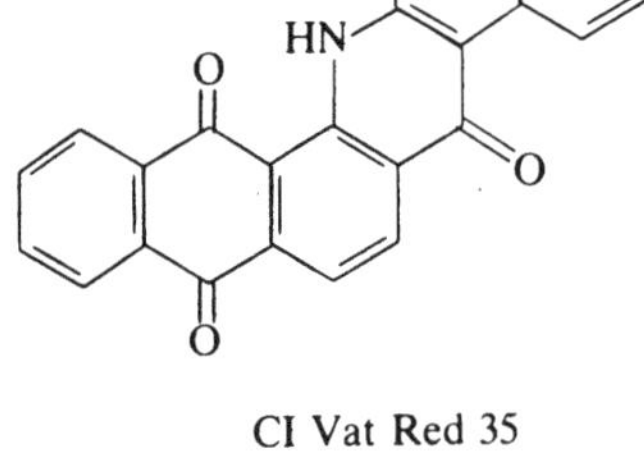

CI Vat Red 35
74

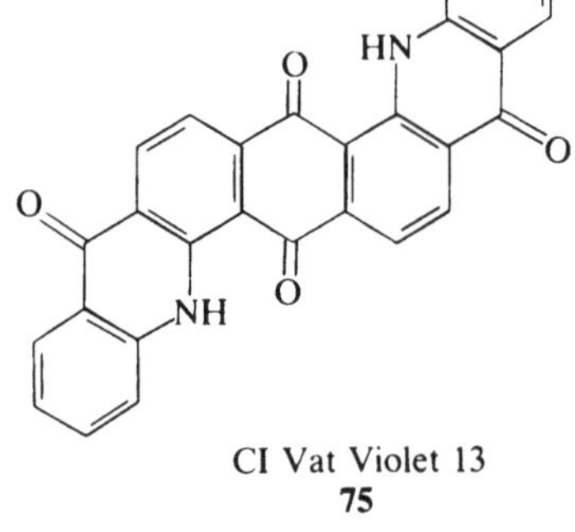

CI Vat Violet 13
75

B. Fused Ring Polycyclic Vat Dyes

A considerable number of vat dyes belong to the fused ring polycyclic type, containing carbonyl groups. The color and constitution and theory of color of these structures has received surprisingly little attention, but has been discussed by Gregory and Gordon.[33] The dyes are classified by the suffix anthrone. The derivatives of benzanthrone are probably the most important.

Pyranthrone (**76**; R = R′ = H), discovered by Scholl in 1905, is an orange vat dye which can be prepared by fusing a mixture of 2,2′-dimethyl-1,1′-bianthraquinonyl with sodium sulfide and ethanolic potassium hydroxide in an autoclave followed by oxidation of the leuco compound.[41] Bromination of the product in chlorosulfonic acid produces the 4,12-dibromopyranthrone, which is an orange (Cibanone Golden Orange 2R, **76**; R = R′ = Br) of good light fastness.

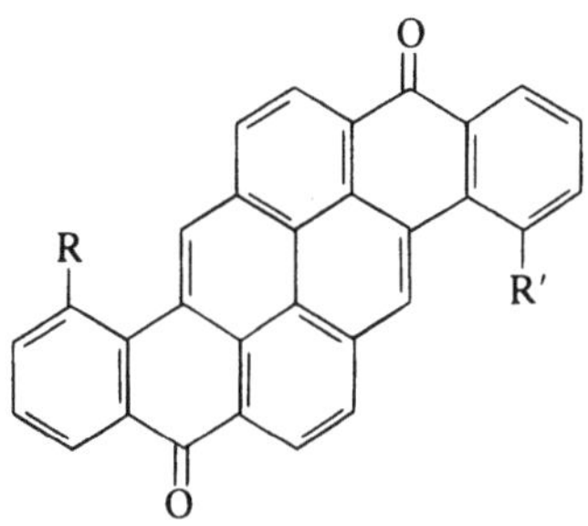

R,R′ = H; CI Vat Orange 9
R,R′ = Br; CI Vat Orange 2
76

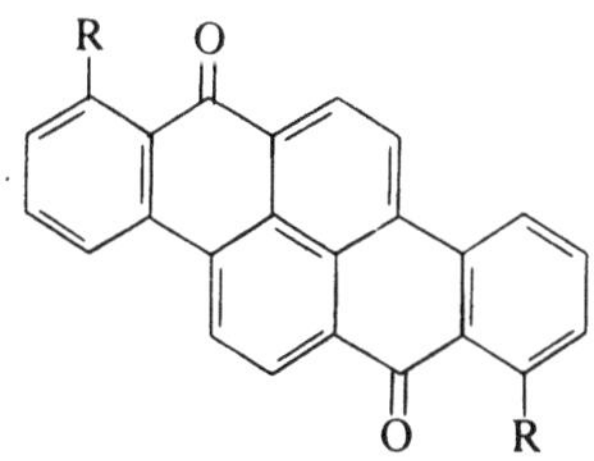

R = H; CI Vat Yellow 4
R = Br; CI Vat Orange 1
77

Indanthrene Golden Yellow GK (**77**; R = H) is an example of the dibenzopyrenequinone type formed in this case by cyclization of 1,5-dibenzoylnaphthalene using either sodium *m*-nitrobenzenesulfonate or a mixture of aluminum and sodium chloride as the oxidizing agent. Bromination (**77**; R = Br) improves the fastness of the dye and gives Indanthren Golden Yellow RK.

Orange dyes are provided by the halogeno derivatives of anthanthrone (**78**; R = H). The substitution may be dichloro or dibromo and the halogenation is carried out on the anthanthrone ring rather than in the intermediate stages. The anthanthrone ring is synthesized from 1-aminonaphthalene-8-carboxylic acid by

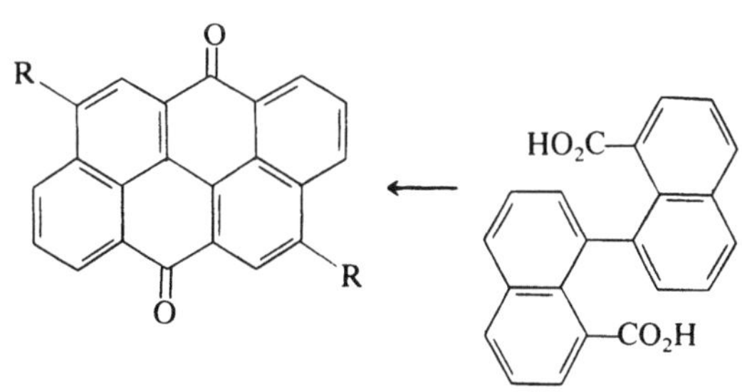

78 **79**

diazotization and reaction with cuprous chloride, copper powder, and base to produce the bisnaphthalenyl compound **(79)**, which is then ring closed in concentrated sulfuric acid at 35 °C.

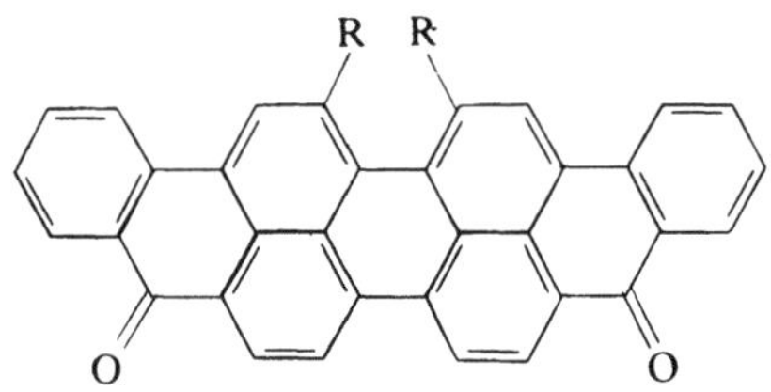

R = H; Violanthrone or Dibenzanthrone
R = OCH_3; Caledon Jade Green XBN
CI Vat Green 1
80

The most important of the dyes of this type are derivatives of benzanthrone, particularly violanthrone and isoviolanthrone, and include one of the classic vat dyes, Caledon Jade Green XBN (**80**; R = OCH_3). Benzanthrone is oxidized by air in the presence of potassium hydroxide and sodium acetate to form 4,4′-dibenzanthronyl, which is treated with manganese dioxide in sulfuric acid. At 20 °C the resultant compound is dibenzanthrone or Violanthrone (**80**; R = H), but treatment at 35 °C results in carbonyl groups at carbons 16 and 17 which are reduced to hydroxyl groups and subsequently methylated, with dimethyl sulfate, to produce Caledon Jade Green XBN. The dye was first prepared in 1920 by Davies, Thomson, and Thomas[56] at ICI and is a beautiful brilliant green color, a rarity, which, as in the case of Malachite Green, arises due to two absorption bands in the visible spectrum with λ_{max} values at 680 and 395 nm.[33] Detailed accounts of the chemistry of violanthrone and its derivatives have been published.[52]

A related group of dyes has been derived from perylene-3,4,9,10-tetracarboxylic acid and naphthalene-1,4,5,8-tetracarboxylic acid. For example, condensation of the former with two molar equivalents of a primary aromatic or an aliphatic amine in aqueous pyridine[57] gives the general structure **81**. CI Vat Red 23 is the dye where R = Me and in CI Vat Red 29 R = 4-methoxyphenyl. A similar condensation with the naphthalene analog is usually carried out with an *ortho*-phenylenediamine. The synthesis with 1,2-phenylenediamine produces *cis* and *trans* isomers, **82** (CI Vat Red 15) and **83** (CI Vat Orange 7), which can be

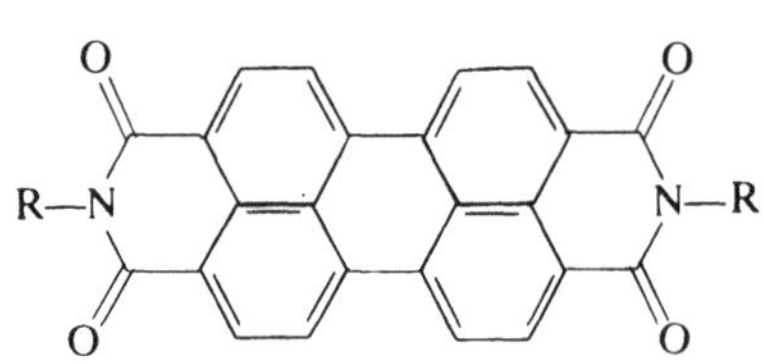

R = Me; CI Vat Red 23
R = 4-MeOPh; CI Vat Red 29
81

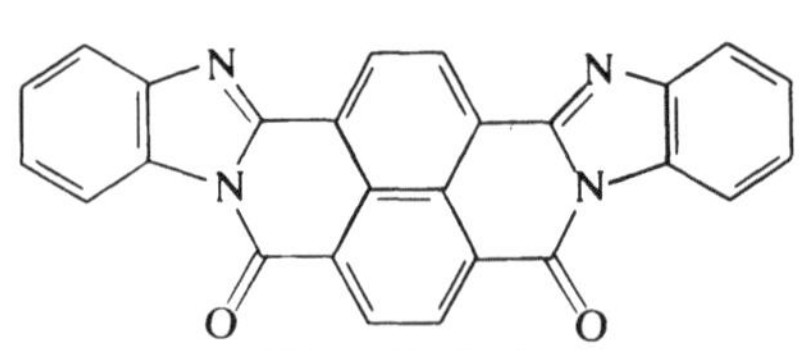

CI Vat Red 15
82

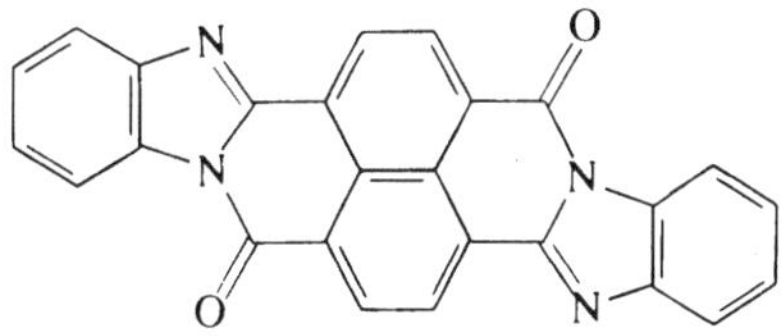

CI Vat Orange 7
83

separated using a treatment developed by IG Farbenindustrie.[44] The mixture of the two dyes is listed as CI Vat Red 14.[48]

C. Indigoid Vat Dyes

No account of vat dye chemistry would be complete without a mention of indigo **(60)** whose structure was elucidated by Baeyer in 1883. Synthetic routes devised by Baeyer,[58] Sandmeyer and Heumann (Section IX.A) soon followed and in 1897 a commercial process based on the caustic fusion of *o*-carboxyphenylglycine was introduced. The process was superseded by that outlined in Scheme 1.[44] Thus, aniline is condensed with formaldehyde-sodium bisulfite and the product is reacted with sodium cyanide to give *N*-cyanomethylaniline. Hydrolysis then gives sodium phenylglycinate, which is cyclized in molten sodium/potassium hydroxide containing sodamide. The resulting indoxyl is air-oxidized to indigo **(60**, CI Vat Blue **1)**.

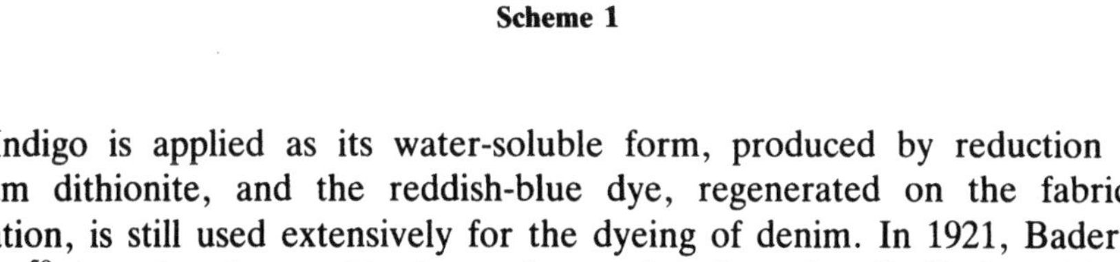

84

Scheme 1

Indigo is applied as its water-soluble form, produced by reduction with sodium dithionite, and the reddish-blue dye, regenerated on the fabric by oxidation, is still used extensively for the dyeing of denim. In 1921, Bader and Sunder[59] introduced a stable leuco form of indigo, its disulfuric acid ester (Indigosol O, **84**), being best prepared by treating leuco indigo with chlorosulfonic acid in pyridine. After application to the fabric, the dye is again regenerated by oxidation. The cultivation of plants for the extraction of natural indigo has now virtually ceased and the parent molecule and a number of its halogeno derivatives are all produced synthetically. Introduction of halogen atoms can be achieved by total synthesis from halogenoanilines or by halogenation of indigo.

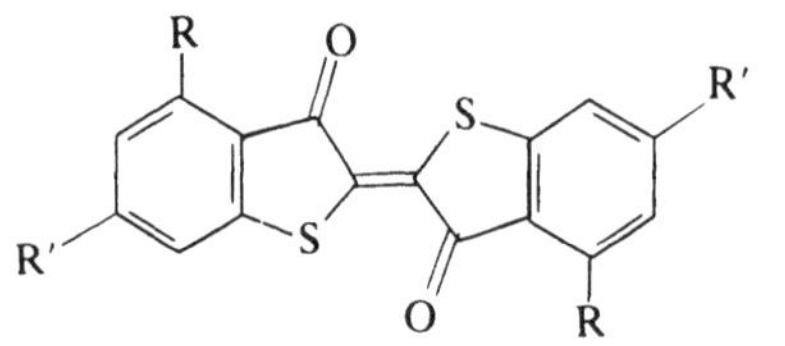

85

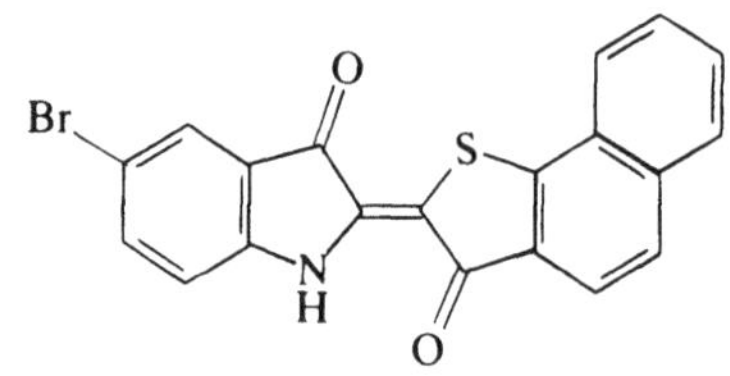

86

Thioindigo (**85**; R = R′ = H, Thioindigo Red B, CI Vat Red 41) and its derivatives are important red dyes discovered by Friedlander in 1906. The most important derivative is Hydron Pink FF (**85**, R = CH_3, R′ = Cl; CI Vat Red 1) which is synthesized in 7 stages from *o*-toluidine hydrochloride.[44,60] A few vat dyes containing both an indigoid and a thioindigoid residue, such as Indanthren Printing Black BL (CI Vat Black 1, **86**), are produced, giving mainly violet, brown, gray, and black hues.

VI. SULFUR DYES

Wood,[61] writing in 1976, indicated that there have been very few books or articles exclusively devoted to sulfur dyes and that, in modern textbooks, authors either neglect them, or give outdated or incorrect information. Sulfur dyes are indeed the enigma of the dye industry in that, throughout the world, they have enormous usage, albeit somewhat less in the more technologically developed West than in the East or Africa. Yet, as noted by Wood, accounts of their chemistry, for reasons which will become apparent, are sparse indeed. The major work on dye chemistry, the eight authoritative volumes of Venkataraman,[42] accords only 75 pages to sulfur dyes, a class which, in 1941, had among its number the dye Sulfur Black T which accounted for 10% of the total dye production in the United States. In the world survey published in 1981[5] sulfur dyes still held the largest market share, at 18%, of any dye class. In the West and Japan, usage on cellulosics is probably about 15% with direct dyes and reactive dyes sharing approximately 65%.

Many dyes, of course, contain sulfur atoms but the present definition is restricted to those dyes that dissolve in and are reduced by sodium sulfide solution. Air oxidation regenerates the insoluble dye molecules. The dyes are, almost invariably, dull hues and the blacks and blues predominate, although it is possible to produce yellow dyes and full reds. The reason that accounts of sulfur dye chemistry are limited is because, for the most part, their absolute structures are uncertain. Since the first commercial production of a sulfur dye, Cachou de Laval, in 1861, obtained by heating bran and sawdust with sodium polysulfide at 300 °C, with concomitant evolution of hydrogen sulfide, this area of chemistry has remained part art and part science.

The process of heating a variety of sulfur dye intermediates with either sodium polysulfide or sodium sulfide and sulfur is known as thionation and intermediates subjected to the process are usually nitrophenols, aminophenols, diphenylamines, and indophenols. Indophenols can be produced by condensation of a 4-aminophenol with an amine having a free *para*-position or a mixture of a phenol having a free *para*-position and a *para*-phenylenediamine in the presence of an oxidizing agent (equation 6). Alternatively, 4-nitrosodialkylanilines may be condensed with phenol derivatives.

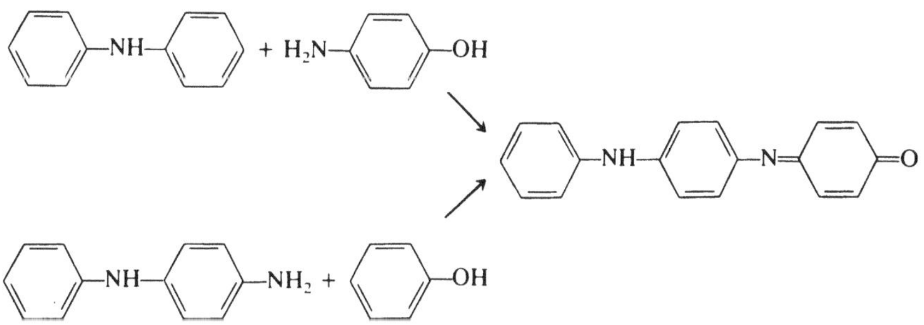

(6)

The yellow sulfur dyes were originally made by thionation of *m*-toluylenediamine in the presence of benzidine, in this case Immedial Yellow D. Increasing the temperature of thionation from 190 to 250 °C results in Immedial Orange C. Thus, it can be seen that the chemistry is rather imprecise. This aspect of the subject has precluded the interest of academic researchers and, consequently, comparatively little effort has been expended on structure determination. The yellow dye Immedial Supra Yellow GWL, prepared by thionation of a mixture of benzidine with dehydrothio-*p*-toluidine and *p*-toluidine, has been shown[62] to consist mainly of **87**.

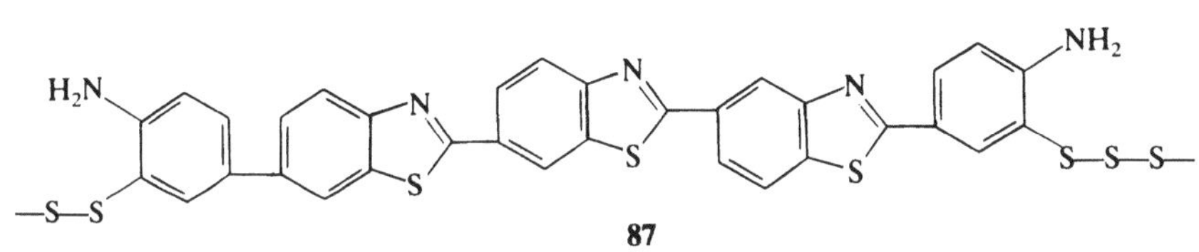

87

Good red and violet colors are unavailable but the major impact of sulfur dyes is in black and blue hues.[63] The blue dyes are conveniently prepared by thionation of indophenols. For example, thionation of 2,4-dinitro-4′-hydroxydiphenylamine produces Immedial Direct Blue G. Greener blues are produced from naphthalene containing indophenols.

It has been shown[64] that Immedial Pure Blue, which forms a crystalline complex with sodium bisulfite, contains the thiazine ring structure in **88**. The compound is isosteric with a number of vat dyes, e.g., indanthrone, and hence its hue and affinity for cellulosic fibers is to be expected.

Sulfur Black T is prepared by thionation of 2,4-dinitrophenol with sodium polysulfide[65] and such is its commercial importance that it has been the subject of a number of investigations to determine its chemical constitution. Hiyama[66] assigned structure **89** to the dye, although when sulfurizing is carried out at

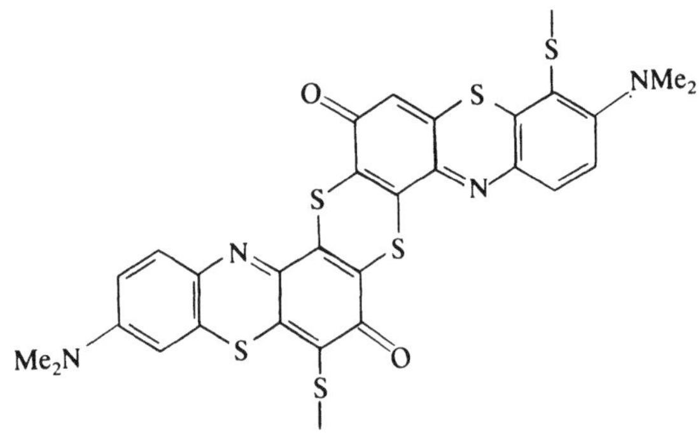

88

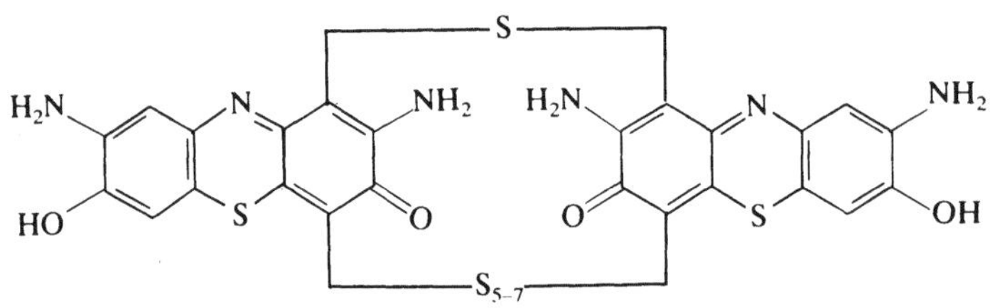

89

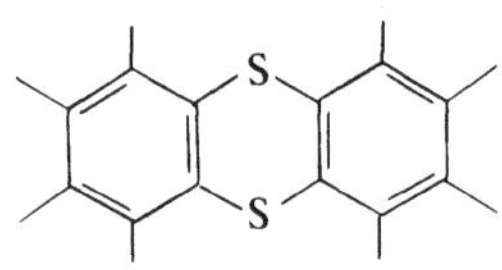

90

temperatures in excess of 110 °C sulfur also enters the molecule at the 5 position *ortho* to the hydroxy group.

Other Sulfur Black dyes have also received investigative attention[67] and thiazone units are frequently proposed in blue and black dyes.[68] The presence of thianthrene units **(90)** has also been established.[69]

A. Sulfurized Vat Dyes

Sulfurized vat dyes are not true sulfur dyes in that they are applied by the vatting process using sodium dithionite. They are, however, produced by the thionation process and are of unknown structure.[70] The first such dye was produced in 1909 by Cassella under the name Hydron Blue R (CI Vat Blue 43) and is obtained by condensation of 4-nitrosophenol with carbazole in sulfuric acid and thionation of the resultant indophenol derivative **(91)** with sodium polysulfide in butanol.[71] A greener blue, Hydron Blue G (CI Vat blue 42), is similarly produced using *N*-ethylcarbazole.

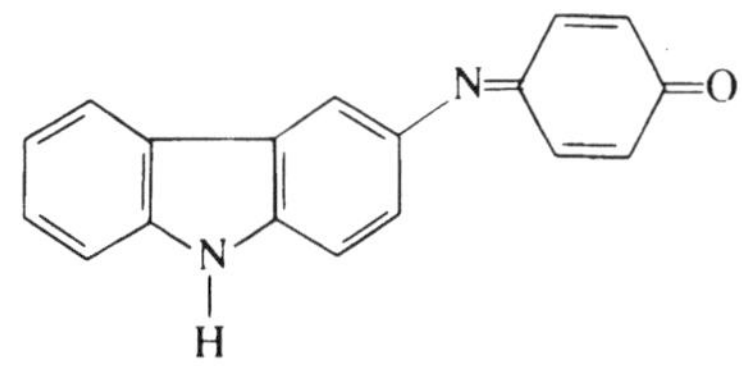

91

VII. SYNTHESES OF REACTIVE DYES

Examples of synthetic procedures for the more important dyes and dye classes are described in this and the following sections. In certain cases the dyes exemplified are not recent inventions but as, for example, in the case of indigo they are still commercially important and as such merit their inclusion.

A. Analysis of Dyes and Intermediates

The majority of commercial dyes (60%) are based on azo systems and the azo dyes for cellulosic fibers, with the exception of azoic dyes, almost invariably contain sulfonic acid groups. The dyes are isolated from aqueous reaction media by "salting out," that is, the water-soluble dye molecules are displaced from solution by addition of salt. Consequently the isolated dyes are never pure compounds as can be achieved with other dye classes, such as the disperse dyes.

The common method to estimate the purity of such a dye sample is by the Knecht reduction technique, which involves the reduction of the azo group with titanous chloride or titanous sulfate in acid solution.[60] The method can also be applied to other dye classes such as triarylmethane, indigo, thiazine, and nitro dyes.

Frequently the coupling components and diazonium components are, also, not pure compounds and may be pastes or contaminated with inorganic salts. In addition it should be noted that sulfonic acid groups can crystallize with five molecules of water of crystallization which must be taken into account when calculating molecular weights. Diazonium components can be estimated by titration with a standard solution of sodium nitrite and coupling components by titration with a standard solution of a diazonium salt, such as 4-chlorobenzenediazonium chloride.

B. Methods of Diazotization

As a general rule, for each gram molecule of aromatic amine to be diazotized two and a half gram molecules of mineral acid are used. One gram molecule is required to generate a molecule of nitrous acid, one to form the amine acid salt (usually hydrochloric acid), and one half a gram molecule to maintain acidity, thus preventing diazoamino compound formation. A slight molar excess of sodium nitrite is usually used and, on completion of reaction, the slight excess decomposed by addition of a little urea or sulphamic acid. Diazonium salts are generally unstable and are prepared at 0–5 °C and immediately reacted further. Aromatic amines containing sulfonic acids are often not soluble in aqueous acid and are most conveniently diazotized by the reverse method. This involves dissolution of the aminesulfonic acid in neutral solution with the required amount of sodium nitrite and running the solution into aqueous hydrochloric acid at 0–5 °C. Discovered by Griess in 1860 this extremely important chemical reaction has been the subject of a number of detailed publications.[72]

C. Coupling Methods

Coupling components fall into four major types. Phenols and naphthols are usually reacted in aqueous alkaline solution, the rate of reaction with the phenolic substrate being much faster than the potential competing hydrolysis. The coupler is dissolved using the minimum amount of sodium hydroxide and sodium carbonate is added to neutralize the acid in the diazonium salt solution, the amount being calculated such that conversion into sodium hydrogen carbonate is affected, thus preventing foaming due to carbon dioxide evolution. Similar conditions apply to reaction with enolizable keto compounds or active methylene compounds. Compounds which fall into this class, important in the production of greenish-yellow dyes, are the open chain compounds such as the acetoacetates and the arylides, acetoacetylamino compounds, and more importantly heteroyclic couplers; pyrazolone and pyridone derivatives.

Aromatic amines, on the other hand, are coupled in weakly acid solution and conditions vary depending on the solubility of the particular compound. Primary, secondary, and tertiary amines are all capable of coupling reactions but in reactive dye chemistry the amino group is usually reacted further either by subsequent diazotization and coupling or by condensation with the reactive group. Consequently tertiary amines are rarely encountered. The amino coupling component is usually added to the diazonium salt, ideally in solution, neutral or acid, and the pH is slowly raised by addition of sodium acetate. Coupling is generally slow in strongly acidic solution but high pH leads to diazoamino formation. The presence of electron-donating groups such as methyl and methoxy, which increase relative electron density at the coupling carbon, increases reaction rate. The reaction rate may also be increased by addition of alcohol.

Diazonium salts react in the *ortho*- and *para*-positions to amino or hydroxyl (phenols and naphthols) groups. In the case of 2-naphthols coupling is rapid, in alkaline solution, in the 1-position. The situation is similar with amino-substituted 2-naphthols such as J-acid which also reacts in the 5-position, *ortho* to the amino group, in acidic solution. In the case of 8-amino-1-naphthol-3,6-disulfonic acid, coupling is *ortho* to the hydroxyl group in alkaline solution and *ortho* to the amino group in acid solution.

D. Synthetic Examples of Reactive Dyes

1. 5-(2′,3′-Dichloro-6′-quinoxalinecarbonylamino)aniline-2-sulfonic acid → 1-(2′,5′-Dichloro-4′-sulfophenyl)-3-methyl-5-pyrazolone (Levafix E Type; Bayer)[73]

2,4-Diaminobenzenesulfonic acid (12 g) in water (100 ml) at 35 °C is adjusted to pH 7.5–8.0 and 2,3-dichloro-6-quinoxalinecarbonyl chloride (16.3 g) is added. The slurry is stirred 4 h at 35–40 °C and pH 7.5 maintained using dilute sodium hydroxide. The precipitate is filtered and the solid redissolved in water (1000 ml) at 50 °C, clarified, cooled to 5 °C, and slowly added to a solution of 5 *M* sulfuric acid (15 g) and 5 *M* sodium nitrite (11 g) in water (50 ml) and the reaction stirred at 5 °C for 4 h. The excess of nitrous acid is decomposed using sulphamic acid and the reaction is allowed to settle. The supernatant liquor is decanted and to the remaining slurry is added a solution of 1-(2′,5′-dichloro-4′-sulfophenyl)-3-methyl-5-pyrazolone (16.1 g) in water (45 g) at pH 7. The reaction is stirred at 0–5 °C and the pH is raised to 4.5 by addition of sodium acetate. On completion of coupling the pH is raised to 8.2 with 2 *M* sodium hydroxide, the temperature increased to 40–45 °C, and sodium sulfate slowly added to precipitate the product, which is collected by filtration. The cake is washed with 20% sodium sulfate solution and dried at 60 °C to give the yellow reactive dye.

2. 3-Amino-4-methoxy-1-methylbenzene-6-β-hydroxyethylsulfonesulfoester sodium salt → 1-(4′-Sulfophenyl)-3-methyl-5-pyrazolone (Remazol type; Hoechst)[74]

4-Amino-5-methoxy-2-methyl-(β-hydroxyethylsulfonyl)benzene (24.5 g) is stirred and heated in 60% sulfuric acid (16.2 g), and the resultant mixture containing the acid sulfoester stirred with ice (150 g) and concentrated hydrochloric acid (15 ml). At 0–5 °C a solution of sodium nitrite (7.2 g) in water (30 ml) is added dropwise in 15–20 min and stirred 10 min. The excess of nitrous acid is removed with a little sulfamic acid and 1-(4′-sulfophenyl)-3-methyl-5-pyrazolone (25.4 g) is added and sodium hydrogen carbonate is added slowly to adjust the pH to 6–7 and complete the coupling reaction. The product is displaced from solution by addition of potassium chloride and filtered and dried *in vacuo* to give a bright yellow dye.

3. 4-Dichlorotriazinylaniline-2,5-disulfonic acid → 6-Hydroxy-4-methyl-1-(4′-sulfobenzyl)pyrid-2-one (Procion MX type; ICI)[32]

Cyanuric chloride (4.63 g) in acetone (25 ml) is added in a thin stream to water (50 ml) and ice (50 g), with good stirring, and a solution of 1,4-phenylenediamine-2,5-disulfonic acid (6.7 g) in water (50 ml) at pH 5 and 0–5 °C is added. The pH is maintained at 4–5 by addition of 2 *M* sodium carbonate as required. After stirring for 1 h concentrated hydrochloric acid (15 ml) is added and, at 0–5 °C, diazotization completed by dropwise addition of a solution of sodium nitrite (3.46 g) in water (25 ml). Excess nitrous acid is deomposed by addition of a little sulfamic acid and the solution is added to a solution of 6-hydroxy-4-methyl-1-(4′-sulfobenzyl)pyrid-2-one (7.4 g) in water (100 ml) at 0–5 °C and pH 7. The solution is stirred for 1 h at pH 6–7 and buffered by addition of potassium dihydrogen phosphate (8 g) and disodium hydrogen phosphate (4 g) in water (50 ml). The product is precipitated by addition of 25% wt/vol of potassium chloride, filtered, and washed with saturated potassium chloride solution (50 ml).

The filter cake is mixed with potassium dihydrogen phosphate (2 g) and disodium hydrogen phosphate (1 g) and dried *in vacuo* at room temperature.

4. 2,4,8-Acid → Dichlorotriazinyl-*m*-toluidine (Procion Yellow MX-R, CI Reactive Yellow 4; ICI)[75]

Concentrated hydrochloric acid (28 ml) is added to a solution of 2-aminonaphthalene-4,8-disulfonic acid sodium salt (34.7 g) and sodium nitrite (7 g) in water (300 ml) while maintaining the temperature at 0–5 °C and stirred for 30 min. Excess nitrous acid is removed by addition of a little sulfamic acid and *m*-toluidine (10.7 g) dissolved in concentrated hydrochloric acid (10 ml) and water (150 ml) at 0–5 °C is added, and the pH is raised to 4 by addition of sodium acetate. The resultant dye is salted out and filtered, washed, and redissolved in water (700 ml) at pH 7 with the addition of sodium hydroxide solution.

Cyanuric chloride (18.5 g) is dissolved in acetone (50 ml) and poured in a fine stream into ice (200 g) and water (300 ml). To this suspension the monoazo dye solution from above is added at 0–5 °C and the pH is maintained at 5.5 to 6.5 by addition of 2 *M* sodium carbonate solution. The dye is isolated by salting to 15% w/v with sodium chloride, filtering and drying the paste under vacuum after thorough mixing with disodium hydrogen phosphate (4 g) and potassium dihydrogen phosphate (8 g) as mixed buffer.

5. 2,4,8-Acid → 5-Chloro-2-methylsulfonyl-4-methylpyrimidin-6-yl-*m*-toluidine (Levafix P Type; Bayer)[76]

The solution of monoazo dye as prepared in the above example is mixed with 2-methylsulfonyl-4,5-dichloro-6-methylpyrimidine (24 g) and stirred vigorously. The temperature is maintained at 65 °C and the pH maintained at 6 to 6.5 by addition of sodium carbonate solution until the reaction is complete. The resultant dye is salted out by addition of salt (80 g) and filtered and dried at 80 °C *in vacuo*.

6. 2,4,8-Acid → 2,4,5-Trichloropyrimidin-6-yl-*m*-toluidine(Drimarene and Reactone Type; Sandoz, Geigy)[77]

To the solution of monoazo dye as prepared in the above example is added, at 50 °C, 2,4,5,6-tetrachloropyrimidine (22.8 g) and the whole is stirred vigorously at 50–60 °C for 6 h. The hydrogen chloride liberated is neutralized by addition of dilute sodium carbonate solution so as to maintain the pH at 5–6. The dye is isolated and dried *in vacuo* at 40 °C.

7. 2,4,8-Acid → *N*-(2,3-Dichloro-6-quinoxalinocarbonyl)-*m*-toluidine (Levafix E Type; Bayer)[73]

The azo dye-base 2,4,8-acid → *m*-toluidine (42 g) prepared as described above is stirred to dissolve in water (700 ml) at 40 °C and pH 7.5–8.0, and powdered 2,3-dichloro-6-quinoxalinecarbonyl chloride (30 g) added. The mixture is stirred 12 h at 35–40 °C, maintaining pH at 7.5–8.0 by addition of 2 *M* sodium carbonate solution. The resultant solution is filtered and sodium sulfate (110 g) added and the solution cooled to precipitate the product, which is filtered and washed with 12% sodium sulfate solution.

8. 2,4,8-Acid → 2-Amino-4-chlorotriazinyl-*m*-toluidine (Procion H, Cibacron Type; ICI, Ciba)

To the solution of the dichlorotriazinyl dye as described above (before addition of sodium chloride to isolate the dye) is added 10% aqueous ammonia (58 ml), the temperature raised to 35–40 °C, and the reaction is stirred for 2 h. The dye is salted out similarly to the Procion MX type.

9. 2 moles (2,4,8-Acid → Dichlorotriazinyl-*m*-toluidine) condensed with *para*-phenylenediaminesulfonic acid (Procion H-E Type; ICI)

To the solution of the dichlorotrazinyl dye as described above (before addition of salt) is added 1,4-phenylenediamine-2-sulfonic acid (8 g) at 0–5 °C and pH 6 and the temperature of the reaction is raised to 35–40 °C and held 2 h. The pH is maintained at 5.5–6.5 by addition of 2 *M* sodium carbonate solution. On completion of reaction the product is salted out in a similar manner to the Procion MX example.

10. 2,4,8-Acid → 2,4-Dichloropyrimidinyl-5-carbonylamino-3′-methylbenzene (Reactofil Type; Geigy)[78]

To the solution of the monoazo dye 2,4,8-acid → *m*-toluidine in water (400 ml) is added, dropwise in 1 h at 0–5 °C, a solution of 2,4-dichloropyrimidine-5-carboxylic acid chloride (22.2 g) in acetone (100 ml). The pH of the reaction is maintained at 6.5–7 by the simultaneous addition of 2 *M* sodium carbonate solution. The product is precipitated by salting, washed with sodium chloride solution, and vacuum dried.

11. Orthanilic Acid → Dichlorotriazinyl-J-acid (Procion MX Type, CI Reactive Orange 1; ICI)[34]

A solution of cyanuric chloride (18.5 g) in acetone (50 ml) is poured into a vigorously stirred mixture of ice (400 g) and water (400 ml). At 0–5 °C a solution of J-acid sodium salt (26.1 g) in water (480 ml) is added over a 40 min period and the pH is maintained at 5–6 by addition of dilute sodium carbonate solution. The mixture is stirred for 30 min to complete reaction.

The diazo solution is produced by addition of a solution of sodium nitrite (6.55 g) in water (20 ml) to a mixture of aniline-2-sulfonic acid (16.45 g) in water (200 ml) and concentrated hydrochloric acid (S.G. 1.18; 18 ml) at 0–5 °C. Excess nitrous acid is removed and the suspension poured into the coupler solution at 0–5 °C. Sodium acetate (40 g) is added and the mixture stirred for 90 min at 0–4 °C and sodium carbonate added gradually to raise the pH to 7. Sodium chloride (20% w/v) is added and after stirring for 30 min at 0–5 °C the product filtered, washed with 20% brine and acetone, and dried *in vacuo* after mixing the paste with disodium hydrogen phosphate (2 g) and potassium dihydrogen phosphate (4 g).

12. Aniline → 2,4-Dichlorotriazinyl-H-acid (Procion Red MX-5B, CI Reactive Red 2; ICI)[14]

Cyanuric chloride (20.3 g, 0.11 gmol) is dissolved by stirring in acetone (100 ml) and reprecipitated by pouring rapidly into ice (300 g) and water (300 g). Concentrated hydrochloric acid (0.5 ml) is added and at 0–5 °C an alkaline (pH 8.5) solution of H-acid disodium salt (equivalent to 36.6 g at 100% pure) in water (160 ml) is added slowly over 60 min. The reaction mixture is stirred 30 min at 0–5 °C to complete the reaction. Simultaneously, aniline (9.3 g) is dissolved in water (100 ml) and 10 *M* hydrochloric acid (25 ml) and cooled to 0–5 °C. Diazotization is effected by dropwise addition of a solution of sodium nitrite (7.2 g) in water (25 ml) maintaining a slight excess of nitrous acid as shown by a positive starch–iodide reaction. The excess is removed by addition of a little sulfamic acid. The benzenediazonium hydrochloride solution is then added rapidly at 0–5 °C to the coupler solution and sodium acetate (50 g) is added. The reaction mixture is then stirred 20 h at 0–5 °C and sodium carbonate is added to raise the pH to 8.5. Sodium chloride equivalent to producing a 20% solution is added and the precipitated dye is filtered after 30 min. The dye cake is mixed intimately with sodium dihydrogen phosphate (2 g) and disodium hydrogen phosphate (2 g) and is dried in vacuum at 30 °C.

13. 2-Aminophenol-4,6-disulfonic acid → 2,4-Dichloropyrimidine-5-carbonyl-J-acid, copper complex (Reactofil Type; Geigy)[24]

The copper complex of the monoazo dye is prepared by diazotization of 2-aminophenol-4,6-disulfonic acid and coupling with J-acid in alkaline medium in a similar manner to example 11 (omitting the reactive group) and copper is introduced by heating in solution with molar quantities of copper sulfate and sodium acetate and salting the product from solution.

The copper complex produced (51.7 g) is dissolved in water (500 ml), the pH is adjusted to 7.5, and the solution cooled to 0–5 °C. A solution of 2,4-dichloropyrimidine-5-carbonyl chloride (22.2 g) in acetone (150 ml) is added and the pH is maintained at 6.5–7 by addition of 2 *M* sodium carbonate solution. On completion of reaction as judged by constant pH the product is precipitated by addition of sodium chloride, filtered, and dried *in vacuo*. The dye is a rubine color.

14. 2-Amino-1-naphthol-4,8-disulfonic acid → Dichlorotriazinyl-H-acid copper complex (Procion MX Type; ICI)[79]

Concentrated hydrochloric acid (28 ml) is added to a solution of 2-naphthylamine-4,8-disulfonic acid disodium salt (34.7 g) and sodium nitrite (7 g) in water (300 ml) while maintaining the temperature at 0–5 °C, and stirred for 30 min. Excess nitrous acid is removed with a little sulfamic acid and the diazo is added to a suspension of *N*-acetyl-H-acid (36.1 g) in water (250 ml at pH 3 and 0–5 °C. The pH is raised to 4–5 by addition of sodium acetate and stirring continued at 0–5 °C until reaction is complete. The dye is salted to 15% with sodium chloride and the product collected by filtration and redissolved in water (1200 ml) at 40 °C. Acetic acid (1 ml), sodium acetate crystals (30 g), and 1 molar copper sulfate solution (100 ml) are added, and 6% hydrogen peroxide solution (230 ml) is added dropwise over 2 h at 40 °C. On completion of reaction the product is displaced from solution by addition of salt and filtered. The acetyl group is hydrolyzed by heating the product in 3% sodium hydroxide solution (1000 ml) at 90 °C for 1 h. The solution is cooled to 20 °C and pH adjusted to 6 by addition of hydrochloric acid and the product isolated by addition of sodium chloride and filtering. The product is dried.

Cyanuric chloride (19 g) in acetone (50 ml) is added to ice-water (400 ml) and at 0–5 °C a solution of monoazo copper complex (74 g) in water (1000 ml) is added and the pH maintained at 5–7 by addition of 2 *M* sodium carbonate. On completion of reaction the product is isolated by addition of salt and filtration. The dye is pasted with disodium hydrogen phosphate (3 g) and potassium dihydrogen phosphate (6 g) and dried in vacuum to give a reddish-blue dye.

15. 2-Amino-1-naphthol-4,8-disulfonic acid → 2-Amino-4-chlorotriazinyl-H-acid copper complex (Procion Blue H-5R,CI Reactive Blue 13; ICI)

To the solution of the dichlorotriazinyl dye described above (before isolation by salting) is added 10% aqueous ammonia (50 ml) and the reaction is stirred for 2 h at 40 °C. The resultant dye is isolated by salting and filtration and is dried to give a reddish-blue dye.

16. 1-Amino-4-[4′-(*N*′-ethyl-*N*′-(2″,3″-dichloroquinoxaline-6″-ylcarbonylamino)-3′-sulfoanilino)]anthraquinone-2-sulfonic acid disodium salt (Levafix E Type, Cavalite Type; Bayer)[73])

Sodium hydrogen carbonate (29 g) and 5-amino-2-ethylaminobenzenesulfonic acid (22.3 g) are stirred in water (160 ml) at 60 °C for 15 min and cooled to 40 °C. Copper sulfate (0.9 g) and copper powder (0.4 g) are added followed by, slowly over a 2 h period, the sodium salt of bromaminic acid (1-amino-4-bromoanthraquinone-2-sulfonic acid; 39.8 g). The blue slurry is stirred at 40 °C for 2 h, clarified, and the product displaced from solution by careful addition of sodium chloride to 14% w/v. The product is filtered and thoroughly washed with 15% sodium chloride solution and redissolved in water (1500 ml) at 30–35 °C and pH 7.6. A xylene (60 g) solution of 2,3-dichloroquinoxaline-6-carbonyl chloride (24.2 g) is added and, at 30–35 °C, the pH is maintained at 5.5–8 by addition of dilute sodium hydroxide solution. On completion of the reaction, usually 3 h, the solution is clarified by filtration and salted to 4% w/v with sodium chloride and 6% w/v with sodium sulfate. The reaction is stirred 14 h and filtered to give, after drying, a bright blue dye.

17. 2,3-Dichloroquinoxaline derivative of copper phthalocyanine (Levafix E; Cavalite Type; Bayer)[73]

Chlorosulfonic acid (160 g) is cooled to 0–5 °C and copper phthalocyanine (23 g) is added at below 25 °C and stirred for 15 min. The reaction is heated over 90 min to 135 °C and held for 3.5 h. The solution is cooled to room temperature and drowned into a mixture of ice and water at below 5 °C (care!). The precipitate is collected and washed with cold 1% hydrochloric acid. The copper phthalocyanine system is substituted primarily in the 3, 3′, 3″, and 3‴ positions by a mixture of sulfonic acid and sulfonylchloride groups.

The wet cake is slurred in ice-water (1200 g) and 2,4-diaminobenzenesulfonic acid (22.5 g) is added, and the pH is adjusted to 7.5 with ammonia. The mixture is heated to 40 °C and stirred at 40 °C and pH 7.5–8, maintained by addition of ammonia, for 4 h, or until pH is constant. The reaction is acidified to pH 2 with hydrochloric acid and the product filtered and washed with dilute hydrochloric acid.

The resultant dye-base is reslurried in water (1500 ml) at pH 7.5–8 at 40 °C and a solution of 2,3-dichloroquinoxaline-6-carbonyl chloride (21 g) in xylene (45 ml) is added with vigorous stirring. The pH is maintained at 7.5–8 with dilute sodium hydroxide until pH is constant (4 h) and the solution is clarified and sodium sulfate (100 g) added. The solution is salted to 16% w/v with sodium chloride and the product filtered, washed with 10% sodium chloride solution, and dried. The resultant turquoise dye should contain approximately one molecule of the reactive system per dye molecule.

VIII. SYNTHESIS OF DIRECT DYES

The principles of diazotization, coupling, and analysis discussed for reactive dyes are generally applicable to direct dye chemistry. Traditionally a number of dyes using known bladder carcinogens have been manufactured. Benzidine and β-naphthylamine and their non-water-soluble derivatives are notable examples. None of the syntheses described uses such compounds. Weights used in synthetic examples assume 100% purity.

A. Synthetic Examples of Direct Dyes

1. 4,4′-Diaminostilbene-2,2′-disulfonic acid ⇉ Phenol (2 mol) (Brilliant Yellow, CI Direct Yellow 4)

4,4′-Diaminostilbene-2,2′-disulfonic acid (37.0 g; 0.1 gmol) is dissolved in water (200 ml) and 10 *M* hydrochloric acid (50 ml) added followed by ice (200 g) to cool to 0–5 °C. Tetrazotization is achieved by dropwise addition of sodium nitrite (14.4 g) in water (60 ml) at 0–5 °C. Excess nitrous acid is decomposed by addition of a little sulfamic acid. The resultant diazonium salt solution is added over a 1 h period to a solution of phenol (20.9 g) in water (200 ml), sodium hydroxide (8.8 g), and sodium carbonate (23.3 g) at 0–5 °C and the reaction stirred overnight. The solution is then heated to 70 °C and the pH adjusted to 8 by careful addition of 10 *M* hydrochloric acid. The product is precipitated by addition of sodium chloride equivalent to 12% weight/volume and the dye is filtered and vacuum dried at 90–95 °C. The expected yield including inorganic salts is 84 g.

2. 4,4′-Diaminodiphenylurea-3,3′-dicarboxylic acid coupled twice to 2 g · mol equivalents of 87.5 mol% acetoacet-2-anisidine-4-sulfonic acid and 17.5 mol% acetoacetanilide-4-sulfonic acid (Benzo Fast Copper Yellow GGL, CI Direct Yellow 69)[80]

2-Anisidine-4-sulfonic acid (12.8 g) and aniline-4-sulfonic acid (2.25 g) are stirred in water (100 ml) and dissolved by addition of sodium carbonate (3.8 g). The solution is cooled to 0–5 °C by addition of ice (100 g) and diketene (7.25 g) is added dropwise over 1 h and the solution held at 0–5 °C for 1 h.

Simultaneously, 4,4′-diaminodiphenylurea-3,3′-dicarboxylic acid (12 g; 0.036 g · mol) is dissolved in sodium hydroxide (3 g) in water (75 ml) and cooled to 0–5 °C by addition of ice (100 g) and sodium nitrite (5.2 g) added. 10 *M* hydrochloric acid (18 ml) is added quickly and the reaction is stirred for 90 min at 0–5 °C, and the excess of the nitrous acid decomposed by addition of a little sulfamic acid.

To the coupler solution described above sodium carbonate (15 g) is added, and, at 0–5 °C, the tetrazo solution is added in 15 min and the reaction mixture heated slowly to 20–25 °C in 45 min and stirred overnight. The reaction is then heated to 40–45 °C and held 3 h. 20% sodium chloride solution (125 g) is added during 2 h at 45–50 °C and the product filtered and vacuum dried at 90 °C. The expected yield including inorganic salts is 40 g.

3. Aniline → Carbonyl-J-acid ← 4-Aminoacetanilide (Benzo Fast Scarlet 4BS, CI Direct Red 23)[80]

Aniline (16.8 g; 0.18 g · mol) is charged into a solution of 10 *M* hydrochloric acid (45 ml) in water (250 ml) and ice (250 g) and diazotized by dropwise addition of sodium nitrite (12.5 g) in water (30 ml).

Simultaneously, 4-aminoacetanilide (27 g) is charged into a solution of 10 *M* hydrochloric acid (45 ml) in water (250 ml) and ice (250 g) and similarly diazotized.

Carbonyl-J-acid (95 g) is dissolved in water (1200 ml) at room temperature by addition of sodium hydroxide solution (40%; 30 ml). Ice (150 g) is added to cool to 10 °C and acidified by addition of 31% hydrochloric acid (30 ml). Sodium carbonate (60 g) is then added and the diazonium salt solutions prepared above are run in simultaneously over 1 h at 10 °C. The mixture is stirred for 1 h and then heated to 70 °C and the dye precipitated by addition of 12% sodium chloride. The product is filtered and vacuum dried at 95 °C to give an expected yield, including inorganic salts, of 220 g.

4. 4-Aminoazobenzene-4′-sulfonic acid → 2-Benzoyl-J-acid (Sirius Red 4B, CI Direct Red 81)[80]

4-Aminoazobenzene-4′-sulfonic acid (27.7 g) is dissolved in water (350 ml) at pH 7 at 70–80 °C and the solution run into a solution of 10 *M* hydrochloric acid (25 ml) in water (200 ml) while simultaneously adding sodium nitrite solution (7.2 g in 30 ml) and cooling by addition of ice (150 g). Final temperature is 15–20 °C. The mixture is stirred for 2 h and the pH adjusted to 2.5.

The coupler solution is prepared by dissolving 2-benzoylamino-5-hydroxynaphthalene-7-sulfonic acid (36.0 g) in water (300 ml) at pH 8.5. Ice (120 g) is added following by 20% sodium acetate solution (75 ml). 10 *M* hydrochloric acid is then added until the solution is acid to litmus paper.

The diazonium salt solution is run in and the mixture stirred at 10–15 °C for 3 h. Sodium chloride (10% w/v) is added and stirring continued overnight. The solution is heated to 65 °C and made alkaline by addition of sodium carbonate (7.9 g). The product is filtered and vacuum dried at 95 °C.

5. Aniline → J-acid → J-acid (Benzo Rubine 6BS; CI Direct Red 16)[80]

Aniline (9.3 g) is diazotized in water (180 ml), ice (100 g), and 10 *M* hydrochloric acid (25 ml) at 0–5 °C with sodium nitrite solution (7.2 g in 30 ml water).

J-acid (23.9 g) is dissolved in water (200 ml) at 40 °C by addition of sodium carbonate to pH 8. The solution is cooled with ice (100 g) and sodium carbonate

(28 g) is added. At 0–5 °C the above diazonium salt solution is run in over 30–45 min. After approximately half the diazo has been added sodium hydroxide (30%; 6 ml) is added followed by the remainder of the diazo. The coupling is stirred overnight.

The mixture is then neutralized with 10 *M* hydrochloric acid, diluted to 1500 ml, and heated to 60 °C. Sodium nitrite (7.2 g) is added and the solution is run slowly into water (100 ml), ice (200 g), and 10 *M* hydrochloric acid (25 ml) maintaining the temperature below 20 °C in an ice bath. The reaction is stirred for 1 h and the excess of nitrous acid is removed with a little sulfamic acid. The resultant diazo solution is added over 2 h to a solution prepared by stirring J-acid (23.9 g) in water (200 ml), at pH 8.5, at 40 °C and subsequently cooled by addition of ice (100 g). The temperature is kept below 25 °C by addition of ice. The pH is maintained at 8.5 by simultaneous addition of sodium carbonate solution. The coupling is stirred overnight and sodium chloride (5% w/v) added, and the product filtered and dried in vacuum at 95 °C.

6. Copper Complex of 4,4′-Diaminobiphenyl-3,3′-dicarboxylic acid coupled twice to 2 moles 3-carboxyphenyl-J-acid (Sirius Supra Blue 3 RL, CI Direct Blue 93)[80]

4,4′-Diaminobiphenyl-3,3′-dicarboxylic acid (11.8 g) is stirred in water (120 ml) and dissolved by addition of sodium hydroxide (40% solution; 29 ml). Ice (200 g) and 30% hydrochloric acid (60 ml) are added and at 0–5 °C a solution of sodium nitrite (6 g) in water (25 ml) added dropwise in 15 min.

3-Carboxyphenyl-J-acid is dissolved in water (200 ml) at 40 °C by addition of sodium carbonate (10 g). The solution is neutralized with hydrochloric acid. Ice (150 g) is added followed by sodium hydrogen carbonate (72 g). The tetrazo solution is then run in as quickly as possible and stirring continued for 10 min to complete reaction. Sodium carbonate (24 g) is added and the coupling is heated to 70 °C and filtered. The filter cake is stirred in water (30 ml), diluted to 1000 ml, and heated to 90 °C. A solution of copper sulfate (22.6 g) in water (60 ml) at 75 °C is mixed with 25% ammonia solution (36 ml) and added to the dye solution. The temperature is maintained for 1 h and the product precipitated by addition of 5% w/v sodium chloride. The pH is adjusted to 8 by addition of hydrochloric acid and the product filtered and dried. The expected yield is 47 g.

7. H Acid → Cresidine → Phenyl-J-acid (Sirius Supra Blue F3R; CI Direct Blue 67)[80]

1-Amino-8-*O*-phenylsulfonylnaphthalene-3,6-disulfonic acid (33.1 g) is dissolved in water (200 ml) at pH 7 by addition of sodium carbonate. 10 *M* hydrochloric acid (18 ml) is added and the solution cooled to less than 10 °C by addition of ice (100 g), and diazotization completed by the dropwise addition of sodium nitrite (5.1 g) in water (20 ml). The reaction is stirred 1 h and the excess of nitrous acid removed by addition of a little sulfamic acid.

2-Methoxy-5-methylaniline (10.3 g) is dissolved at 80 °C in water (200 ml) and 10 *M* hydrochloric acid (10 ml) and cooled with ice (100 g) to 25 °C. The solution is run into the diazo solution, and sodium acetate solution (20%; 50 ml) added to raise the pH to 3. After stirring for 45 min the mixture is further cooled by addition of ice (100 g) and acidified with 10 *M* hydrochloric acid (15 ml) and diazotized by dropwise addition of sodium nitrite (5.5 g) in water (20 ml). The reaction is stirred for 1 h and sodium chloride (8% w/v) added. The precipitated diazonium salt is filtered and the diazo paste stirred in water (200 ml) at 0–5 °C.

Phenyl-J-acid (20.5 g) is dissolved in water (300 ml) and 25% ammonia (6 ml) and, after cooling with ice (150 g), a further portion of 25% ammonia (30 ml) is added. At 0–5 °C the diazo from above is run into the coupler solution and stirred to complete reaction.

The coupling is then heated to 75–80 °C and, after addition of 30% sodium hydroxide solution (81 ml), held at 75–80 °C for 20 min. The completed hydrolysis is then run quickly into water (100 ml) and ice (200 g). 10 *M* hydrochloric acid (90 ml) is added to lower the pH to 8, and the batch is treated with sodium chloride (2% w/v). The product is filtered and dried and the expected weight yield of blue dye is 40 g.

8. 4-Aminobiphenyl-3-sulfonic acid → 1-Amino-2-ethoxynaphthalene-6-sulfonic acid → Benzoyl-H-acid (Brilliant Benzo Fast Green GL, CI Direct Green 13)[80]

4-Aminobiphenyl-3-sulfonic acid (7.2 g) is dissolved at pH 7–7.5 in water (80 ml) at 70 °C and the solution run into a mixture of 10 *M* hydrochloric acid (8 ml) and ice (50 g). Sodium nitrite (2 g) in water (8 ml) is added dropwise at 18–20 °C and stirred 1 h. 1-Amino-2-ethoxynaphthalene-6-sulfonic acid (9.3 g) is added and stirred 10 min and 20% sodium acetate solution (45 ml) added to raise the pH to 4.5. The mixture is stirred overnight at 18 °C and then 40% sodium hydroxide solution (8 ml) added. The solution is treated with sodium chloride (90 g) followed by sodium nitrite (3 g) in water (10 ml). This is quickly followed by 31% hydrochloric acid (25 ml) and the temperature allowed to rise to 35 °C and held at 35 °C for 15 min. The precipitated diazonium salt is filtered and the press paste stirred in water (30 ml) and ice (30 g).

Benzoyl-H-acid (11.8 g) is dissolved at pH 7–8 in water (50 ml) by addition of sodium carbonate at 70 °C. The solution is cooled to 0 °C by addition of ice (100 g), and pyridine (100 ml) added followed by the diazo suspension. The reaction is stirred at 0–5 °C overnight and the product precipitated by addition of sodium chloride (15% w/v). After stirring for 2 h the product is filtered and dried to give the green dye with expected yield of 40 g including inorganic salts.

IX. SYNTHESIS OF VAT DYES

Vat dyes are, as previously mentioned, frequently difficult to prepare and require particular experimental care. Several of the dyes are among the oldest known synthetic dyes, but nevertheless interest in their chemistry remains and hence examples of typical synthetic methods are included.

A. Synthetic Examples of Vat Dyes

1. Indanthrone (e.g., Navinon Blue RSN; Indanthren Blue RS; CI Vat Blue 4)[53,60]

To a stainless steel reactor, fitted with a stirrer and a lid, is charged sodium hydroxide (65 g) and potassium hydroxide (130 g) and the mixture heated to 220 °C to form a melt. At 220 °C fused sodium acetate (55 g), fluxing agent, is added and the mixture stirred in a nitrogen atmosphere. 2-Aminoanthraquinone (100 g) is then added at 180 °C over 20 min followed by a mixture of sodium hydroxide (65 g), potassium hydroxide (130 g), and sodium nitrate (14 g) over 1 h. The mixture is maintained at 201 °C for 1.5 h and then poured quickly into cold water (2500 ml). The aqueous mixture is heated to 48 °C and treated with sodium hydrosulfite (100 g), and is held 2 h during which time the leuco form crystallizes. The crystals are collected and washed copiously with dilute sodium hydrosulfite solution. The filter cake is then stirred in water (300 ml), treated with sodium hydroxide (5 g), and at 60 °C air is blown through the solution with good agitation and the dye precipitated, filtered, washed, and dried to give Indanthrone in 58% yield.

2. Flavanthrone (CI Vat Yellow 1)[54]

1-Bromo-2-aminoanthraquinone (10 g), finely divided copper powder (2.1 g), hydroquinone (0.25 g), and dimethylformamide (22 ml) are stirred for 2 h at 90–95 °C and then for 3 h at 110–115 °C. The suspension is filtered and the crude product washed with warm dimethylformamide (25 ml). The filter cake is resuspended in 20% hydrochloric acid (250 ml) and a solution of sodium chlorate (12 g) in water (60 ml) added dropwise to the suspension at 90 °C, and the mixture stirred at 90 °C for 2 h. The resultant suspension is filtered hot and the dye cake thoroughly water washed until neutral. The product is dried to give flavanthrone in 80% yield.

3. 1,2-Benzo-5,6-phthaloylacridone (Indanthren Red RK, CI Vat Red 35)[55]

Tobias acid (2-naphthylamine-1-sulfonic acid, 10.5 g), 1-chloroanthraquinone-2-carboxylic acid (16.2 g), sodium borate (hydrated, 28.5 g), and boric acid (4.5 g) are stirred in nitrobenzene (75 ml). The mixture is heated to 115 °C and water vapor is evolved. The reaction is held at 115 °C for approximately 1 h until the vigorous evolution of water vapor has ceased. The vapors evolved should be condensed and collected in a Dean and Stark trap. The mixture is then heated slowly to 200 °C and maintained at 200–210 °C for 2 h. Chloroacetic acid (0.5 g) is then added and the reaction continued for a further 2 h at 200–210 °C. The product is isolated by cooling and filtering. The filter cake is washed with nitrobenzene and steam-distilled to remove nitrobenzene, again filtered, water washed, and dried to give 1,2-benzo-5,6-phthaloylacridone.

Note. The treatment with chloroacetic acid is to convert any 2-naphthylamine byproduct into a water-soluble, noncarcinogenic derivative. The process is claimed as an improvement over the earlier processes, which were discontinued because of their use of 2-naphthylamine. Tobias acid is manufactured by sulfonation of 2-naphthol followed by Bucherer amination, thus avoiding carcinogenic intermediates.

4. 16,17-Dimethoxyviolanthrone (Caledon Jade Green XBN, CI Vat Green 1)[81]

a. Benzanthrone. Anthraquinone (30 g) is stirred in 96% sulfuric acid (210 g) and water (20 ml) at 125 °C. To this mixture iron powder (22.8 g) and glycerol (16.6 g) are carefully added, the vigorous exotherm being controlled by adjusting the addition rate and by external cooling such that the temperature does not exceed 140 °C (ideally 135–138 °C). The reaction mixture is held at 135–140 °C for 1 h and poured carefully into water (1000 ml) and the precipitated product is filtered and water washed. The filter cake is slurried in dilute sodium hydroxide, filtered, washed, and dried to give benzanthrone (28.8 g). The product may be crystallized as yellow needles, mp 170 °C, from acetic acid or purified by vacuum sublimation.

b. 4,4′-Dibenzanthronyl. Benzanthrone (20 g) is stirred and heated in isobutyl alcohol (38.5 g) with potassium hydroxide (50 g) and sodium acetate (6 g) at 112 °C for 4 h. On completion of reaction the mixture is poured carefully into water (100 ml) and stirred 15 min. The two layers are then separated, the upper layer being retained and oxidized at 50 °C by treatment with 12% sodium hypochlorite solution (8 g) in water (60 ml). Isobutyl alcohol is removed by steam distillation and the precipitated 4,4′-dibenzanthronyl is filtered and dried. The yield is usually 92%.

c. 16,17-Dihydroxyviolanthrone. 4,4′-Dibenzanthronyl (15 g) is dissolved in 96% sulfuric acid (400 g) and water (56 ml) and oxidized at 25–30 °C for 4 h by addition of manganese dioxide (19.2 g). The resultant mixture is poured slowly into aqueous sodium bisulfite solution (10%; 2000 ml) and heated at reflux for 1 h. The suspension is cooled, filtered, and thoroughly washed with water to give 16,17-dihydroxyviolanthrone.[56]

d. 16,17-Dimethoxyviolanthrone. 16,17-Dihydroxyviolanthrone (10 g) is suspended in dry nitrobenzene (100 ml) and sodium carbonate (10 g) added followed by dimethyl sulfate (10 g). There is an exothermic reaction and heating is continued, and the mixture is held at reflux for 3 h. The mixture is cooled and a solution of sodium carbonate (10 g) in water (200 ml) added. The nitrobenzene is removed by steam distillation. The product is collected by filtration. The product may be purified by dissolving in boiling nitrobenzene (1 g/10 ml) and filtering from undissolved material. The filtrate is concentrated by vacuum distillation to small bulk and allowed to cool. The product which crystallizes is filtered and washed with ethyl alcohol.

5. 4,12-Dibromopyranthrone (Cibanone Golden Yellow 2R, CI Vat Orange 2)[82]

Potassium hydroxide (9 g) is stirred in isobutyl alcohol (72 ml) at reflux for 1 h and cooled to 70 °C. 2,2′-Dimethyl-1,1′-dianthraquinonyl (15 g) is added portionwise and the reaction heated at reflux for 3 h. A solution of disodium hydrogen phosphate (1.2 g) in water (150 ml) is added and the solution cooled to 30 °C. Air is drawn through the reaction for 9 h. The reaction is distilled to a pot temperature of 101–102 °C to remove isobutanol and the orange mass diluted by addition of water (100 ml) and disodium hydrogen phosphate (0.6 g). The mixture is stirred at 60 °C for 1 h and filtered and washed neutral with water to give pyranthrone in quantitative yield (described as Indanthrene Golden Orange G).

Pyranthrone (10 g) is added to chlorosulfonic acid (80 g; care!) followed by iodine (0.15 g) and bromine (3.75 g). The mixture is heated to 58 °C and held for 2 hours and cooled to 30 °C. Sulfuric acid (40 g) is run into the reaction and stirring continued until hydrogen chloride evolution ceases. The mixture is then added slowly to water (900 g) and the precipitated 4,12-dibromopyranthrone filtered and dried. The expected yield is 12.5 g and 90% of theory.

6. Anthraquinoneacridone Vat Dye (Indanthrene Violet FFBN, CI Vat Violet 13)[82]

(The first stage of this synthesis requires an autoclave.) 1,5-Dichloroanthraquinone (24 g), anthranilic acid (75 g), 50% aqueous potassium hydroxide solution (60 g), water (90 ml), isobutanol (12 g), magnesium oxide (6.4 g), and cupric oxide (1.7 g) are charged to a steel autoclave and heated for 1 h at 150 °C. The reaction mass is cooled back to 50 °C and added to an excess of 5% aqueous sulfuric acid and the isobutyl alcohol is removed by distillation. The precipitate of the so-called acridone acid is filtered, washed acid free, and dried in quantitative yield (41.5 g).

The "acridone acid" (41.5 g) is added to a mixture of sulfuric acid (41.5 g) and chlorosulfonic acid (62.5 g) at 15–20 °C and the mixture stirred at 28 °C for 1 h. 75% Sulfuric acid (27.5 g) is added slowly (foaming!) and the reaction stirred 30 min. Water (44 g) is added slowly (care!) and the reaction mixture allowed to stand overnight.

The product which crystallizes is collected by filtration and immediately (without washing) redissolved in sulfuric acid (600 g) at 15 °C and then drowned into water (2400 ml). The precipitate of Indanthrene Violet FFBN is filtered and washed acid free and dried in expected yield of 73% (28 g).

7. Dianthrimidepyrrole Vat Dye (Indanthren Olive R, CI Vat Black 27)[82]

a. 1,1′-Dianthrimide. 1-Aminoanthraquinone (19.4 g), 1-chloroanthraquinone (21.6 g), sodium carbonate (9.6 g), and sodium acetate (2.2 g) are stirred in nitrobenzene (180 ml) and the mixture is heated over 5 h to 205–210 °C. The mixture is cooled back to 190 °C and copper powder (0.54 g) is added and the reaction held 1 h. The reaction is then heated and nitrobenzene (100 ml) distilled.

The remainder of the nitrobenzene is distilled under vacuum and the residue digested in 5% aqueous hydrochloric acid (400 ml) at reflux for 15 min. The suspension is cooled and the product filtered, washed neutral, and dried to give 1,1′-dianthrimide in expected yield of 95% (37.2 g).

b. 4,4′-Diamino-1,1′-dianthrimide. Sulfuric acid (70 g) and oleum (20%, 130 g) are stirred together and boric acid (6 g) is added, and the mixture heated to 70 °C for 30 min and cooled to 20 °C. 1,1′-Dianthrimide (30 g) is added and the reaction mixture is heated to 60 °C and held for 90 min. To the resultant solution is added mixed acid (28% nitric acid; 44 g) at 20–25 °C over a period of 4–6 h. The reaction mixture is stirred for 2 h and drowned into water (1200 ml), and stirred for a further 1 h. The precipitated product (4,4′-dinitro-1,1′-dianthrimide) is filtered, washed acid free, and resuspended in water (150 ml) and 30% sodium hydroxide (54 g) and stirred 15 min. A solution of sodium sulfide (38.4 g) in water (130 ml) is added and the reduction completed by stirring at 95 °C for 1 h. The reaction is diluted to 1600 ml with water and the product filtered and thoroughly water washed and dried to give 4,4′-diamino-1,1′-dianthrimide in expected yield of 84% (26.6 g).

c. 4,4′-Dibenzoylamino-1,1′-dianthrimide. 4,4′-Diamino-1,1′-dianthrimide (18 g) is stirred in dry *o*-dichlorobenzene (85 ml) and heated to 125 °C. Benzoyl chloride (18 ml) is run in over 90 min and the reaction heated to 150 °C over 2 h and held 3 h at 150 °C. Sodium carbonate (2.4 g) is added and the solvent is removed under vacuum to give crude 4,4′-dibenzoylamino-1,1′-dianthrimide in 84% expected yield.

d. Indanthren Olive R (Ring closure to dianthrimidepyrrole). 4,4′-Dibenzoylamino-1,1′-dianthrimide (15 g) is added to sulfuric acid (100 g) over 8 h at 25–30 °C and stirred 2 h at 30 °C. The resultant solution is run slowly into a solution of sodium chlorate (1.25 g) in water (450 ml). The reaction is heated to 80 °C over 30 min and filtered and thoroughly washed with water, and dried to give Indanthren Olive R in expected 90% yield.

8. Dibenzopyrenequinone (Indanthren Golden Yellow G, CI Vat Yellow 4)[82]

a. 1,5-Dibenzoylnaphthalene. Aluminum chloride (110 g) is added slowly to benzoyl chloride (77 g) allowing the temperature to rise to 80–90 °C. The mixture is cooled to 63 °C and naphthalene (26.3 g) added portionwise with care, at 60–65 °C. (Foaming may be a problem.) The reaction is then held 8 h at 70 °C and then added slowly to water (100 ml) and the precipitate filtered and thoroughly washed acid free. The wet filter cake is added to chlorobenzene (230 g) and the water removed by azeotropic distillation. The dry solution is cooled to 15 °C and the crystals of 1,5-dibenzoylnaphthalene collected by filtration. Further purification may be achieved by recrystallization from chlorobenzene. Expected yield is 46% of theory (33 g).

b. Dibenzopyrenequinone. A melt of aluminum chloride (160 g) and sodium chloride (38.5 g) is prepared at 120 °C. 1,5-Dibenzoylnaphthalene (37.5 g) is added carefully portionwise over 1 h during which time air is bubbled through the melt. The temperature is allowed to rise to 160 °C. On completion of the addition, aluminum chloride (90 g) is added while air bubbling is maintained. The temperature is held at 160–170 °C by cooling as necessary. After 2.5 h at 160–170 °C the melt is added carefully to water (1000 ml). Hydrochloric acid (12 g) is then added and the temperature raised to 100 °C and held 30 min. The reaction is cooled and cold water (500 ml) added, and the product filtered and thoroughly washed acid free. The product is purified by reslurrying in water (1200 ml) and hydrochloric acid (12 g) at 100 °C, cooling, filtering, washing, and drying to give 34.2 g of dibenzopyrenequinone in 92% yield.

c. Dibromodibenzopyrenequinone (Indanthren Golden Yellow RK, CI Vat Orange 1). To the melt of the dibenzopyrenequinone (before drowning into water) bromine (33 g) is added slowly dropwise at 160 °C. (Reference indicates 20 h but this is unrealistic for 33 g.) The resultant melt is drowned out and the dye isolated as the dibromodibenzopyrenequinone.

9. Indigo (CI Vat Blue 1)

There are many syntheses of indigo. Two are described here. The first is the classical Heumann Synthesis, and the second is a simpler, more recent laboratory procedure.

a. Heumann Synthesis.[83,60] Anthranilic acid (68.5 g) is "pasted" with a small amount of water (15–20 ml) and dissolved in a solution of sodium hydroxide (20 g) in water (40 ml). A solution prepared from chloroacetic acid (47.25 g) in water (100 ml) and sodium carbonate (26.5 g) is added and the reaction mixture held at 40 °C for 100 h. The precipitated phenylglycine-2-carboxylic acid sodium salt is collected, washed with a small amount of cold water, and dried. Yield is expected to be 81 g at 75%.

A mixture of completely dehydrated sodium hydroxide (12.5 g), dehydrated potassium hydroxide (12.5 g), calcium oxide (3.75 g), and dry sodium phenylglycine-2-carboxylate (12.5 g) is ground in a ball mill with total exclusion of moisture, and the resultant mixture is heated under vacuum for 2 h at 150 °C and 6 h at 230–235 °C. The resultant fusion mixture containing sodium indoxylcarboxylate is dissolved in water (1000 ml) at 70 °C and a vigorous stream of air is passed through the reaction mixture to precipitate the indigo, which is filtered and washed with water. The precipitate is resuspended in dilute hydrochloric acid (5%; 100 ml) and heated under reflux for 15 min. The suspension is cooled and filtered and the precipitate washed with water and dried to give indigo in an expected yield of 6 g (80%).

b. Harley–Mason Synthesis.[84] 2-Nitrobenzaldehyde (5 g) and nitromethane (2.3 g) are stirred in methanol (15 ml) and treated slowly at 0 °C with a solution of sodium methoxide prepared by dissolving sodium (0.9 g) in methanol (10 ml). The reaction is stirred 12 h at 0 °C and the resultant sodium salt of 2-nitro-1-*o*-nitrophenylethyl alcohol collected and washed with ether. The salt (3 g) is dissolved in water (50 ml) and 2 N sodium hydroxide (15 ml) added. Sodium dithionite (6.5 g) is then added slowly with stirring and a thick precipitate of indigo is formed at once. Air is drawn through the solution for 15 min to complete the oxidation of any leuco-component and the indigo (1.51 g; 90%) is collected. The product may be purified by vacuum sublimation.

10. Thioindigo (CI Vat Red 41)[60]

Anthranilic acid (68.5 g) is added to a solution of hydrochloric acid (33%, 78.5 g) in water (500 ml) and cooled to 0 °C with good agitation, and a solution of sodium nitrite (34.5 g) in water (75 ml) is added slowly to diazotize the amine. Simultaneously, flowers of sulfur (11.1 g) are added to a solution of sodium sulfide (85 g) in water (100 ml) and the mixture heated under reflux until the solution is clear. The resultant solution is cooled and ice (500 g) is added followed by the diazonium salt solution, portionwise, in 10 min. The mixture is stirred for 2 h and acidified to pH 1 with concentrated hydrochloric acid. The precipitate of thio and dithiosalicyclic acids is collected and thoroughly washed with water. The precipitate is resuspended in water (500 ml) and sodium carbonate (25.8 g) added gradually, and the solution heated to reflux. The pH must at this stage still be acid to litmus. Reduction is completed by addition of Béchamp iron (400 g) and holding at 95 °C for 2 h. The reaction course is followed by thin layer chromatography and on completion the solution is made alkaline by addition of 35% sodium hydroxide (60 g). At 95 °C a solution of chloroacetic acid (52 g) and sodium carbonate (29 g) in water (150 ml) is added. The mixture which must be kept alkaline during the reaction is held at 90 °C for 30 min, cooled, and allowed to stand. The supernatant liquor is decanted from the iron and the product precipitated by slow addition of 33% hydrochloric acid (150 g) with good agitation. The product is filtered and washed free of acid and dried at 80 °C to give 2-carboxyphenylthioglycollic acid (expected yield 80%, 85 g).

A mixture of 2-carboxyphenylthioglycollic acid (50 g) and sodium hydroxide (100 g) is ground in a ball mill for 24 h and the resultant mixture is placed under vacuum and heated to 205 °C over 2 h and held for 24 h. The resultant 3-hydroxythionaphthene is dissolved in water (1000 ml) and treated slowly with concentrated sulfuric acid (50 g). Flowers of sulfur (40 g) are added and the mixture stirred at 95 °C for 3 h. The resulting dye is filtered, water thoroughly, and dried at 80 °C to give thioindigo (25 g) in expected 70% yield.

X. SYNTHESIS OF SULFUR DYES

The structures of sulfur dyes remain largely undetermined and indeed are probably indeterminate, being mixtures. The most important process in the synthesis of sulfur dyes is thionation and this process is exemplified.

A. Synthetic Examples of Sulfur Dyes

1. CI Sulfur Red 10[68]

4-Hydroxydiphenylamine (25 g), sodium sulfide (90 g), sulfur (50 g), and water (70 ml) are heated together under reflux at 145–150 °C for 34 h with good stirring. The mixture is cooled and stirred in water (1000 ml) and a stream of air is passed through the liquid for 24 to 48 h to precipitate the dye, which is collected by filtration and thoroughly washed with warm water. The product is dried and ground and successively Soxhlet-extracted with water and carbon disulfide, each for 24 h, and finally dried to constant weight.

2. CI Sulfur Blues 1,3,4,5,11[45]

4-(2′,4′-Dinitroanilino)phenol (25 g) is added as a 75% paste, over 2 h at 90–92 °C to a solution of sodium sulfide (90 g), sulfur (50 g), and water (70 ml) and heated at reflux for 3 h. Sodium chloride sufficient to raise the reflux temperature to 109–110 °C is added and this temperature is maintained for 10 h. The reaction mixture is cooled to 45 °C and diluted with water (200 ml) and the product filtered and dried at 70 °C.

3. Hydron Blue R (CI Vat Blue 43)[82]

95% Sulfuric acid (97 g) is cooled to 0 °C and carbazole (8 g) and anthracene (0.08 g) added slowly portionwise. The resultant solution is stirred and cooled to −28 °C. The cooled solution is added slowly at −23 °C to a solution prepared by dissolving 4-nitrosophenol (5.92 g) in sulfuric acid (43 g) at −10 °C and cooling to −27 °C. The addition takes 2 h and is followed by a reaction period of 1 h. The resultant mixture is added slowly to a mixture of water (300 ml), ice (120 g), and iron powder (5.1 g). Further ice up to 200 g may be used to maintain the temperature at 0 °C. The reaction is stirred for a further 6 h and the mixture heated to 20 °C. The precipitate of 3-(4′-hydroxyphenylamino)carbazole is filtered and thoroughly washed with water and dried. The expected yield is 13 g, which is quantitative.

A mixture of butanol (40 ml), sulfur (10.8 g), 60% sodium sulfide (8.55 g), and 3-(4′-hydroxyphenylamino)-carbazole (9 g) is heated under reflux for 36 h. The reaction is stirred in water (60 ml) and 20% sodium chloride solution (20 g) and the butanol removed by distillation. Sodium sulfide (1 g) is added and the mixture stirred to dissolve sulfur. The product is collected by filtration and washed sulfide-free with 10% sodium chloride solution and resuspended in water (100 ml), acidified to pH 1 with hydrochloric acid, and stirred at 50 °C for 15 min. The precipitate is filtered, thoroughly washed, and dried to give Hydron Blue R.

4. Sulfur Black T (CI Sulfur Black 1)[60]

2,4-Dinitrochlorobenzene (35 g) is suspended in water (60 ml) and heated to 90 °C with stirring and 35% sodium hydroxide (40 g) added dropwise over 2 h.

Heating is continued until complete conversion to 2,4-dinitrophenol is achieved. The suspension of sodium dinitrophenolate is cooled to 40–45 °C and mixed with a solution formed by dissolving sodium sulfide (62.5 g) and sulfur (25 g) in water (62.5 g). The temperature is raised to 60 °C and the volume adjusted to 300 ml, and the temperature carefully raised over 3 h to 105 °C. The mixture is then heated under reflux without stirring for 30 h and a further amount of water (300 ml) added. Air is drawn through the reaction mixture at 60–65 °C for 1 h and the precipitated product filtered and dried at 70 °C to give 35 g of dye.

ACKNOWLEDGMENTS

The author would like to take this opportunity to express his gratitude to Dr. G. Hallas and Dr. C. V. Stead for their invaluable suggestions and for reading the whole of the manuscript. The author would like to thank Ms Marie Williams who typed the majority of the script, and his wife Margaret for assistance in additional typing and proofreading.

REFERENCES

1. J. S. Ward, *Rev. Prog. Color. Relat. Top. 14,* 98 (1984).
2. I. D. Rattee, *Rev. Prog. Color. Relat. Top. 14,* 50 (1984).
3. I. D. Rattee, *Endeavour 20,* 154 (1961).
4. W. E. Stephen, *Chimia 19,* 261 (1965).
5. K. Aeberhard, *Textilveredlung 16,* 442 (1981).
6. S. Fujioka and S. Abeta, *Dyes and Pigments 3,* 281 (1982).
7. P. Rys and H. Zollinger, in: *The Theory of Coloration of Textiles,* Chap. 8, Reactive Dye-Fibre Systems, The Dyers Company Publication Trust (1975).
8. C. F. Cross and E. J. Bevan, *Researches on Cellulose 1,* 34 (1895).
9. I.G. Farbenindustrie, British Patents 259,634 (1925); 458,684 (1936).
10. D. H. Peacock, *J. Soc. Dyers Colour 42,* 53 (1926).
11. D. N. Kursanov and P. A. Solodkov, *Zh. Prikl. Khim. SSSR 16,* 351 (1943) [*J. Appl. Chem. USSSR 16,* 36 (1943)].
12. F. R. Alsberg, R. N. Heslop, I. D. Rattee, W. E. Stephen, W. J. Marshall, R. W. Speke, C. D. Weston, and ICI, British Patent 798,121 (1958).
13. W. F. Beech, *Fibre-Reactive Dyes,* Logos, London (1970).
14. R. N. Heslop, W. E. Stephen, and ICI, British Patent 838,337 (1957).
15. C. V. Stead, *Dyes and Pigments 3,* 161 (1982).
16. D. R. Waring and ICI, German Patent 2,244,537 (1973) [*CA 78,* 148961 (1973)].
17. E. P. Sommer, *Am. Dyest. Rep. 47,* 895 (1958).
18. H. Zimmermann, *Melliand Textilber. 39,* 1026 (1958).
19. J. Heyna, *Angew. Chem. 74,* 966 (1962).
20. E. Siegel, in: *The Chemistry of Synthetic Dyes* (K. Venkataraman, ed.), Vol. VIII, p. 1, Academic Press, London (1978).
21. G. Hallas, in: *Developments in the Chemistry and Technology of Dyes* (J. Griffiths, ed.), p. 31, Blackwell Scientific, Oxford (1984).
22. M. Capponi, E. Metzger, and A. Giamara, *Am. Dyestuff Reptr. 50,* 23 (1961).
23. H. F. Andrew, V. D. Poole, and ICI, British Patent 917,780 (1963) [*CA 61.* 10816 (1963)].
24. Geigy, Belgian Patent 644,495 (1964) [*CA 63,* 11743 (1965)].
25. H. S. Bien, E. Klauke, and Bayer, British Patent 1,169,254 (1969) [*CA 72,* 122901 (1970)].

26. K-H. Schundehutte, K. Trautner, and Bayer, U.S. Patent 3,853,840 (1974) [*CA 83,* 195209 (1975)].
27. G. A. Gamlen, C. Morris, D. F. Scott, H. J. Twitchett, and ICI, British Patent 937,182 (1963) [*CA* 60, 4282 (1964)]; M. Miujumoto and R. Parham, *A.A.T.C.C. Int. Conf. and Exhib.*, 153 (1986).
28. B. L. McConnell, L. A. Graham, and R. A. Swidler, *Text. Res. J. 49,* 458 (1979).
29. K-H. Schundehutte, in: *The Chemistry of Synthetic Dyes* (K. Venkataraman, ed.), Vol. 6, p. 211, Academic Press, New York (1972).
30. ICI French Patent 1,198,036 (1959).
31. W. E. Stephen and ICI German Patent 1,019,025 (1957) [*CA 53,* 18496 (1959)].
32. P. W. Austin, A. Crabtree, and ICI, German Patent 2,150,598 (1972) [*CA 77,* 116022 (1972)].
33. P. F. Gordon and P. Gregory, *Organic Chemistry in Colour,* Springer-Verlag, Berlin (1983).
34. F. R. Alsberg, I. D. Rattee, W. E. Stephen, W. J. Marshall, R. W. Speke, C. D. Weston, and I.C.I., British Patent 797,946 (1958).
35. H. R. Hensel and G. Lutzel, *Angew. Chem., Int. Ed. Engl. 4,* 312 (1965).
36. K. G. Kleb, E. Siegel, and K. Sasse, *Angew. Chem., Int. Ed. Engl. 3,* 408 (1964).
37. D. R. Waring, P. Gregory, and ICI, German Patent 2,209,837 (1972) [*CA* 78, 73626 (1973)].
38. I.C.I., French Patent 1,471,782 (1967) [*CA 68,* 3910 (1968)].
39. A. H. M. Renfrew, *Rev. Prog. Color. Relat. Top. 15,* 15 (1985).
40. Geigy, Belgian Patent 650,328 (1965) [*CA 63,* 13454 (1965)].
41. E. N. Abrahart, *Dyes and their Intermediates, 2nd edn.*, Edward Arnold, London (1977).
42. K. Venkataraman, *The Chemistry of Synthetic Dyes,* Academic Press, New York (1952).
43. C. V. Stead, *Rev. Prog. Color. Relat. Top. 6,* 1 (1975).
44. R. L. M. Allen, *Colour Chemistry,* Nelson, London (1971).
45. H. E. Fierz-Davis and M. Matter, *J. Soc. Dyers Colour. 53,* 424 (1927).
46. Ciba, British Patent, 554,463 (1943) [*CA* 39, 1296 (1945)].
47. C. C. Cook, *Rev. Prog. Color. Relat. Top. 12,* 73 (1982).
48. *Colour Index,* Revised 3rd Edition, The Society of Dyers and Colourists (1975).
49. F. M. Rowe and C. Levin, *J. Soc. Dyers Colour. 40,* 218 (1924); F. M. Rowe and E. Levin, *J. Soc. Dyers Colour. 41,* 354 (1925); *42,* 82 (1926).
50. R. L. Desai and T. N. Mehta, *J. Soc. Dyers Colour. 54,* 422 (1938).
51. Hoechst, Belgian Patent 654,559 (1965) [*CA 64,* 19874 (1966)].
52. K. Venkataraman and V. N. Iyer, in: *The Chemistry of Synthetic Dyes* (K. Venkataraman, ed.), Vol. V, p. 132, Academic Press, London (1971).
53. S. Balasubramanian, R. Shantarum, and L. K. Doraiswamy, *Br. Chem. Eng. 12,* 377 (1967).
54. Ciba-Geigy, British Patent 1,278,914 (1972) [German Patent 2,105,286 (1971); *CA* 76, 4898 (1972)].
55. R. S. Barnes, W. Smith, and ICI, British Patent 691,118 (1953) [*CA 47,* 10237 (1953)].
56. A. H. Davies, R. F. Thomson, J. Thomas, and ICI, British Patent 181,304 (1922) [U.S. Patent 1,531,260 (1925); *CA 20,* 114 (1926)].
57. R. L. Walker and du Pont, U.S. Patent 3,340,264 (1967) [*CA 68,* 88196 (1968)].
58. D. R. Waring, in: *Comprehensive Heterocyclic Chemistry* (O. Meth-Cohn, ed.), Vol. 1, p. 317, Pergamon, Oxford (1984).
59. M. Bader and C. Sunder, U.S. Patent 1,448,251 (1923) [*CA 17,* 2195 (1923)].
60. H. E. Fierz-David and L. Blangey, *Fundamental Processes of Dye Chemistry,* Interscience, New York (1949).
61. W. E. Wood, *Rev. Prog. Color. Relat. Top. 7,* 80 (1976).
62. J. Marek and D. Markova, *Collect. Czech. Chem. Commun. 27,* 1533 (1962).
63. C. Heid, K. Holoubeck, and R. Klein, *Melliand Textilberichte 54,* 1314 (1973).
64. W. Zerweek, H. Ritter, and M. Schubert, *Angew. Chem.* (*Ausg. A*) *60,* 141 (1948).
65. D. G. Orton, in: *The Chemistry of Synthetic Dyes* (K. Venkataraman, ed.), Vol. VII, p. 1, Academic Press, London (1974).
66. H. Hiyama, *J. Chem. Soc. Japan Ind. Chem. 51,* 97 (1948).
67. H. H. Hodgson, *J. Soc. Dyers Colour. 40,* 330 (1924).
68. W. N. Jones and E. E. Reid, *J. Am. Chem. Soc. 54,* 4393 (1932).

69. H. Hiyama, *Yuki Kagako no Shimpo, Progress of Synthetic Sulphur-Containing Dyes,* p. 199, Osaka Municipal Tech. Res. Inst., Osaka (1959).
70. K. H. Shah, B. D. Tilak, and K. Venkataraman, *Proc. Indian Acad. Sci. 28A,* 111 (1948).
71. B.I.O.S. *983*, 43, 53–58, 71–74.
72. K. H. Saunders, *The Aromatic Diazo Compounds and Their Technical Applications,* Arnold, London (1949); H. Zollinger, *Azo and Diazo Chemistry,* Interscience, New York (1961).
73. E. Kissa and du Pont, U.S. Patent 3,184,284 (1965) [*CA 63,* 4454 (1965)].
74. J. Heyna, A. Bauer, K. Berner, and Hoechst, German Patent 1,150,163 (1963) [*CA 60,* 1867 (1964)].
75. ICI French Patent 1,143,176 (1957).
76. Bayer, Netherlands Patent 6,516,117 (1966) [*CA 66,* 11849 (1967)].
77. Sandoz, British Patent 916,094 (1963) [*CA 59,* 8914 (1963)].
78. Geigy, Belgian Patent 644,495 (1964) [*CA 63,* 11743 (1964)].
79. Ciba, British Patent 949,711 (1964) [*CA 63,* 711 (1965)].
80. B.I.O.S., 1548 (1946).
81. B.I.O.S., 987.
82. FIAT, II, 1313.
83. B.A.S.F., Brisith Patent 127,178 (1901).
84. J. Harley-Mason, *J. Chem. Soc.*, 2907 (1950).

4

Dyes for Polyester Fibers

L. SHUTTLEWORTH and M. A. WEAVER

I. INTRODUCTION

Dyes traditionally used for natural fiber coloration were water-soluble or solubilized prior to use and required salt linkages or van der Waals forces to impart affinity for the fiber. The dyes may be insolubilized within the natural fiber after being applied in a water-soluble form, or they were made to react chemically with the fiber to prevent their removal.

The development of cellulose acetate as a "synthetic" textile fiber posed many problems to the colorist because available dyes were inadequate for its coloration. The Ionamines were an early series of dyes used for the coloration of cellulose acetate. They functioned by slow hydrolysis of the water-soluble species, which created in the dyebath a small amount of a water-insoluble derivative that colored the fiber. The development of disperse dyes—that is, dyes insoluble in water and applied from aqueous dispersion rather than from solution—was a revolutionary solution to the problem of coloration of synthetic fiber. The dyes were essentially insoluble in water and were prepared for application by being ground, in the presence of dispersing agents, to microscopically fine particles of the order of a few microns and, then, by pan drying the resultant suspension. The resulting readily dispersible solid could then dye the more hydrophobic acetate fiber by partitioning into the fiber from low dyebath concentrations.

The development of disperse dyes led to an upsurge in new chemistry, resulting in novel intermediates which were unsuitable for the synthesis of water-soluble dyestuffs. The structural features affecting dyeing and fastness properties needed to be reconsidered; smaller, more compact dye molecules

L. SHUTTLEWORTH • Research Laboratories, Photographic Products Group, Eastman Kodak Company, Rochester, NY 14650. M. A. WEAVER • Retired from Research Laboratories, Eastman Chemicals Division, Eastman Kodak Company, Kingsport, TN 37662.

needed to be invented, which would cover the whole gamut of the color spectrum.

The advent of polyamide and polyester fibers, which were even more hydrophobic than cellulose acetate fibers, increased the need for disperse dyes. Many of the dyes used successfully to color cellulose acetate were found to be unsuitable for polyester coloration because of the compact crystalline nature of the polyester fiber. Diffusion of disperse dyes into the polymer matrix was difficult and occurred slowly compared to the diffusion of disperse dyes into cellulose acetate. The ability to use smaller dye molecules and higher temperatures with polyester fiber, than was possible for cellulose acetate without causing deformation, alleviated the problem to some extent. Carriers were found to be effective in swelling the polyester fiber and in aiding the dye penetration and diffusion. Also, the development of machinery to allow the dyeing process to proceed at up to 130 °C under pressure helped to speed up the process.

Drastic conditions, such as the use of carriers or high temperature, imposed further dye structure requirements in relation to hydrolytic stability, reduction stability, and general dye reactivity. These requirements restricted the options available to the dye chemist in designing dye molecules. For example, residual carriers remaining in the fibers decreased the lightfastness of colored materials. Also, the continuous dyeing of polyester and polyester cellulosic blends by using dry heat and high temperatures in the THERMOSOL process created a need for new disperse dyes which would withstand these high-temperature conditions because disperse dyes are nonionic and show a tendency to sublime at elevated temperatures. Today, the use of polyester fabrics in automotive upholstery has created a need for disperse dyes of exceptionally high lightfastness to withstand extended exposure to light.

The popularity in the 1970s of continuous filament polyester knits led to the introduction of a new coloration technique for this material. Because of the material's dimensional instability and its particular popularity as a fashion fabric, the technique of heat transfer printing became popular. Intricate designs could be transferred thermally by dye sublimation from a paper support to the polyester fabric. This led to a requirement for dyes of high sublimation efficiency and led to a reexamination of many dyes rejected earlier because of their poor sublimation fastness.

The development of the chemistry of disperse dyes has largely mirrored the requirements of the end uses of polyester fiber and the techniques used to color it. The innovative abilities of the dyestuff chemist have been challenged in the invention of new chemical structures that cope with these changes. By way of intermediates, chromophoric systems and reaction methods, novel chemical techniques were developed to satisfy the aforementioned needs and, also, to satisfy the need for cost-effective products in today's competitive world.

It is the aim of this chapter to describe in detail some of the more recent chemical techniques used to develop disperse dyes for the effective coloration of polyester fiber.

II. ANTHRAQUINONE DYES

A. Preparative Index

Synthetic Method 1

Synthetic Method 2

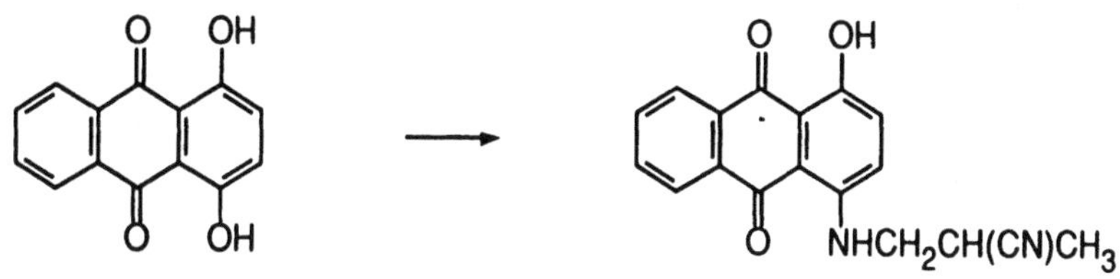

Synthetic Method 3

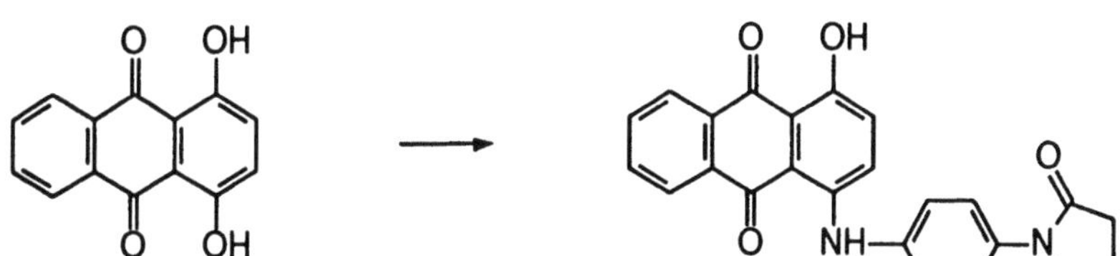

Synthetic Method 4

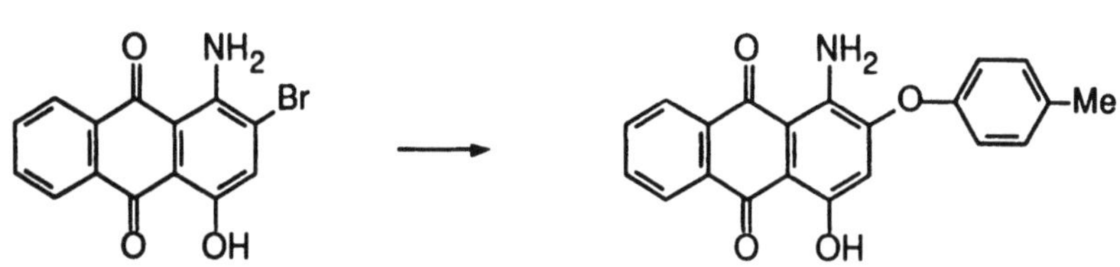

Synthetic Method 5

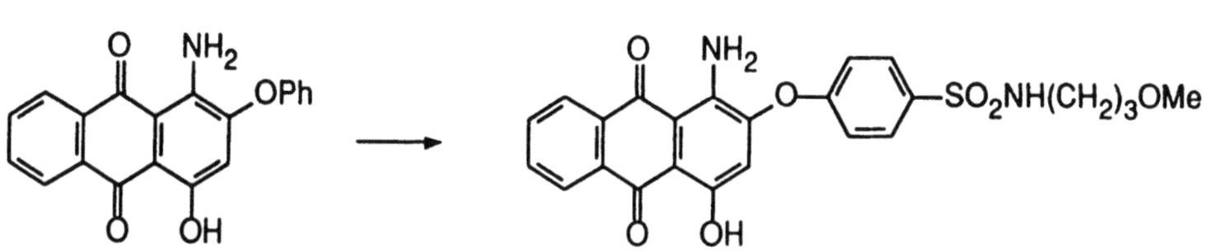

Synthetic Method 6

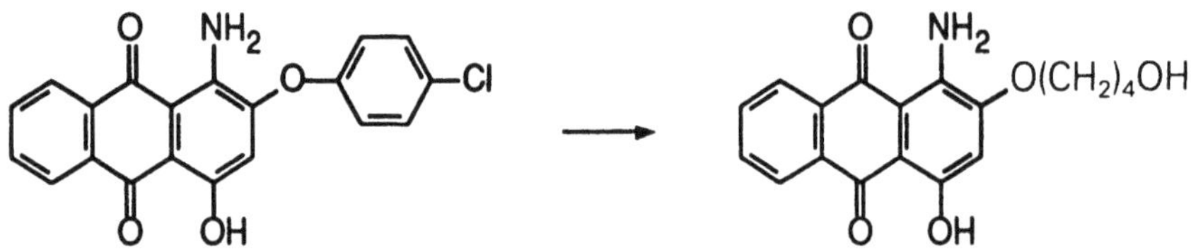

Synthetic Method 7

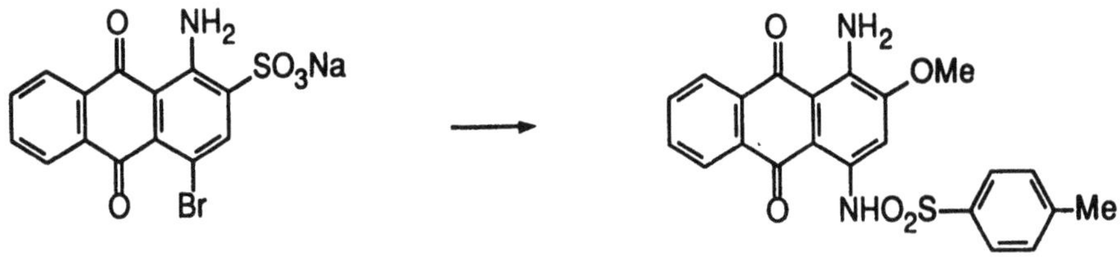

Synthetic Method 8

Synthetic Method 9

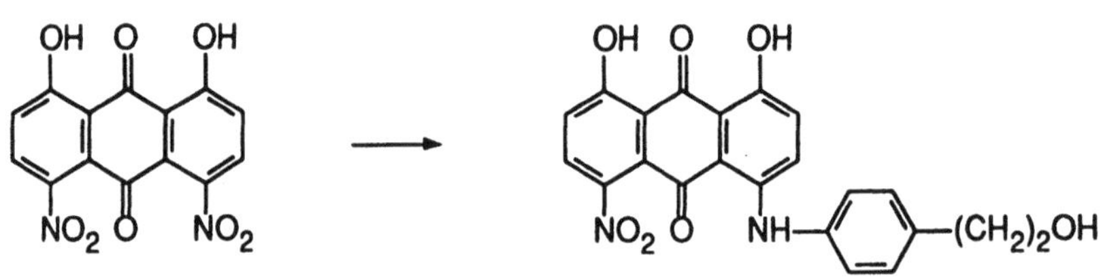

Synthetic Method 10

Synthetic Method 11

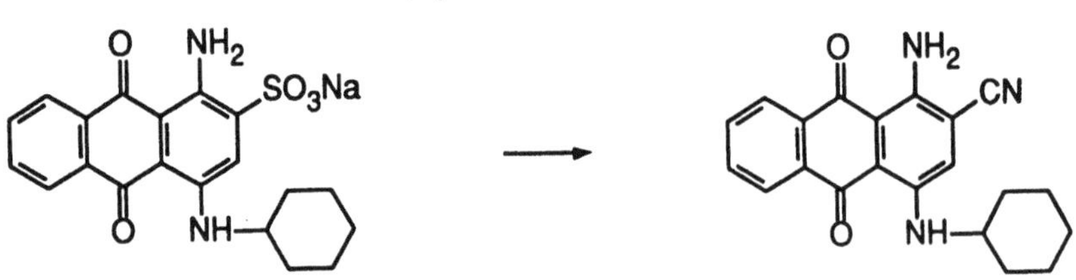

Synthetic Method 12

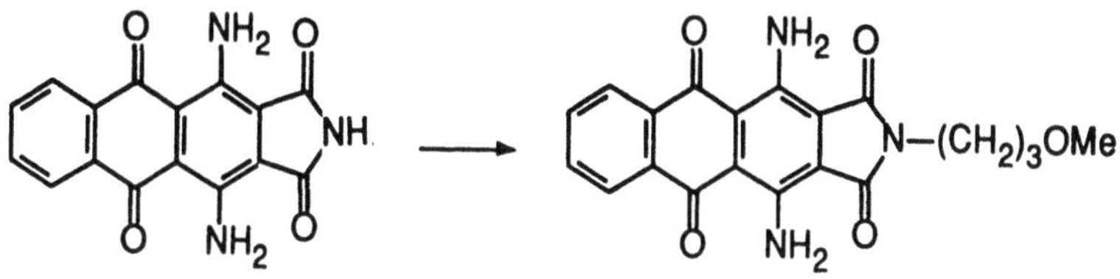

B. Discussion

The anthraquinone chromophoric system has been used extensively for many years for the preparation of textile dyes. It is possible to produce the complete color gamut from the chromophore, but yellow or orange dyes are often better produced by using alternative chemistry from the point of view of both performance and economics. This is true in the case of disperse dyes in general and dyes for polyesters in particular. Red, violet, blue, and turquoise blue shades are still produced commercially by using anthraquinone chemistry because of the particularly attractive performance characteristics afforded by the chromophore, from the standpoint of both dyeability and fastness properties. Many of the brightest hues, ranging from red through blue, when coupled to high lightfastness, can be generated from this chromophore.

Extended polycyclic quinone systems have been used for the coloration of other fiber types, such as cellulose fibers, to produce high wetfastness dyes. In general, these systems are not used extensively for polyester coloration because of the difficulties encountered with fiber penetration in relation to the method of coloration.

Until the 1950s when heterocyclic azo dyes based on 2-amino-5-nitrothiazole were discovered, blue disperse dyes were almost invariably based on the anthraquinone nucleus. This was only partly true in the case of red disperse dyes; but, since that time, much work, as reflected by the extent of patent activity, has been done to replace anthraquinones in the field of both red and blue disperse dyes. However, despite the effort by dye chemists over the years, anthraquinone dyes have retained their premier position in this field.

The hue of anthraquinone dyes is essentially controlled by the presence of electron donors, such as amino and hydroxyl groups in the 1-, 4-, 5-, and 8-positions of the nucleus because, in general, the other substitution positions have a lesser influence on overall hue but are often used to influence fastness or dyeing properties. In general, the anthraquinone compounds containing hydroxy groups are more lightfast on cellulose acetate and polyester than the anthraquinone compounds containing amino groups, while the reverse is true for polyamides. The substitution of hydroxyl groups in the 1- and 4-positions of the anthraquinone nucleus gives the orange derivative quinizarin which is of no utility as a dye in itself but is a very important intermediate for a class of 1,4-disubstituted anthraquinone dyes. The utility of quinizarin lies in its ability to form (equation 1) a leuco derivative that reacts readily with amines to introduce

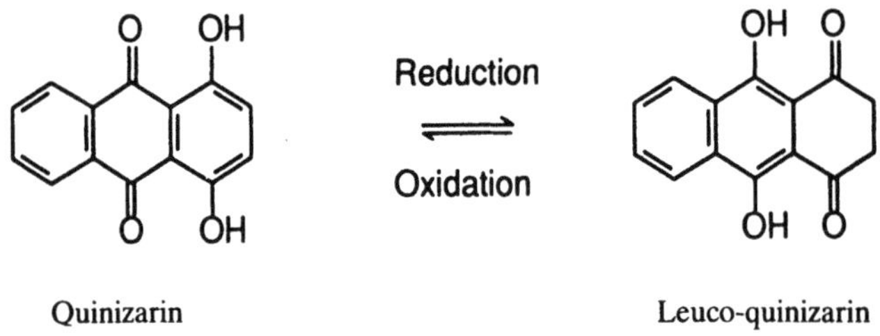

(1)

the amine function in the 1- and/or 4-positions with a corresponding bathochromic shift and an increase in extinction coefficient.

Condensation of leuco quinizarin with various aliphatic amines leads to a generic class of dyes classified as CI Disperse Blue 3. These are often mixtures chosen to optimize dyeing properties. The dyes were originally developed for acetate fibers, but they suffer from poor colorfastness due to "burnt gas fumes." This phenomenon was first noticed as a "reddening" of the dyed material caused by the acetate fiber absorption of trace amounts of atmospheric oxides of nitrogen. These oxides originally resulted from burning gas, but now result from electric sparking in air from motor emissions and other kinds of discharge. The color change presumably results from a nitrosation process on the dye molecule within the fiber and is probably best related to a problem in storing the dyed fabric.

The problem appears to be specific to acetate fibers because CI Disperse Blue 3 types encountered some success on nylon carpet because of their high lightfastness. However, the dyes had little success on polyester fiber because of their very poor light and sublimation fastness. The advent of heat transfer printing led to a reintroduction of CI Disperse Blue 3 for polyester coloration. In this process, the high sublimation rate and attractive hue made the dye desirable, and the poor lightfastness could be tolerated to some extent because the heat-transfer printing technique was largely used in production prints for fashion outlets where long-term life was of lesser importance than that for nylon carpet.

Synthetic Method 1[1] (1,4-Dialkylaminoanthraquinone-CI Disperse Blue 3 Type Dye) (Procedure quoted verbatim from U.S. Patent 2,211,943).

"A mixture of 50 parts of leuco quinizarine (obtained by reducing quinizarine in neutral or slightly alkaline aqueous suspension with sodium hydrosulfite at 60 to 80°), 325 parts of methyl alcohol and 50 parts of a 33 percent aqueous solution of monoethylamine is agitated and boiled vigorously for 1 hour in a flask with a suitable reflux condenser. The resulting condensation product which is formed in the mixture is mainly the leuco derivative of 1-mono-ethylamino-4-hydroxyanthraquinone.

"30 parts of a 38 percent aqueous solution of mono-methylamine are then added to the mixture and the resulting mixture is boiled for about one and a half hours. The condensing reaction results in the formation of leuco 1-mono-ethylamino-4-mono-methylaminoanthraquinone. To convert the leuco compound into its corresponding quinoid form, 20 parts of nitrobenzene are added to the mixture which is then boiled for a further two hours and then cooled to about 30° whereby 1-mono-ethylamino-4-mono-methylaminoanthraquinone is precipitated. The mixture is filtered and the crystalline residue on the filter is washed with about 120 parts of methyl alcohol and then dried in air or vacuo."

Substitution of only one of the hydroxyl groups of quinizarin with an alkylamine or arylamine gives dyes whose shades vary from pink to violet depending upon the substitution. 1-Hydroxy-4-alkylaminoanthraquinones tend to be more hypsochromic and to be of use primarily on cellulose acetate because their lightfastness is, in general, inferior to the arylamino derivatives on polyester fibers. In contrast, the 1-hydroxy-4-arylamino derivatives tend to be preferred on polyester because of their increased affinity to the fiber.

The dyes can be prepared from quinizarin and/or leuco quinizarin with the arylamine, but care must be taken to avoid an excess of amine which would lead to 1,4-disubstitution. The use of boric acid as a catalyst, particularly with arylamino derivatives, facilitates the condensation, especially in cases of steric inhibition. Dyes of this type have been commercialized for cellulose acetate and for polyester; one example of each follows.

Synthetic Method 2[2] (1-Hydroxy-4-alkylaminoanthraquinone) (Procedure quoted verbatim from U.S. Patent 2,659,739).

"16.8 Grams (.07 mol) of quinizarin and 7.3 grams (.03 mol) of leuco quinizarin were placed in 300 cc. of *n*-butyl alcohol and heated to boiling with stirring. Then 8.6 grams of β-cyanopropylamine (0.102 mol) dissolved in 50 cc. of *n*-butyl alcohol were added slowly over a period of 2 hours. The reaction mixture was refluxed for 6 hours longer, cooled overnight to 0°C. −5°C., and filtered, and the product collected on the filter was pressed as dry as possible. The dye cake was then dried at 60°C. to obtain 25.5 grams (83% of theory) of 1-hydroxy-4-β-cyanopropylamino-anthraquinone which melted at 149°C.–150°C."

This dye, which has good lightfastness and gasfastness, produces violet shades when dyed on cellulose acetate materials. Substitution in the 4-position by arylamino tends to yield dyes which are a little more bathochromic in shade and which have better lightfastness on polyester fibers than dyes with alkylamino in the 4-position. The arylamino substitutent is often further substituted in order to optimize fastness and dyeing performance towards the use for which the dye is designed. The following example gives a typical preparation of a polyester dye of this type.

Synthetic Method 3[3] (1-Hydroxy-4-arylaminoanthraquinone) (Procedure quoted verbatim from U.S. Patent 3,279,880).

"A mixture of 10.5 grams of 1,4-dihydroxyanthraquinone, 6.5 grams of leuco 1,4-dihydroxyanthraquinone, 4 grams of boric acid, 15 grams of *N*-(*p*-aminophenyl)pyrrolidone and 200 cc. of isopropyl alcohol was refluxed together with stirring, for 20 hours. The reaction mixture resulting was cooled and 100 cc. of water were added with stirring. After one hour of stirring the solid which precipitated was recovered by filtration and washed well with water. The filter cake was reslurried with 3000 cc. of 7% aqueous sodium hydroxide, and the resulting mixture was heated to boiling over a period of 40 minutes and then boiled for 5 minutes. The dye compound formed was collected by filtering the hot reaction mixture, and the dye cake was washed with water until the filtrate became colorless."

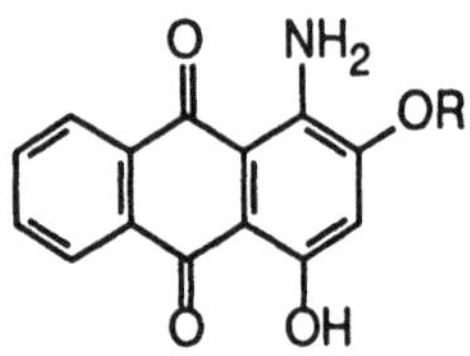

R = Alkyl or substituted alkyl
or aryl or substituted aryl

1

The dye formed has excellent affinity for polyester fibers, giving a blue shade with fastness to light, washing, and sublimation.

A group of compounds which provides bright red and pink shades on polyester materials is based on 1-amino-4-hydroxy-2-alkoxy (or aryloxy) anthraquinone **(1)**.

1-Amino-4-hydroxy-2-phenoxyanthraquinone (CI Disperse Red 60) has long been used as a workhorse red dye for polyester. Its low energy of application, high lightfastness, and relatively good economics make it particularly attractive for a wide range of applications. The following preparation is typical of dyes of this type.

Synthetic Method 4[4] (CI Disperse Red 60 Type Dye) (Procedure quoted verbatim from U.S. Patent 3,189,398).

> "The dye may be made by slowly adding 344 parts of pulverized potassium carbonate over a 2-hour period to 2400 parts of *p*-cresol at 50–55°C. The temperature is then raised to 80–85°C. and held at that point for an hour; 800 parts of 1-amino-2-bromo-4-hydroxyanthraquinone are added and the batch is heated 5 hours at 150–155°C. At this point the progress of the reaction may be checked by chromatographic analysis of a sample. The mass is drowned in about 30,000 parts of approximately 1N sodium hydroxide, stirred 6 to 7 hours and heated one hour at 60°C. It is then filtered and washed with cold water until Brilliant Yellow paper no longer gives an alkaline test in the wash water. The dye may be kept damp or used as a pulp or it may be dried at about 60–65°C. Yield of the crude product is about 98% of theory."

Dyes of this type have somewhat poor fastness to sublimation and, as such, are used in heat transfer printing of polyester. However, in order to improve fastness to sublimation for the regular dyeing of polyester while retaining most of the dyeing properties, the phenyl ring of the 2-phenoxy substituent can be chlorosulfonated, and the chlorosulfonyl derivatives can then be reacted with various amines.

Synthetic Method 5[5] (Sulfonamido Substituted Red 60) (Procedure quoted verbatim from U.S. Patent 3,299,103).

> "20 Parts of 1-amino-2-phenoxy-4-hydroxyanthraquinone are introduced in a finely divided form into 80 parts of chlorosulfonic acid within 30 minutes at 20° to 25°C. After stirring for two hours at this temperature, the mixture is poured onto ice, the reaction material sharply filtered off by suction and washed with ice-water until it has a slightly acid reaction. The moist filtered material is then introduced into 80 parts of γ-methoxypropylamine while cooling and then stirred for about two to four hours at

room temperature. After standing for 12 to 15 hours, the reaction product is filtered off by suction, washed with methanol and water and dried."

Dyes where the 2-substituent is alkoxy rather than aryloxy have been proposed. Substitution of the alkoxy function with polar groups leads to an increase in sublimation fastness on polyester materials, and this has been used commercially in the production of red dyes for polyester. The dyes are advantageously prepared indirectly via the aryloxy compound rather than by direct substitution of the nuclear halogen with alkoxide.

Synthetic Method 6[6] (1-Amino-4-hydroxy-2-hydroxyalkoxyanthraquinone) (Procedure quoted verbatim from U.S. Patent 3,694,467).

"50 Parts of 1-amino-2-(4-chlorophenoxy)-4-hydroxyanthraquinone is introduced while stirring into a mixture consisting of 150 parts of butanediol-(1,4) and 6 parts of potassium hydroxide at 100°C. The reaction mixture is heated at 135°C for 2 1/2 hours while a weak current of nitrogen is passed over the mixture. The reaction mixture is allowed to cool, 150 parts of methanol is added, and the whole is slightly acidified with acetic acid and allowed to stand for some hours. The dye is filtered off, washed with methanol and then water until the wash has a neutral reaction, and dried. 41 parts of a dye having a melting point of 161° to 162°C (from butanol) is obtained."

Red dyes for polyester, where the 4-hydroxyl group has been replaced by a sulfonamido function, have been described and commercialized. They are claimed to have outstanding fastness to light and to sublimation, and to produce similar shades to the hydroxyl-substituted dyes. Their advantage over the hydroxyl-substituted derivatives is that they are often found to be less sensitive to metal contamination because they are less prone to complexation with heavy metals. The dyes can be made directly from bromamine acid as described in the following procedure.

Synthetic Method 7[7] (1-Amino-2-methoxy-4-toluenesulfonamidoanthraquinone) (Procedure quoted verbatim from U.S. Patent 3,087,773).

"10 Grams of $CuSO_4{\cdot}5H_2O$, 300 grams of 1-amino-4-bromoanthraquinone-2-sodium sulfonate, 150 grams of *p*-toluenesulfonamide and 55 grams of K_2CO_3 were added with stirring to 7 liters of water. The reaction mixture thus obtained was heated to 97°C. over a period of about 2 hours and then held at 96–98°C. for 12 hours while stirring. The reaction mixture was then cooled to 30°C. and filtered. The 1-amino-4-*p*-toluenesulfonamidoanthraquinone-2-potassium sulfonate collected on the filter was washed with 3 liters of 2% KCl and then with 2 cold water washes. Upon drying to a constant weight of 110–120°C. a yield varying from 310–320 grams is obtained. An amount of 1-amino-4-bromoanthraquinone-2-sodium sulfonate equivalent to 240 grams of the free acid was employed. This compound can be purchased as "Bromamine Acid" and assays 80–85% free acid.

"250 Grams of flake KOH were dissolved in 500 grams of methyl alcohol with cooling: 102 grams of 1-amino-4-*p*-toluenesulfonamidoanthraquinone-2-potassium sulfonate were added at 60–65°C. over a period of 1/2 hour with vigorous stirring. The reaction mixture thus obtained was stirred at moderate reflux (87–89°C) for 4 hours and then drowned in 2800 cc. of cold water with stirring and filtered. The blue potassium salt collected on the filter and was washed 6 times with hot water. The wet cake weighing about 300 grams was added to 2500 cc. of water at 80–90°C. with good stirring, neutralized to a pH of 5–7 with acetic acid and filtered. The 1-amino-2-methoxy-4-*para*-toluenesulfonamidoanthraquinone collected on the filter was washed with hot water until neutral. A yield of about 72 grams dry basis, was obtained."

A number of blue dyes for polyester has been developed based on the 1,4,5,8-tetrasubstituted anthraquinone system. The dyes are often based on diaminochrysazin or diaminoanthrarufin as intermediates. To date, dyes of this type offer the most attractive blue dyes for polyester because of their excellent working properties, coupled with high fastness, particularly to light.

One modern high-volume workhorse blue dye for polyester is CI Disperse Blue 56 and it is based on this system. The dye has excellent application properties and economy, and offers reddish so neutral blue shades on polyester with acceptable lightfastness for most applications. Sublimation fastness, as may be expected from the low energy of application, is not good, and so its application is restricted to those areas where sublimation fastness is not a primary factor for consideration. The dye is prepared by the bromination of diaminodihydroxyanthraquinones and an example of the type of procedure which can be used follows.

Synthetic Method 8[8] (CI Disperse Blue 56) (Procedure quoted verbatim from U.S. Patent 2,990,413).

"27 Parts of 4,8-diamino-1,5-dihydroxyanthraquinone and 10 parts of boric acid are dissolved in 250 parts of sulfuric acid monohydrate and 0.1 part of iodine and 10 parts of bromine are added thereto. The mixture is then stirred at 50°C. until the bromine is used up. The reaction mixture is then pressed onto ice, the precipitate is filtered off by suction and washed neutral; 35 parts of a blue dyestuff are thus obtained which contains about 25 percent of bromine."

The diaminodihydroxyanthraquinones can be made from dinitroanthrarufin and dinitrochrysazin or mixtures thereof. It has been found that replacement of one of the nitro groups of dinitrochrysazin with substituted anilines leads to a series of blue dyes which have particularly good gasfastness on acetate fiber, coupled with high lightfastness. The dinitrochrysazin compounds are slightly greener and have greater lightfastness than those from dinitroanthrarufin. These dyes also exhibit excellent performance on polyester fibers. The second nitro group is resistant fo further substitution by arylamines, but it can be reduced to produce brighter and more bathochromic dyes with slightly better dyeing properties. The preparation of dyes of this type is exemplified as follows.

Synthetic Method 9[9] (1,8-Dihydroxy-4-nitro-5-arylaminoanthraquinone) (Procedure quoted verbatim from U.S. Patent 2,641,602).

"50 Parts of *p*-aminophenylethyl alcohol, 50 parts of 4,5-dinitro-1,8-dihydroxyanthraquinone and 175 parts of *n*-amyl alcohol are placed in a suitable reaction vessel and refluxed together, with stirring for 6 hours. The reaction mixture is then cooled and 500 parts of 5% hydrochloric acid are added thereto. The *n*-amyl alcohol is removed from the reaction mixture by steam distillation and the reaction mixture remaining is filtered. The product obtained on the filter is washed with water and ground in a ball mill with 200 parts of 2% aqueous sodium carbonate solution. The grind is filtered, washed neutral with water and dried at 60 °C. 53 parts of a product are obtained."

Diaminoanthrarufin derivatives which have a hydroxyphenyl or lower alkoxyphenyl substituent in the 2-position, *ortho* to the α-hydroxyl substituent, have been commercialized as blue dyes for polyester materials. The dyes (CI Disperse Blue 73) have excellent buildup and provide desirable bright neutral blue shades with good lightfastness. The preparation of these dyes is best achieved from diaminoanthrarufin disulfonic acid by first substituting one of the sulfonic acids with the alkoxyphenyl group and then by removing the second sulfonic acid by reduction.

Synthetic Method 10[10] (CI Disperse Blue 73) (Procedure quoted verbatim from U.S. Patent 1,652,584).

"20 parts of *p*-diaminoanthrarufin-2,6-disulfonic acid and 10 parts of boric acid are dissolved with heating in 400 parts of 96% sulfuric acid. After complete solution has been obtained and the mass is cooled to about 10°C, 10 parts of anisol are added. The color of solution quickly changes from blue to brownish red indicating the formation of the addition product. The solution is then poured into 2000 parts of water and the yellow solution obtained heated to about 90–95°C; a beautiful crystalline precipitate soon starts to separate, which precipitate is formed by displacement of a sulfo group by the anisol radicle, and by the concurrent escape of the so liberated sulfur dioxide. This heating is continued for about 2 or 3 hours, when the precipitate does not increase further, the reaction mass is then allowed to cool and, the precipitate is filtered off and washed to neutrality with a little water. A beautiful crystalline pyridine salt is obtained by dissolving the product in pyridine and precipitating with alcohol."

"10 parts of the pyridine salt of the *p*-diamino-hydroxy-phenyl anthrarufin sulfonic acid are stirred up with 280 parts by weight of 20% ammonia, and then 2000 parts of water are added; 5 1/2 parts of 86% sodium hydrosulfite are added to the so obtained blue solution. The solution immediately becomes orange and now contains a hydrocompound distinguished from the parent material by the addition of 2 hydrogen atoms in the molecule. This new product can be isolated by salting it out from the above solution. The reaction mass is now slowly heated on the water bath, the color changing gradually to violet and a new compound, formed by the splitting off of a sulfo group, separates as beautiful fine needles. The same reaction also occurs by prolonged standing at ordinary temperatures. The new compound is filtered off, after the precipitation is complete; it is washed with slightly acidified water and then with pure water. The product can be purified by dissolving the above wet precipitate in pyridine and precipitating it with water; it is obtained in beautiful crystals. Recrystallizing the dry product from nitrobenzol also yields the product in crystalline form."

1,4-Disubstituted anthraquinones have been described earlier and, although of use on acetate fiber, have found little utility on polyester because of poor lightfastness. The introduction of an electron-withdrawing group into the 2- and/or 3-position adjacent to the amino substituent leads to a significant increase in lightfastness on polyester materials and effects a bathochromic shift. The lightfastness of CI Disperse Blue 3 types has been improved for the heat transfer printing of polyester by the introduction of the cyano group into the 2-position of a 1,4-diaminoanthraquinone derivative **(2)**.

This reaction is normally carried out by displacement of the 2-sulfo group.

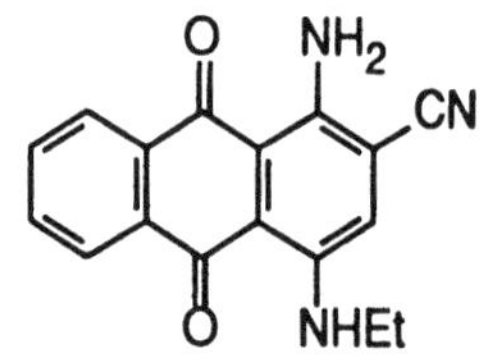

2

The sulfo derived dye can be made by using standard procedures from bromamine acid. The following procedure describes the type of chemistry which can be used.

Synthetic Method 11[11] (1-Amino-2-cyano-4-alkylaminoanthraquinone) (Procedure quoted verbatim from U.S. Patent 1,938,029).

> "20 Parts of the sodium salt of 1-amino-4-hexahydroanilidoanthraquinone-2-sulfonic acid are heated with 40 parts of potassium cyanide and 1000 parts of water in a closed vessel for six to eight hours to a temperature of 110–115°C., while stirring. After cooling, the crystalline product having separated is filtered by suction, washed with hot water, dried, and transformed into its brownish red sulfate by means of 12 parts its weight of sulfuric acid (65%) at a temperature of 40–45°C. The sulfate forms small orange crystals which are filtered and decomposed with water at 30°C. A blue compound is thus obtained crystallizing from glacial acetic acid in the form of blue needles and melting at 211–212°C."

Bright turquoise dyes are produced if both the 2- and 3-positions are substituted with withdrawing groups. The dicyano substituents are effective; but of more utility in commercial applications is the *N*-substituted cyclic imide system (CI Disperse Blue 60). Beautiful, clean, greenish-blue shades on polyester can be produced from dyes of this type. Lightfastness and build-up of these types of dyes are not ideal, and the dyes are expensive to produce; nevertheless, they have found widespread utility for the production of bright turqouise and greens on polyester fiber.

The dyes can be produced by a number of methods that involve some unique chemistry. The following example serves to describe the type of chemistry used.

Synthetic Method 12[12] (*CI Disperse Blue 60*) (Procedure quoted verbatim from U.S. Patent 2,753,356).

> "A mixture of 40 parts of 1,4-diaminoanthraquinone-2,3-dicarboximide, 14 parts of gamma-methoxypropylamine, 158 parts of methanol and 240 parts of nitrobenzene is heated for 16 hours at 175°C. After cooling to room temperature, the desired compound is isolated by filtration, washing with methanol and finally with water, and drying. The yield of 1,4-diamino-anthraquinone-*N*-gamma-methoxypropyl-2,3-dicarboximide is 43 parts (87% of theoretical) and nitrogen analysis shows 11.80% N (theory = 11.08% N). When the same process is carried out at a reaction temperature of 185°C., the product has a N-content of 11.25%."

III. METHINE DYES

A. Preparative Index

Synthetic Method 13

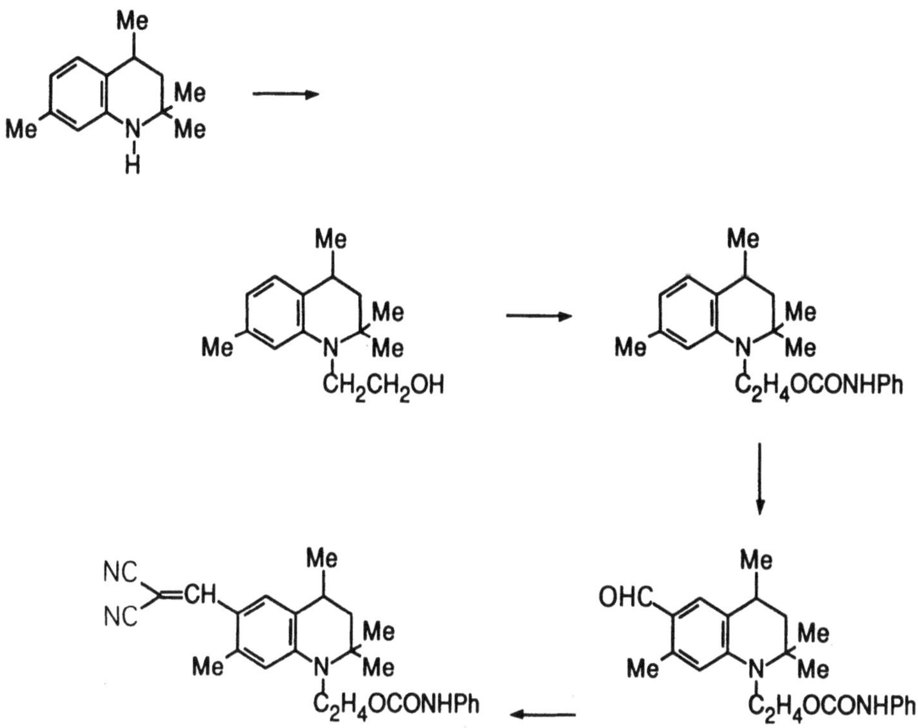

Synthetic Method 14

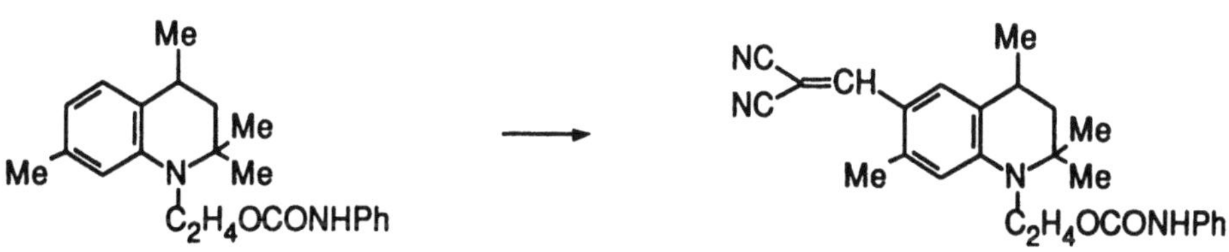

Synthetic Method 15

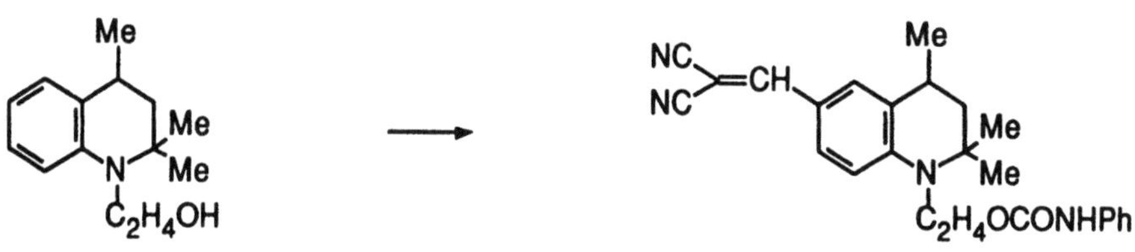

Synthetic Method 16

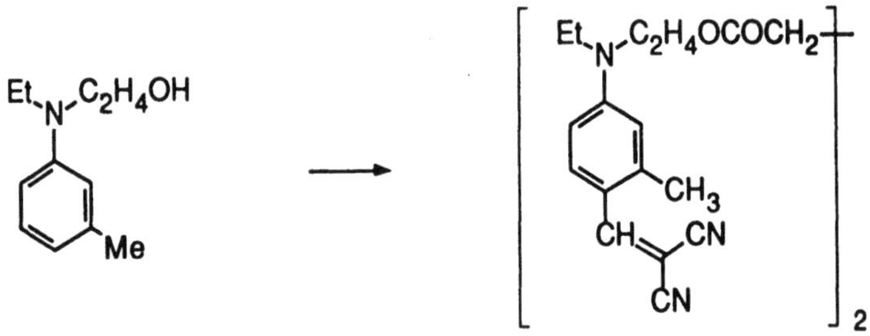

Synthetic Method 17

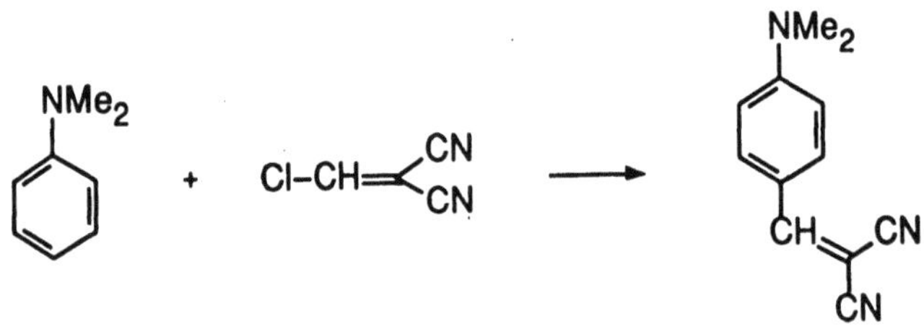

Synthetic Method 18

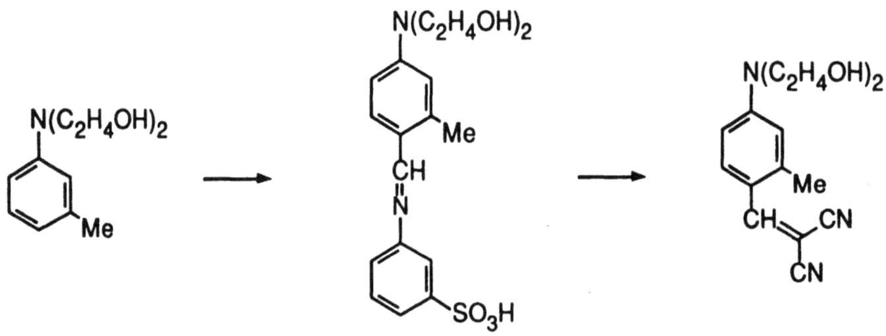

Synthetic Method 19

Synthetic Method 20

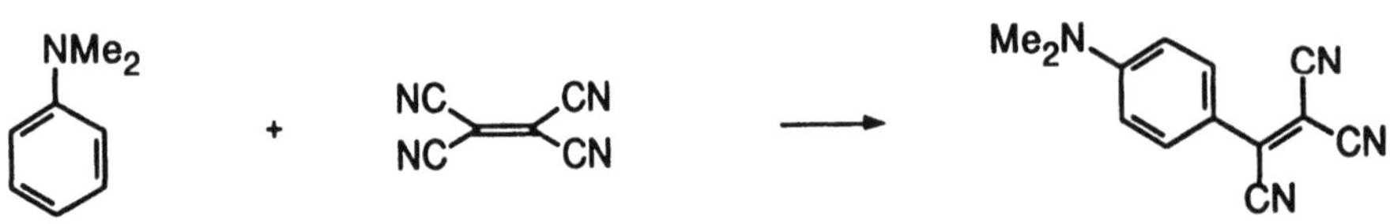

Synthetic Method 21

Synthetic Method 22

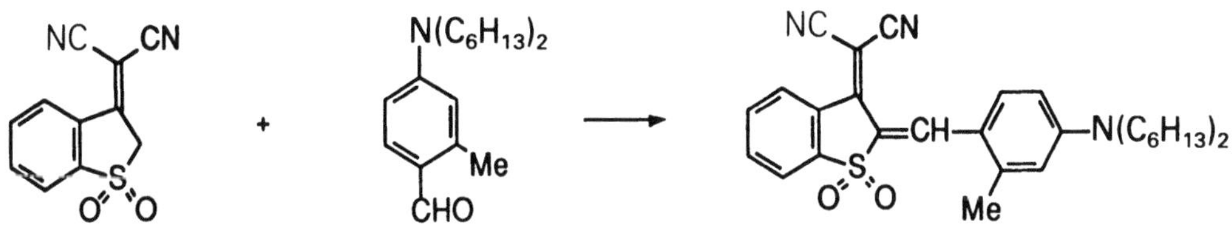

Synthetic Method 23

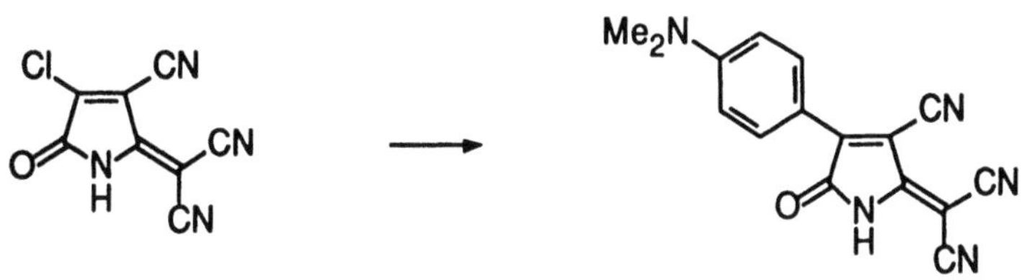

B. Discussion

The methine or arylidine class of dyes has found widespread use for the coloration of acetate and polyester textiles. Until recently, this class of dyes has been limited to the production of very bright lemon yellow shades which are characterized by high lightfastness, particularly on polyester. The dye class as a whole has good dyeing properties; but, in general, the dyes have relatively poor hydrolytic stability, particularly at high pH. Consequently, dyeings at high temperatures, such as for polyesters, may result in some loss of shade depth because of hydrolysis. It is this characteristic of poor hydrolytic stability, along with that of poor lightfastness, which renders the methine class unsuitable for polyamide.

Conventional methine dyes are characterized by the presence of the methine group from which the class takes its name **(3)**. They are normally made by condensation of an aromatic aldehyde containing a *para*-donor group (normally

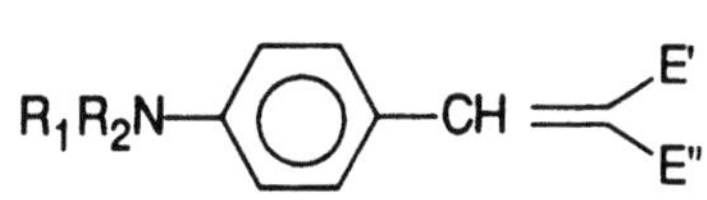

3

substituted amino) with an activated methylene compound that contains the necessary electron-withdrawing functions to give the required charge separation in the final dye. Commercial dyes normally use cyanoacetic acid derivatives as the active methylene components. Malononitrile tends to yield dyes which are slightly bathochromic and have higher extinction coefficients and better lightfastness than alkylcyanoacetates.

The dyes produced are used in the preparation of Kelly green and similar shades on polyester where their brightness can be used advantageously. Because of their relatively small molecular size, the dyes' sublimation fastness can be relatively low for certain polyester applications. Some of the more recent methine chemistry reflects attempts to overcome problems associated with pH instability and to improve sublimation fastness. Also, attempts have been made to extend the rather limited shade range and to use this class of dyes' very high extinction coefficients.

Methine dyes are normally prepared from a formylated aniline derivative by condensation with an active methylene compound. Formylation of the aniline derivative is usually achieved by Vilsmeier reaction by using phosphorus oxychloride and *N,N*-dimethylformamide. The following example of dye preparation uses these standard conditions and is generally applicable to similar systems. This example also describes the introduction of the phenylurethane moiety into the aniline portion to provide improved fastness to light, washing, and sublimation.

Synthetic Method 13[13] (Vilsmeier Formylation and Condensation) (Procedure quoted verbatim from U.S. Patent 3,247,211).

"a. *N-β-Hydroxyethyl-2,2,4,7-tetramethyl-1,2,3,4-tetrahydroquinoline.* 113 g. of 2,2,4,7-tetramethyl-1,2,3,4-tetrahydroquinoline, 33 g. of ethylene oxide and 450 cc. of ethanol were heated in an autoclave at 180°C. for 1 hr. The ethanol was distilled off, and 200 cc. of hexane added. The resulting white solid was filtered off, washed with hexane and dried. M.P. 86–88°C.

"b. *Preparation of the carbamate.* 11.6 g. of the product from (a), 5.95 g. of phenyl isocyanate and 100 cc. of benzene were refluxed 2 hr. and the solvent distilled off. The product was a viscose oil.

"c. *Preparation of the aldehyde.* 17.6 g. of the product of (b) was dissolved in 20 cc. of dimethyl formamide and cooled in the ice bath. 5.5 g. of $POCl_3$ was added slowly below 30 °C. The mix was then heated 1 hr. on the steam bath and poured onto 200 cc. cracked ice. The mix was made basic with NaOH, the product coming down as a thick greenish oil.

"d. *Dye preparation.* 7.6 g. of (c), 1.32 g. of malononitrile, 3 drops of piperidine and 30 cc. of ethanol were refluxed 1 hr. The mix was chilled, filtered and the product washed and air dried. It melts at 169–171°C., and imparts fast yellow shades to Dacron and Kodel polyester fibers and cellulose acetate fibers."

Attempts to simplify the process for the preparation of methine dyes have been reported. The following synthesis of the dye described above produces the dye in high yield and purity without isolation of the formyl derivative.

Synthetic Method 14[14] (Non-isolation of Formylated Amine) (Procedure quoted verbatim from U.S. Patent 3,917,604).

"To a solution of 175 g. of 1-(2-phenylcarbamoyloxyethyl)-2,2,4,7-tetramethyl-1,2,3,4-tetrahydroquinoline in 300 ml. DMF at 45°C. is added 88.0 g. of phosphorus oxychloride and the reaction mixture is heated to 80–85°C. and held at that temperature for one hour to complete formation of the intermediate compound. The solution of the intermediate compound is cooled to 45°C. and added over a 1 hour period to a mixture of 800 ml isopropanol, 46.5 g. malononitrile and 186 g. sodium acetate. The mixture then is heated to 60–65°C. and held at that temperature for 2 hours. To this mixture is added 625 ml. of water over a 1 hour period while maintaining the temperature at 60–65°C. The reaction mixture then is cooled to 30°C., held for 1 hour and the methine product is filtered off, washed with isopropanol, cold water, and hot water, and then dried. The assay yield of dye is 96.5%."

An ingenious procedure for the preparation of this type dye has been reported whereby the complete synthesis can be carried out without isolation of intermediates. Furthermore, cyanoacetamide can be used and dehydrated *in situ* rather than by using the more expensive malononitrile, which is illustrated in the following method.

Synthetic Method 15[15] (One-step Procedure) (Procedure quoted verbatim from U.S. Patent 4,180,663).

"*N*-(β-Hydroxylethyl)-2,2,4-trimethyl-1,2,3,4-tetrahydroquinoline (10.9 g., 0.050 m.) was dissolved in 40 ml. of dimethylformamide. Phenylisocyanate (6.54 g., 0.055 m.) was added and the mixture was heated on a steam bath for two hours. After cooling to 20°C., phosphorus oxychloride (8.4 g., 0.055 m.) was added dropwise at 20°–30°C. and the reaction mixture was heated at 40°–50°C. for $1\frac{1}{4}$ hour. The mixture was cooled to 20°C. and cyanoacetamide (4.6 g., 0.055 m.) and dimethylformamide (10 ml.) was added followed by the dropwise addition of phosphorus oxychloride (5.9 g., 0.038 m.) at 20°–30°C. The reaction mixture was stirred at 20°–30°C (some external cooling necessary) for four hours and was added dropwise to a mixture of sodium acetate (40 g., 0.49 m.) in isopropanol (100 ml.) at 30°–45°C. After addition was complete, the temperature was held at 50°–60°C. for one hour. External heating was removed and water (80 ml.) was added dropwise with stirring. The mixture was cooled to 20°C., and the solid was collected by filtration, washed with isopropanol, cold water, and hot water, and air-dried. The yield of dye was 17.4 g. (0.042 m., 84.5%) as a yellow orange microcrystalline powder."

In order to achieve good fastness to sublimation while still utilizing the small "volatile" methine chromophore, either the molecular weight or the polarity of the molecule was increased. In the above example, the molecular weight is increased by introduction of the phenyl moiety, and the urethane group tends to increase polarity and to reduce the tendency of the molecule to sublime.

Another effective method which has been used successfully commercially is dimerization of the molecule. Here the molecular weight is increased; hence, the molecule's tendency to sublime is reduced without reducing the effective color yield per unit weight of dye, as would be the case when introducing a bulky polar grouping. An example of such a dye follows.

Synthetic Method 16[16] (Dimeric Methine Dye) (Procedure quoted verbatim from U.S. Patent 3,879,434).

"a. An amount of 34.8 g. diethylsuccinate, 71.6 g. *N*-ethyl-*N*-β-hydroxyethyl-*m*-toluidine and 5 drops of titanium tetraisopropoxide are stirred and refluxed together for

3 hr., allowing ethanol to be removed. The reflux temperature rose from 170°C. to 210°C. during this period and about 90% of the theoretical amount of ethanol is collected. The product is allowed to cool and a portion is converted to the bis-aldehyde.

"b. An amount of 22.0 g. of bis[2-(*N*-ethyl-*m*-toluidino)ethyl]succinate is dissolved by stirring in 20 ml. of dry dimethylformamide. The solution is cooled and 10 ml. of phosphorus oxychloride added below 30°C. After heating 1 hr. on a steam bath, the reaction is drowned on an ice–water mixture. This mixture is then made basic with 50% NaOH, resulting in an oily product. The aqueous portion is decanted off and the bis-aldehyde taken up by heating with 75 ml. of ethanol. On cooling, the product crystallizes and is collected by filtration, washed with ethanol and air dried. The product melts at 85–90°C. and one recrystallization from ethanol gives pure bis-aldehyde melting at 95–97°C.

"c. An amount of 4.97 g. bis[2-(*N*-ethyl-4-formyl-*m*-toluidino)ethyl]succinate, 1.32 g. malononitrile, 100 ml. ethanol and 5 drops of piperidine are heated together on a steam bath for 30 minutes. The product crystallizes on cooling and is collected by filtration. The crude product is heated to boiling in 100 ml. of ethanol, allowed to cool to about 50°C., filtered, washed with ethanol, and air dried. The yellow dye obtained melts at 152°–157°C and dyes polyester fibers yellow shades of outstanding light and sublimation fastness."

The methods so far described for the preparation of methine or arylidine dyes have all involved the condensation of an aldehyde with an active methylene compound, usually malononitrile. Alternative methods have been described which avoid the formylation and condensation reactions. One such method, as shown below, uses dicyanovinyl derivatives for the introduction of the dicyanovinyl group to form the methine dye.

Synthetic Method 17[17] (Methine Dye Using 1-Chloro-2,2-dicyanoethylene) (Procedure quoted verbatim from U.S. Patent 4,006,178).

"24.2 Parts of dimethylaniline are dissolved in 100 parts of anhydrous ether. A solution of 11.2 parts of 1-chloro-2,2-dicyanoethylene in 50 parts of anhydrous ether is added dropwise at 20°C over the course of 30 min, whilst stirring. The reaction mixture is boiled for 1 hour under reflux and is left to stand overnight at room temperature. The red-brown crystals are filtered off and washed with water until neutral. After drying, 14.7 parts (75%, relative to the 1-chloro-2,2-dicyanoethylene employed) of the required dye are obtained; m.p. 179°C."

Another method for synthesis of methine dyes suitable for the preparation of dyes containing hydroxyl groups, that would be reactive under normal formylation condtions, uses the Schiff's base of the coupler. This reacts with malononitrile in the presence of a weak base to give the corresponding methine dye as shown in equation (2). The synthesis described in detail follows.

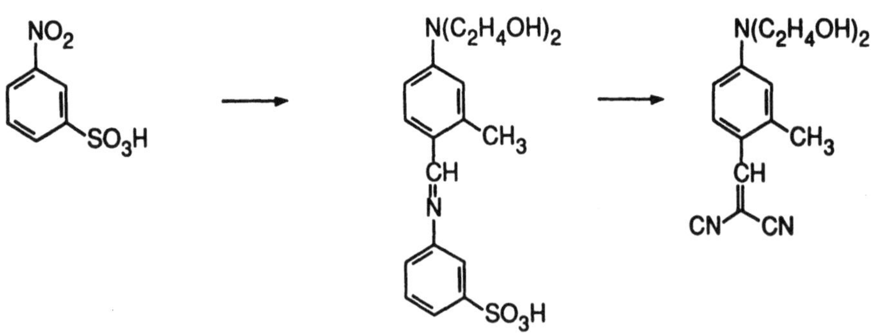

(2)

Synthetic Method 18[18] (Methine Dye via Schiff's Base) (Procedure quoted verbatim from U.S. Patent 2,798,090).

"123 Parts of nitrobenzene were slowly added with good agitation to 460 parts of 20% oleum at 50°C. By the time addition of the nitrobenzene was complete the temperature of the reaction mixture had risen to 75°C.–80°C. Following the addition of the nitrobenzene, the reaction mixture was heated very slowly to 95°C. and maintained at 95°C. for two hours, after which it was allowed to cool to room temperature.

"214 Parts of the *meta*-nitrobenzene sulfonic acid reaction mixture prepared as described above were added to 545 parts of water and then 49 parts of *N,N*-di-β-hydroxyethyl-*m*-toluidine were dissolved in the resulting mixture. The reaction mixture thus obtained was cooled to 20°C. and 27 parts of 36% aqueous formaldehyde were added thereto. 56 Parts of iron filings were then added to the reaction mixture during approximately two hours while maintaining the temperature between 16°C. and 20°C. After the iron addition was complete, the reaction mixture was stirred for four hours at 16°C–20°C, and then filtered after decanting the reaction mixture from any unreacted iron. 4′-(di-β-hydroxyethylamino)-2′-methylbenzalaniline-3-sulfonic acid was recovered as an orange solid, washed well with water and dried. 85.5 Parts (90% yield) of dried product were obtained."

"378 Parts of the 4′-(di-β-hydroxyethylamino)-2′-methylbenzalaniline-3-sulfonic acid were placed in 3200 parts of water and 92 parts of sodium bicarbonate were added. The reaction mixture thus obtained was heated slowly to 60°C. and stirred until solution was complete. 73 Parts of malononitrile were then added to the reaction mixture after adding two parts of wetting agent. The reaction mixture thus obtained was stirred for 16 hours without heating, and the product which formed was recovered on the filter by filtration. The yellow dye obtained on the filter was washed well with cold water and then dried. 220 parts (81% of theory) of 4-(di-β-hydroxyethylamino)-2-methylbenzylidene-malononitrile were obtained. It melts at 138–140°C."

The methine dyes which have been commercialized to date are almost exclusively yellow shades, usually greenish-yellow. However, in recent years, some efforts have been made to extend the color gamut of the methine or arylidene chromophoric system into red and blue shades. This has resulted in some interesting chromophores and in interesting potential for further work.

If malononitrile in the above series of dyes is replaced with 3-cyano-5-phenyl-2-furanone, the hue of the dyes changes from yellow to magenta. 3-Cyano-5-phenyl-2-furanone can be readily synthesized from phenacyl bromide and cyanoacetic acid.[19] Unfortunately, a decrease in lightfastness accompanies the deepening of hue.

Synthetic Method 19[20] (Cyanofuranone Methine Dye) (Procedure quoted verbatim from U.S. Patent 3,661,899).

"Dissolved in 300 ml of toluene were the following: 25 grams (0.168 mole) of *p*-dimethylaminobenzaldehyde, 31.0 grams (0.168 mole) of *p*-dimethylaminobenzaldehyde, 31.0 grams (0.168 mole) of 3-cyano-4-phenyl-2(5H)-furanone, 0.5 ml. of piperidine and 0.5 ml. of glacial acetic acid. The solution was refluxed for 10 minutes during which time 2.8 ml. of water was removed azeotropically. The mixture was cooled and the product was filtered, washed with benzene and dried, giving 47.8 grams of shiny blue-gray crystals of 5-*p*-dimethylaminobenzal-3-cyano-4-phenyl-2(5*H*)-furanone. One crystallization with filtration with 300 ml of chlorobenzene gave 41.9 grams of the product melting at 235–237.5°."

Replacement on the methine proton with an electron-withdrawing group also has a profound effect on the hue of methine or arylidene dyes. Replacement of the methine proton with the cyano group has been widely studied and this replacement results in a shift in hue from yellow to magenta when the other substituents are identical.

Synthetic Method 20[21] (Tricyanovinyl Dye from Tetracyanoethylene) (Procedure quoted verbatim from U.S. Patent 2,889,335).

"A solution of 10 parts of tetracyanoethylene and 178 parts of dry tetrahydrofuran is treated with 19.3 parts of *N,N*-dimethylaniline. A deep blue color forms immediately. The mixture is refluxed on a steam table and the tetrahydrofuran is boiled off. The residue consists of *N,N*-dimethyl-*p*-tricyanovinylaniline in the form of a bright blue crystalline solid, weight 16 parts, sparingly soluble in ethyl alcohol. The solid is washed by suspending with 143 parts of diethyl ether and recrystallized from ethyl alcohol."

The magenta dye may be prepared alternatively by the addition of cyanide to the conventional methine dye; oxidation of the resultant product follows.

Synthetic Method 21[21] (Tricyanovinyl Dye from Methine Dye) (Procedure quoted verbatim from U.S. Patent 2,889,335).

"1. To 180 parts of 50% aqueous ethanol are added 20 parts of *p*-dimethylaminobenzalmalononitrile and 13 parts of potassium cyanide. The mixture is warmed and stirred on a steam table for 3–4 minutes to complete solution. The solution is filtered and diluted with 200 parts of water containing 21 parts of acetic acid. The precipitate of 4-(alpha, beta, beta-tricyanoethyl)-*N,N*-dimethylaniline is collected on a filter, washed with water and dried to give 21 parts by weight (92% yield) of a product with an indicated melting point of 125–130°C. as determined with a preheated Fisher block. A sample for analysis is recrystallized twice from 60% aqueous ethanol after treatment with Darco and dried *a vacuo* at 80°C.; the purified sample gave M.P. 138°–9°C.

"2. To 210 parts of glacial acetic acid and 20 parts of 4-(alpha, beta, beta-tricyanoethyl)-*N,N*-dimethylaniline is added 44 parts of lead tetraacetate. The solution is stirred and heated to 100°C. for two hours. Immediately the reaction is accompanied by the formation of a deep red color characteristic of the dye. An additional 52 parts of acetic acid is added to wash down the sides of the flask, and the dark red solution is allowed to cool slowly to room temperature. Lustrous dark blue needles of 4-tricyanovinyl-*N,N*-dimethylaniline separate and are collected on a filter and washed several times with cold acetic acid and ether, the dry weight being 7.8 parts (39% yield)."

Arylidene dyes, which have an even more powerful electron acceptor than those previously discussed, have been prepared, and the hues of these dyes can be extended to give remarkably blue arylidene dyes with simple aniline type donors. Only recently, a dye of this type has been commercialized for dyeing of polyester textiles, and extremely bright turquoise shades appear on the fiber when the dye is used.

Synthetic Method 22[22] (Blue Methine Dyes) (Procedure quoted verbatim from U.S. Patent 4,281,115).

"a. 20 Parts 3-oxo-2,3-dihydro-1-benzothiophene-1,1-dioxide, 8 parts malonic acid dinitrile and 0.2 parts of a mixture (1:5) of piperidine and glacial acetic acid are dissolved in 250 parts of anhydrous ethanol and heated to 60° for 6 hours with stirring. A further 4 parts of malonic acid dinitrile are added and the mixture is stirred further for 16 hours at 60°. Subsequently, the mixture is cooled to 5°, and the product is filtered, washed with ice-cold ethanol and dried in vacuo.

" b. 12 Parts of 3-*N,N*-di-n-hexyltoluidine are dissolved in 50 parts of dry dimethylformamide, reacted with 7 parts phosphorous oxychloride with cooling and subsequently heated at 60° for 4 hours. The cooled solution is added dropwise at room temperature with stirring to a suspension of 10 parts of the product obtained as described under (a,) above and 15 parts anhydrous sodium acetate in 200 parts absolute alcohol. The mixture is stirred overnight at room temperature, and the precipitated dye is filtered, washed with ice-cold absolute alcohol, slurried in 500 parts ice water, filtered again, washed with ice-cold water and dried in vacuo."

The dicyanovinyl function is demonstrated in these dyes to confer a profound bathochromic shift; alkyl chains present on the aniline function are effective at increasing the pH stability. They are, in fact, true methine dyes in that the methine proton is still present. It was demonstrated earlier that substitution of this proton with an electron-attracting function has a profound effect on the absorption maximum of these arylidene dyes. A series of dyes has been described which make use of both of these hue effects, that is, the incorporation of the dicyanovinyl function and, also, substitution of the methine proton by an electrophilic species. These dyes use the 5-cyanomethylene-2-oxo-3-pyrroline system as the electron-withdrawing function and are capable of producing blue and cyan hues with aniline-type donors. The pyrrolines[23,24] are readily made from available precursors as in equation (3).

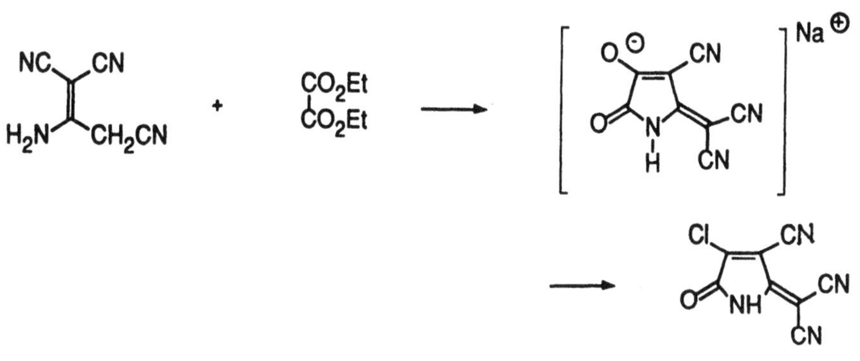

(3)

Synthetic Method 23[24] (Dicyanovinylpyrroline Dyes) (Procedure quoted verbatim from U.S. Patent 3,013,013).

" 1. A solution of 132 parts of malononitrile dimer [$C(CN)_2{=}C(NH_2)CH_2CN$, Carboni *et al., J. Am. Chem. Soc. 80,* 2838–40 (1958)] and 160 parts of diethyl oxalate in 793 parts of absolute methanol is added quickly to a solution of 108 parts of sodium methoxide in 595 parts of absolute methanol. The solution becomes warm and turns yellow. It is stirred at room temperature for two hours. About one third of the methanol is removed by distillation. The mixture is diluted with two volumes of dry benzene. The yellow precipitate which forms is collected by filtration to yield 203 parts of the disodium salt of 4-cyano-5-dicyanomethylene-3-hydroxy-2-oxo-3-pyrroline. This salt is purified by recrystallization from 90/10 n-propyl alcohol/water.

"The disodium salt of 4-cyano-5-dicyanomethylene-3-hydroxy-2-oxo-3-pyrroline is dissolved in a minimum of water and treated with a molar excess of aqueous hydrochloric acid. The dihydrate of the corresponding monosodium salt is obtained as a bright yellow precipitate which is collected by filtration and purified by recrystallization from 90/10 n-propyl alcohol/water."

2. "50 Parts of the dihydrate of the monosodium derivative of 4-cyano-5-dicyanomethylene-3-hydroxy-2-oxo-3-pyrroline is suspended in 157 parts of acetonitrile, and 60 parts of oxalyl chloride is added. The mixture is stirred at reflux for one hour. The mixture is then cooled, and the precipitate which forms is collected by filtration, washed with acetonitrile and dried under reduced pressure to obtain 36 parts (80% yield) of crude 3-chloro-4-cyano-5-dicyanomethylene-2-oxo-3-pyrroline in the form of buff colored crystals. It is purified by sublimation at 180°C and 0.1 mm pressure to give colorless crystals."

3. "To a solution of 15 parts of 3-chloro-4-cyano-5-dicyanomethylene-2-oxo-3-pyrroline in 2250 parts of ethyl acetate is added 48 parts of *N,N*-dimethylaniline. The mixture is allowed to stand at room temperature for two hours. A precipitate of 20 parts of 4-cyano-5-dicyanomethylene-3-(*p*-*N,N*-dimethylaminophenyl)-2-oxo-3-pyrroline is recovered by filtration. It is recrystallized from dimethylformamide."

IV. NITRODIPHENYLAMINE DYES

A. Preparative Index

Synthetic Method 24

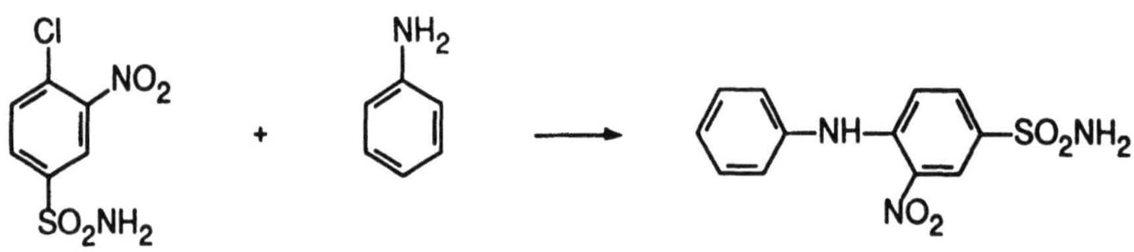

Synthetic Method 25

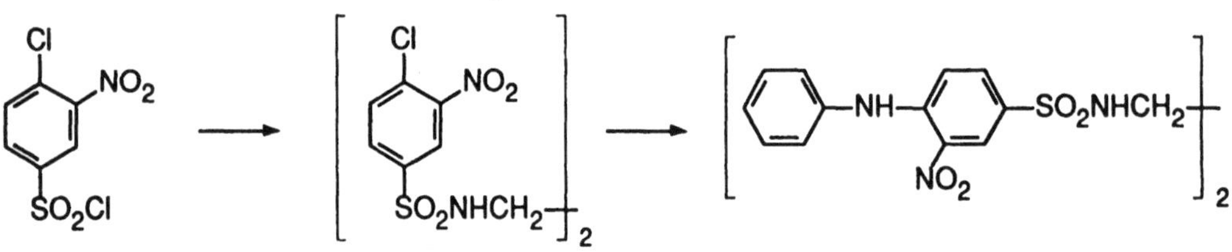

B. Discussion

The nitrodiphenylamine dye class was originally developed for acetate fiber because the inexpensive nature of the dye class made it attractive for the production of exclusively bright yellow shades. However, methine dyes have largely replaced nitrodiphenylamines for the coloration of polyester fiber. One of the principal reasons for the replacement is the comparatively low extinction coefficient of nitrodiphenylamine dyes to that of methine dyes. However, nitrodiphenylamine dyes have retained a niche in the armory of dyes available for the polyester coloration, principally because of their ability to produce yellow shades of very high lightfastness on the fiber. Universally, the dyes of this class have poor lightfastness on polyamide.

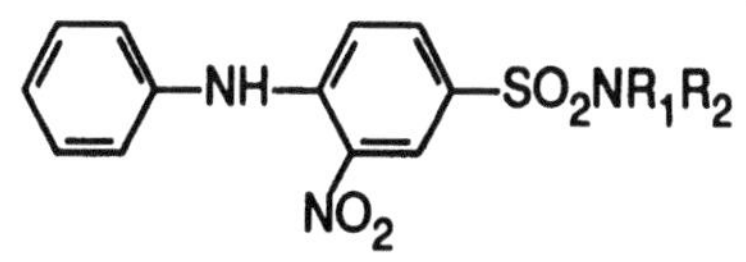

4

The dyes of particular interest on polyesters also have a sulfonamido substituent *para* to the amine in the ring containing the nitro group **(4)**. The nitro substituent is always found *ortho* to the amine, because this forms part of the chromophoric system and results in improved lightfastness when compared to the *para* nitro-substituted compounds. A typical example of a commercial nitrophenylamine dye follows.

Synthetic Method 24[25] (Sulfonamide Substituted Nitrodiphenylamine) (CI Disperse Yellow 33) (Procedure quoted verbatim from U.S. Patent 2,422,029).

"In a flask fitted with a mechanical stirrer were placed 92.4 g. (1.1 moles) of sodium carbonate and 102.3 g. (1.1 moles) of aniline. The stirrer was started and the mixture in the flask heated to 125°C. To the hot mixture were added portionwise 236 g. (1 mole) of 4-chloro-3-nitrobenzene sulfonamide, so that the reaction does not become too vigorous. After the addition, heating was continued for 6 hours at 125° to 135°C. The reaction mixture was permitted to cool and then was steam distilled to remove unreacted aniline. The yellow dye remaining with the aqueous still liquors was filtered off, washed with water and dried in air. It melted at 178 to 180°C."

By reacting with 4-chloro-3-nitrobenzenesulfonyl chloride and 2 equivalents of aniline, the *N*-phenylsulfonamide derivative of the above dye (CI Disperse Yellow 42) can be made. The dye has found significant application in the dyeing of polyester textiles in lightfast yellow shades.

The difference in reactivities of the two halogen atoms in 4-chloro-3-nitrobenzenesulfonyl chloride can be utilized by first substituting the chlorosulfonyl halogen and, then, by substituting the aromatic halogen with aniline to prepare the dye. This procedure can be utilized to prepare dimeric nitrophenylamine dyes by using a diamine in the first condensation and aniline in the second condensation. An example of this procedure follows.

Synthetic Method 25[26] (Dimeric Nitrodiphenylamine) (Procedure quoted verbatim from U.S. Patent 3,729,493).

"To a solution of 12 grams of ethylenediamine and 67.2 grams of sodium bicarbonate dissolved in 400 ml. of water at 40°C., 108.4 grams of 3-nitro-4-chlorobenzenesulfonyl chloride (90% assay) is added portionwise. The reaction mixture is then heated to 60–65°C. for two hours, cooled to 15°C. and filtered. The filter cake is washed with water and slurried in a mixture of 80 ml. isopropanol and 160 ml. water. Then, 16.8 grams of sodium bicarbonate, 36 grams of aniline and 10 drops of piperidine are added, and the reaction mixture is heated to reflux and held for 18 hours. The product is stirred at 15–20°C., filtered and washed. A yield of 79 g of product are obtained."

The purpose of dimerization, as discussed with methine dyes, is to increase molecular weight and, hence, fastness to sublimation without reducing tinctorial power. The dye prepared by the above method dyes cellulose esters and polyesters in yellow shades characterized by lightfastness and sublimation.

V. AZO DYES

A. Preparative Index

Synthetic Method 26

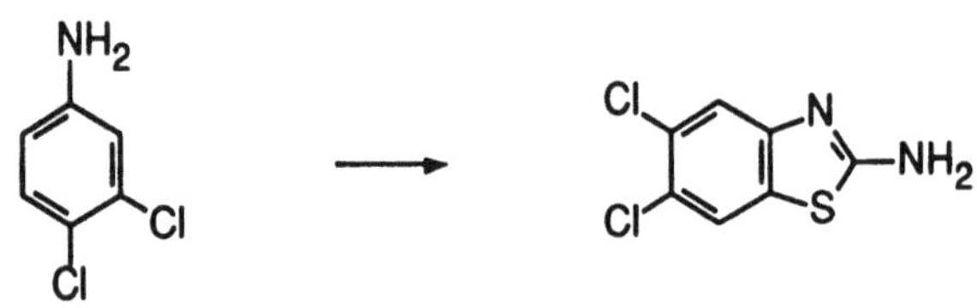

Synthetic Method 27

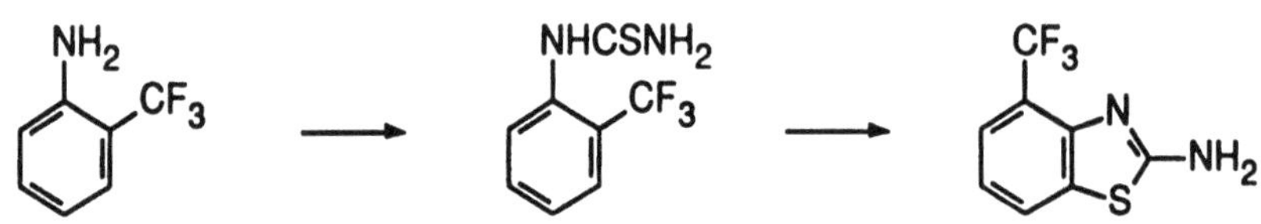

Synthetic Method 28

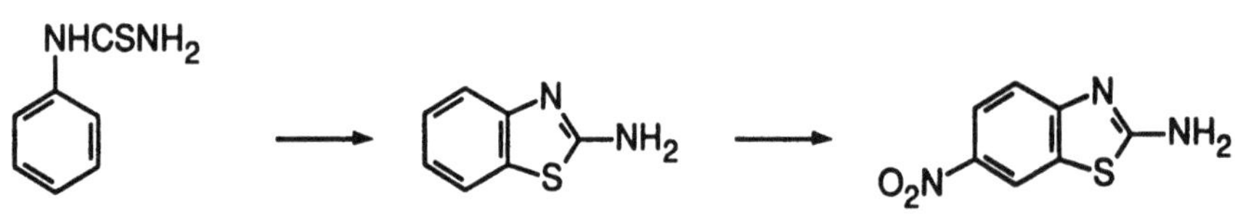

Synthetic Method 29

Synthetic Method 30

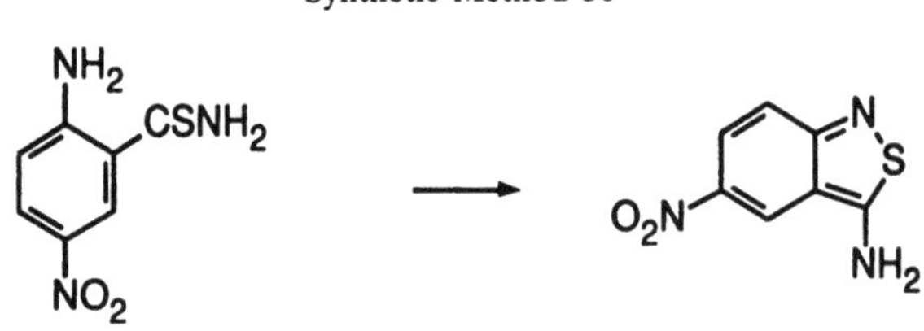

Synthetic Method 31

Synthetic Method 32

Synthetic Method 33

Synthetic Method 34

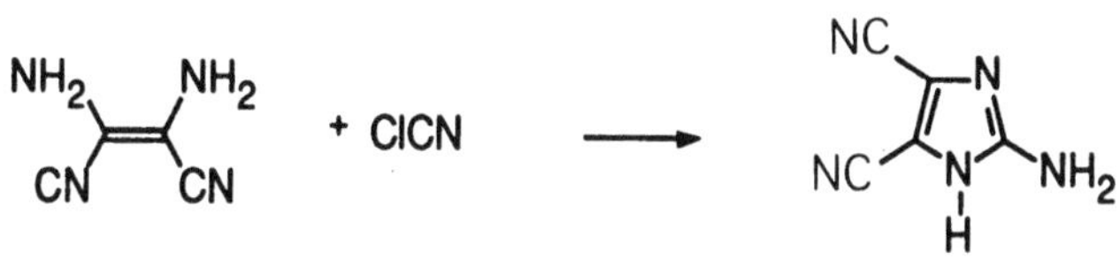

Synthetic Method 35

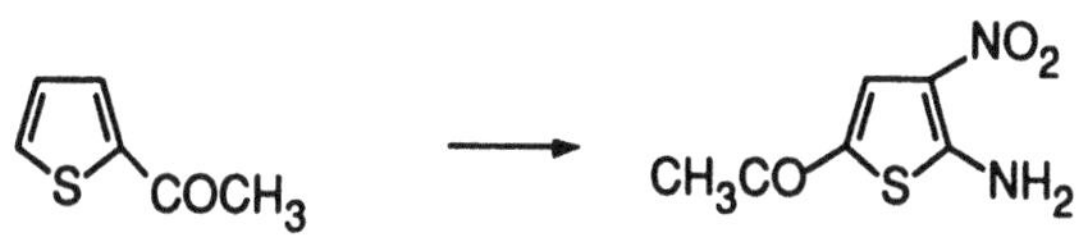

Synthetic Method 36

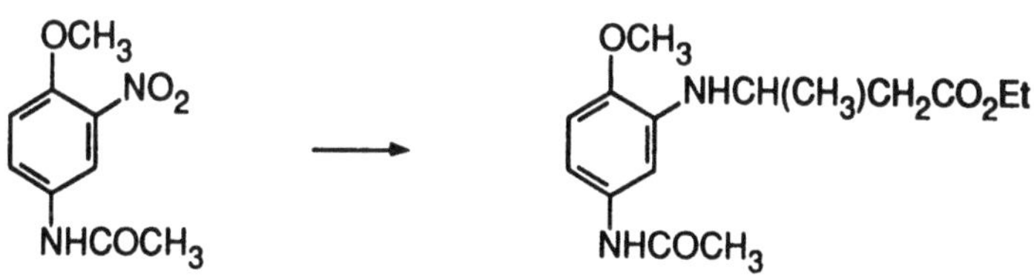

Synthetic Method 37

Synthetic Method 38

Synthetic Method 39

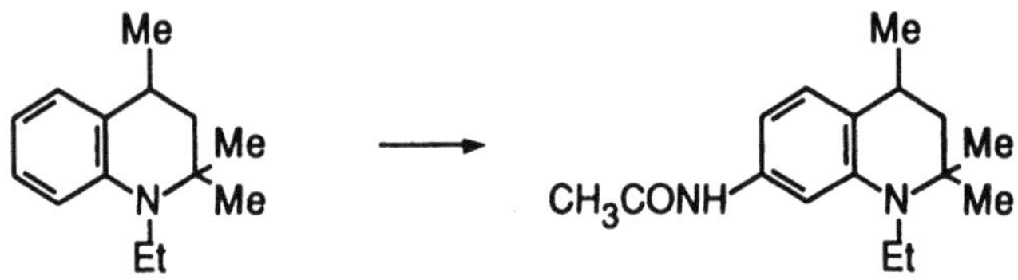

Synthetic Method 40

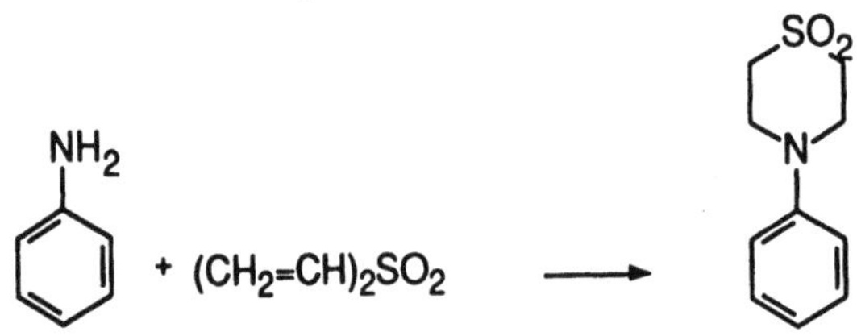

Synthetic Method 41

Synthetic Method 42

Synthetic Method 43

Synthetic Method 44

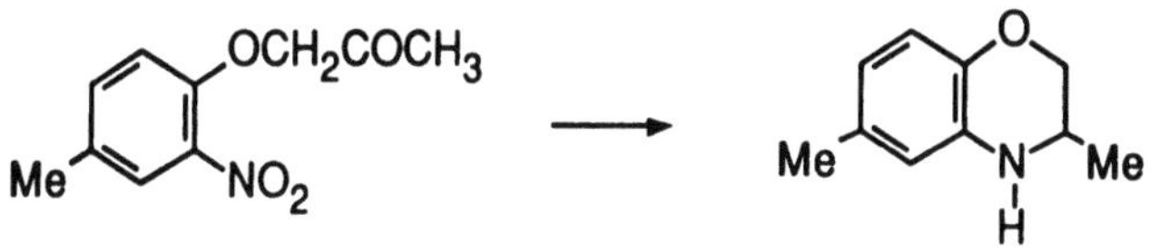

Synthetic Method 45

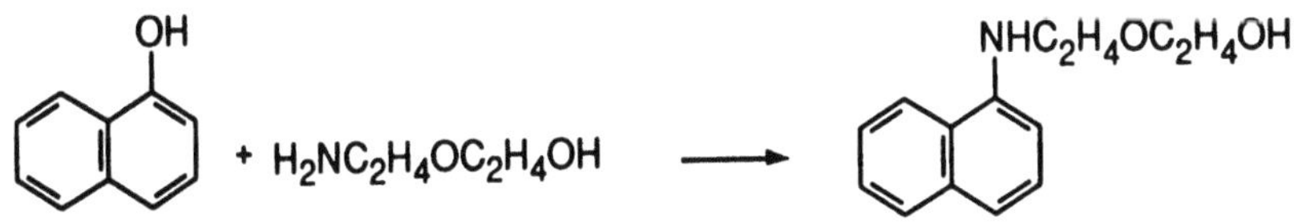

Synthetic Method 46

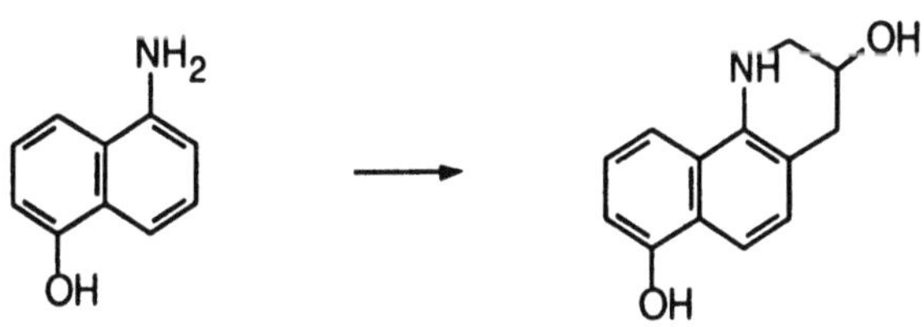

Synthetic Method 47

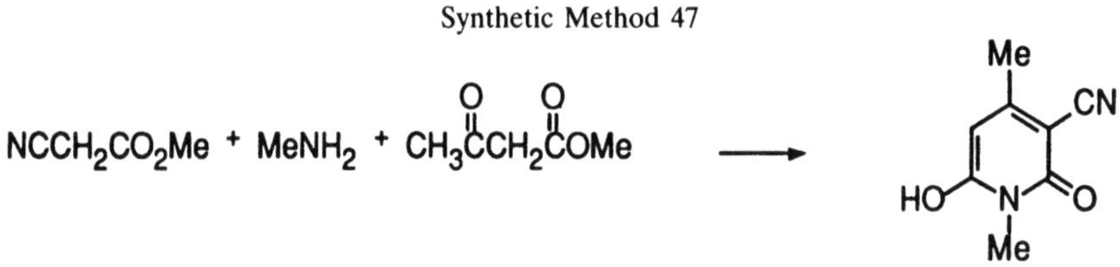

Synthetic Method 48

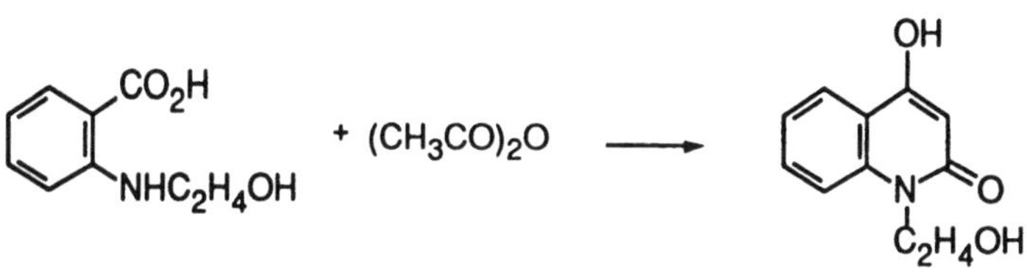

Synthetic Method 49

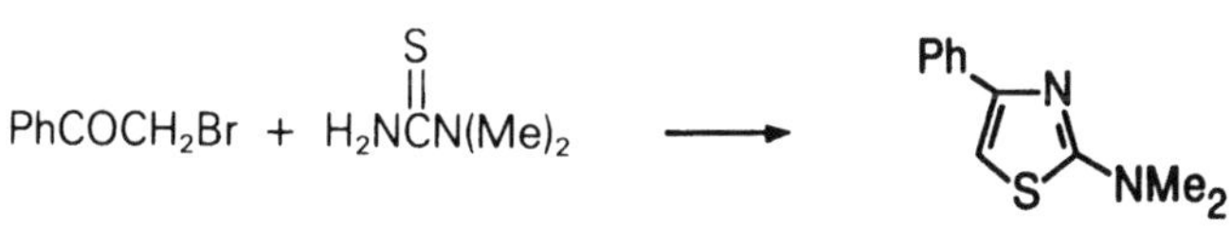

Synthetic Method 50

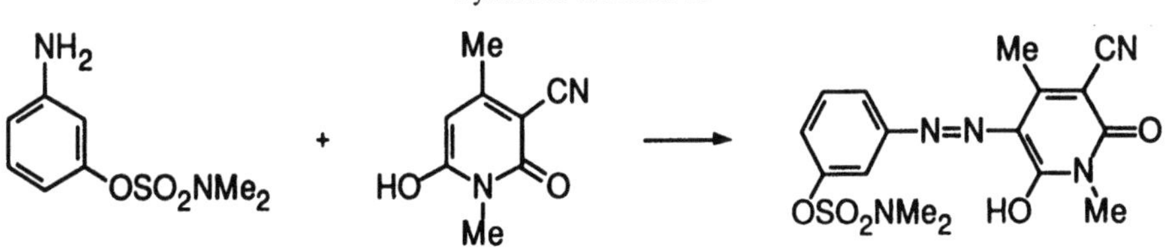

Synthetic Method 51

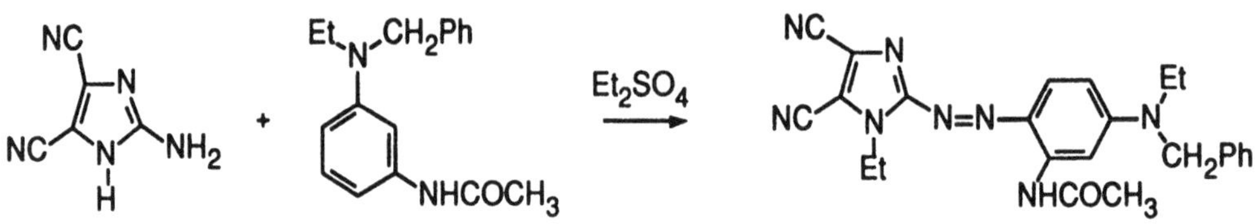

Synthetic Method 52

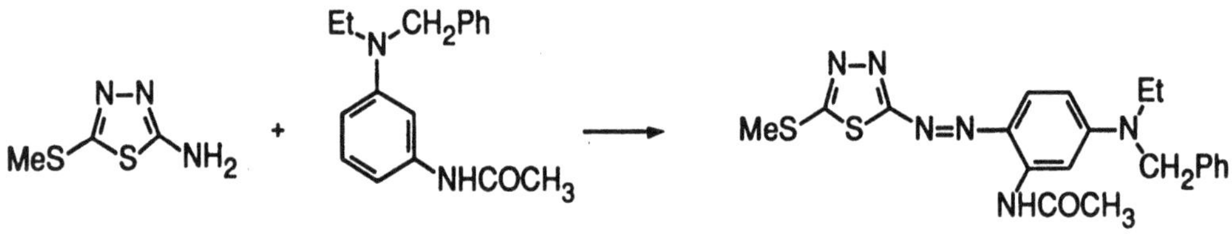

Synthetic Method 53

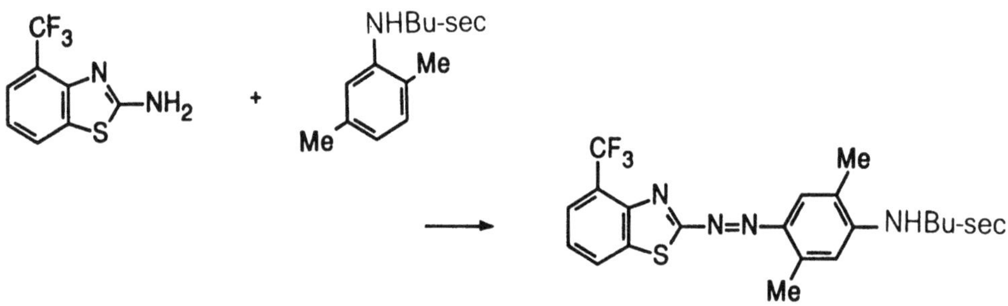

Synthetic Method 54

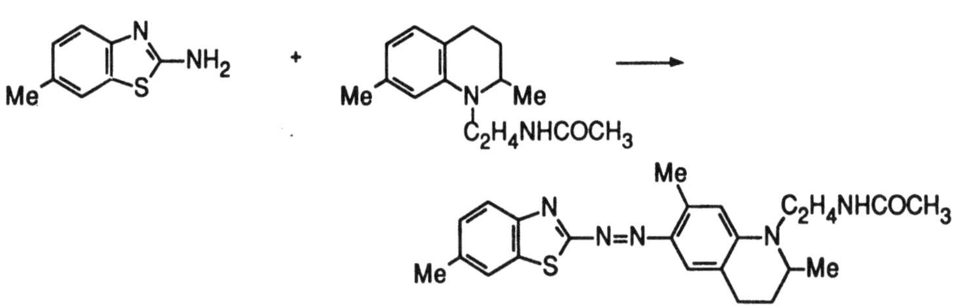

Synthetic Method 55

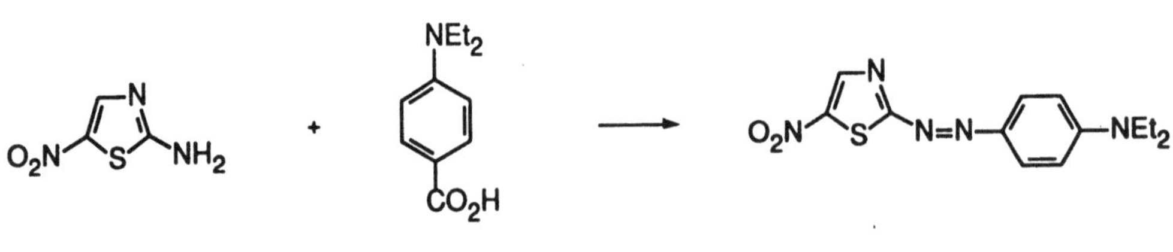

Synthetic Method 56

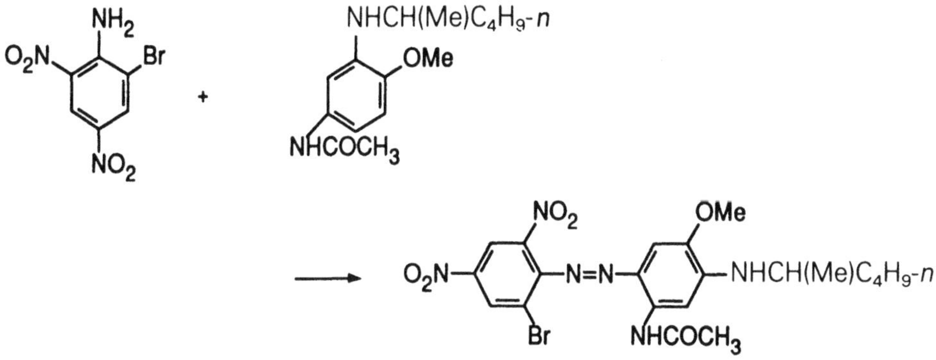

Synthetic Method 57

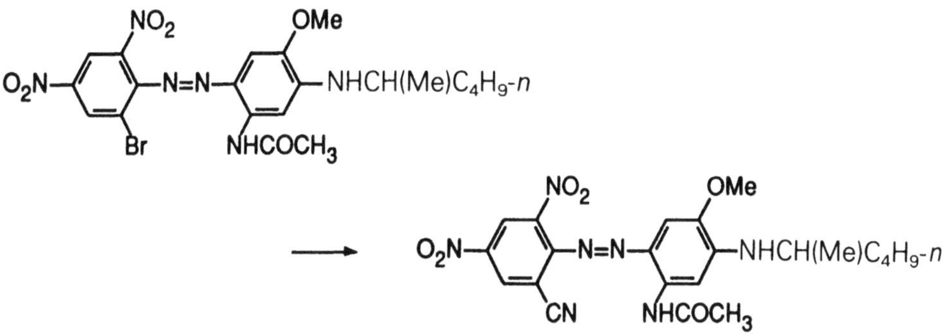

Synthetic Method 58

Cu_2Br_2 + NaCN ⟶ $Na[Cu(CN)_2]$ ⟶ Use in cyanide displacement

Synthetic Method 59

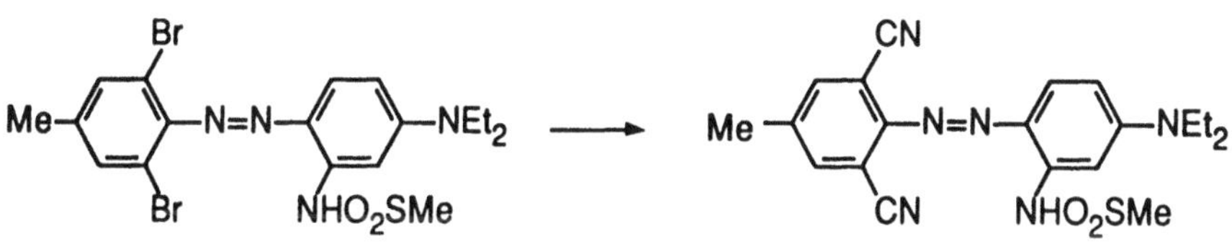

B. Discussion

The diazotization of aromatic amines by using nitrous acid and the coupling of the diazonium compound with active hydrogen species to produce colored compounds has long been accepted by organic chemists. The mechanism of the reaction has been studied in great detail, and by no means is it the purpose of this section to repeat or to extend this concept. Instead, the purpose of this section is to elaborate on how the dyestuff chemist has attempted to utilize this chemistry to produce polyester dyestuffs which meet industrial requirements.

Several methods are available to the dyestuff chemist to adjust the hue, performance (including both application and fastness properties), and economics. For example, structural variations are possible in the amine to be diazotized, and in the coupling component to be used. Also, chemical reactions are possible in the dye itself if it is advantageous to carry them out from the standpoint of synthesis or economics. All of these techniques have been used by the dyestuff chemist to devise colored azo compounds which give performance/economic features.

Blue dyes for synthetic fibers were almost exclusively the domain of anthraquinone derivatives until 2-amino-5-nitrothiazole was diazotized in the 1950s by J. B. Dickey and his co-workers. They found that the diazonium component could be made to yield brilliant blue azo dyes by using aniline and toluidine couplers. These dyes found application on acetate fibers, rather than on polyester, because they provided alternatives to the anthraquinone dyes when gas fume fastness was required. Poor fastness to light and poor hydrolytic stability precluded this class of dyes from any widespread application to polyester. Nevertheless, this class did point the way to widespread research on and subsequent use of diazonium components producing bathochromic dyes on polyester.

Prior to Dickey's discovery, the only blue azo dye of any note had been developed from a benzothiazole diazonium component which used a powerfully donating coupler to produce the dye CI Disperse Blue 15 (**5**). The dye had poor lightfastness on cellulose acetate and polyester fibers. Since that time, coupling components have been widely used to vary the hue and particularly the dyeing performance of polyester fiber dyes.

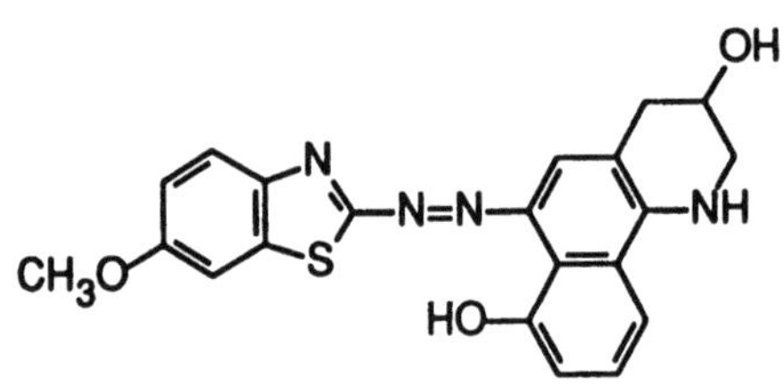

5

Bayer chemists discovered that halogen atoms, particularly bromine atoms, present in phenylazo derivatives are activated to nucleophilic substitution if they are *ortho* to the azo group and this led to the development of yet another way to induce bathochromic shifts in azo compounds. Nucleophilic replacement of *ortho* haloazo derivatives, which are often sterically hindered because of the halogen atom, with the linear and electron-withdrawing cyanide group led to profound bathochromic shifts and the ready development of blue dyes. These dyes had remarkably high extinction coefficients and excellent fastness properties on polyester fibers. This type of reaction on preformed dye has since been widely employed in the manufacture of commercial blue dyes for polyester fibers.

It is the aim of this section to provide details of some of the more interesting and useful developments available for the syntheses of blue azo dyes by nucleophilic substitution. The preparation of azo dyes from such a variety of diazonium and coupling components often requires special conditions, and some of the major commercial diazotization and coupling methods are described by using specific examples of each.

1. Diazonium Components

The earliest diazonium components used commercially were based on substituted aniline derivatives and often contained nitro substituents to yield a bathochromic shift and to improve lightfastness on polyester and cellulose acetate fibers. Many of these derivatives are used today, and some of the dyes derived from them are described later. Often the diazonium components are made by using well-known chemistry which will not be repeated here. The first commercial purpose of heterocyclic amines was the production of red dyes by using the aminobenzothiazole nucleus as a diazonium component. Earlier benzothiazoles were substituted with donor groups but, in order to achieve the necessary fastness requirement for polyester, particularly lightfastness, substitution with electron-withdrawing groups was more desirable.

Some of the earliest dyes containing electron-withdrawing groups were based on 2-amino-6-methylsulfonylbenzothiazole. The 6-position was usually selected for substitution because of the relative ease of synthesis of intermediates, although the 4-position was used in rare cases.

Two main routes are used for the synthesis of aminobenzothiazoles. In one route, an aniline derivative is reacted with thiocyanogen made *in situ* from a thiocyanate and a halogen, usually bromine. The aminobenzothiazole ring is produced directly. An example of this method in the preparation of 2-amino-5,6-dichlorobenzothiazole follows. This intermediate is used commercially for the preparation of bright scarlet dyes for polyester fibers.

Synthetic Method 26[27] (2-Amino-5,6-dichlorobenzothiazole–Thiocyanate Route) (Procedure quoted verbatim from U.S. Patent 3,502,645).

> "5,6-Dichloro-2-aminobenzothiazole may be prepared by adding 600 ml. glacial acetic acid to 81 gms. 3,4-dichloroaniline, adding 81 gms. of sodium thiocyanate thereto, heating to 40°C, holding at 40°–45°C. for 90 minutes, cooling to 10°C., adding dropwise 80 gms. bromine dissolved in 100 ml. glacial acetic acid during 1 hour while stirring at 10°–13°C., stirring 2 hours at 13°–15°C., collecting the precipitate by filtration, slurrying the precipitate in 2.4 liters water at 40°–50°C., adjusting to pH 9.5 by adding 230 gms. 25% aqueous NaOH gradually during 1 hour, cooling to 10°C, collecting the solids by filtration, washing the solids with 150 ml. H_2O at room temperature and drying the resulting 5,6-dichloro-2-aminobenzothiazole at 70°C."

A second route which is commonly used for the preparation of aminobenzothiazoles is to cyclize the corresponding thiourea by using a halogen, commonly bromine. This method is useful if it is desirable to keep the 6-position of the benzothiazole ring unsubstituted after cyclization. The thiocyanogen route described earlier tends to introduce the thiocyanate substitution in the 6-position of the benzothiazole ring if the 4-position on the aniline precursor is unsubstituted. It is particularly useful if a synthesis of a 2-amino-4-substituted benzothiazole is desired. The following examples describe the preparation of 2-amino-4-trifluoromethylbenzothiazole. This diazonium component has been used commercially to produce bright red dyes for heat transfer printing with high lightfastness on polyester and nylon materials. The use of the trifluoromethyl group gives the good thermal transfer required for the process, and the 4-substitution is preferred in this case because the economics of the precursors are more favorable than the other isomers.

Synthetic Method 27[28] (2-Amino-4-trifluoromethylbenzothiazole–Thiourea Route) (Procedure quoted verbatim from U.S. Patent 4,052,379).

"To a stirred solution of 7.9 g. (0.104 m.) of ammonium thiocyanate in 30 ml. of acetone is added at 30°C., 13.25 g. (0.094 m.) of benzoyl chloride. The mixture is stirred under reflux for 10 min, cooled to 50°C., and 15.1 g. (0.094 m.) of 2-aminobenzotrifluoride is added in one portion. The mixture is stirred under reflux for 20 min. A solution of 13.2 g. of sodium hydroxide in 140 ml. of water is added. The mixture is boiled for 15 min. and the clear solution is cooled to 20°C. The pH of the mixture is adjusted to 5 with 21 ml. of conc. HCl and then made just alkaline by the addition of 4 ml. of 88% ammonia. After 30 min. the mixture is cooled to 10°C., and the solid, which separates, is collected by filtration, washed with water and dried. The yield of *o*-trifluoromethylphenylthiourea is 18.8 g. (91%), mp 165°–166°C. To a stirred mixture of 31.5 g. of *o*-trifluoromethylphenylthiourea in 135 ml. of 1,2-dichloroethane is added below 30°C. a solution of 9 ml. of bromine in 65 ml. of 1,2-dichloroethane. The mixture is stirred under reflux for 3.7 hr., cooled to 10°C., and the solid collected by filtration, washed with 1,2-dichloroethane and dried. The solid is stirred into 400 ml. of water, basified with ammonia, filtered and dried. The yield of 2-amino-4-trifluoromethylbenzothiazole is 25.6 g. (82%)."

2-Amino-6-nitrobenzothiazole is an important intermediate in the preparation of CI Disperse Red 144, a major commercial neutral red dye used for polyester materials. One synthesis of this intermediate uses the thiourea method of cyclization previously described, followed by nitration directly in the cyclization medium to yield the 6-nitro derivative.

Synthetic Method 28[29] (2-Amino-6-nitrobenzothiazole) (Procedure quoted verbatim from U.S. Patent 4,363,913).

"A solution of phenylthiourea (152 g., 1.0 mol) in 300 ml. of 98% sulfuric acid at 5–10°C is treated with bromine (8 g.) over a period of 30 minutes. During the bromine addition the temperature is allowed to rise to 12°C., and the evolution of gas begins. The temperature is allowed to gradually rise at such a rate that gas evolution does not become too vigorous, with most of the gas evolution having been completed when the temperature reaches 25°C. The reaction mixture then is warmed to 35°C. and held at that temperature for 2 hours. When the reaction is complete (as determined by thin layer chromatography), nitrogen is passed through the reaction mixture with vigorous stirring for 2 hours. The solution is cooled to 10°C. and stirred. 70 G (1.1 mole) of 98% nitric acid is added over a 45 minute period. Nitrogen is bubbled through the solution prior to and during the nitric acid addition. The solution is allowed to warm to 15°C. and then is poured over 500 g of ice. The product is recovered by filtration and reslurried with 800 ml. water and the mixture made basic with 28% aqueous ammonia. The solid is filtered off, washed with water and dried to give 185 g. (95%) of 2-amino-6-nitrobenzothiazole (m.p. 240°–243°C)."

The first blue monoazo derivatives which achieved these bathochromic hues by virtue of the diazonium component rather than by using special couplers were derived from 2-amino-5-nitrothiazole. Monoazo dyes derived from this amine yielded strong blue shades on acetate fibers from simple aniline and toluidine couplers, and heralded the considerable research effort into heterocyclic azo compounds which followed this discovery. 2-Amino-5-nitrothiazole can be prepared by nitration of 2-aminothiazole as documented by Dickey and his co-workers.

Synthetic Method 29[30] (Nitration of 2-aminothiazole) (See NOTE Below Preparation) [Procedure quoted verbatim from *J. Org. Chem. 20,* 505 (1955)].

"A solution of 25 g. (0.25 mole) of powdered 2-aminothiazole (free from sodium chloride) in 75 ml. (1.41 mole) of 96% sulfuric acid, prepared at 10–15°, was stirred while 11.6 ml. (0.25 mole) of nitric acid (d. 1.49–1.50) was added during 45 minutes so that the temperature did not exceed 12°. The system was maintained at 0° during three hours and at 25° for 12 hours, then drowned in ice-water to a volume of 1.5–2 l. The insoluble material (crude nitramine, insignificant) was filtered off, and the filtrate was neutralized with 150 g. of powdered sodium carbonate. The product was filtered off, reslurried in ice water, refiltered, and dried at 100°, 30.6 g. (84%), m.p. 189–192°. Crystallization from 2 l. of boiling water left 25 g. (70%), m.p. 192–195°, not raised by crystallization from hot ethanol (8 ml. per g.). Crystallization from hot acetic acid (16 ml. per g.) raised the melting point to 195–196°". [NOTE: Hazards have been noted with the procedure.[31]]

Dyes from 2-amino-5-nitrothiazole have achieved significant commercial success for the dyeing of acetate fibers in which gas fastness is of a higher priority than lightfastness.

BASF chemists later found that azo dyes from aminobenzisothiazoles were significantly more bathochromic than dyes from the isomeric aminobenzothiazoles described earlier. In fact, the nitrobenzisothiazole derived dyes were found to be, in general, similar in hue to those from 2-amino-5-nitrothiazoles described earlier by Dickey, but with improved lightfastness on polyester fibers. These dyes have been manufactured and sold successfully for many years for the dyeing of both acetate and polyester.

Several routes to substituted 3-amino-2,1-benzisothiazoles have been reported in the literature.[32] These routes often involve the oxidative cyclization of an *o*-aminothioamide usually by using hydrogen peroxide. An interesting alternative involves the use of 100% sulfuric acid as the oxidant and is described as follows.

Synthetic Method 30[33] (3-Amino-5-nitro-2,1-benzisothiazole) (Procedure quoted verbatim from U.S. Patent 3,981,883).

"49.25 Parts of 5-nitro-2-aminobenzoic acid thioamide are introduced in portions, whilst stirring, into 450 parts of 100% strength sulfuric acid at room temperature. An exothermic reaction takes place and the temperature rises to 40°–50°C. After all the thioamide has been added, the temperature is raised slowly until the evolution of sulfur dioxide starts at from 60° to 70°C. The mixture is maintained at from 60° to 70°C for 2 hours and is then heated to 100°C for 1 hour. The solution is then cooled, poured onto 1,000 parts of ice-water and brought to pH 5–6 with aqueous sodium hydroxide solution. The precipitate is filtered off, washed with water and dried. 47.2 Parts (96% of theory) of 3-amino-5-nitro-2,1-benzisothiazole, having a decomposition point in excess of 250°C, are obtained."

Since the cyclization was completed in high yield in sulfuric acid alone, it was possible to use this solution directly without isolation for diazotization and coupling to produce dye.

Another important class of heterocyclic amines which has been used commercially to produce dyes for polyesters are the aminothiadiazoles, including both 2-amino-1,3,4-thiadiazole and 5-amino-1,2,4-thiadiazole. These systems, which incorporate another nitrogen atom in the hetero ring, produce a series of brilliant red shades on polyester fibers. Some of the dyes produced have been used to substitute for the anthraquinone red and pink dyes, which for so long have dominated this market area. Azo dyes derived from diazotized 2-amino-5-ethylthio-1,3-thiadiazole have been successful in replacing red and pink anthraquinone dyes. This intermediate is prepared by first preparing the aminothiadiazole thiol and then by alkylating the derivative. The following is a useful procedure for the preparation of this type of intermediate.

Synthetic Method 31[34] (2-Amino-5-mercapto-1,3,4-thiadiazole) (Procedure quoted verbatim from U.S. Patent 2,759,947).

"A mixture of 198 parts of concentrated hydrochloric acid and 50 parts of a 1/2% solution of hypophosphorous acid is diluted to a total of 800 parts by volume with water. This acidic solution is heated to a temperature of about 98°C. and 40 parts of *N,N'*-bis-(thiocarbamyl)-hydrazine is added. The mixture is heated between 100°C–104°C. (reflux) until the reaction is substantially complete. After cooling to room temperature, the crystals which form are removed by filtration, washed with water, and dried at about 90°C. A yield of 25.7 grams (72.5% of theoretical) of 2-amino-5-mercapto-1,3,4-thiadiazole is obtained. An ultra-violet analysis (at 310 millimicrons) showed the product to be 98.3% pure. The product melted between 229.5°C. and 229.7°C., uncorrected."

The alkylation of this intermediate can be achieved by a number of methods. The following procedure serves as an example.

Synthetic Method 32[35] (2-Amino-5-cyanomethyl-1,3,4-thiadiazole) (Procedure quoted verbatim from U.S. Patent 4,528,368).

"A mixture of 2-amino-5-mercapto-1,3,4-thiadiazole (66.5 g, 0.50 mol.) and ethanol (500 ml) is heated to reflux, and chloroacetonitrile (41.1 g, 0.55 mol.) is added dropwise at reflux. Heating is continued for four hours and then about 400 ml of distillate is removed using a Dean–Stark trap. The clear solution is cooled to about 75°C and water (150 ml) is added, allowing the temperature to drop to about 50°C. After adjusting the pH to about seven by adding slowly a solution of Na_2CO_3, the mixture is cooled to room temperature and the gray product is collected by filtration and washed with water. The moist filter cake is recrystallized from 300 ml of ethanol to yield the desired product which melts at 163°–164°C."

The isomeric 5-amino-1,2,4-thiadiazole tends to yield dyes which are a little more bathochromic than the 1,3,4-isomer but which still retain the brilliance of shade. The following procedure describes the synthesis of a typical diazonium component of this type.

Synthetic Method 33[36] (5-Amino-3-ethylthio-1,2,4-thiadiazole) (Procedure quoted verbatim from U.S. Patent 3,221,006).

> "30 Grams of sodium thiocyanate were dissolved in 250 ml. of methanol with good stirring. To this were added with stirring 54.5 grams of 2-ethyl-2-thiopseudourea hydrobromide (*Org. Syn. Coll.*, Vol. III, p. 440). After chilling to −14°C., this solution of sodium thiocyanate and 2-ethyl-2-thiopseudourea hydrobromide was treated simultaneously from two separate dropping funnels containing solutions of (1) bromine (48 grams) in 140 ml. of methanol and (2) sodium methylate in methanol, prepared from sodium (13.8 grams) in 150 ml. of methanol. The reaction mixture was stirred and cooled, and the additions were at such a rate as to keep the temperature at −5°C. and took 1 hour. About 1/7 of the sodium methylate solution was added first, after which the two solutions were added simultaneously. The reaction was stirred 1 hour more without cooling after the addition was complete. The reaction mixture was slightly alkaline and was carefully neutralized with concentrated HCl. It was then filtered from a small amount of precipitated NaBr. The methanol was removed *in vacuo,* and the resulting solid residue was triturated and washed twice with water and air dried. The crude product weighed 39.5 grams (81.7%) and melted at 74–92°C. Crystallization from 100 ml. of water yielded 30 grams of product, M.P. 90–5°C., softening point 60°C. Recrystallization from benzene-hexane yielded 26.2 grams of 2-amino-3-ethylthio-1,2,4-thiadiazole in the form of white needles melting at 92–6°C."

An interesting heterocycle is the aminoimidazole having two cyano substituents. This is one of the few nonsulfur-containing heterocyclic diazoniums which provides dyestuffs that have excellent performance on polyester. Dyes produced are bright red in shade with very good lightfastness on polyester. The following route is acceptable for a laboratory preparation of the amino derivative.

Synthetic Method 34[37] (2-Amino-4,5-dicyanoimidazole) (Procedure quoted verbatim from U.S. Patent 3,770,764).

> "To a mixture of 10 g. of diaminomaleonitrile, 20 ml. of cyanogen chloride and 50 ml. of acetonitrile was added 20 ml. of BF_3 etherate at 0°–10°C. A solution formed after 1.5 hours. The excess cyanogen chloride and part of the acetonitrile were removed under reduced pressure, and the solid hydrochloride of 2-amino-4,5-dicyanoimidazole (2.5 g.) was collected on a filter. On treatment with water, the hydrochloride released the parent 2-amino-4,5-dicyanoimidazole (0.78 g.)." [NOTE: "A small amount of 2-amino-4,5-dicyanoimidazole was treated with aqueous HCl then KNO_2. A precipitate of 2-diazo-4,5-dicyano-2*H*-imidazole formed. This substance exploded at 148°C."]

Aminothiophene derivatives have been widely studied as diazonium components for polyester dyes. Dickey in his classic work in the 1950s on heterocyclic diazonium derivatives observed that 2-amino-3-nitro-5-acetylthiophene readily produced blue monoazo dyes when diazotized and coupled to simple aniline and toluidine couplers. The dyes were known to have excellent performance on both acetate and polyester materials but were never commercialized because an economic route to the heterocyclic amine could not be found.

Later, in the 1960s, Karl Gewald discovered a new route to aminothiophene derivatives with both a high and varied degree of substitution.[38] This discovery spurred much research effort, as evidenced by the profusion of patent literature, but still little was commercialized from the apparently fruitful area. It remained for ICI chemists to commercialize the first polyester dye based on Gewald's synthetic procedures. They were able to prepare 2-amino-3,5-dinitrothiophene by using modified Gewald chemistry and to commercialize the first dye based on the thiophene azo chemistry. The route used involved the reaction of mercaptoacetaldehyde dimer with cyanoacetic acid to produce an aminothiophene having a 3-carboxy group. Nitration of this intermediate, after protection of the amino group, led to decarboxylation and formation of the 3,5-dinitroaminothiophene (equation 4).[39]

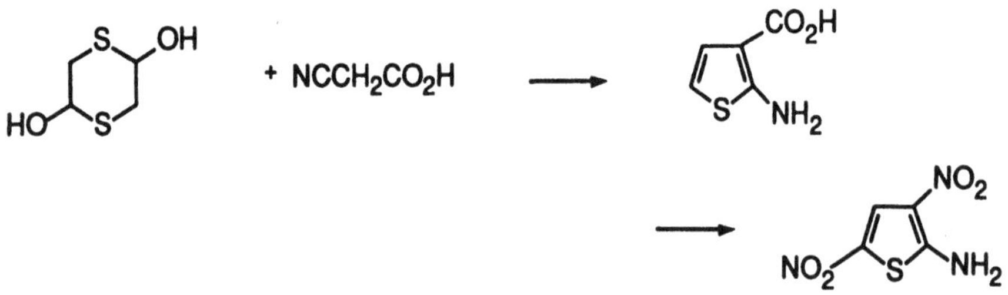

(4)

Recent work on the preparation of 2-amino-3-nitro-5-acetylthiophene by using modifications to Dickey's original procedure appears to have overcome the synthetic and economic difficulty experienced earlier with this route. Dyes from the 5-acetyl analog instead of from the 5-nitro analog, as described previously, are less bathochromic but have significantly better lightfastness.

Synthetic Method 35[40] (2-Amino-3-nitro-5-acetylthiophene) (Procedure quoted verbatim from U.K. Patents 2,092,128A and 2,078,713A).

"2-Acetylthiophene (25.2 g) (0.2 mole) and hydroxylamine sulfate (16.8 g) (0.102 mole) were stirred together at room temperature in acetic acid (80 g). A solution of sodium hydroxide (20 g NaOH + 40 ml H_2O) was added dropwise, allowing the temperature to rise gradually from the heat of reaction. After addition was complete, the mixture was refluxed for 3 hours and then cooled to 30°C.

"A solution of bromine (35.2 g) (0.22 mole) in acetic acid (10 g) was added dropwise at 30–40°C. After the addition, the mixture was stirred further at 50°C. for 15 minutes to complete the reaction and then cooled back to about 30°C. The mixture was then drowned into cold water, filtered, and the solid product washed with water and dried. The yield of the brominated product was 35.6 g (80.9%).

" 2-Bromo-5-acetylthiophene oxime (35.2 g) (0.16 mole) was dissolved in concentrated sulfuric acid (80 ml) at 5–10°C. A mixture of fuming nitric acid (95–98%) (10.6 g, 0.16 mole) in concentrated sulfuric acid (10 ml) was added dropwise at less than 10°C. The temperature, after the addition, was allowed to rise to room temperature and the system stirred there for 60 minutes. Paraformaldehyde (2.0 g) was then added carefully and the mixture stirred at room temperature for 15 minutes. This was followed by the addition of a further amount (15.0 g) of paraformaldehyde at about 20–26°C. Stirring was continued at this temperature for 2–3 hours further, and then the mixture was drowned into water, stirred for 30 minutes and filtered. The solid 2-bromo-3-nitro-5-acetylthiophene was washed with water and dried. The yield was 35.3 (88.25%).

"2-Bromo-3-nitro-5-acetylthiophene (1.0 g) was added to a solution of hexamethylenetetramine (1.0 g) in water (50 ml). The reaction mixture was refluxed for 1 hour, cooled and filtered. The product 2-amino-3-nitro-5-acetylthiophene was washed with a little water and dried in air. The yield was 0.65 g (83%) mp. 220–222°C."

2. Coupling Components

The principal type of coupling component used in the production of azo dyes for polyester is the aromatic amine coupler, although there are some notable exceptions which will be discussed later. Substitutions on the amine nitrogen can be used to vary application properties and/or fastness properties. For example, alkylated amines tend to result in more bathochromic dyes than do the dyes derived from unalkylated amines, as may be expected in terms of the nitrogen basicity. Secondary substitution on the alkyl group tends to have a less dramatic effect on hue and is used more often to vary application properties and fastness of the resultant dye.

Ring substitution of the aniline coupler with donor groups *ortho* to azo tends to result in a more bathochromic shift. Ring substitution in the *meta* position to the azo group (*ortho* to the amine group) often results in a hypsochromic shift, particularly if the amine is dialkylated as a result of steric interactions leading to a loss of planarity in the charge-separated species. Indeed, coupling of *ortho*-substituted dialkyl aniline derivatives can often be sluggish and may lead to formation of impurities that sometimes result in partial dealkylation of the amine.

The introduction of the acetamido group into an aniline coupler, *meta* to the amine groups and *ortho* to the subsequent azo group, is of major importance in the consideration of modern azo dyes for polyester fibers. Somewhat unexpectedly from a theoretical standpoint, this group increases the extinction coefficient, produces a bathochromic shift, and increases the lightfastness, sometimes dramatically. These are very desirable features both from a standpoint of technical performance and from consideration of the economics of a polyester dye. The widespread use of this substitution pattern and variation of it in dyes intended for polyester is an indication of its general applicability.

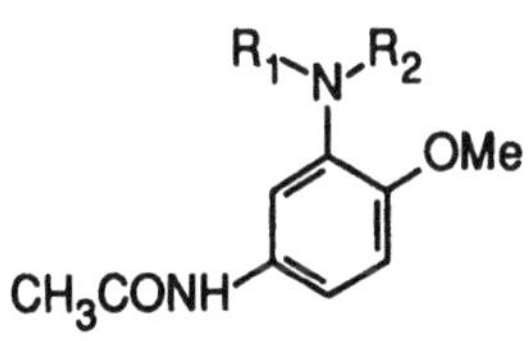

6

The incorporation of a methoxy group *para* to the acetamido function and *ortho* to the amine groups extends the hue effects of the acetamido function and leads to couplers **(6)** widely used in the production of blue, particularly navy-blue, azo dyes. The intermediates can be made by using standard chemistry and are available commercially from a number of suppliers. The most widely used derivative is the coupler where R1 and R2 are —CH_2CH_2OAc **(6)**; but if R1 is hydrogen and R_2 is alkyl, then navy-blue dyes produced from 2,4-dinitro-6-bromoaniline have similar hues but have increased extinction coefficients and, hence, better potential economics. The following preparation describes couplers of this type.

Synthetic Method 36[41] (Reductive Alkylation of 2-Nitro-4-acetamidoanisole) (Procedure quoted verbatim from U.S. Patent 4,210,586).

"A mixture of 105.0 g. (0.5 mole) of 2-nitro-4-acetamidoanisole, 67.0 g. (0.5 mole) of ethylacetoacetate, 550 ml of isopropyl alcohol, 10.0 g. of 5% Pt/C and 3.0 g. of *p*-toluenesulfonic acid is treated in an autoclave at 165°C. and 1,000 psi of hydrogen until uptake of hydrogen ceases. The solvent and catalysts are removed. Upon standing, 143.8 g. (98%) of ethyl-3-(2′-methoxy-5′-acetamidoanilino)butyrate is obtained."

Alkylated aniline derivatives where the alkyl group is cyclized onto the *ortho* position of the phenyl ring to create bicyclic or tricyclic structures, such as julolidine **(7)** and tetrahydroquinoline **(8)**, react with diazonium components to give dyes which are considerably more bathochromic than the normal aniline.

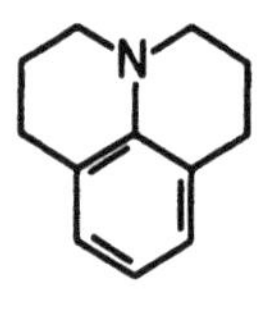

7

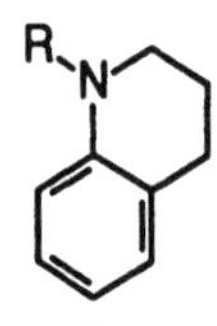

8

Lightfastness of dyes derived from julolidine is almost invariably poor, and so the coupler is of little use in preparing bathochromically shifted dyes for polyester. However, lightfastness of tetrahydroquinoline derived dyes is much better, and the intermediate has found utility for the synthesis of polyester dyes. The intermediate is usually prepared from an aniline or toluidine derivative by first converting the aniline or toluidine derivative to the dihydro compound,[42] followed by catalytic reduction (equation 5).

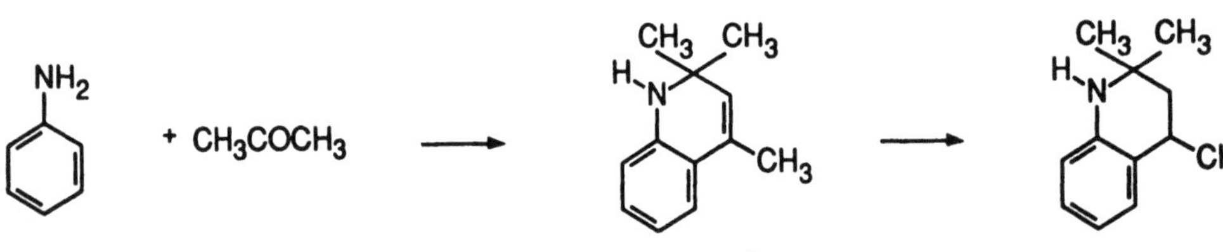

(5)

Another synthetic procedure which has been used to prepare the highly bathochromic 8-methoxy derivative is described below.

Synthetic Method 37[43] (Reduction of 2,5-Dimethyl-8-methoxyquinoline) (Procedure quoted verbatim from U.S. Patent 4,400,318).

"A mixture of the above 2,5-dimethyl-8-methoxyquinoline (260 g, 1.3 m), isopropanol (600 ml), and Raney nickel catalyst (25.0 g) was subjected to 1000 psi hydrogen for

3 hrs in a stainless steel autoclave. The reaction mixture from the autoclave was filtered to remove the catalyst. Essentially a quantitative yield of product was obtained, after removal of the isopropanol under vacuum, and the product was used without further purification."

Alkylation of nitrogen is often used to increase lightfastness and to achieve a further bathochromic shift. The following example gives a procedure for the ethylation of a tetrahydroquinoline which is suitable for laboratory use and which can be modified for alkylation of other amine couplers.

Synthetic Method 38[43] (2,7-Dimethyl-*N*-ethyltetrahydroquinoline) (Procedure quoted verbatim from U.S. Patent 4,400,318).

"A mixture of 2,7-dimethyl-1,2,3,4-tetrahydroquinoline (48.3 g, 0.3 m), potassium carbonate (41.4 g, 0.3 m), and iodoethane (100 g) was heated at reflux for 4 hrs. Water (200 ml) was added and then the organic layer was extracted using chloroform (100 ml). The organic layer was separated and dried over sodium sulfate. Chloroform and excess iodoethane were distilled off to leave essentially pure product (44 g), which was used without further purification."

A combination of the tetrahydroquinoline coupler together with the acetamido function has been used commercially to produce lightfast bathochromic blue dyes when combined with appropriate diazonium components. The 7-acetamidotetrahydroquinoline may be prepared by using the following procedure.

Synthetic Method 39[44] (7-Acetamido-1-ethyl-2,2,4-trimethyl-1,2,3,4-tetrahydroquinoline) (Procedure quoted verbatim from U.S. Patent 3,635,941).

"1-Ethyl-2,2,4-trimethyl-1,2,3,4-tetrahydroquinoline (96.5 g.) is added slowly to 500 ml. of concentrated H_2SO_4 at about 5°C. Then a solution of 33 ml. concentrated HNO_3 and 33 ml. concentrated H_2SO_4 is added dropwise at 0–5°C. The reaction is stirred 0.5 hr longer after the addition is completed and then drowned in ice–water mixture. After the mixture is made basic with concentrated ammonium hydroxide, the product is taken up in hexane and toluene and washed with water. The solvent is evaporated to leave the 7-nitro compound as a semi-solid mass. The nitro compound (105 g.) is dissolved in 1250 ml. of 2B alcohol and hydrogenated in the presence of Raney nickel at 75°C. and about 1500 p.s.i. hydrogen pressure. The catalyst is filtered off and the filtrate evaporated to yield 7-amino-1-ethyl-2,2,4-trimethyl-1,2,3,4-tetrahydroquinoline in the form of a black viscous oil. The 7-amino compound (21.8 g.) is dissolved in 30 ml. of acetic acid. Acetic anhydride (10 ml.) is added and the reaction is allowed to stand for 1 hr. and then drowned in water. The product, 7-acetamido-1-ethyl-2,2,3-trimethyl-1,2,3,4-tetrahydroquinoline, solidifies on standing, is collected by filtration and is recrystallized from water–methanol solution; M.P.: 123–127°C."

Side-chain substitution on the nitrogen atom was used principally to affect dyeing properties and/or fastness properties, rather than to effect hue shifts. The following examples describe a number of methods used for the introduction of various side-chain substituents on couplers.

Synthetic Method 40[45] (*N*-Phenylthiomorpholine Dioxide) (Procedure quoted verbatim from U.S. Patent 3,383,379).

"46.5 g (0.5 m) aniline and 59.0 g. (0.50 m) divinylsulfone were stirred and heated on the steam bath for three hours. The mixture was cooled and 750 cc. 6% HCl was added. Steam was blown through until the mixture boiled and solution was almost complete. The above product crystallized out on cooling. The white solid was filtered off, washed with water and dried. M.P. 118–120°C."

Synthetic Method 41[46] (*N*-Ethyl-*N*-*β*-succinimidoethyl-*m*-toluidine) (Procedure quoted verbatim from U.S. Patent 3,493,556).

"An amount of 89 g. (0.5 M) of *N*-*β*-aminoethyl-*N*-ethyl-*m*-toluidine and 50 g. (0.5 M) of succinic acid anhydride is mixed intimately and gradually heated until an exothermic reaction begins. When the temperature begins to fall, heat is applied and the temperature is held at 130–140°C for one hour. The mixture is then drowned in one liter of water and allowed to cool. The yield is 99 g. of product melting at 81.5–82.5°C."

Synthetic Method 42[43] (Hydroxyethylation of a Tetrahydroquinoline) (Procedure quoted verbatim from U.S. Patent 4,400,318).

"2,5-Dimethyl-8-methoxy-1,2,3,4-tetrahydroquinoline (189 g., 1.0 m) was reacted with ethylene oxide (52.0 g.) in an autoclave at 160°C. for 10 hrs. The reaction mixture from the autoclave was distilled under vacuum to yield 169 g. (72%) of product which boiled at 150–138°C. at 1.7/0.5 mm Hg. A small amount of material was slurried in hexane, and after filtering and drying, the product melted at 65–67°C."

Synthetic Method 43[47] (3-Acetamido-*N,N*-dibenzylaniline) (Procedure quoted verbatim from U.S. Patent 3,765,830).

"A mixture of 15.0 g. 3′-aminoacetanilide, 38.0 g. benzyl chloride and 25 ml. *N,N*-dimethylformamide is heated and stirred for 2 hr. at 95°–105°C. Then 10.1 g. triethylamine is added, and the reaction is heated for another hour at 110–125°C. After drowning the reaction in water, the product 3-acetamido-*N,N*-dibenzylaniline is collected by filtration, washed with water and air dried; m.p. 140–142°C."

The aforementioned examples represent only a few of the many methods used to modify fastness properties by variation of side-chain substituents on the coupler. These methods, with the exception of the hydroxyethylation technique, are used to improve sublimation fastness. Hydroxyalkyl substitution, as compared to alkyl substitution, frequently induces a deterioration in lightfastness on polyester and often the hydroxy group is protected by acetylation in polyester dyes to achieve adequate lightfastness. However, the hydroxyalkyl substituent is effective in improving dyeing properties for acetate fiber dyes.

Couplers containing polyether-type substituents have been found to aid the dyeing properties of polyester and acetate without significant deterioration in lightfastness.

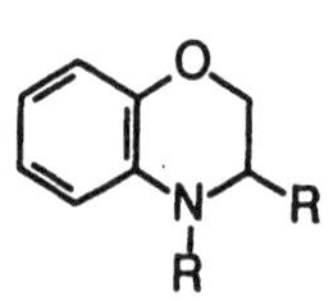

9

Another useful technique to produce more bathochromic dyes by variation of coupler moiety involves the benzomorpholine system **(9)**. As with tetrahydroquinoline, cyclization of the *N*-alkyl substituent onto the aniline ring in the *ortho* position, in this case through oxygen, leads to a bathochromic shift. Benzomorpholines, however, have not found widespread commercial utility, principally because of the lack of a good economic commercial procedure for their synthesis. A preparation suitable for laboratory use of the ring system follows.

Synthetic Method 44[48] (3,6-Dimethylbenzomorpholine) (Procedure quoted verbatim from U.S. Patent 3,453,270).

> "An amount of 209 g. of 2-nitro-4-methylphenoxyacetone in 220 cc. absolute ethanol was reduced over Raney Ni at 1700–1800 p.s.i. hydrogen pressure with temperature raised over 2 hours to 167°C. The solution was cooled, filtered, and distilled to yield 104 g. of the benzomorpholine; B.P. 95–98°C/2 mm."

The derivative may be alkylated, to give useful couplers, by using some of the methods already described.

Couplers which yield even more bathochromic dyes can be found in the naphthylamine derivatives. The fusion of a second phenyl ring to the derivative leads to an extension of the conjugate system and to a profound bathochromic shift from the corresponding aniline dyes. However, the system is essentially restricted to monoalkylated 1-naphthylamines because dialkylation leads to steric effects which are evidenced by slow coupling and by impurities as described earlier in relation to *ortho*-substituted aniline couplers. Because of their good economics and bathochromic hue, naphthylamine derivatives have found utility in the preparation of navy-blue dyes and black components. Lightfastness of the series cannot be rated as particularly high on polyester. Alkylation of 1-naphthylamine or reductive alkylation of 1-nitronaphthalene can be used to prepare these derivatives. Still another route is given by the following example.

Synthetic Method 45[49] (*N*-hydroxyethoxyethyl-1-naphthylamine) (Procedure quoted verbatim from U.S. Patent 4,324,721).

> "A mixture of 144 parts of 1-naphthol, 100 parts of water, 210 parts of diglycolamine ($H_2NC_2H_4OC_2H_4OH$) and 12.8 parts of sulfur dioxide was heated at 135 ± 5°C for 20 hours. Aqueous sodium hydroxide was added in an amount sufficient to form the water soluble sodium salt of any unreacted 1-naphthol and the resulting mixture heated at 50° to 60°C for 1 hour. The crude product was extracted with benzene and the benzene extract water washed. Upon removal of the benzene, 190.5 parts (77.5% yield) of *N*-(β-hydroxyethoxyethyl)-1-naphthylamine were obtained which had purity of 93.8%."

One of the earliest blue azo disperse dyes, CI Disperse Blue 15 **(5)**, used the particularly bathochromic tetrahydronaphthoquinoline system. It is formed by cyclization of an *N*-alkylated naphthylamine in a manner similar in principle to conventional tetrahydroquinolines. This type of coupler has not yet found any application for polyester dyes.

Synthetic Method 46[50] (3,7-Dihydroxytetrahydronaphthoquinoline) (Procedure quoted verbatim from British Patent 470,640).

"A mixture of 160 parts of 1-amino-5-oxynaphthalene, 200 parts of n-butyl alcohol and 110 parts of epichlorohydrin is heated in an atmosphere of carbon dioxide, while slowly stirring to 75°C., and is kept at this temperature for 1 1/2 hours. The mass is then brought to the boil and boiled for 9 hours. The feebly coloured crystalline magma thus obtained is cooled in an atmosphere of carbon dioxide, stirred with 250 parts of a mixture of alcohol and acetone in the proportions of 1:1 and quickly filtered. The solid matter is washed with a small quantity of the same mixture and finally with ether and dried. 130 parts of a nearly colourless hydrochloride of *py*-3-oxytetrahydro-7-oxy-α-naphthoquinoline are obtained; this amounts to 51% of the theory.

"The base obtained from the hydrochloride and recrystallized from dilute alcohol melts at 186–187°C."

Heterocyclic compounds have been utilized as coupling components in preparing azo dyes for polyester, in addition to diazonium components, but, according to the volume of patent literature, less emphasis seems to have been expended toward their use as couplers. However, one series of heterocyclic couplers which have achieved spectacular success consists of the so-called pyridone couplers. Azo dyes from these couplers on polyester produce bright yellow dyeings with particularly high lightfastness, excellent dyeing, and other fastness properties. One important feature of these dyes is their particularly high extinction coefficient which, together with the ready availability of raw materials for the coupler and the dye's ease of synthesis, makes for an economic combination of properties. A suitable procedure for the synthesis of pyridone couplers follows.

Synthetic Method 47[51] (1,4-Dimethyl-3-cyano-6-hydroxy-2-pyridone) (Procedure quoted verbatim from U.S. Patent 4,284,782).

"99 Parts of cyanoacetic acid methyl ester are added at 5°–10°C., while stirring to 162 parts of a 44.1% strength aqueous solution of monomethylamine. Stirring is continued for one hour. 133.5 parts of acetoacetic acid methyl ester are run in, and the mixture is heated at 90°C. for 4 hours in an autoclave, a maximum pressure of 1.3 bars being generated. 296 Parts of a 27% strength sodium hydroxide solution are then added to the reaction solution, and 33 parts of monomethylamine (46.2% of theory, relative to material employed) are distilled at 85°C into a receiver charged with approx. 42 parts of ice water, so that an approx. 44% strength aqueous solution of monomethylamine, suitable for further reactions, is formed. The 1,4-dimethyl-3-cyano-6-hydroxypyrid-2-one is then precipitated by being run into a mixture of 100 parts of ice and 217 parts of a 61% strength sulfuric acid. The product is filtered off, washed with water and dried. Yield: 155.8 parts (95% of theory) with a melting point of 285°C."

Variations of this synthesis are possible which incorporate other substituents in the pyridone ring. A similar type of coupler with a fused benzene ring, a quinolone derivative, has also been used commercially to produce yellow dyes. This derivative can be synthesized as follows.

Synthetic Method 48[52] (*N*-Hydroxyethyl-4-hydroxy-2-quinolone) (Procedure quoted verbatim from U.S. Patent 2,529,924).

> "36.2 Grams of *N*-β-hydroxyethylanthranilic acid are added to 60 grams of acetic anhydride and the resulting solution is heated under reflux with stirring at about 130°C. for 30 minutes. The reaction mixture is then poured into 300 cc. of 10% sodium hydroxide solution and agitated to effect complete solution. The reaction product is precipitated by adding hydrochloric acid until the cold mixture is acid to congo red. The *N*-β-hydroxyethyl-4-hydroxy-quinolone-2 is recovered in good yield by filtration, washed with cold water and dried."

Other heterocyclic couplers have been evaluated for producing polyester dyes. Heterocyclic amines which are alkylated have been coupled in an analogous manner to substituted anilines. Dyes having good performance have been produced in the laboratory from these types, but to date they have had very limited commercial success. The following preparation describes the preparation of *N,N*-dialkylated-2-aminothiazole, which has been coupled to produce blue polyester dyes that have good dyeing and fastness properties.

Synthetic Method 49[53] (2-(*N,N*-Dimethylamino)-4-phenylthiazole) (Procedure quoted verbatim from U.S. Patent 3,770,719).

> "A solution of 15.6 g. of *N,N*-dimethylthiourea in 100 ml. of absolute ethanol is treated cautiously with 29.8 g. of phenacyl bromide by portionwise addition of the solid. After the mild exothermic reaction has subsided, the reaction is completed by heating for 2 hrs. at reflux. Upon cooling, the solid is filtered off to give a 98% yield of 2-(*N,N*-dimethylamino)-4-phenylthiazole hydrobromide."

3. Dyes

Azo dyes for polyester are normally of the monoazo type, although a number of disazo compounds have reached commercial status. Normally, the azo dye is required to be small and compact in order to achieve an acceptable diffusion rate into polyester. Hence, multiple azo derivatives, as may be observed in many direct dyes for cellulose, are not common. Rather, hue shifts are obtained from the use of highly substituted diazonium components to achieve charge separation rather than extended conjugation.

The use of highly negatively substituted diazonium components and the use of heterocyclic diazonium components mean that special conditions are often necessary to achieve satisfactory diazotization. The highly negatively substituted amines are often weakly basic; hence, the commonly described method of diazotization by using dilute acid and sodium nitrite is unsuitable except in the case of very simple amines. Often, the highly substituted and heterocyclic diazonium components formed are very active and couple rapidly but, sometimes if coupling is insufficiently rapid, special coupling conditions may be necessary to speed up coupling and/or reduce the tendency of the diazonium component to decompose.

This section is concerned with polyester dye formation from diazotizing and coupling. Also, polyester dye modification is discussed in terms of preformed dyes reacting primarily by cyanide displacement, which has achieved considerable commercial success in the preparation of primarily blue phenyl azo dyes for polyester. Specific working examples from the patent literature are described which exemplify some of the methods used.

Often, highly negatively substituted aniline diazonium components and many heterocyclic diazonium components are unstable in the aqueous acidic conditions used for the diazotization of simple aniline derivatives. Hence, anhydrous conditions are necessary for the diazotization of these amines. The use of nitrosyl sulfuric acid in an organic acid has been found to be effective for the diazotization of many of these amines. Nitrosyl sulfuric acid may be purchased as a 40% solution in sulfuric acid. Alternatively, it may be purchased in solid form from chemical suppliers, or generated *in situ* from sodium nitrite and concentrated sulfuric acid. The organic acid normally used for this type of diazotization is acetic acid but, because this material solidifies at a temperature normally above that for diazotization, the addition of propionic acid is necessary to depress the freezing point of acetic acid to an acceptable level. Normally, a mixture of 1 part propionic acid to 5 parts acetic acid is used and is commonly referred to in the literature as "1:5 acid." Other ratios may be more desirable for certain applications. This method can be used almost universally for laboratory diazotizations of both basic and weakly basic amines, but commercial utility is limited by the disposal problems associated with large quantities of organic acids.

For various reasons, such as lack of solubility of sulfate or other salts, certain amines do not effectively react under the above conditions and special conditions are more suitable in these cases. For instance, the diazonium component from 2,4-dinitro-6-cyanoaniline appears particularly difficult to form and to couple cleanly; therefore, dyes of this and related types are more effectively produced by first preparing the dinitrobromoazo dye and then displacing the bromine atom on the preformed dye with cyanide. This procedure is discussed later.

Table 1 includes some of the principal types of amines used in the preparation of azo dyes for polyesters and gives some of the best methods found to diazotize them. The list is not intended to indicate the only methods which can be used, but rather to assist in indicating the type of diazotization conditions which may be expected to give good results with a particular type of amine.

The following examples, mostly of commercial polyester dyes, serve to illustrate some of the different diazotization and coupling techniques available. Simple aniline derivatives are usually best prepared by using dilute acid for diazotization. The following example describes such a diazotization in which the diazonium component is coupled to a pyridone-type coupler described previously. This type of dye is used commercially as a bright lemon yellow dye of high extinction coefficient. The dye has excellent fastness to light and is found to be superior to the methine dye class because of its excellent pH stability. This type dye is particularly valuable in the preparation of bright green shades for polyester when combined with a blue dye with the correct shade and fastness.

Table 1. Conditions for Amine Diazotization

Diazotizable amine	Preferred conditions
Monosubstituted Anilines	1:5 Acid or dilute HCl
p-Nitroaniline	Dilute H_2SO_4
p-Chloroaniline	60% Acetic acid
Disubstituted Anilines	1:5 Acid or dilute HCl
2,4-Dichloroaniline	60% Acetic acid
2,5-Dichloroaniline	60% Acetic acid
Trisubstituted Anilines	1:5 Acid
2,6-Dichloro (or bromo)	
-4-nitroaniline	Conc sulfuric acid
2,4,6-Trinitroaniline	Conc sulfuric acid
2-Cyano-4,6-dinitroaniline	70% Sulfuric acid
2 Bromo-6-cyano-4-nitroaniline	70% Sulfuric acid
Benzothiazoles	50% H_2SO_4 or 60% HOAc
2-Amino-6-nitrobenzothiazole	85% Phosphoric acid
Thiazoles and Thiadiazoles	1:5 Acid
2-Amino-5-nitrothiazole	Conc H_2SO_4
Thiophenes	1:5 Acid

Synthetic Method 50[54] (Aniline Diazo Component + Pyridone Coupler) (Procedure quoted verbatim from U.S. Patent 3,957,749).

"21.6 Parts 3-amino-phenyl *N,N*-dimethylsulfamate are dissolved in 300 parts of water with the addition of 36 parts of 30% hydrochloric acid and diazotized at 0° to +5°C with a solution consisting of 7,7 parts sodium nitrite dissolved in 50 parts of water. The diazo solution is allowed to run into a solution consisting of 18,0 parts 1-methyl-3-cyano-4-methyl-6-hydroxy-2-pyridone dissolved in 400 parts of water, 18,2 parts of sodium hydroxide solution (33° Bé) and 36.0 parts of sodium acetate; during the coupling process the reaction temperature is kept at 0 to +5° by the addition of 500 parts of ice. The coupling process being terminated, the dyestuff so formed is sucked off, washed with water until neutral and finally dried. The dyestuff constitutes a yellow powder dissolving in concentrated sulfuric acid with a reddish yellow color. It has a melting point of 196° to 197°."

Neutralization of the coupling solution in the above example is necessary to speed the coupling rate to an acceptable level, because of the relatively weak nature of the diazonium component. Anilines with more highly negative substituents and the heterocyclic amines are not likely to require buffering to achieve a satisfactory coupling rate. However, the weakly basic nature of these amines normally requires special diazotization conditions. An exception to this generality lies in the diazotization of 2-amino-4,5-dicyanoimidazole described earlier.

Synthetic Method 51[55] (Dicyanimidazole Diazo Component + Acetamidoaniline Coupler) (Procedure quoted verbatim from U.S. Patent 4,097,475).

"2-Amino-4,5-dicyanoimidazole (6.65 parts, 0.050 mole) was suspended in 165 parts of water and 15.9 parts of concentrated hydrochloric acid (0.135 mole) at 0°–5°C. 5N Sodium nitrite solution (10.66 parts, 0.052 mole) was added dropwise at 0°–5°C., and a positive nitrite test was maintained for 30 minutes. Excess nitrite was destroyed with sulfamic acid. This suspension of diazonium zwitterion was used directly in the coupling step.

"A solution of 14.8 parts (0.055 mole) of 3-(*N*-benzyl-*N*-ethylamino)acetanilide in 40 parts of methanol and 40 parts of acetic acid was added dropwise over 15 minutes at 0°–5°C to the diazonium zwitterion suspension. The reaction was allowed to warm to room temperature over 4 hours, the solids collected by filtration, washed acid free with water and dried to give 20.5 parts (99% yield) of product.

"The above dry product (8.24 parts, 0.02 mole) was suspended in 125 parts of acetone. A solution of 7.52 parts (0.055 mole) of potassium carbonate in 125 parts of water was added and the reaction stirred for 5 minutes until solution was complete. Diethylsulfate (15.4 parts, 0.10 mole) was added and the reaction mixture stirred at 25°–30°C. for 6 hours. The solids were isolated by filtration, washed with water and dried to give 7.9 parts (90.0% yield) of red product. Recrystallization from isopropanol provided an analytically pure sample, $a_{max.}$ 141 liters g.$^{-1}$ cm.$^{-1}$ at $\lambda_{max.}$ 527 mμ."

The above heterocyclic azo system is exceptional among heterocyclic azo dyes in that the imidazole nitrogen atom must be alkylated after azo coupling in order to get useful dyes having good lightfastness. Also, an organic solvent, such as methanol, used in the coupling reaction has been observed to increase coupling rate. Unfortunately, the organic solvent also increases the decomposition rate of the diazonium component and, therefore, a balance based on specific circumstances must be made to be useful. The presence of the acetamido group on the coupler leads to faster coupling, more bathochromic hue, and improved lightfastness over the dye than without it. These dyes produce bright red shades with excellent fastness, particularly to light, on polyester.

Most heterocyclic amines require diazotization under anhydrous conditions, and by far the most effective method to do this commercially is by the use of nitrosyl sulfuric acid in mixed propionic and acetic acids in the ratio of 1:5 or 2:5. Coupling of these diazonium solutions is usually rapid and can normally be achieved under aqueous acid conditions, with or without buffering by weakly basic salts, such as ammonium or sodium acetates.

Synthetic Method 52[56] (1,3,4-Thiadiazole Diazo Component + Acetamidoaniline Coupler) (Procedure quoted verbatim from U.S. Patent 3,639,384).

"Sodium nitrite (0.72 g.) is added portionwise to 5 ml. of conc. H_2SO_4. The solution is cooled and 10 ml. of 1:5 acid is added below 15°C. The mixture is cooled further, and 1.47 g. of 2-amino-5-methylthio-1,3,4-thiadiazole is added, followed by 10 ml. 1:5 acid, all below 5°C. After stirring for 2 hrs. at 0–5°C., the diazonium solution is added to a chilled solution of 2.68 g. of 3-acetamido-*N*-benzyl-*N*-ethylaniline in 100 ml. of 1:5 acid below 5°C. The reaction is kept cold and ammonium acetate added until the coupling mixture is neutral to Congo Red test paper. After allowing to couple 1 hr. at about 5°C., the reaction mixture is drowned in water. The product is collected by filtration, washed with water and dried in air. The dye produces bright red shades on polyester fibers and has good lightfastness and resistance to sublimation."

The isomeric 1,2,4-thiadiazole derivative can be diazotized and coupled in a similar manner, to give dyeings on polyester which are a little more bathochromic than the 1,3,4-isomer. These dyes are characterized by their brightness of shade coupled to high fastness, and they have been commercialized to compete with the anthraquinone red and pink dyes described earlier. Because of their relatively high tinctorial strength, they offer good economics and lack of metal sensitivity as advantages over the high volume anthraquinone red dyes.

Another important series of heterocyclic azo dyes is based on the benzothiazole nucleus. The diazonium components from variously substituted 2-aminobenzothiazoles generally produce, with commonly used couplers, red to rubine dyes. They are, in general, less bright than the thiadiazole-azo derivatives, although a few exceptions exist which have specific applications. 2-Aminobenzothiazoles commonly diazotize readily in strong mineral or organic acid solutions which use nitrosyl sulfuric acid as the diazotizing agent. If the acid strength is too high, insoluble sulfates form. If it is too low, the amine precipitates, and the nitrosyl sulfuric acid decomposes. In practice, 50 to 70% aqueous sulfuric or acetic acids are most advantageous. The use of phosphoric acid, either alone or in combination with other acids, is advantageous in particularly difficult cases. Occasionally, 1:5 acid itself gives satisfactory results, depending upon the specific substitution of the benzothiazole nucleus.

Synthetic Method 53[57] (2-Amino-4-CF_3-benzothiazole + Aniline Coupler) (Procedure quoted verbatim from U.S. Patent 4,052,379).

"Sodium nitrite (0.72 g., 0.01 mole) was added slowly in portions with stirring to concentrated sulfuric acid (5 ml.) at below 70°C. The mixture was warmed to 70°C. for 5 minutes and the resultant solution cooled to room temperature. Concentrated sulfuric acid (14 ml.) was added to a stirred suspension of 2-amino-4-trifluoromethylbenzothiazole (2.18 g., 0.01 mole) in water (24 ml.). The mixture was cooled to 0°C. and the solution of sodium nitrite in sulfuric acid was added while the temperature was kept below 3°C. Diazotization was allowed to proceed for 2 hours at 0°C., and the diazo solution was then added slowly to a well-stirred mixture of 2,5-dimethyl-*N-sec*-butylaniline (2.2 g., 0.0125 mole), anhydrous sodium acetate (12 g.) and ethanol (60 ml.) while the temperature was kept below 6°C. Coupling was allowed to proceed for 30 minutes at 6°C., and the mixture was then poured into cold water and stirred for 5 minutes. The solid dye was filtered off, washed with water and dried."

An example follows of a benzothiazole derived from diazotization of the amino derivative by using 60% aqueous acetic acid.

Synthetic Method 54[58] (2-Amino-6-methylbenzothiazole + Tetrahydroquinoline Coupler) (Procedure quoted verbatim from U.S. Patent 3,998,801).

"To 150 g. of 60% acetic acid containing 10 g. of concn. H_2SO_4 is added 8.2 g. (0.05 mole) of 2-amino-6-methylbenzothiazole at room temperature. The solution is cooled to −5°C. and a solution of 3.6 g. $NaNO_2$ in 20 ml. conc. H_2SO_4 is added below

0°C. The reaction mixture is stirred at −5° to 9°C. for 1.5 hours. The coupler *N*-(2-acetamidoethyl)-2,7-dimethyl-1,2,3,4-tetrahydroquinoline (0.005 mole) is dissolved in 20 ml. of 15% H_2SO_4 and the solution chilled in an ice bath. To the chilled coupler solution is added a 0.005 mole aliquot of the diazonium solution. The coupling mixture is treated with ammonium acetate to a pH of 3–4 and allowed to stand for one hour. The azo compound product is precipitated by addition of water, collected by filtration, washed with water and dried in air. The compounds are recrystallized or reslurried in methanol or ethanol for purification."

The presence of a donor group in the benzothiazole ring, as in the above example, tends to reduce the lightfastness on polyester, but has the opposite effect on polyamide. This dye is hence claimed to produce a fast red shade on polyamide fibers.

Certain diazonium components, particularly those from some heterocyclic amines, tend to produce low yields of dye upon coupling. This appears to be true regardless of the coupling medium employed although, of course, there are differences. This is often accompanied by reduced purity of the dye formed also, presumably as a result of diazo decomposition. In some cases the situation can be improved by using displacement coupling, which entails blocking the coupling position of the coupler with certain displaceable groups. The groups which appear to be most effective are —COAlk, —CO_2H, and —CHO. The diazonium component which is particularly known for producing dyes in low yield is that developed originally by Dickey in producing blue monoazo dyes, namely, 2-amino-5-nitrothiazole. This diazo component also appears to be one which responds well to displacement coupling. The following example describes the technique in relation to the diazotization and coupling of 2-amino-5-nitrothiazole.

Synthetic Method 55[59] (Displacement Coupling) (Procedure quoted verbatim from U.S. Patent 4,247,458).

"Sodium nitrite (2.2 g) was dissolved in conc sulfuric acid (15 ml) at 70°C. and the mixture cooled below room temperature. A mixture of propionic acid and acetic acid (30 ml) in the ratio 1:5 was added, maintaining the temperature below 20°C. The mixture was cooled to <5°C. and 2-amino-5-nitrothiazole (4.35 g) was added, followed by a further portion of 1:5 acid (30 ml). Diazotization was allowed to proceed for two hours at <5°C., and the resulting diazonium solution was added dropwise to a solution of 4-diethylaminobenzoic acid in dilute sulfuric acid. The dye was precipitated immediately and was filtered and washed with water. Thin layer chromatography indicated the dye to be in a very pure state. The yield was 6.5 g (71%)."

In a similar preparation, substituting *N,N*-diethylaniline for diethylaminobenzoic acid in the coupling just discussed, the dye was produced in very low yield and was extremely impure. After considerable work, the best conditions found for preparing the dye from *N,N*-diethylaniline coupler were those in which ethanol was the coupling solvent. All other solvents appeared to yield very impure products. Even with ethanol as the coupling solvent, the best yield obtainable was 4.0 g (44%), and then the purity was lower than that of the dye produced in 71% yield by the method described above.

Because of its high volatility, the dye just described has found commercial application in heat transfer printing for the coloration of polyester textiles in violet shades. the dye is particularly useful in the production of black shades owing to the high density of color produced. Lightfastness on polyester is excellent.

The production of navy-blue and black shades by exhaust dyeing represents one of the major uses for azo dyes in polyester. The azo dye CI Disperse Blue 79 **(10)** represents probably the highest volume of azo dye manufactured for use on polyester because of the large quantities needed for the production of blacks.

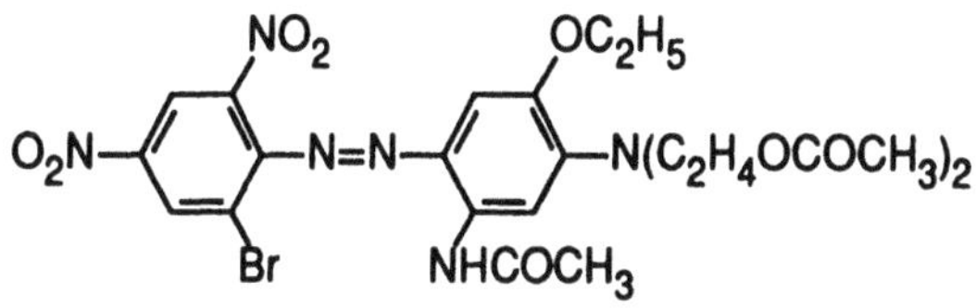

10

The following procedure serves to illustrate the method used for the production of CI Disperse Blue 79 type dyes. The use of a monoalkyl substituent on the coupler nitrogen, rather than the use of the bis-acetoxyethyl grouping, appears to cause an increase in the extinction coefficient of the dye while the navy-blue shade is still maintained.

Synthetic Method 56[60] (Diazotization and Coupling of 2-Bromo-4,6-dinitroaniline) (Procedure quoted verbatim from U.S. Patent 4,076,706).

"To a solution of nitrosyl sulfuric acid, from 18.0 grams of $NaNO_2$ and 125 ml of concentrated H_2SO_4, is added 100 ml of 1–5 acid (1 part propionic and 5 part acetic) below 20°C. The solution is cooled to 0°–5°C. and 65.5 grams (0.25 mole) of 2-bromo-4,6-dinitroaniline dissolved in 200 ml of concentrated H_2SO_4 is added dropwise below 5°C. The reaction is allowed to stir at 0–5°C. for 2 hours to complete diazotization.

"To a solution of 26.4 g (0.1 mole) of 2-(2′-methoxy-5′-acetamidoanilino)hexane in 75 ml of 1–5 acid is added 250 ml (0.1 mole) of the above diazo solution with stirring below 20°C. The reaction system is allowed to stand for one hour, poured into 800 ml of water, the solids collected by filtration, washed with water, and air-dried to yield 53.7 g (98%) of the dye."

Bayer chemists discovered that a halogen atom when present in the *ortho* position to the azo group of a phenyl azo dye is particularly labile, especially if the phenyl ring is further activated by electron-attracting groups, such as the nitro group. This situation is, of course, commonly encountered in azo dyes for polyester. Under these circumstances, the *ortho* substituent is readily replaced by a nucleophile. If the nucleophiles are also strong acceptors, then large bathochromic shifts are observed upon substitution into the phenyl azo dye. Of particular commercial interest is the introduction of the cyano substituent, because the linear nature of the group introduces little or no steric interaction in the *ortho* position. The replacement of, for an example, an *o*-bromo substituent with a cyano group results in a profound bathochromic shift, an increase in the brightness of the resulting dye because of the release of steric hindrance usually caused by the bulky *o*-bromo substituent, and an increase in lightfastness.

An acetamido function present on the coupler portion of the azo dye, *ortho* to the azo group, also appears to facilitate the displacement of the halogen atom. Indeed, ideally substituted azo dyes commonly and rapidly undergo cyanide displacement at ambient temperatures under the right reaction conditions. The reaction conditions normally reported involve treating the *o*-haloazo dye with cuprous cyanide in dimethylformamide solvent. The usually high crystalline cyano dye can be precipitated with water.

The navy-blue dye reported above is, in fact, ideally substituted to undergo cyanide displacement and, indeed, this has been done commercially to produce a greenish-blue dye for polyester. The dye has been used in conjunction with lemon yellow dyes to produce bright green shades with good lightfastness and working properties. The preparative conditions used for cyanide displacement are given below.

Synthetic Method 57[60] (Cyanide Displacement to Produce Blue Dyes) (Procedure quoted verbatim from U.S. Patent 4,076,706).

"A solution of sodium dicyanocuprate [$NaCu(CN)_2$] in dimethylformamide is added with stirring at 95°–100°C. to a solution of 52.1 g. (0.098 moles) of the above dye in 133 ml. of dimethylformamide. After addition is complete, the reaction is heated at 95°–100°C. for two hours. To the reaction is added 400 ml of isopropyl alcohol. The mixture is allowed to cool to room temperature, collected by filtration, washed well with water and air dried to yield 42.4 g. (91%) of the dye."

The above example describes the use of sodium dicyanocuprate instead of cuprous cyanide for the displacement. In commercial production, the use of cuprous cyanide presents certain problems. Excess of cuprous cyanide is necessary to achieve displacement effectively, and this results in excess cyanide ion in the effluent and excess copper in the dye and effluent. Excess copper in the dye results in hue shifts and dye instability if the azo dye happens to be used in combination with certain anthraquinone dyes that are sensitive to metals. The use of sodium cyanocuprate is very efficient in displacement and requires little or no excess. Also, free cyanide ion is not present in the sodium cyanocuprate method, and any remaining copper tends to be solubilized on addition of water. The following example describes in detail the preparation and use of sodium dicyanocuprate.

Synthetic Method 58[61] (Sodium Dicyanocuprate—Use in CN Displacement) (Procedure quoted verbatim from British Patent 1,438,374).

"In a 2 litre, three-necked flask are placed 2-ethoxyethanol (651.7 g., 700 ml) and with stirring 5-acetamido-4-(2′-bromo-4′,6′-dinitrophenylazo)-*N*-cyclohexyl-2-methoxyaniline (79.8 g.), followed by warming to 50–60°C. The cyanide solution is prepared in a separatory funnel by the addition of solid cuprous bromide (11.67 g.) to a dimethylformamide solution (141 g., 150 ml.) containing sodium cyanide (8.08 g.) predissolved in 25 ml. water. The mixture is shaken vigorously for several minutes during which time the cuprous bromide dissolves yielding a light green solution accompanied by a 10°C. temperature rise. Care is taken to avoid any physical contact with the solution. After the complex solution is prepared, it is added rapidly to the slurry of the azo compound at 50–60°C. Little or no temperature response is observed during the addition. The reaction temperature is then raised to 90–95°C. and held for one and one half hours. Completion of the reaction is verified by TLC. After the reaction is completed, 180 ml. of water is added to the mixture which is then cooled to 25°C. The product is filtered off, washed with 150 ml. of 50% 2-ethoxyethanol in water, then washed with water and dried. The yield of 5-acetamido-4-(2′-cyano-4′,6′-dinitrophenylazo)-*N*-cyclohexyl-2-methoxyaniline is 68.5 g. (94.6% of theory), which is free from azo byproducts and contains less than 400 ppm. copper."

Although it is commonly used to prepare blue phenyl azo dyes, cyanide displacement chemistry may be also used to prepare other shades, for which the alternative route may be difficult or expensive. The following example describes the use of this chemistry to produce a bright red dye which has been commercialized for polyester textiles coloration.

Synthetic Method 59[62] (Cyanide Displacement Producing Red Dyes) (Procedure quoted verbatim from U.S. Patent 4,452,609).

"A solution of 26.5 g of 2,6-dibromo-4-methylaniline in 125 ml of 60% strength sulfuric acid is diazotized with 17 ml of a 42% strength nitrosyl sulfuric acid at 38–40°C. The mixture is subsequently stirred for 60 min at 40°C., and the solution of the diazo component is then added at 0°–5°C., to a solution or suspension of 24.2 g of 3-diethylamino-methanesulfonanilide and 2 g of amidosulfonic acid in 200 ml of water. The pH value is increased to 3 by adding sodium acetate. The precipitate is filtered off and washed several times with water. Yield; 45.5 g.

"10.36 g of the 3-methanesulfonamido-4-(2′,6′-dibromo-4′-methylphenylazo)-*N,N*-diethylaniline thus obtained are dissolved or suspended in 50 ml of dimethylformamide, together with 2.35 g of zinc cyanide and 0.2 g of copper cyanide, and the mixture is warmed to 100°C. for 30 minutes. The product is precipitated with 50 ml of water and 7.5 g of the dicyanoazo dyestuff are filtered off. It dyes polyester fibers in a bluish-tinged red shade with good fastness properties, in particular fastness to light, sublimation and wet processing."

VI. MISCELLANEOUS CLASSES

A. Preparative Index

Synthetic Method 60

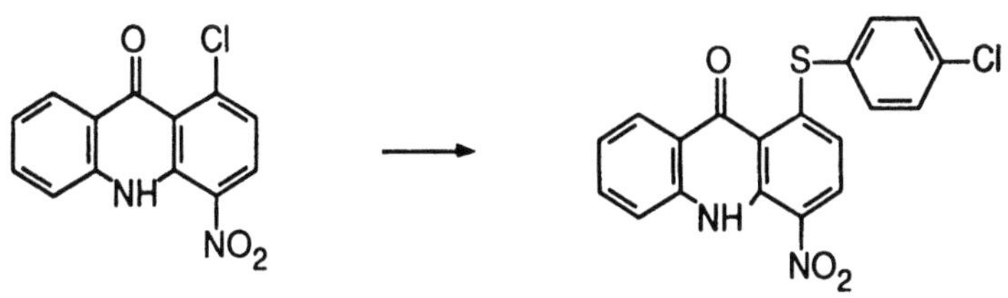

Synthetic Method 61

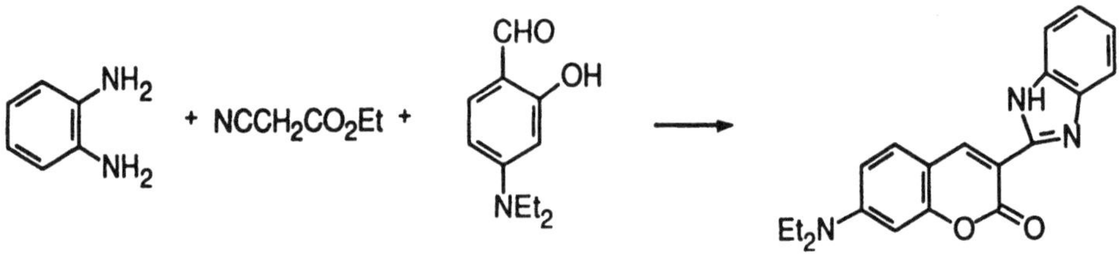

Synthetic Method 62

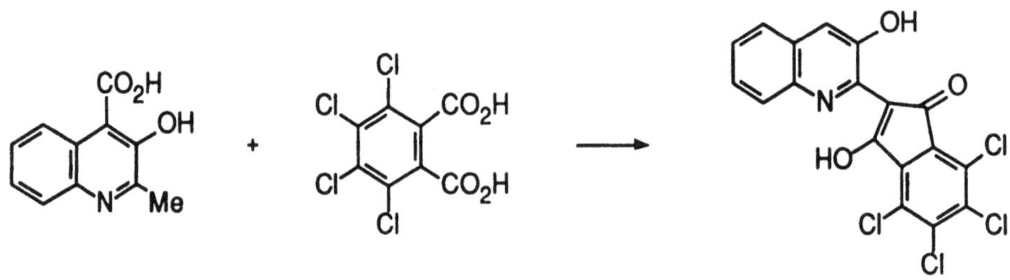

Synthetic Method 63

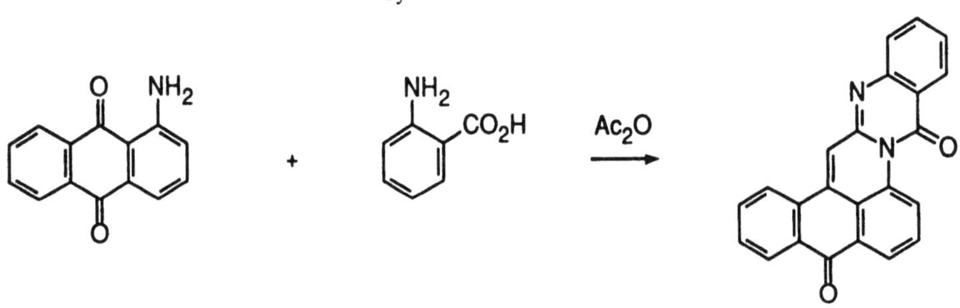

B. Discussion

Several disperse dyes for polyester have been described from various other chromophoric systems. Sometimes these dyes have found commercial application because of a particularly advantageous property. Often the chromophore can be related to one of the more commonly used chromophoric systems. One such type is the nitroacridone system. Yellow dyes **(11)**, claiming to have particularly good dyeing properties and to produce strong fast dyeings on polyester, have been reported from this system. The dyes may be regarded as structurally related to the nitrodiphenylamines already described, in which further cyclization of the diphenylamine moiety has formed the acridine ring. The preparation of one such dye is described below.

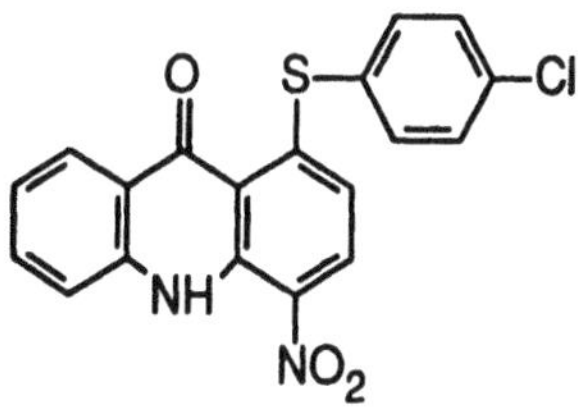

11

Synthetic Method 60[63] (Nitroacridones) (Procedure quoted verbatim from U.S. Patent 3,541,099).

> "12 Grams of anhydrous potassium carbonate are added to a dispersion of 41.2 g. of finely pulverized 1-chloro-4-nitroacridone (M.P. 250–251°) and 24.9 g. of 4-chlorothiophenol in 300 ml. of dimethylsulfoxide, the addition being made at room temperature (20°) while stirring well. The temperature rises to 45°. The yellow slurry which forms after a short time is slowly heated to 90–95° and stirred for 1 hour at this temperature. The mixture is then cooled to 50° and 1500 ml. of water are added. The precipitated yellow reaction product is stirred for another 20 minutes and then filtered off. It is washed with hot water and then dried. The yield is 57.4 g."

The preparation of the intermediate chloronitroacridone has been described.[64]

Another series of dyes which is closely related to the methine dye class is that of the coumarin dyes. These dyes give bright fluorescent-yellow shades on polyester, but have only limited lightfastness. The dyes can be prepared in a similar fashion to the methine dyes already described. A 4-formylaniline derivative, which in this case has a hydroxyl group *ortho* to the formyl group, is condensed with an activated acetonitrile compound, as is the case for the methine class of dyes. However, the *o*-hydroxyl group cyclizes, in the case of the coumarin dyes, onto the nitrile function to yield a derivative which, upon hydrolysis, forms the coumarin compound (equation 6).

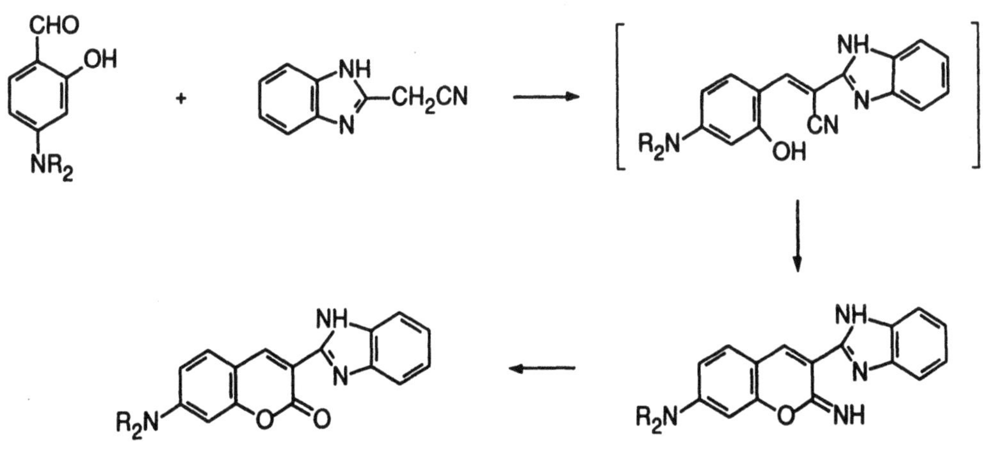

(6)

A one-step alternative synthesis of coumarin dyes of this type has been reported which starts with *o*-phenylenediamine and ethyl cyanoacetate and making benzimidazoyl acetic acid *in situ*, which is used in place of the nitrile. This acid is allowed to react with a 3-hydroxy-4-formylaniline derivative as described earlier. The synthesis is shown in the following procedure.

Synthetic Method 61[65] (Coumarin Dye—One Step Synthesis) (Procedure quoted verbatim from U.S. Patent 3,933,847).

"A mixture of 11.0 parts of *o*-phenylenediamine and 12.5 parts of ethyl cyanoacetate in 100 parts of 50% sulfuric acid was heated under reflux while stirring until *o*-phenylenediamine disappeared from the system, during which it took about 10 hours.

"To the reaction mixture containing 18 parts of the resulting 2-benzimidazolyl acetic acid were further added 19.7 parts of 2-hydroxy-4-diethylaminobenzaldehyde, and the mixture was heated at 90° to 95°C for 1 hour while stirring. After the reaction was completed, the reaction mixture was adjusted under cooling to 3 to 4 of pH with 20% aqueous caustic soda solution. Then the precipitated solids were filtered, washed thoroughly with water and dried to give 27 parts of the dye."

For some time, quinoline yellow has been known as a dye for natural fibers. The presence of a hydroxyl group on the quinoline ring capable of hydrogen bonding to the phthalone carbonyl group has been known to markedly increase the lightfastness of these dyes. If these dyes are not sulfonated but are applied to polyester from aqueous dispersions, a very useful class of polyester dyes, the quinophthalones, results. These dyes are typified by neutral yellow shades on polyester which have very high lightfastness. The excellent low energy, dyeing properties coupled with good economics makes the quinophthalone neutral yellow dye, CI Disperse Yellow 54, an excellent workhorse. This high-volume dye is used in combination with the anthraquinone red, CI Disperse Red 60, and the anthraquinone blue, CI Disperse Blue 56. The dye has better pH stability than the methine dyes and has a more neutral hue, making it more desirable for general applications. A synthetic procedure for the preparation of CI Disperse Yellow 54 is outlined below (equation 7).

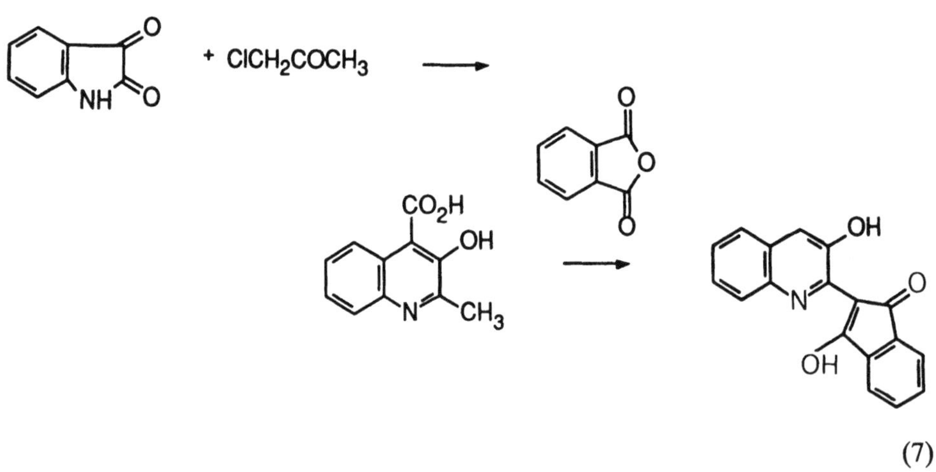

(7)

The sublimation fastness of CI Disperse Yellow 54 is not desirable for certain applications; however, the sublimation fastness has been improved by halogenation of the dye to give CI Disperse Yellow 64, which is more suitable for these type applications but where good dyeing properties are still retained.

A preparation procedure for a halogenated derivative of CI Disperse Yellow 54 is described below, and the synthesis can be adapted for the preparation of CI Disperse Yellow 54 itself.

Synthetic Method 62[66] (Quinophthalone Preparation) (Procedure quoted verbatim from U.S. Patent 3,399,028)

"40 Parts of 3-hydroxyquinaldine-4-carboxylic acid, 63 parts of 3,4,5,6-tetrachlorophthalic acid and 600 parts of dichlorobenzene are thoroughly mixed and heated to 110°C. In the course of 2 hours the temperature is raised to 170°C. while stirring the melt and distilling off the water formed. Upon elimination of the water the melt is held at 175–180°C. for a period of 15 hours. The reaction mixture is then permitted to cool to 60°C. and the batch is suction-filtered to yield a cake comprising 3,4,5,6-tetrachloro-3′-hydroxyquinophthalone from which the residual *o*-dichlorobenzene is removed by steam distillation. The resultant product is washed with water until salt free and dried at 100°C. The dye, 3,4,5,6-tetrachloro-3′-hydroxyquinophthalone, is obtained in good yield and in the form of yellowish-brown crystals with a chlorine content of 32.9% and melting point above 350°C."

Although hydrophobic in nature and of potential utility as disperse dyes, condensed polycyclic compounds (such as those often found used as vat dyes for cellulose fibers) do not find much application on polyester. The reason probably lies in their large molecular size and, hence, poor diffusion rate into the compact polyester fiber structure. However, a few such derivatives have found commercial utility in the dyeing of polyester particularly via the THERMOSOL process. An example of such a dye, CI Disperse Yellow 77, is given below **(12)**. This dye is claimed to impart brilliant yellow shades to polyester with excellent light, wash and sublimation fastness.

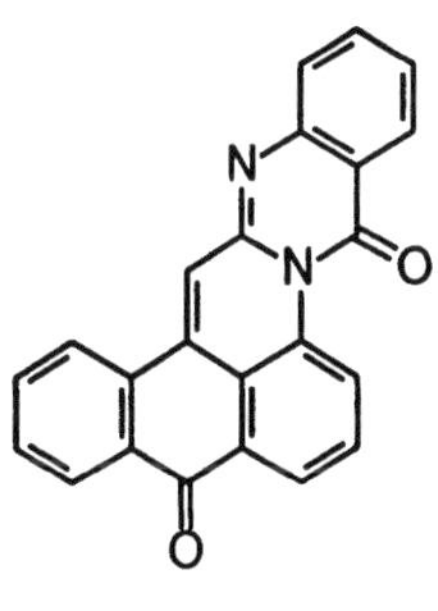

12

Synthetic Method 63[67] (CI Disperse Yellow 77 Type) (Procedure quoted verbatim from U.S. Patent 3,037,836)

"22.3 Parts of 1-aminoanthraquinone and 200 parts of acetic anhydride are heated to 100°C., and the glacial acetic acid formed is distilled off under slight vacuum. When the acylation is completed, 25 parts of zinc chloride and 15 parts of anthranilic acid are added, and the glacial acetic acid formed is again distilled off at 100–110°C. After completion of the condensation (1 to 2 hours), the mixture is cooled, filtered off with suction and washed with acetic acid anhydride until the discharge shows a clear yellow color. The residue is briefly boiled in a 10% sodium acetate solution, filtered off with suction and washed neutral. The dyestuff dissolves in concentrated sulfuric acid with a ruby-red color, melts at 293–296°C., and, after pasting with sulfuric acid, is ground with sulfite waste liquor."

REFERENCES

1. National Aniline & Chem. Co., U.S. Patent 2,211,943 (1940).
2. Eastman Kodak Co., U.S. Patent 2,659,739 (1953).
3. Eastman Kodak Co., U.S. Patent 3,279,880 (1966).
4. Interchemical Corp., U.S. Patent 3,189,398 (1965).
5. BASF, U.S. Patent 3,299,103 (1967).
6. BASF, U.S. Patent 3,694,467 (1972).
7. Eastman Kodak Co., U.S. Patent 3,087,773 (1963).
8. Farbenfabriken Bayer, U.S. Patent 2,990,413 (1961).
9. Eastman Kodak Co., U.S. Patent 2,641,602 (1953).
10. Grasselli Dyestuff Corp., U.S. Patent 1,652,584 (1927).
11. General Aniline Works, U.S. Patent 1,938,029 (1933).
12. DuPont, U.S. Patent 2,753,356 (1956).
13. Eastman Kodak Co., U.S. Patent 3,247,211 (1966).
14. Eastman Kodak, Co., U.S. Patent 3,917,604 (1975).
15. Eastman Kodak, Co., U.S. Patent 4,180,663 (1979).
16. Eastman Kodak, Co., U.S. Patent 3,879,434 (1975).
17. Ciba-Geigy, U.S. Patent 4,006,178 (1977).
18. Eastman Kodak, Co., U.S. Patent 2,798,090 (1957).
19. Eastman Kodak, Co., U.S. Patent 3,468,912 (1969).
20. Eastman Kodak, Co., U.S. Patent 3,661,899 (1972).
21. DuPont, U.S. Patent 2,889,335 (1959).
22. Sandoz, U.S. Patent 4,281,115 (1981).
23. R. A. Carboni, D. D. Coffman, E. G. Howard, *J. Am. Chem. Soc. 80,* 2838 (1958).
24. DuPont, U.S. Patent 3,013,013 (1961).
25. Eastman Kodak, Co., U.S. Patent 2,422,029 (1947).
26. Eastman Kodak, Co., U.S. Patent 3,729,493 (1973).
27. Martin Marrietta, U.S. Patent 3,502,645 (1970).
28. Eastman Kodak, Co., U.S. Patent 4,052,379 (1977).
29. Eastman Kodak, Co., U.S. Patent 4,363,913 (1982).
30. J. B. Dickey, E. B. Towne, and G. F. Wright, *J. Org. Chem. 20,* 505 (1955).
31. L. Silver, *Chem. Eng. Prog. 63,* 44 (1967).
32. R. F. Meyer, B. L. Cummings, P. Bass, and H. O. J. Collier, *J. Med. Chem. 8,* 515 (1965); J. Gray and D. R. Waring, *J. Heterocycl. Chem. 17,* 65 (1980).
33. BASF, U.S. Patent 3,981,883 (1976).
34. American Cyanamid, U.S. Patent 2,759,947 (1956).
35. Eastman Kodak, Co., U.S. Patent 4,528,368 (1985).
36. Eastman Kodak, Co., U.S. Patent 3,221,006 (1965).
37. DuPont, U.S. Patent 3,770,764 (1973).
38. K. Gewald, *Angew. Chem. 73,* 114 (1961); *Chem. Ber. 98,* 3571 (1965).
39. ICI, British Patent 1,394,368 (1975).
40. Eastman Kodak Co., U.K. Patent 2,092,128A (1982); U.K. Patent 2,078,713A (1982).
41. Eastman Kodak, Co., U.S. Patent 4,210,586 (1980).

42. *Organic Syntheses*, Coll. Vol. 3, 329.
43. Eastman Kodak, Co., U.S. Patent 4,400,318 (1983).
44. Eastman Kodak, Co., U.S. Patent 3,635,941 (1972).
45. Eastman Kodak, Co., U.S. Patent 3,383,379 (1968).
46. Eastman Kodak, Co., U.S. Patent 3,493,556 (1970).
47. Eastman Kodak, Co., U.S. Patent 3,765,830 (1973).
48. Eastman Kodak, Co., U.S. Patent 3,453,270 (1969).
49. DuPont, U.S. Patent 4,324,721 (1982).
50. I. G. Farben., British Patent 470,640 (1937).
51. Cassella, U.S. Patent 4,284,782 (1981).
52. Eastman Kodak, Co., U.S. Patent 2,529,924 (1950).
53. Eastman Kodak, Co., U.S. Patent 3,770,719 (1973).
54. Cassella, U.S. Patent 3,957,749 (1976).
55. DuPont, U.S. Patent 4,097,475 (1978).
56. Eastman Kodak, Co., U.S. Patent 3,639,384 (1972).
57. Eastman Kodak, Co., U.S. Patent 4,052,379 (1977).
58. Eastman Kodak, Co., U.S. Patent 3,998,801 (1976).
59. Eastman Kodak, Co., U.S. Patent 4,247,458 (1981).
60. Eastman Kodak, Co., U.S. Patent 4,076,706 (1978).
61. Eastman Kodak, Co., British Patent 1,438,374 (1976).
62. Bayer, U.S. Patent 4,452,609 (1984).
63. Ciba-Geigy, U.S. Patent 3,541,099 (1970).
64. H. B. Nesbit, *J. Chem. Soc.*, 1372 (1933).
65. Sumitomo Chem. Co., U.S. Patent 3,933,847 (1976).
66. Toms River Chem. Corp., U.S. Patent 3,399,028 (1968).
67. Bayer, U.S. Patent 3,037,836 (1962).

5

Dyes for Polyacrylonitrile

P. GREGORY

I. INTRODUCTION

Until Perkin's historic discovery of the first synthetic dye, Mauveine, in 1856, all of the colorants were of natural origin. However, before the end of the nineteenth century, synthetic dyes had displaced virtually all the natural dyes. This was not the case with the textile fibers; the naturally occurring fibers such as cotton, wool, and silk dominated until well into the twentieth century. It was not until the period 1930–1955 that the three major synthetic fibers, namely polyamide (nylon), polyester (e.g., Terylene), and polyacrylonitrile (e.g., Dralon), were discovered and introduced to the market place.

Acrylonitrile was used extensively during the Second World War as a copolymer with butadiene to produce "nitrile rubber," used to augment the depleted supplies of natural rubber. This surfeit of acrylonitrile in the 1940s, especially in the USA, no doubt prompted chemists at du Pont to investigate the polymer derived from this cheap and plentiful monomer and in 1953 du Pont launched the first polyacrylonitrile fiber, Orlon.[1]

Polyacrylonitrile, or acrylic fibers as they are known, are prepared by addition polymerization of acrylonitrile (equation 1). The pure homopolymer is difficult to dye but acrylic polymers containing small amounts of anionic centers, such as sulfonic acid or carboxylic acid groups, can be dyed readily with dyes bearing a positive charge, viz. cationic dyes. The cationic dyes are attracted to and then anchored to the fiber by ionic bonds (equation 2). The anionic centers arise either from the polymerization inhibitors or from small amounts of copolymer added deliberately to introduce the anionic sites.

Other manufacturers quickly followed du Pont's lead and launched their own polyacrylonitrile fibers. Table 1 lists the major brands of polyacryonitrile fibers and the anionic centers present.

P. GREGORY • Fine Chemicals Research Centre, ICI Colours and Fine Chemicals, Hexagon House, Blackley, Manchester M9 3DA, England.

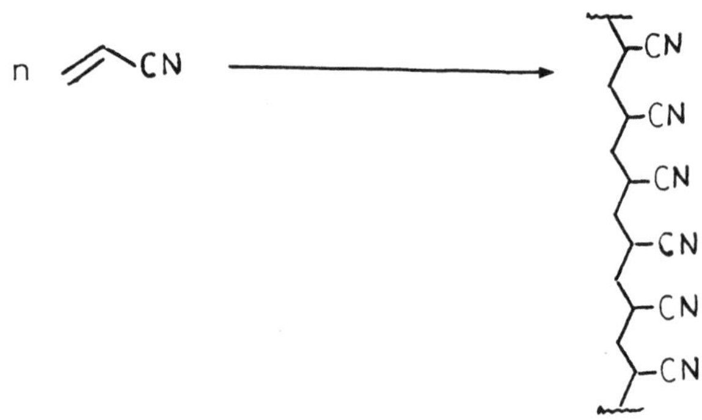

(1)

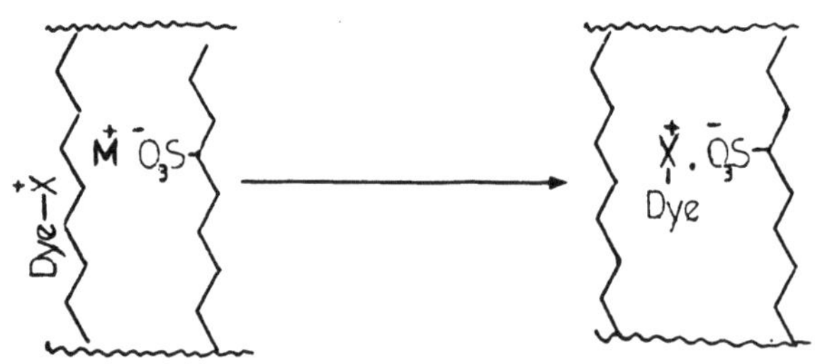

(2)

The first commercial dyes used for dyeing this new fiber were classical basic (cationic) dyes such as polymethine, triphenylmethane, and oxazine dyes. These dyes are among the brightest and strongest dyes known but had not found use on textile fibers because of their poor fastness properties, in particular their low fastness to light. Surprisingly, selected dyes of these types displayed remarkably good fastness to light on polyacrylonitrile fibers. Further research unearthed new classes of cationic dyes having excellent all-round properties with the result that the present cationic dyes for polyacrylonitrile are the most technically excellent of all the dyes. Indeed, since polyacrylonitrile fibers are often used as a cheaper substitute for wool, it is commonplace to find long-life furnishings such as carpets, curtains, furniture suites, and even tents and awnings composed of dyed acrylic fibers.

Disperse dyes (see Chapter 4) have also been used to dye polyacrylonitrile, but they are of little importance nowadays (only pale depths of shade can be achieved) and therefore they are not discussed further in this chapter.

The cationic dyes used for dyeing polyacrylonitrile fibers may be divided into

Table 1. Major Types of Polyacrylonitrile Fibers

Orlon (du Pont)	SO_3H groups
Dralon (Bayer)	
Acrilan (Chemstrand)	
Courtelle (Courtaulds)	CO_2H groups

two main types:

1. Those in which the positive charge is localized on one atom, usually a nitrogen atom (localized or pendant cationic dyes).
2. Those in which the positive charge is delocalized over the entire molecule (delocalized cationic dyes).

Each of these types is discussed, beginning with the pendant cationic dyes.

II. PENDANT CATIONIC DYES

These tend to be largely azo or anthraquinone disperse dyes modified for polyacrylonitrile fibers by the introduction of a pendant cationic group, typically pyridinium or trimethylammonium. Consequently, the dyes display a similar level of brightness and tinctorial strength to the corresponding azo and anthraquinone disperse dyes from which they are derived. As a class, the pendant cationic dyes are much less important than the brighter and stronger delocalized cationic dyes. Therefore, less attention is devoted to pendant cationic dyes. For convenience, the dyes are discussed according to the three major color areas of yellow, red, and blue.

A. Yellow Dyes

1. Miscellaneous Chromogens

Cationic versions of known classes of disperse yellow dyes have been used for dyeing polyacrylonitrile. Thus, fluorescent yellow dyes of both the styryl **(1)** and coumarin type **(2)** have been used. Cationic versions of the quinophthalone dyes **(3)** and nitro dyes **(4)** have also been used to dye polyacrylonitrile.[2] Dye types **1–4** are of no importance nowadays.

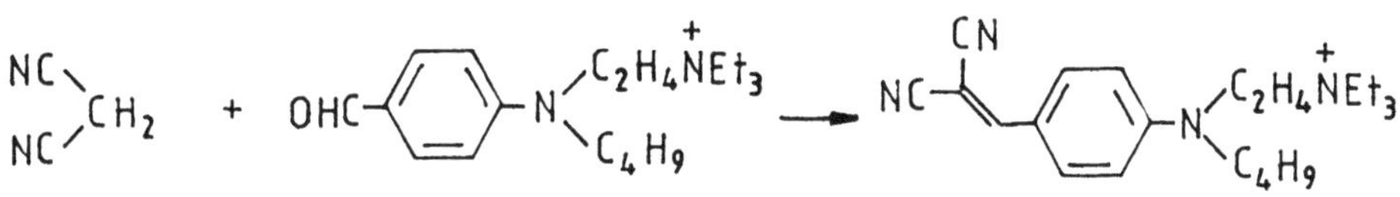

1

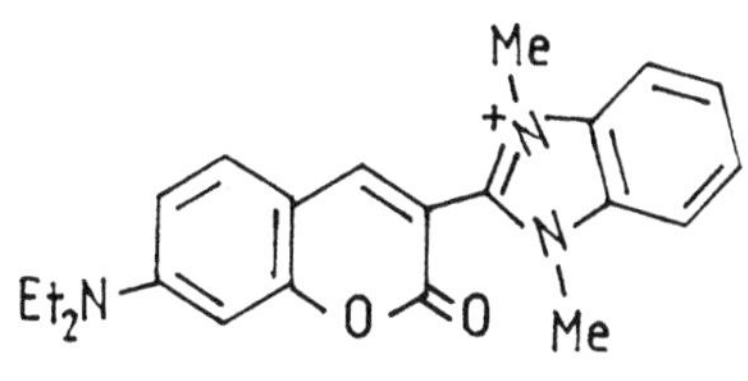

2

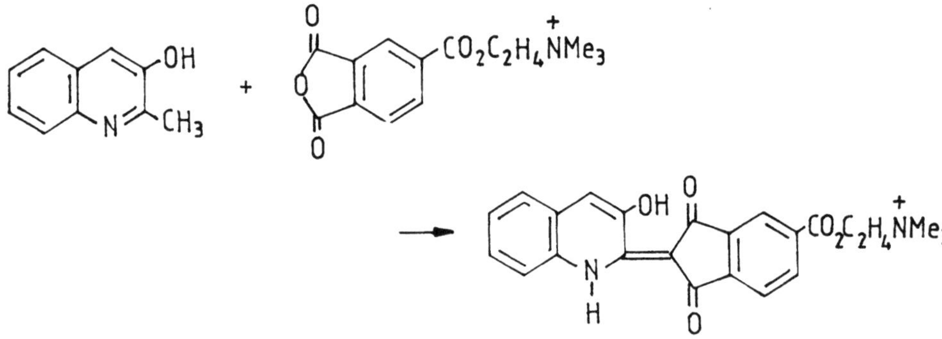

3

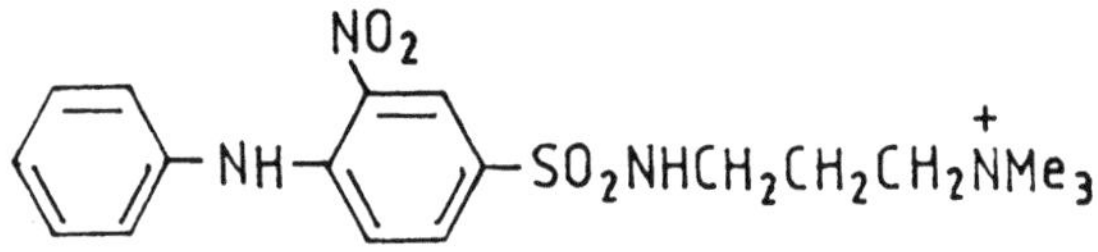

4

2. Azopyridone Dyes

Cationic versions of the azopyridone yellow dyes that have made such an impact in other dye types, such as disperse dyes for polyester and reactive dyes for cotton, have also been devised. Both ICI and Sandoz introduced dyes of this type. ICI[3] introduced greenish-yellow dyes such as structure **5**, prepared as shown in Scheme 1, while Sandoz[4] introduced golden yellow dyes such as **6** (Scheme 2). In both cases the cationic group is present in the pyridone coupling component.

Unlike the azopyridone yellow dyes for polyester, these cationic pyridone yellow dyes have not gained the same prominence for polyacrylonitrile. The reason is that superior dyes, in terms of both brightness and strength, exist in which the positive charge is delocalized over the entire molecule (see Section III.C).

B. Red Dyes

The only important red dyes containing a pendant cationic group are those derived from the carbocyclic azo class of dyes. Red dyes from anthraquinones, such as **7**, are just not cost effective against the azo dyes, especially those containing a delocalized positive charge such as the diazahemicyanines (Section III.D).

The pendant cationic group in the carbocyclic azo dyes can be located in either the diazo component or the coupling component. The old Janus dye **(8)**, derived from *m*-aminophenyltrimethylammonium chloride, and the du Pont dye **(9)**, derived from 4-aminophenacyltrimethylammonium chloride, prepared as shown in Scheme 3,[5] are examples of dyes containing a pendant cationic group in the diazo component. These types of dyes are of no interest nowadays.

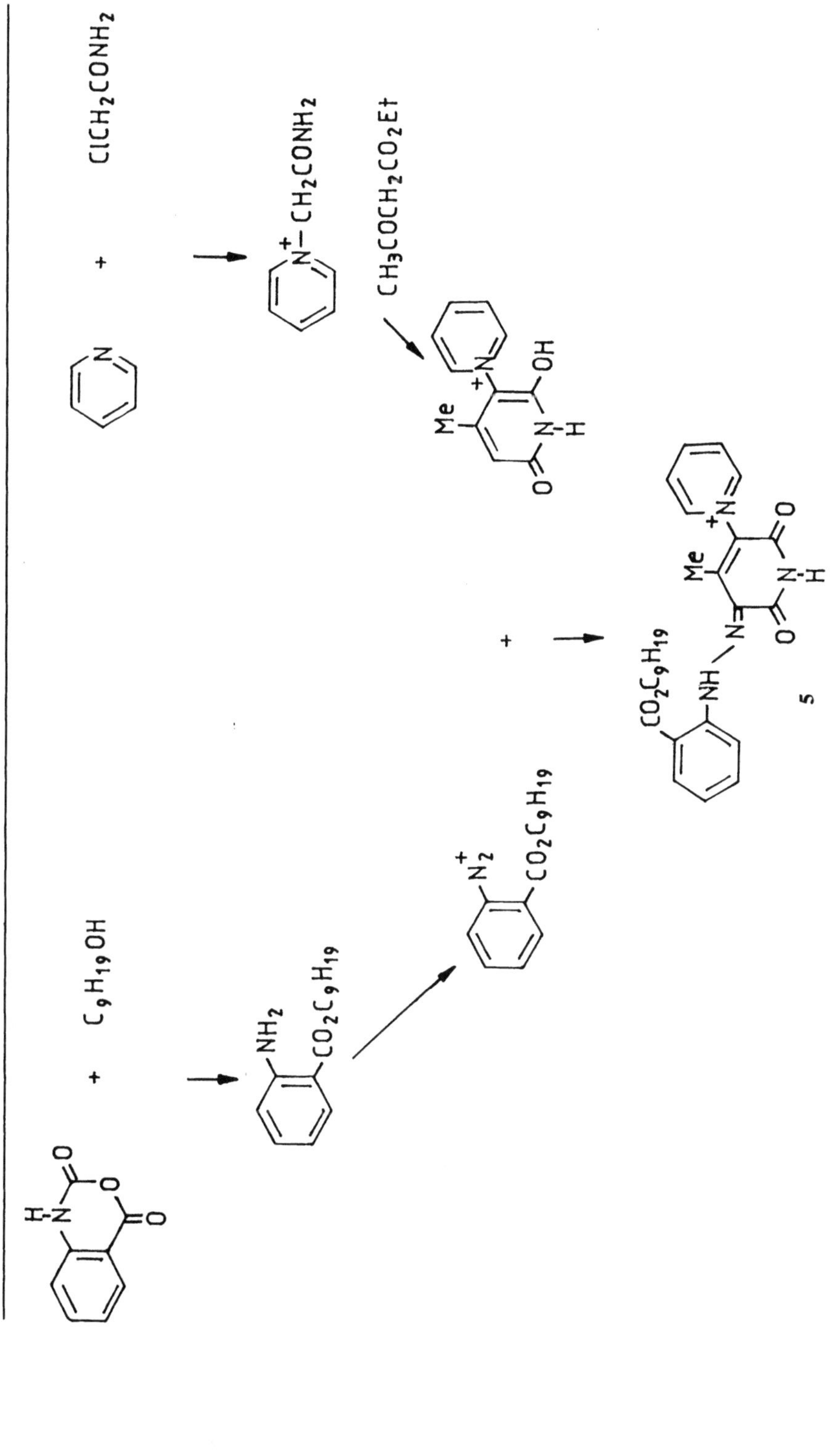

5

Scheme 1

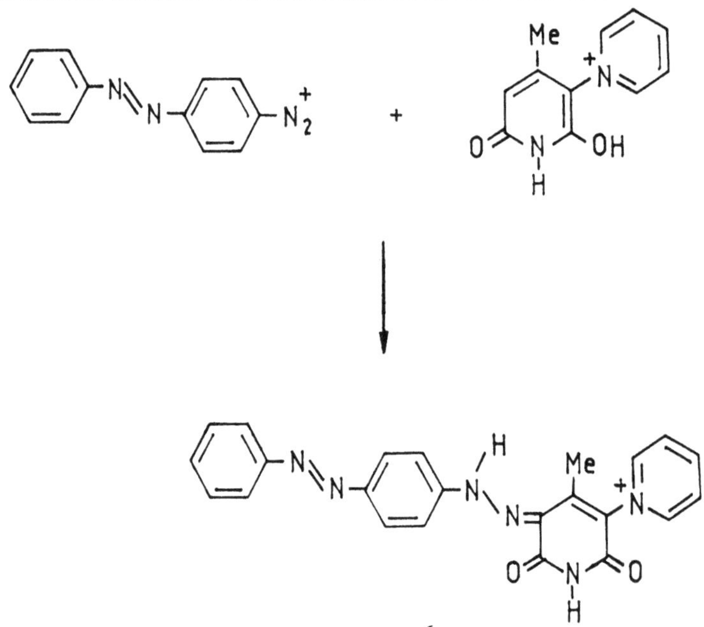

6

Scheme 2

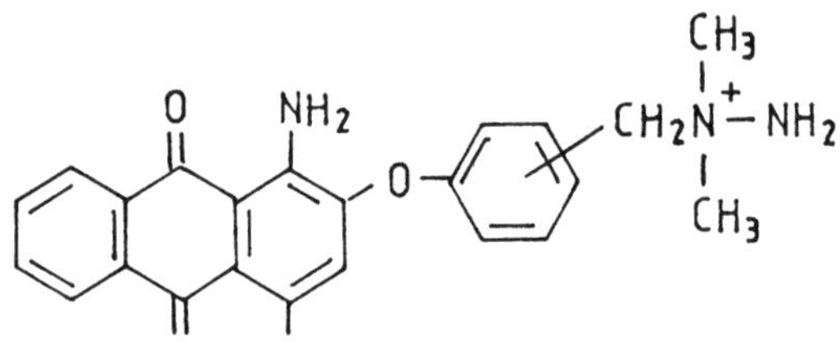

7

The most important red pendant cationic dyes are those in which the cationic group is located in the coupling component. CI Basic Red 18 types **(10)** are the most important dyes. This cost effective, dull, brownish-red dye tends to be the workhorse dye for producing the duller red and tertiary shades such as black. Several types of cationic group have been employed in this dye **(10a–d)**. (The

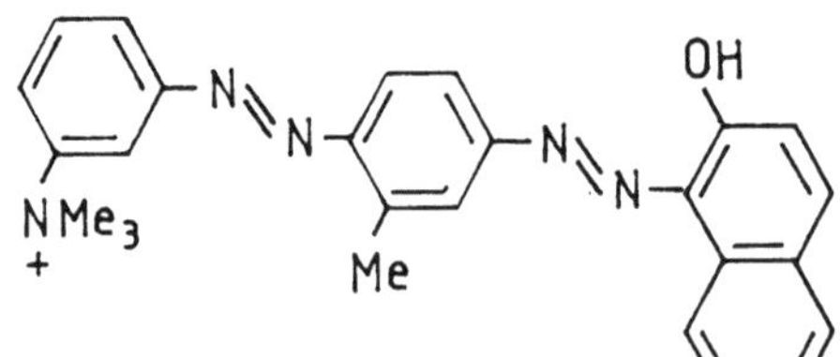

8

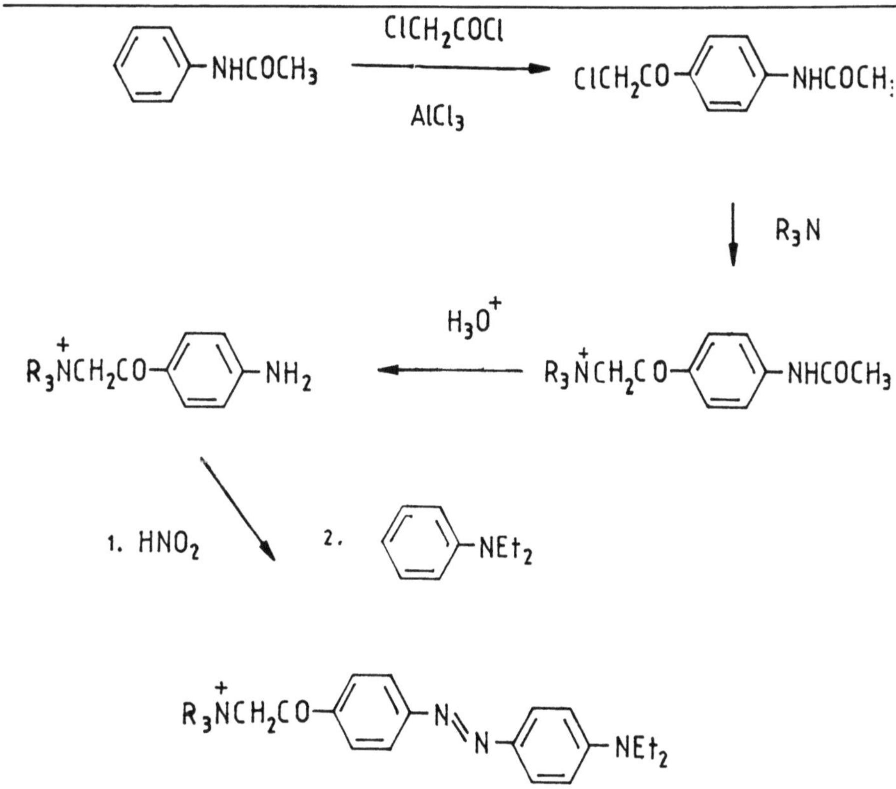

9

Scheme 3

colors and strengths are the same; the different cationic groups just affect properties such as solubility.) The most prevalent group is pyridinium **(10b)** followed by trimethylammonium **(10a)**. A 4-methylpyridinium group has also been patented by ICI[6]; the resultant dye **(10c)** is claimed to have high crystallinity (easier filtration) and better aqueous solubility than **(10a)** or **(10b)**. It is therefore ideal for both powder brands and aqueous-based liquid brands. Sandoz[7] cleverly introduced a novel cationic group, the 1,1-dimethylhydrazinium group **(10d)**. Dyes based upon this grouping have been withdrawn because of toxicity concerns about 1,1-dimethylhydrazine. Scheme 4 illustrates the synthesis of CI Basic Red 18 types.

Another useful dye is CI Basic Orange 30:1 **(11)**. The additional chlorine atom *ortho* to the azo linkage causes enforced nonplanarity of the phenyl ring; this produces the hypsochromic shift from red to yellow-brown but also causes an undesirable drop in tinctorial strength.[8]

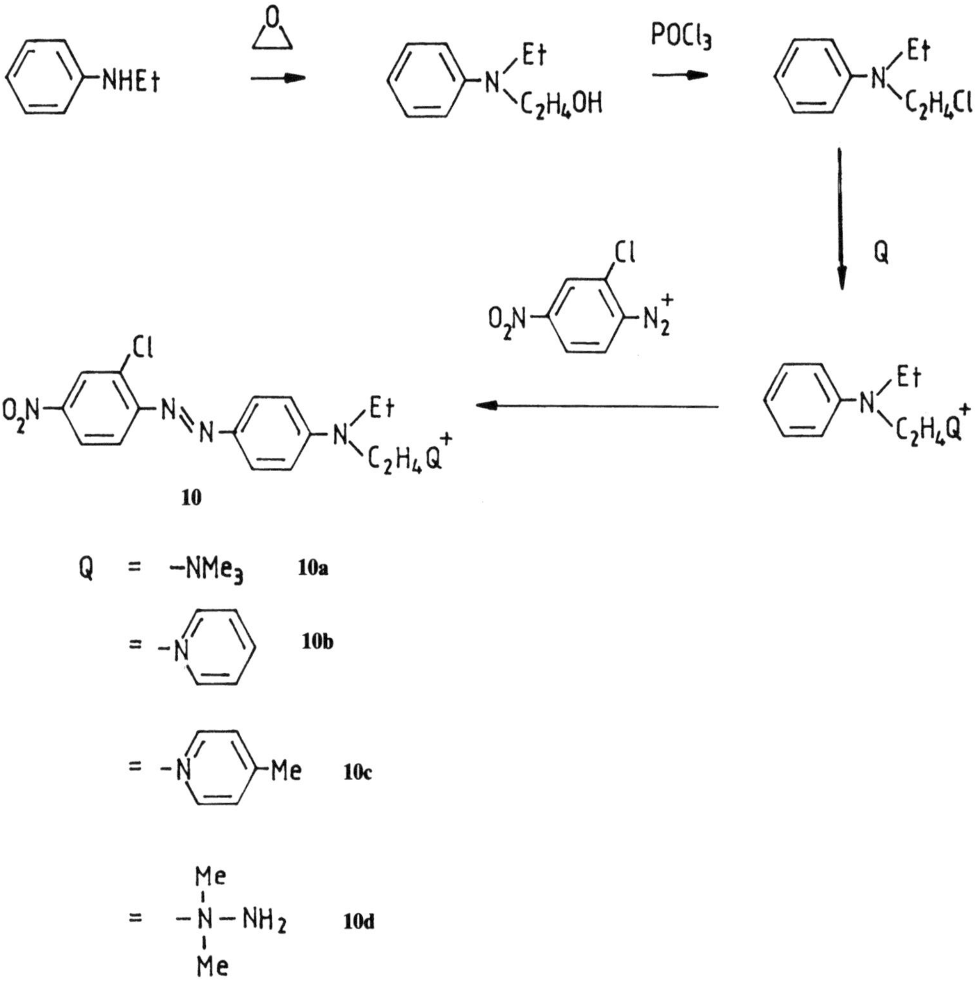

10

Scheme 4

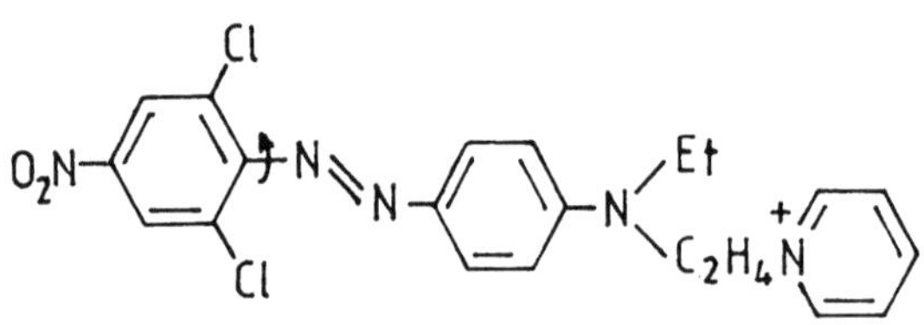

11

C. Blue Dyes

Pendant cationic blue dyes belong to the azo and anthraquinone classes. As is the case with the yellow and red dyes, pendant cationic blue dyes are of much less importance than the blue dyes containing a delocalized positive charge (Section III.E).

1. Azo Dyes

In contrast to the azo reds, the pendant blue dyes tend to be derived from heterocyclic diazo components. CI Basic Blue 119 **(12)**, derived from an aminonitrothiazole, is a typical dye.

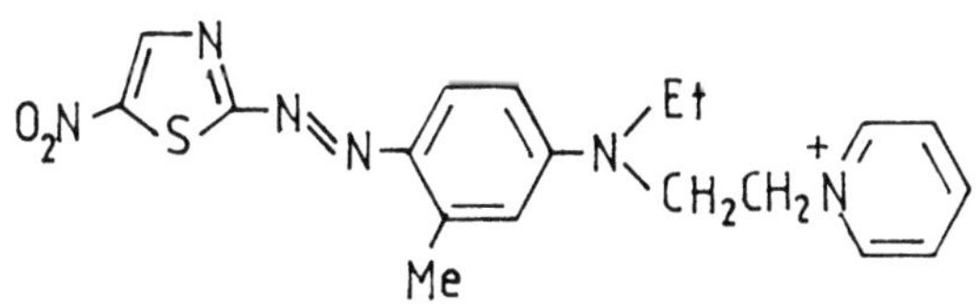

12

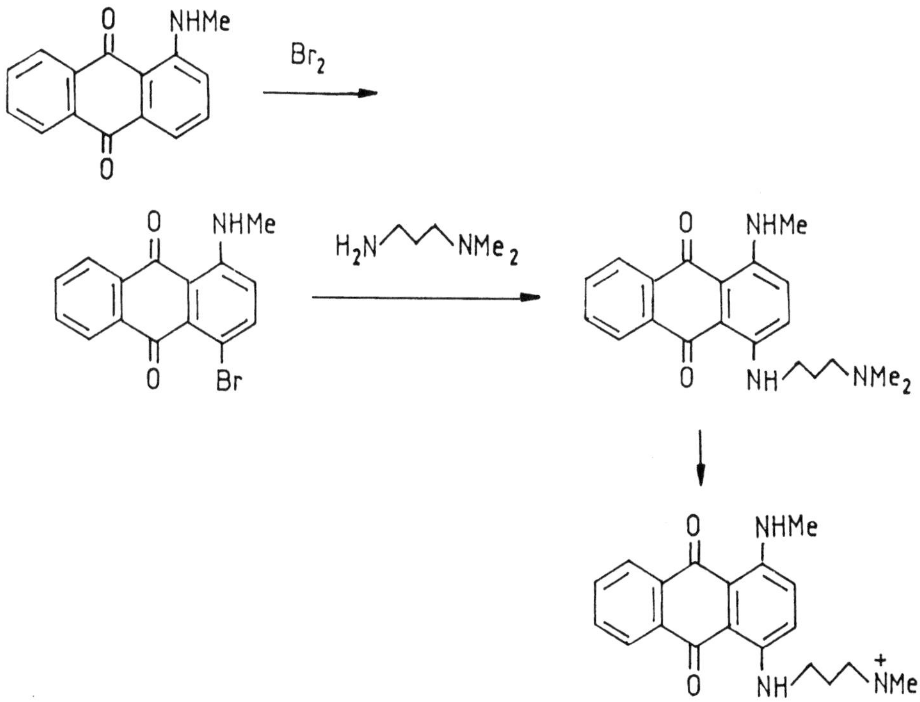

13

Scheme 5

2. Anthraquinone Dyes

1,4-Bisalkylaminoanthraquinones provide the important pendant cationic blue dyes. Perhaps the most important dye is CI Basic Blue 22 **(13)**. This dye is prepared by the sequence of reactions shown in Scheme 5.

The inherent low tinctorial strength of anthraquinone dyes combined with the rather protracted reaction sequence diminish the cost effectiveness of this dye. It is used because it is technically excellent, including high fastness to heat treatments and to dyebath hydrolysis (see Section III.E.5).

The bis-cationic dye **(14)** uses cheaper intermediates and a "one-pot" reaction sequence (Scheme 6). Consequently it is much more cost effective than **(13)**. However, its dyeing properties are not quite as good.

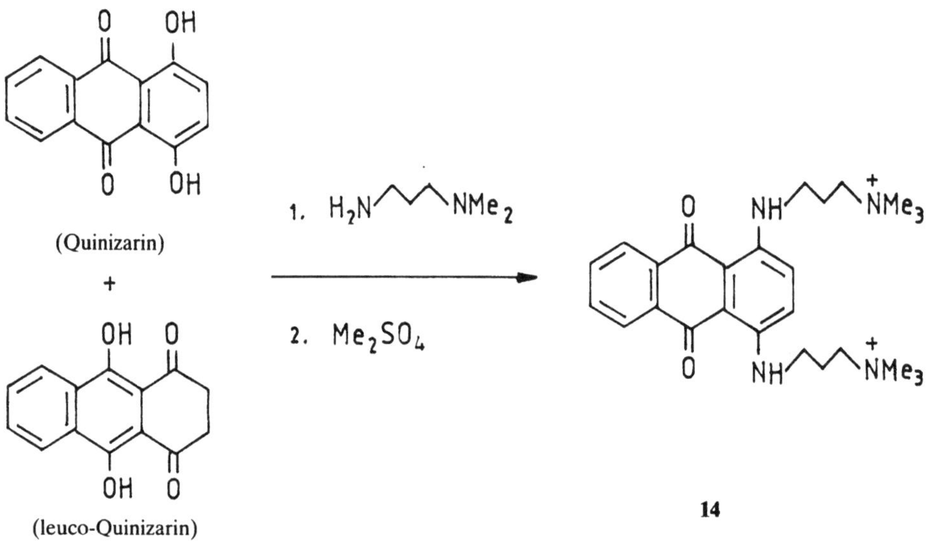

Scheme 6

D. Miscellaneous Cationic Groups

All the commercial cationic dyes for polyacrylonitrile contain a quaternized nitrogen atom. However, cationic dyes in which the positive charge is centered upon sulfur (sulfonium) and phosphorus (phosphonium) dyes have been patented. Typical dyes are **(15)** and **(16)**. A more detailed account of pendant cationic dyes has been given by Baer.[2]

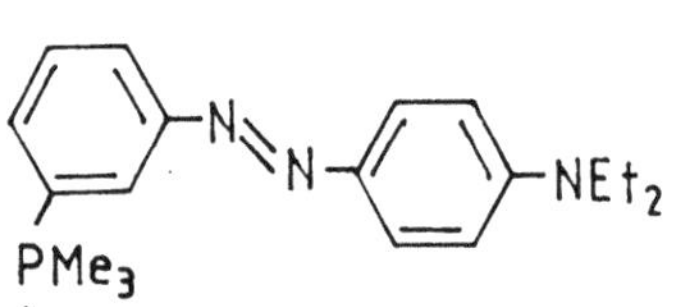

15

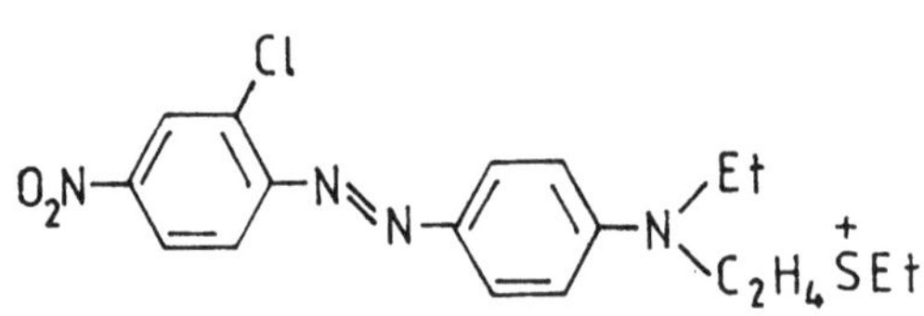

16

III. DELOCALIZED CATIONIC DYES

A. Comparison with Pendant Dyes

In contrast to the localized or pendant cationic dyes, in which the positive charge is centered on one atom (usually nitrogen), in delocalized cationic dyes the positive charge is delocalized over many atoms. This seemingly insignificant change has a profound effect upon the properties of the dyes. Thus, the delocalized cationic dyes are inherently much stronger and brighter than their pendant counterparts. Indeed, many of the delocalized dyes have molar extinction coefficients greater than 50,000, with a number above 100,000. Their enhanced brightness is a function of their structural similarity to cyanine dyes which, in turn, are closely related to the "odd-alternant hydrocarbon" molecules used as models by Dewar for his Perturbational Molecular Orbital (PMO) theory of color.[9] This relationship results in the first excited state of the molecule having a similar structure to the ground state which, in turn, results in a narrower absorption curve which is manifested as brightness.[10]

These key advantages of higher tinctorial strength, which results in improved cost effectiveness, and a high level of brightness do not in themselves ensure commercial success. Many dyes possessing these attributes have been known for many years without ever achieving the commercial impact of weaker and duller dyes such as the azos and anthraquinones. The reason is simple. The dyes must also possess suitable fastness properties, particularly fastness to light. Almost without exception, delocalized cationic dyes exhibited poor light fastness on the substrates available prior to the discovery of polyacrylonitrile and this restricted their use to applications where high light fastness was not required, e.g., wrapping paper and toilet paper. However, the advent of polyacrylonitrile fiber changed the scenario completely and opened up a new lease of life for delocalized cationic dyes. Very surprisingly, many of the known delocalized cationic dyes displayed good light fastness on polyacrylonitrile and this promoted research into other systems.[2] As a consequence, the present ranges of cationic dyes for polyacrylonitrile are the most technically excellent of any of the dye ranges. They combine exceptionally high tinctorial strength, high brightness, and high fastness properties, a combination of properties extremely difficult to achieve across a complete range of dyes.

The delocalized cationic dyes are now discussed according to the three most important shade areas of yellow, red, and blue. However, protonated azo dyes represent a good transitional point from the pendant cationic dyes to the delocalized cationic dyes. Therefore, these are discussed first.

B. Protonated Azo Dyes

The best example to illustrate a delocalized cationic dye is Chrysoidine **(17)**. In this yellow dye the proton resides on the β-nitrogen atom of the azo group. The positive charge is delocalized (shared) between the azo nitrogen atom, structure **(17A)**, and the two amino groups, structures **(17B)** and **(17C)**.

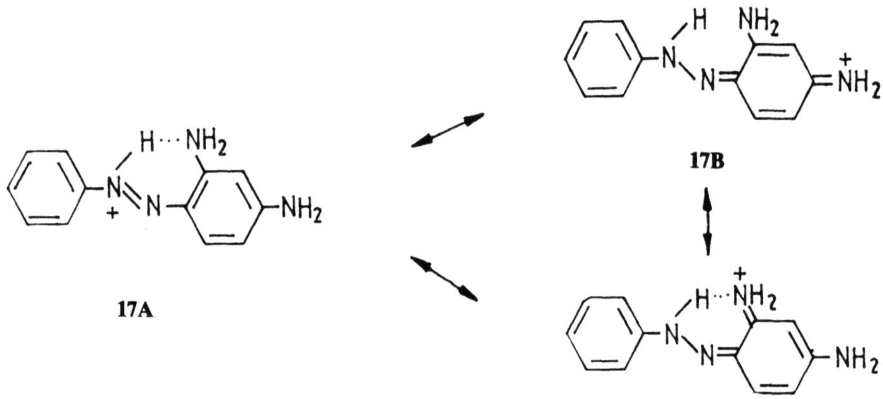

17A

17B

17C

In effect, Chrysoidine exists as the azonium tautomer.[11] However, it is a highly stabilized azonium tautomer because of the intramolecular hydrogen-bonding between the nitrogen atom of the amino group *ortho* to the azo linkage and the hydrogen atom on the azo group. Indeed, it is known that the stabilization of azonium tautomers by groups *ortho* to the azo linkage that are capable of forming intramolecular hydrogen-bonds with the azo hydrogen atom is a general feature. For example, the orange dye **(18)**, incorporating a methoxy group, and the fluorescent magenta dye **(19)**, incorporating an acetyl group, are stable dyes.[12]

Groups in the diazo component also stabilize the azonium tautomer. A good example here is the indicator dye, Methyl Red **(20)**.

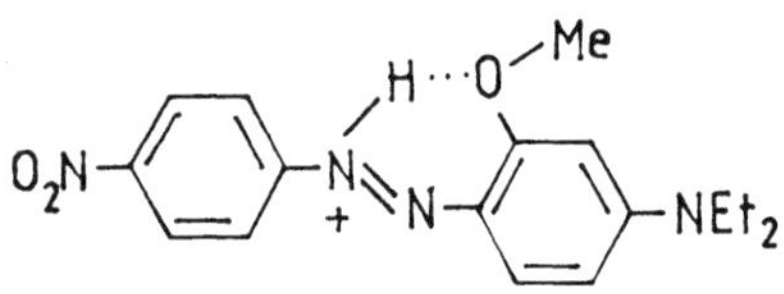

18

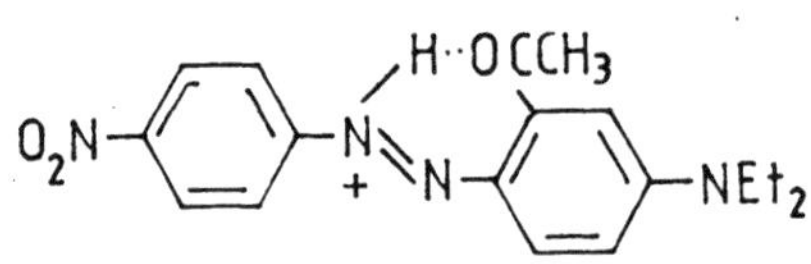

19

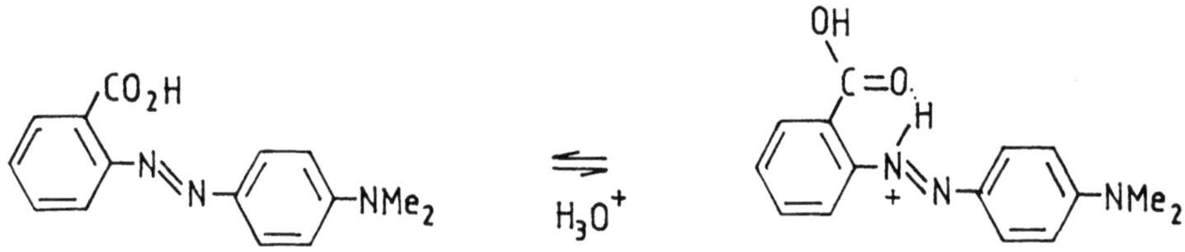

20
Orange
λ_{max} 435 nm
ε_{max} 19,000

Red
λ_{max} 520 nm
ε_{max} 45,000

Table 2. Delocalized Yellow Cationic Dye Systems

Class	General structure	Color
Carbocyanine	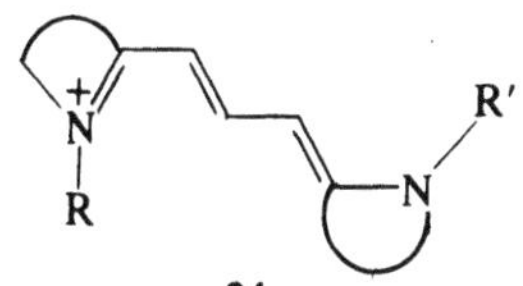 **24**	—
Azacarbocyanine	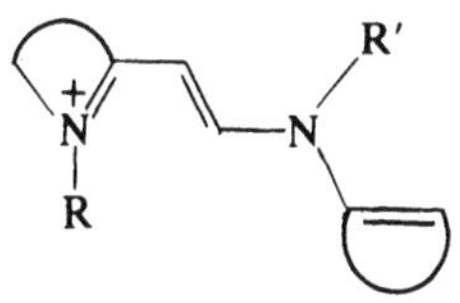 **21**	Greenish-yellow
Diazacarbocyanine	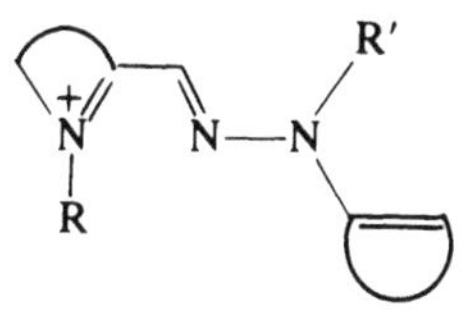 **22**	Yellow-orange
Triazacarbocyanine	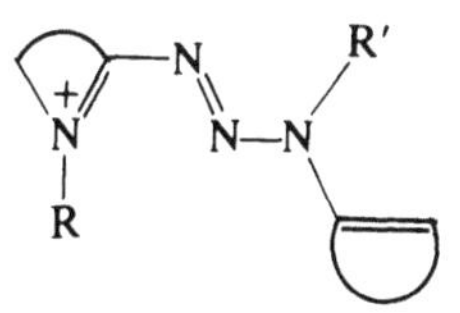 **23**	Yellow

C. Yellow Dyes

There are three noteworthy delocalized yellow cationic dye systems, namely, the azacarbocyanines **(21)**, the diazacarbocyanines **(22)**, and the triazacarbocyanines **(23)**. These are related to the parent carbocyanine molecule **(24)** by replacing one, two, or three CH units with nitrogen atoms;[13] see Table 2.

1. Azacarbocyanines

CI Basic Yellow 11 **(25)**, prepared by condensing Fischer's aldehyde **(26)** with the aniline **(27)**, is one of the leading greenish-yellow dyes for polyacrylonitrile. It provides bright, strong, greenish-yellow shades having good light fastness.

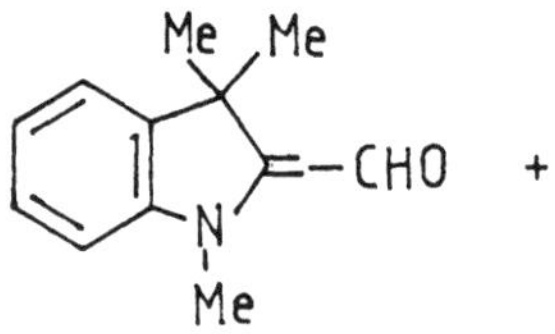

26

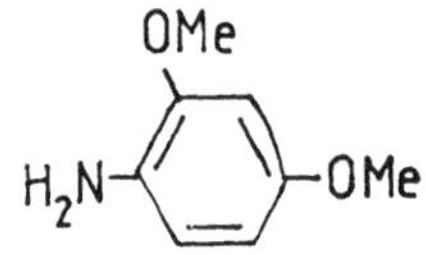

27

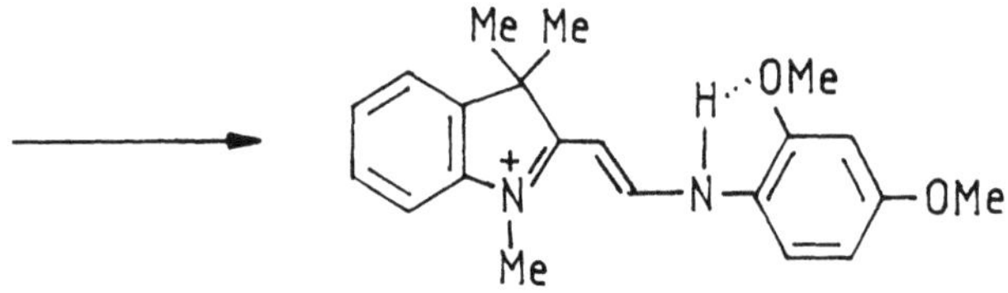

25

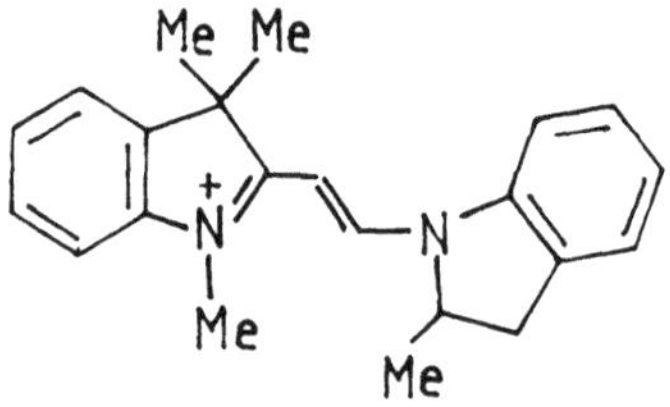

28

By incorporating the nitrogen atom into a heterocyclic ring, a greenish-yellow dye is produced **(28)** which has excellent light fastness.[14]

2. Diazacarbocyanines

Diazacarbocyanines undoubtedly represent the most important class of cationic yellow dyes for polyacrylonitrile. The leading golden yellow dye is CI Basic Yellow 28 **(29)**, which is prepared by methylating the orange dye **(30)**. The simple effect of methylation, discovered by chemists at Bayer,[15] has a dramatic effect on light stability: the *N*-methyl dye **(29)** has vastly superior light fastness to the protonated dye **(30)**, i.e., 6–7 vs 4 at standard depth of shade. The dye **(30)** is prepared[15] by diazotization of *p*-anisidine followed by coupling to Fischer's base **(31)**. The resulting orange dye is then treated first with a base and then with dimethyl sulfate to give **(29)**; see Scheme 7.

3. Triazacarbocyanines

Triazacarbocyanine dyes are a more recent discovery than either the aza- or diaza-carbocyanine dyes. They are less important commercially than their aza or diaza counterparts but commercial dyes do exist, such as the yellow dye **(32)**. Like the diazacarbocyanines, the triazacarbocyanine dyes are prepared by diazotization and coupling except that in this case the coupling takes place on a nitrogen atom rather than a carbon atom; the dyes are normally a mixture of two isomers. Scheme 8 shows the preparative route to **(32)**.[14,16]

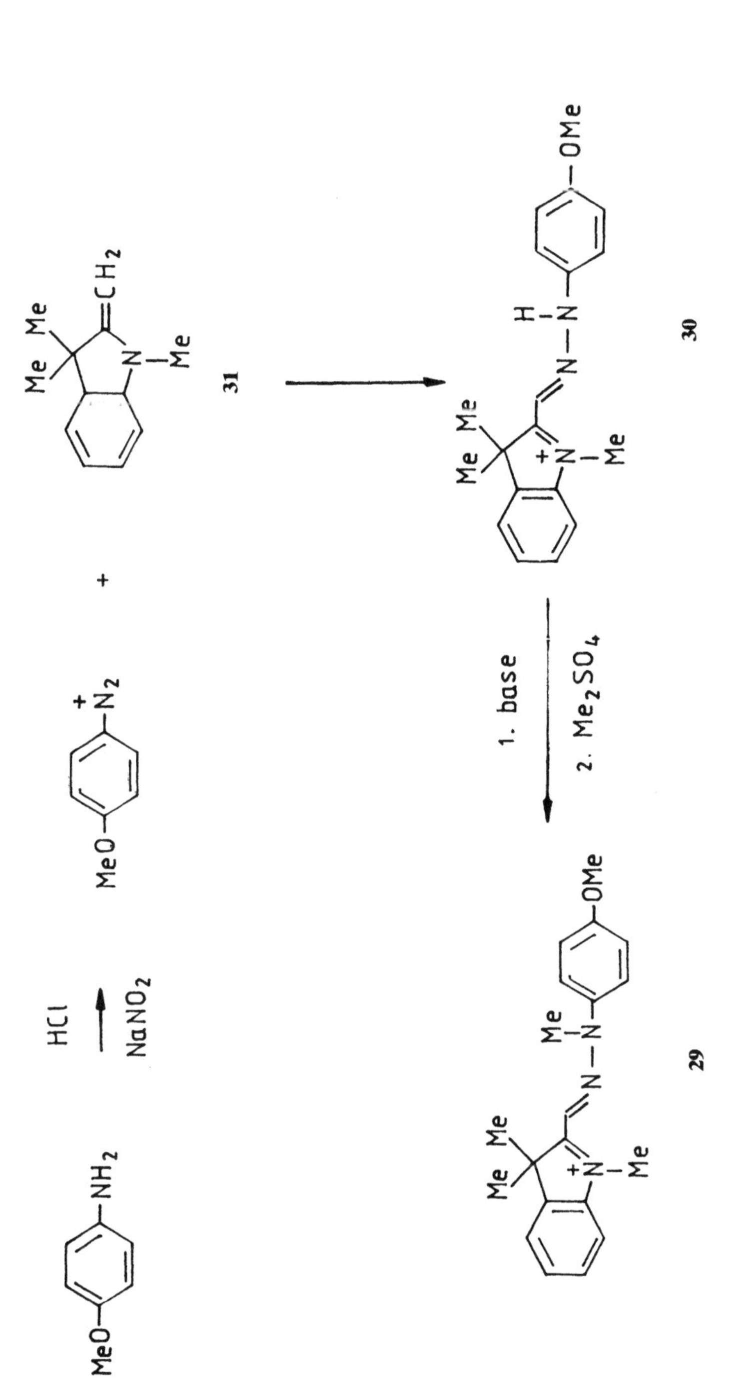
MeO
NH2
HCl
NaNO2
MeO
N2
+
Me
Me
CH2
N
Me
31
Me
Me
+
N
Me
N
N
H
OMe
30
1. base
2. Me2SO4
Me
Me
+
N
Me
N
N
Me
OMe
29

Scheme 7

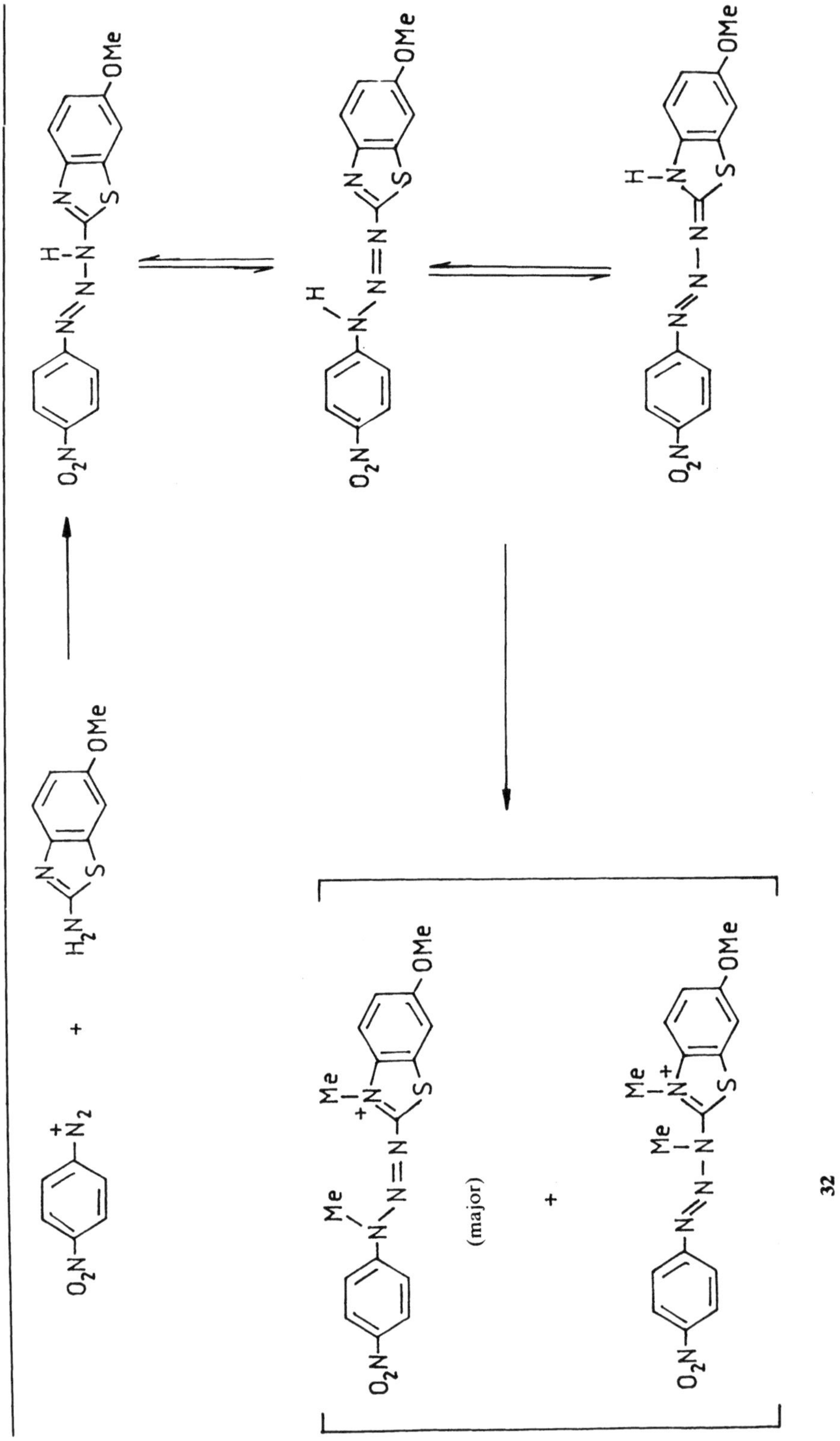

Scheme 8

33 33A

X, Y = CH or N

In the delocalized cationic yellow dyes discussed, the two limiting structures are **(33)** and **(33A)**. For simplicity, only one structure has generally been depicted but it should be realized that the positive charge is delocalized as shown.

D. Red Dyes

The two main types of delocalized cationic red dyes are hemicyanines and diazahemicyanines.

1. Hemicyanines

As the name implies, hemicyanines are half-cyanines in which a dialkylaniline residue replaces "half" of a true cyanine molecule. CI Basic Red 14 **(34)** is a typical cationic red hemicyanine dye. They are prepared by condensation of Fischer's base **(31)** with a 4-formyl dialkylaniline (equation 3).[2]

31 +

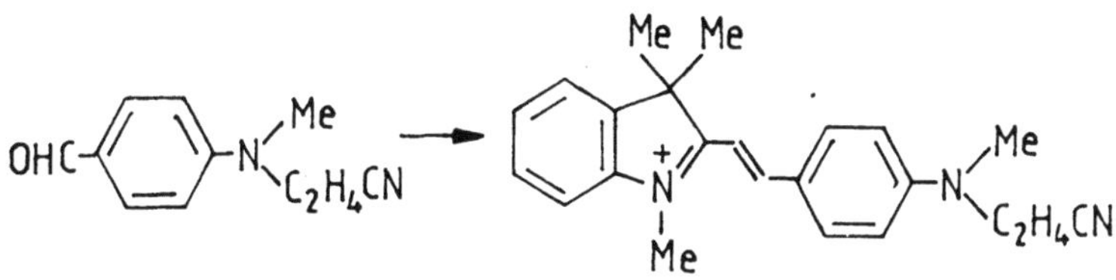

(3)

34

As a class, hemicyanines are exceptionally bright, fluorescent dyes with high tinctorial strength. However, like many fluorescent dyes, they display only poor to moderate light fastness, even on polyacrylonitrile fibers. For this reason, their use has declined in favor of the diazahemicyanine red dyes.

2. Diazahemicyanines

Diazahemicyanine dyes represent the biggest single advance in cationic dyes for polyacrylonitrile. Both the leading red and blue cationic dyes belong to this class. Formally, they are derived from the hemicyanines by replacing the two central CH units by nitrogen atoms to give the basic structure **(35)**. Various heterocycles can be used to complete the ring and these normally contain either nitrogen atoms or nitrogen and sulfur atoms. As seen from Table 3, the heterocycles containing only nitrogen give red dyes while those containing both nitrogen and sulfur give blue dyes.

Table 3. Colors of Some Typical Diazahemicyanine Dyes

Hetero ring	Color	λ_{max}^{MeOH} (nm)
Tetrazole	Red	520
Imidazole-2	Red	522
1,2,4-Triazole-2	Red	531
Indazole-3	Rubine	538
Pyridine-2	Reddish-violet	553
Thiazole-2	Reddish-blue	588
Benzthiazole-2	Reddish-blue	592
3-Methyl-isothiazole-5	Reddish-blue	595
2-Naphthothiazole-2	Blue	619
Benzoisothiazole-3	Green	683[a]

[a] In aqueous acetic acid.

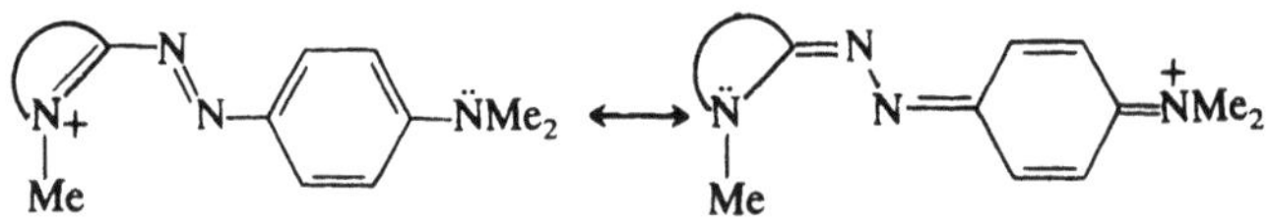

35

Although diazahemicyanine dyes have been known since 1949, they proved difficult to synthesize and were obtained only in poor yield.[2] This is because many amino heterocycles could not be diazotized and coupled successfully owing to competing hydrolysis reactions. For example, strongly acidic conditions (nitrosyl sulfuric acid) are required to effect diazotization of the weakly basic amino heterocycles and this causes protonation of the heterocycle, which in turn renders the diazonium group labile to nucleophilic substitution.[2] The reactions are depicted for the diazotization of 2-aminopyridine; see Scheme 9.

Hunig, in a series of 29 papers from 1957 to 1968, revealed a better route to diazahemicyanines by oxidative coupling.[2] In oxidative coupling, a heterocyclic hydrazone **(36)** is coupled with an aromatic coupling component in the presence of a mild oxidizing agent, e.g., ferricyanide. The reaction medium is generally water plus an organic solvent such as dimethylformamide, methanol, or glycol ether. The fundamental reaction is a four-electron oxidation (equation 4).

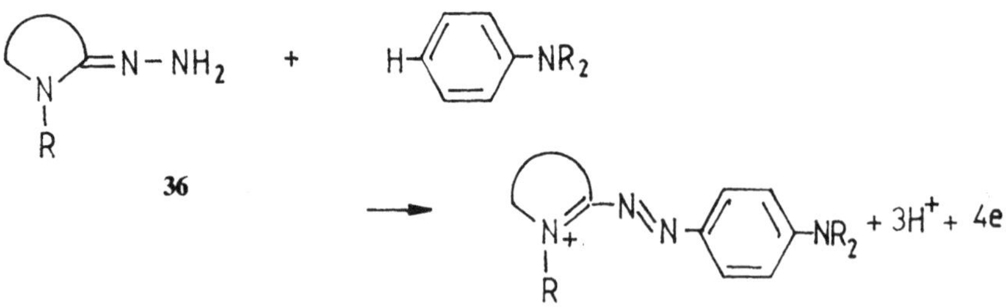

36 (4)

Hunig's discovery was timely, appearing just after the commercial launch of the first acrylic fibers. Had this not been so, Hunig's work may well just have

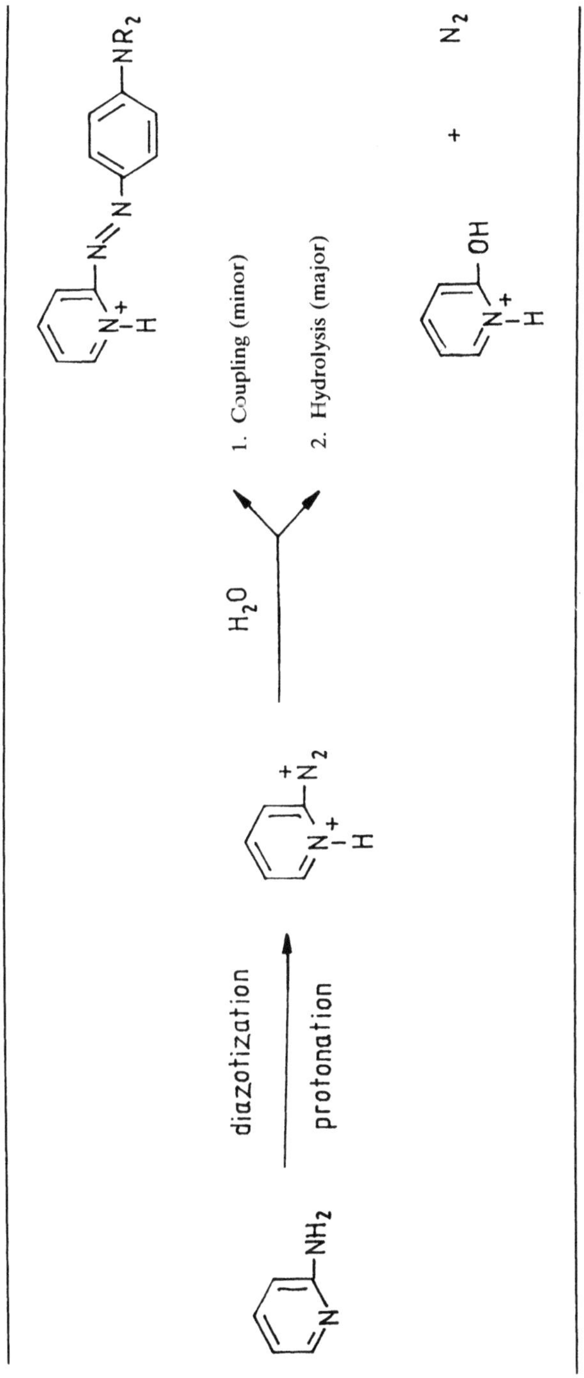
diazotization
protonation
H_2O
1. Coupling (minor)
2. Hydrolysis (major)
NH_2
NR_2
OH
N_2

Scheme 9

been an academic curiosity. In the event, certain diazahemicyanine dyes were found to possess outstanding properties on polyacrylonitrile. They gave bright, strong colors having exceptional fastness properties. Of the many types examined (see Table 3), the best red dyes were those derived from 3-amino-1,2,4-triazole **(37)**. The derived triazolium dyes may be prepared by the oxidative coupling route just described. However, increased knowledge about the diazotization of amino heterocycles has led to improved conditions which allow satisfactory diazotization and most diazahemicyanine dyes are now usually prepared by diazotization and coupling reactions. In the case of the triazolium dyes, the aminotriazole **(37)** is diazotized and coupled to an aniline to give a dye base **(38)**; methylation yields the cationic dye **(39)**. The two leading red dyes are CI Basic Red 22 (**39**; R = R^1 = Et) and especially the yellower homolog (**39**; R = Me, R^1 = CH_2Ph).

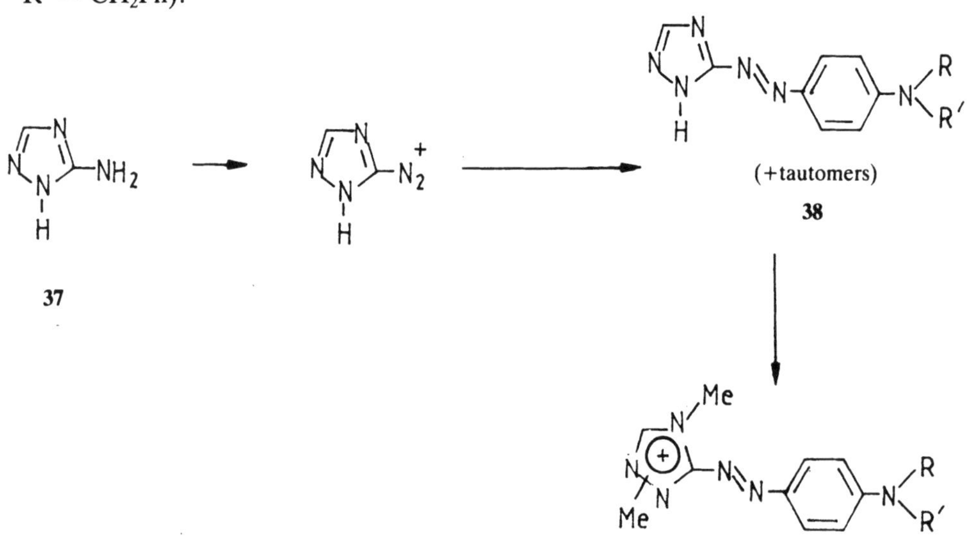

(+tautomers)

38

37

39

An interesting feature of the triazolium dyes is that both commercial products are a mixture of two isomers. Theoretically, three isomers are possible although in practice only two are formed: the bluish-red 2,4-isomer **(40)** and the yellowish-red 1,4-isomer **(41)** in an approximately 6:1 ratio.[17] The 1,2-isomer **(42)** is not formed.

The bathochromic shade of the 2,4-isomer is attributable to the fact that the positive charge is delocalized throughout the molecule, as depicted by the limiting structures **(40)** and **(40A)**. Such delocalization is not possible in the 1,4-isomer; the positive charge is restricted to the triazolium ring—structures **(41)** and **(41A)**.

E. Blue Dyes

The delocalized cationic blue dyes contain the widest variety of structural types and these include oxazines, thiazines, triphenylmethanes, naphtholactams, and diazahemicyanines.

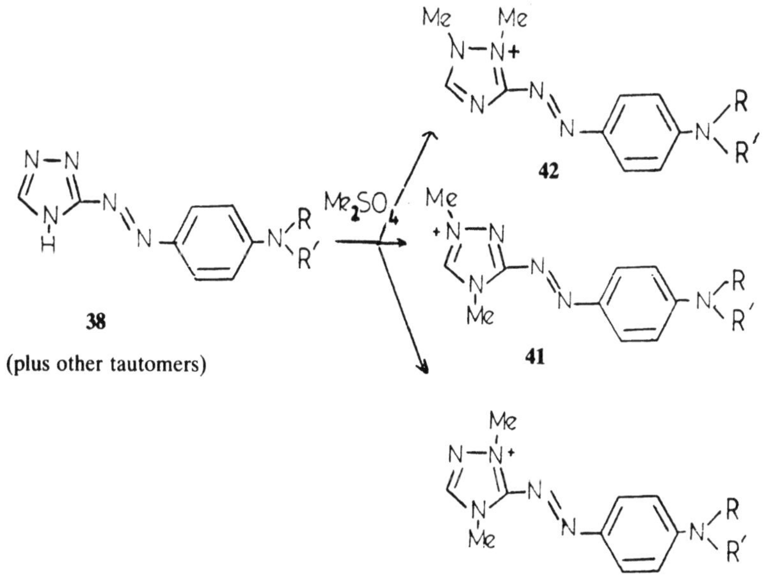

38

(plus other tautomers)

42

41

40

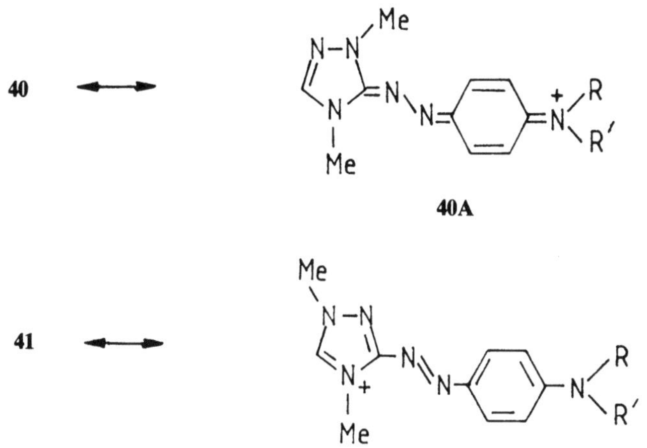

40A

41A

1. Oxazines

Oxazines represent one of the oldest classes of delocalized cationic dyes. They are bright, strong dyes with moderate to good light fastness properties. Oxazines give attractive greenish-blue to turquoise shades, typified by CI Basic Blue 3 **(43)**. The general synthetic route is shown in Scheme 10.

Chemists at BASF discovered that by replacing one of the dialkylamino groups of CI Basic Blue 3 with arylamino groups, dyes of higher light fastness are obtained.[18] Typical dyes are **(44)** and **(45)**.

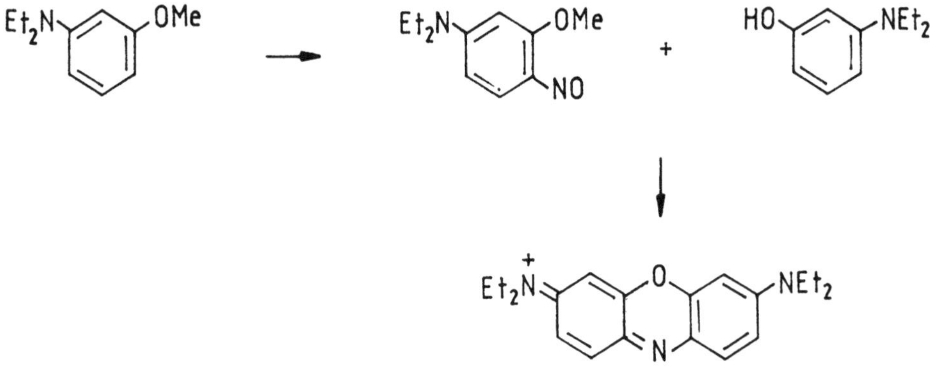

43

Scheme 10

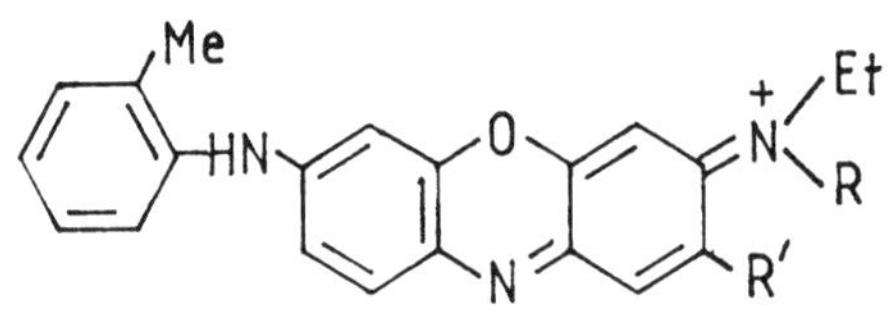

44 R = Et, R' = H
45 R = H, R' = Me

2. Thiazines

The sulfur analogs of the oxazines, namely the thiazines, are little used nowadays. Methylene Blue **(46)**, a dye used extensively for staining bacteria,[19] is the simplest and most common thiazine dye. Its color and strength are similar to the oxazine analog but its light fastness is lower. Surprisingly, the introduction of a nitro group produces a dye, Methylene Green **(47)**, that has excellent light fastness.

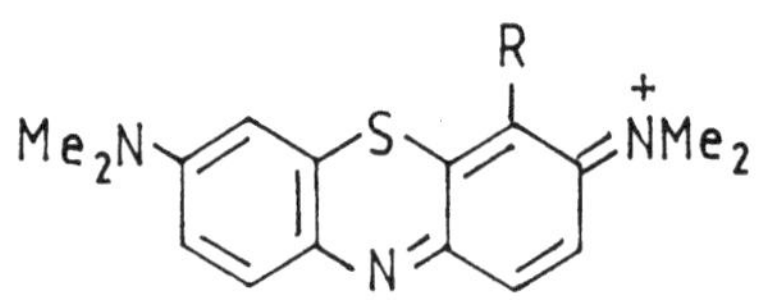

46 R = H
47 R = NO_2

3. Triphenylmethanes

Triphenylmethane dyes are one of the oldest classes of synthetic dyes. They have high tinctorial strength, are reasonably bright, but generally have only moderate light fastness.

The green dyes such as Malachite Green **(48)** are generally used as constituents of blacks, usually in combination with a red dye such as CI Basic Red 18 **(10)**. More recently, blue dyes (such as **49**) have appeared and these are used in self-shades. The general synthetic route to triphenylmethane dyes is shown in Scheme 11.[20]

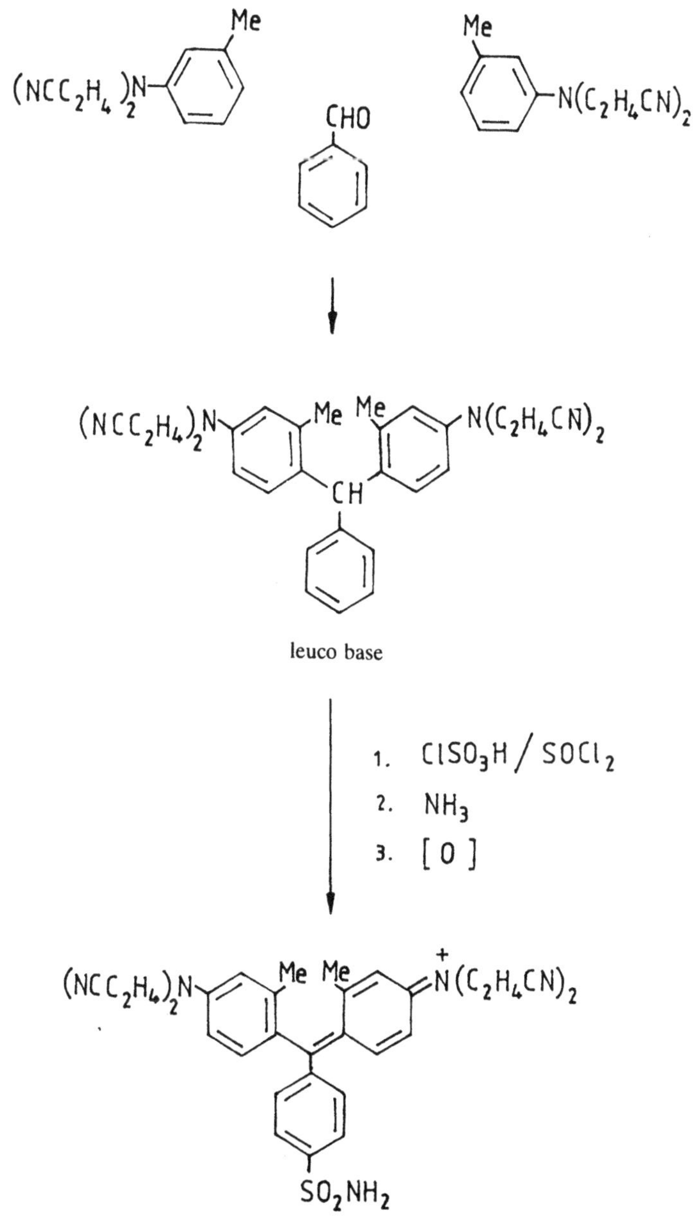

Scheme 11

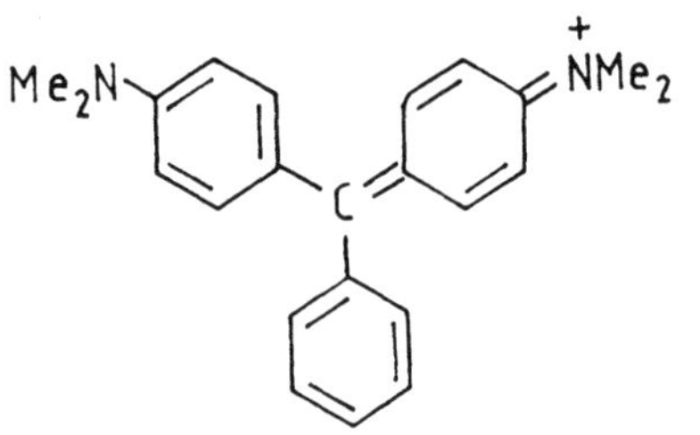

48

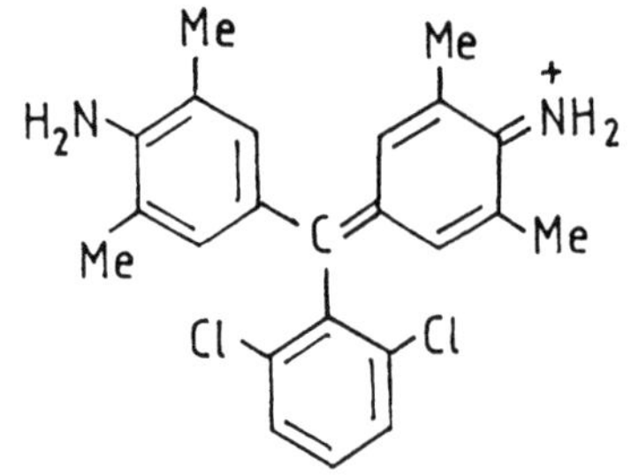

49

4. Naphtholactams

Naphtholactams represent a fairly recent class of blue dye. A typical route to a typical dye **(50)** is shown by Scheme 12.[21]

Naphtholactams are bright, reddish-blue dyes having only moderate tinctorial strength (ε_{max} ~ 25,000–30,000); they display good light fastness and good heat fastness properties.

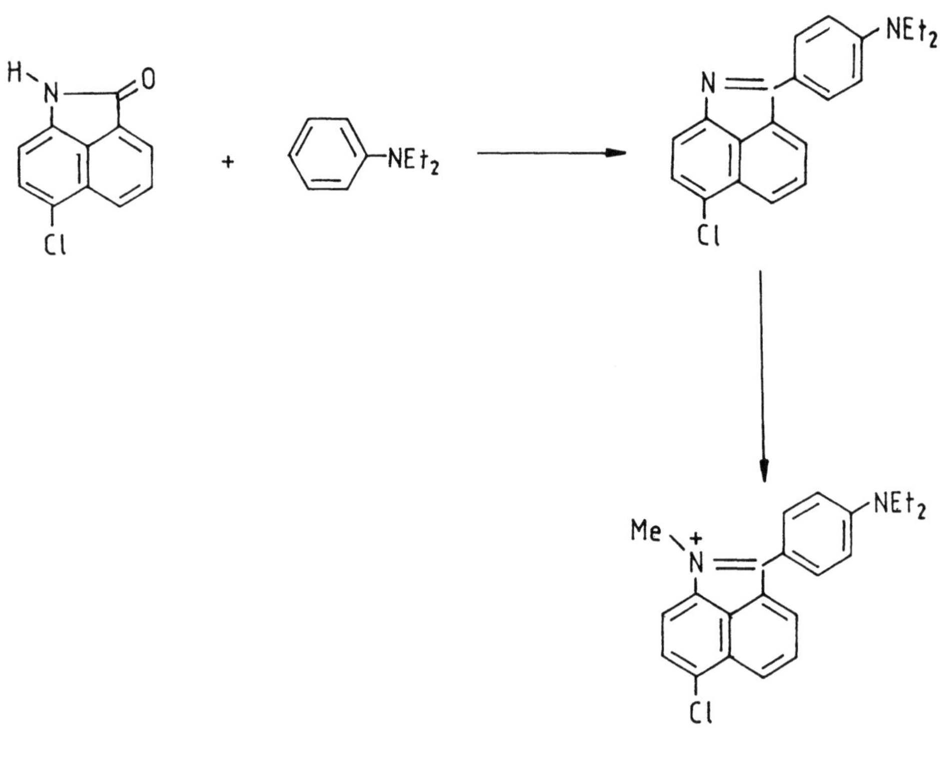

50

Scheme 12

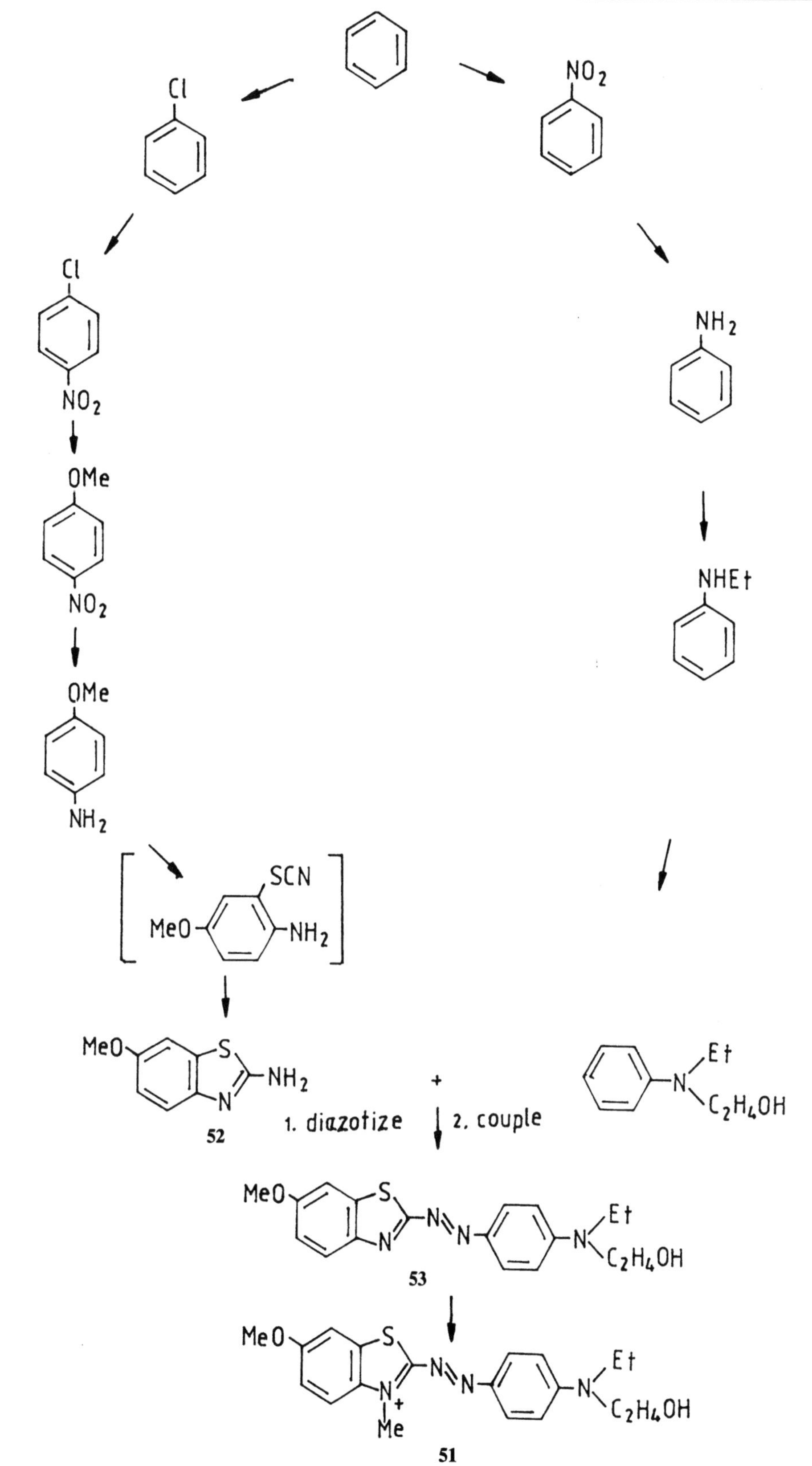
Cl
NO_2
OMe
NH_2
NHEt
SCN
MeO
52
1. diazotize
2. couple
Et
C_2H_4OH
53
Me
51

Scheme 13

5. Diazahemicyanines

As was the case with the red dyes, the diazahemicyanine class again boasts the market leading blue dye, namely CI Basic Blue 41 **(51)**. The dye is prepared from benzene by the sequence of reactions shown in Scheme 13.[22] The final steps involve diazotization of the benzothiazole **(52)** and coupling to an aniline to give the dyebase **(53)**. Quaternization with dimethyl sulfate yields CI Basic Blue 41.

The diazahemicyanine blue dyes are the strongest of the cationic blues typically having ε_{max} values in the range 70,000–90,000, with a broad absorption curve. Hence, the area under the absorption curve, which is the true measure of tinctorial strength, is large. The hues range from reddish blue, e.g. **(54)**, to green, e.g. **(55)**. CI Basic Blue 41 is an attractive mid-blue. The dyes are surprisingly bright in view of their broad absorption curves. The light fastness depends upon the substituents but ranges from good **(54)** to excellent **(51)**.

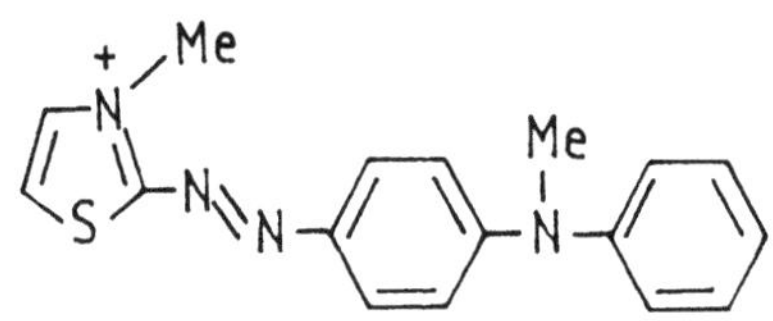

54

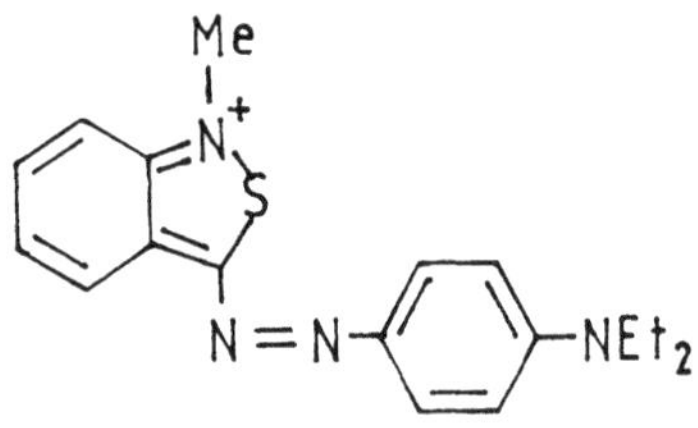

55

Unlike the market-leading yellow and red dyes, CI Basic Blue 41 has a technical defect, its so-called heat fastness. Essentially, this means that the dye has some instability toward post heat treatments such as steaming conditions used to introduce permanent pleats into garments, such as skirts. It is also manifested in the boiling aqueous dyebath where some hydrolysis of the dye takes place.

The relative instability of the blue diazahemicyanine dyes is a direct consequence of the increased delocalization of the positive charge. In these dyes there is a large contribution from the limiting structure **(51A)** to the ground state of the molecule. Since the majority of chemical reactions occur in the ground state of the molecule, it is apparent why CI Basic Blue 41 undergoes hydrolysis. Nucleophiles such as water or, especially, the hydroxide ion attack the electrophilic carbon atom of the immonium group to produce the spiro-type compound **(56)**; the amine group is then eliminated to form the red merocyanine dye **(57)**—this product has been isolated and characterized. The dye **(57)** then degrades further to produce colorless compounds. The reactions are illustrated in Scheme 14.

There have been many attempts to improve the heat fastness of blue diazahemicyanine dyes. Knowing the probable degradation route, these attempts have followed two basic lines:

1. Increase the steric hindrance around the amino group.
2. Decrease the contribution of the limiting structure such as **(51A)**.

For the first approach, more bulky groups have been used on the amino nitrogen atom, such as C_4 to C_6 alkyl, cyclohexyl, and α-branched groups such as

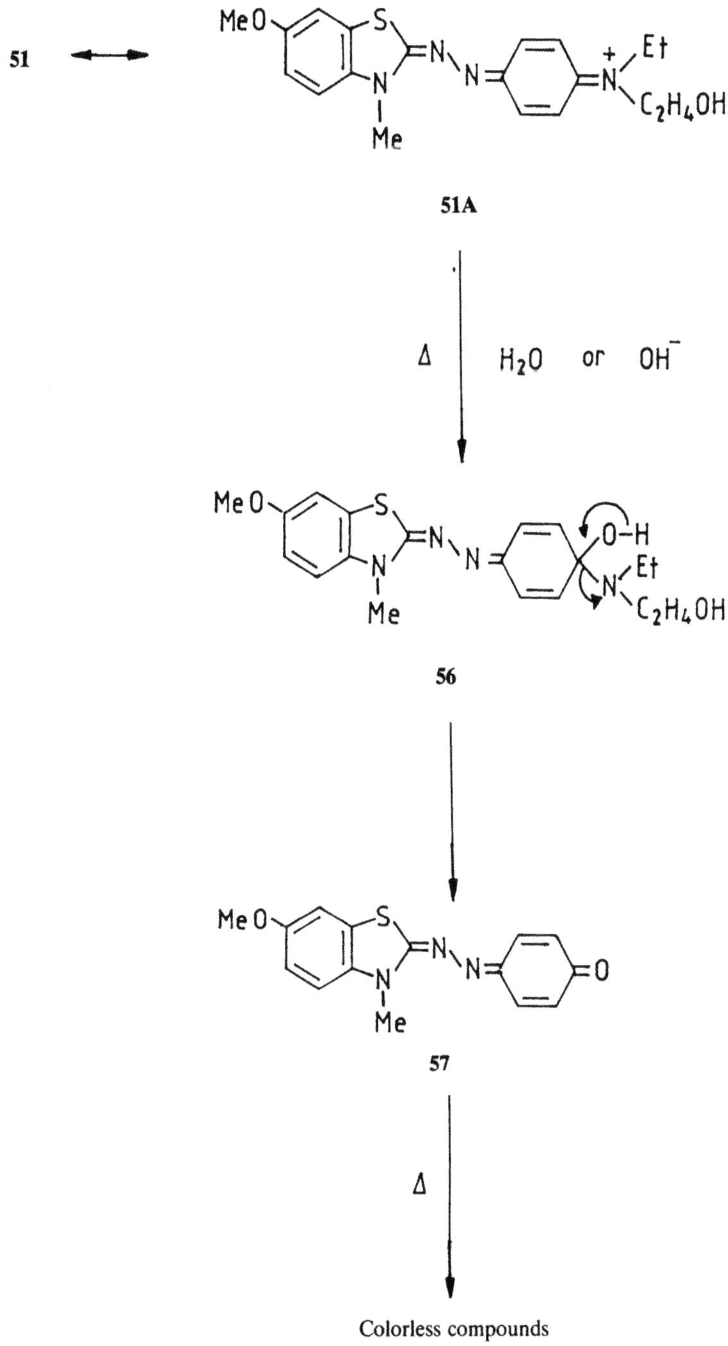
51
MeO
S
N
N
N
Et
C_2H_4OH
Me
51A
Δ
H_2O or OH^-
O–H
56
O
57
Δ
Colorless compounds

Scheme 14

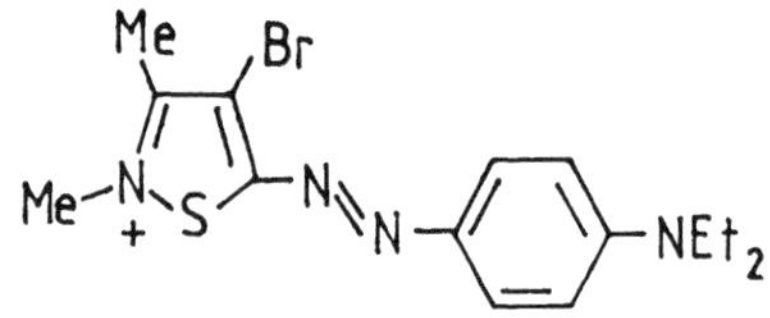

58

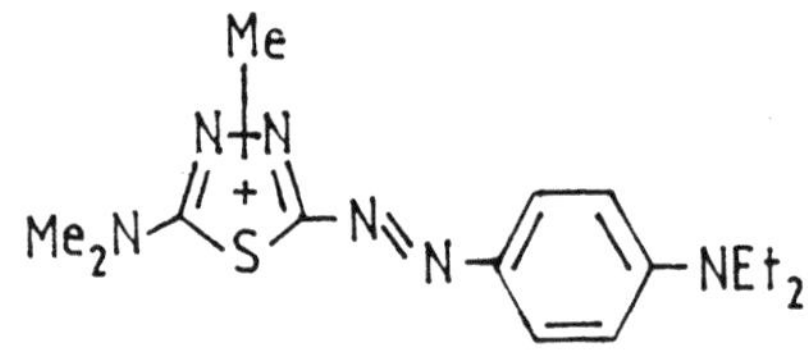

59

isopropyl. In the second approach, which has generally proved more successful, heterocycles other than benzothiazoles have been used. These include isothiazoles, e.g. **58,** and thiadiazoles, e.g. **59**. To date, however, CI Basic Blue 41 still reigns supreme, its many virtues outweighing its minor defect.

IV. SYNTHESIS

In this section the synthesis of selected dyes from the above six sections is given. The selection is based primarily upon commercial importance and, to a lesser extent, upon interesting/novel chemistry. At least one example each of a yellow, red, and blue dye of both the pendant and delocalized cationic types is given.

Where possible, the stoichiometry and yields of the reactions are quoted.

A useful general account of practical dye chemistry is given in the book by Fierz-David and Blangey.[23]

A. Pendant Cationic Dyes

1. Azopyridone Yellow

a. Preparation of the Golden Yellow Dye **(6).** Chloroacetamide is quaternized with pyridine and then reacted with ethylacetoacetate to form a pyridinium pyridone. Coupling of this compound with diazotized 4-aminoazobenzene gives the yellow dye **6**; see Schemes 1 and 2.

b. Quaternization.[3,4] Chloroacetamide (93.5 g; 1.0 *M*) is dissolved in dimethylformamide (200 ml) by warming to 45–50 °C. The solution is cooled to ambient temperature and pyridine (85 ml; 1.05 *M*) is added. The mixture is stirred and heated gently to 100 °C whereupon the clear colorless solution becomes opaque and an exothermic reaction occurs (the temperature rises to about 145 °C). A thick, white crystalline precipitate forms and the exotherm soon subsides. The mixture is stirred and heated at 100–110 °C for a further 1 h to complete the reaction and is then allowed to cool to ambient temperature. The pyridinium acetamide is filtered off, pressed dry, and then slurried in acetone (500 ml) and refiltered. The white solid is dried at 40 °C to give a yield of 157 g (91% of theory).

c. Condensation.[3,4] To a rapidly stirring mixture of ethylacetoacetate (130 g; 1.0 *M*) and pyridinium acetamide (173 g; 1.0 *M*) in methanol (500 ml) is added a solution of sodium hydroxide (40 g: 1.0 *M*) in water (100 ml). After refluxing for 3 h, the mixture is cooled to ambient temperature and the pyridinium pyridone, which is present as a yellow solid, is filtered off to yield 202.5 g (85% of theory). Recrystallization from water gives yellow needles in 65% yield (154 g).

d. Diazotization and Coupling.[3,4] 4-Aminoazobenzene (9.7 g: 0.05 *M*) is stirred thoroughly with ice-water (125 ml) and 10 *M* hydrochloric acid (12.5 ml; 0.125 *M*) is added followed by 1 *M* sodium nitrite solution (50 ml; 0.05 *M*). The reaction mixture is stirred at 0–5 °C for 30 min.

The above diazo solution is then added to a stirred solution of the pyridinium pyridone (120 g; 0.05 *M*) in water (1 1) at 0–5 °C and the mixture stirred for 1 h at this temperature to complete the coupling reaction. The yellow dye **6** is then filtered off, washed with a little cold water, and dried at 40 °C, to yield 13.0 g (60% of theory).

2. Azobenzene Red

a. Preparation of CI Basic Red 18:1.[6,24] CI Basic Red 18:1 **(10b)** is prepared by the diazotization of 2-chloro-4-nitroaniline followed by coupling to *N*-ethyl-*N*-2-pyridiniumethylaniline; see Scheme 4.

b. Preparation of N-*ethyl*-N-2-*pyridiniumethylaniline.* Phosphorus oxychloride (45.2 g; 0.295 *M*) is stirred in a reaction flask at about 40 °C. *N*-Ethyl-*N*-2-hydroxyethylaniline (66 g; 0.4 *M*) is added at such a rate that the temperature increases to about 85 °C; the remainder is then added so that the temperature is maintained at 80–90 °C. On completion of the addition, the reaction mixture is stirred at 95–100 °C for 1.5 h.

Sodium hydroxide liquor (124 g; *ca* 10 *M*; 1.0 *M*) and water (80 ml) are stirred in a 500-ml beaker at 30 °C. The phosphorus oxychloride reaction mixture is then added to the stirred sodium hydroxide solution, ice being added as necessary to keep the temperature between 35 and 40 °C. The addition takes about 30 min and about 200 g of ice are required. The reaction mixture is stirred for a further 30 min and then transferred to a separating funnel. The lower aqueous layer is discarded and the upper layer of *N*-ethyl-*N*-2-chloroethylaniline is weighed and bottled to yield 71.8 g (93% of theory).

N-Ethyl-*N*-2-chloroethylaniline (55 g; 0.3 *M*), pyridine (26.1 g; 0.33 *M*), and water (90 ml) are stirred under reflux for 3 h. The solution is allowed to cool to ambient temperature and filtered to remove any insoluble matter. The clear filtrate (170.6 g, 46% assay) is bottled and used directly in the coupling reaction.

c. Diazotization of 2-chloro-4-nitroaniline and Coupling to N-*ethyl*-N-2-*pyridiniumethylaniline.* 2-Chloro-4-nitroaniline (54.3 g; 63.5% strength; 0.2 *M*) is dissolved in glacial acetic acid (320 ml) by heating to 70 °C and concentrated hydrochloric acid (52 ml; 0.52 *M*) added. The stirred solution is cooled to 0–5 °C and the resulting yellow suspension diazotized by the addition of 2 *M* sodium nitrite solution (104 ml; 0.208 *M*). Ice-water (200 ml) is added and the diazo solution is then filtered from insoluble matter. The excess of nitrous acid is destroyed by the addition of sulfamic acid solution.

The above diazo solution is added to the stirred coupling solution of *N*-ethyl-*N*-2-pyridiniumethylaniline (114 g; 0.2 *M*) at 0–5 °C and the resulting solution stirred for 4 h at 0–5 °C and then for a further 18 h at ambient temperature. The solution is diluted to about 2.5 l with water and the acidity to Congo Red paper removed by the addition of concentrated ammonia solution. Sodium chloride (*ca* 300 g) is then added and the mixture stirred for 15 min. The precipitated CI Basic Red 18:1 is filtered off, pressed dry, and the cake slurried in acetone (2 l), filtered, washed with acetone, and dried to yield 65 g.

3. Anthraquinone Blue

a. Preparation of CI Basic Blue 22. CI Basic Blue 22 **(13)** is synthesized by reaction of 1-methylamino-4-bromoanthraquinone (prepared by bromination of 1-methylaminoanthraquinone) with 1-amino-3-*N*,*N*-dimethylaminopropane followed by quaternization with dimethyl sulfate; see Scheme 5.

b. Condensation reaction.[25] 1-Methylamino-4-bromoanthraquinone (10 g) and copper acetate (0.1 g) are stirred in 1-amino-3-*N*,*N*-dimethylaminopropane (30 ml) at 90–100 °C until there is no further change in the blue color of the solution. Sufficient water is then added to precipitate the dye. The mixture is allowed to cool to ambient temperature and the dye is filtered off, washed with water, and dried.

c. Quaternization reaction.[25] The blue dye (10 g) is dissolved in *o*-dichlorobenzene (200 ml) with stirring at 60 °C. Dimethyl sulfate (4 g) is added dropwise at 60 °C and the precipitated blue needles of CI Basic Blue 22 are filtered off, washed with toluene, and dried.

B. Delocalized Cationic Dyes

1. Azacarbocyanine Yellow

a. Preparation of CI Basic Yellow 11. Fischer's aldehyde **(26)** is condensed with an equimolecular proportion of 2,4-dimethoxyaniline sulfate **(27)** in the presence of alcohol to give CI Basic Yellow 11 **(25)**.

b. Condensation.[26,27] 2,4-Dimethoxyaniline (30.6 g; 0.2 *M*) is stirred with ethanol (80 ml) and 79% strength sulfuric acid (7.2 ml = 12.4 g of 100% sulfuric acid; 0.1 *M*) is added dropwise over about 10 min whereby a white crystalline paste of the half sulfate of 2,4-dimethoxyaniline is formed. Fischer's aldehyde

(40 g of 100%; 0.2 *M*) is added and the mixture stirred at a temperature of 40–45 °C for 1 h.

Water (60 ml) is then added to temporarily precipitate the dye and, after a further 15 min stirring at 40–45 °C, the dye redissolves. The solution is poured into cold water (800 ml) and the dye salted out by the addition of sodium chloride (40 g). After stirring for 2 h, the CI Basic Yellow 11 **(25)** is filtered off, washed with a little 2% brine, and dried at 40–50 °C to yield 91.6 g.

2. Diazacarbocyanine Yellow

a. Preparation of CI Basic Yellow 28 **(29).** *p*-Anisidine is diazotized and coupled to Fischer's base **(31)** to give the orange hydrazone dye **(30)**. This is basified and the resultant dyebase quaternized with dimethyl sulfate to give CI Basic Yellow 28 **(29)**; see Scheme 7.

b. Preparation of the Dyebase.[15] A stirred solution of *p*-anisidine (12.3 g) in water (100 ml) and 10 *M* hydrochloric acid (20 ml) is diazotized by the addition of an aqueous 2 *M* sodium nitrite solution (53 ml) at a temperature of 0 °C. The resulting solution is stirred for 1 h at 0–5 °C, filtered, and then added dropwise to a stirred solution of Fischer's base (18.2 g) in water (100 ml) and 10 *M* hydrochloric acid (20 ml) at 0–5 °C. The mixture is allowed to stir for 30 min then the pH is raised to 5–6 by the addition of 4 *M* sodium acetate solution (100 ml), and the coupling allowed to continue for a further 2 h at 0–5 °C. The product is filtered off and washed with cold water (25 ml). The cake is slurried in 0.5 *M* aqueous sodium carbonate solution (500 ml) to convert the protonated hydrazone form **(30)** into the neutral azo dyebase. The solid is filtered off, washed with water (100 ml), and dried at 60 °C in 54% yield (17.0 g), mp 150–152 °C. After crystallization from methanol, the product melts at 157–158 °C.

c. Quaternization.[28] The crystallized dyebase (30.7 g) is dissolved with heating in toluene (100 ml) and to this stirred solution dimethyl sulfate (12.6 g) is added dropwise. The reaction mixture is then heated on a steam bath and, after a short time, the quaternary dye separates out as yellow crystals. The mixture is heated on the steam bath for a further 3 h, cooled, and the dye filtered off. The methosulfate of CI Basic Yellow 28 **(29)** is thus obtained in a pure form; no further purification is necessary.

3. Triazacarbocyanine Yellow

a. Preparation of the Yellow Dye **(32).**[16] The yellow dye **(32)** is prepared by diazotization of *p*-nitroaniline followed by coupling to 2-amino-6-methoxybenzothiazole. Quaternization of the resulting tautomeric triazene dye with dimethyl sulfate leads to the yellow dye **(32)**, itself a mixture of two isomers; see Scheme 8.

b. Diazotization and Coupling. p-Nitroaniline (13.8 g; 0.1 *M*) is slurried in concentrated hydrochloric acid (30 ml; 0.3 *M*) and then diluted by the addition of water (150 ml). The suspension is stirred at 0–5 °C and 1 *M* sodium nitrite solution (100 ml; 0.1 *M*) is added dropwise. The reaction mixture is stirred at 0–5 °C for 1 h and the excess of nitrous acid removed by the addition of sulfamic acid solution.

The above diazo suspension is poured into a stirred mixture of the aminobenzothiazole (18.0 g; 0.1 *M*) in glacial acetic acid (100 ml) and crushed ice (300 g). The solution is carefully neutralized by the dropwise addition of 10 *M* sodium hydroxide solution, keeping the temperature at 0–5 °C, until the pH of the reaction mixture is 7.0–7.5. After stirring for a considerable time at 0–5 °C, the precipitated yellow triazene dye is filtered off, washed with water, and dried. In a mixture of acetone/ethyl cellosolve 1:1, the dye has an absorption maximum of 419 nm.

c. Quaternization. A suspension of the triazene dye (3.3 g) in dimethyl sulfate (10 ml) is heated at 95–100 °C for 5 min. The warm melt is dissolved in warm water (500 ml) and the solution buffered with sodium acetate until it is only slightly acid (pH 6). A little charcoal is then added and the "solution" filtered. Zinc chloride solution is added to the filtrate to precipitate the yellow dye **(32)** as the zinc chloride double salt. The dye has an absorption maximum in water of 432 nm.

4. Hemicyanine Red

a. Preparation of CI Basic Red 14. Fischer's base **(31)** is condensed with 4-formyl-*N*-methyl-*N*-2-cyanoethylaniline in glacial acetic acid. CI Basic Red 14 **(34)** is isolated as the zinc chloride double salt (equation 3).

b. Condensation.[27,29] Fischer's base (69.2 g; 0.4 *M*) is added to glacial acetic acid (192 ml) with stirring and the temperature rises to about 45 °C. 4-Formyl-*N*-methyl-*N*-2-cyanoethylaniline (75.2 g; 0.4 *M*) is then added. The temperature of the resulting solution is then raised from 35 to 80 °C over 1 h and maintained at 80 ± 2 °C for a further 4 h.

The solution is poured into cold water (4.5 l) with stirring and 10 *M* hydrochloric acid (80 ml) is added followed by a solution of Dispersol OG (4 g) in water (40 ml). The temperature is lowered to 10 °C by the addition of crushed ice (*ca* 1 kg), and then a solution of zinc chloride (120 g) in water (120 ml) is added dropwise over about 2 h. The temperature is maintained at 10 °C by the addition of ice (*ca* 500 g). The reaction mixture is stirred overnight and the dye filtered off and washed with 5% brine (2 l). The cake is dried at 60 °C to give CI Basic Red 14 (**34**, 189 g).

5. Diazahemicyanine Red

a. Preparation of CI Basic Red 22.[30] 3-Amino-1,2,4-triazole **(37)** is diazotized and coupled with *N,N*-diethylaniline and the resulting azo dyebase **(38)** is quaternized with dimethyl sulfate to yield CI Basic Red 22 (**39**; R = R^1 = Et).

b. Diazotization and Coupling. 3-Amino-1,2,4-triazole (2.52 g, 0.03 *M*) is added portionwise to a stirred solution of concentrated sulfuric acid (3.3 ml) in water (12 ml) at 25 °C. Glacial acetic acid (12 ml) is added, the solution cooled to 0–5 °C, and 5 *M* sodium nitrite solution (6.6 ml) is added over 10 min. The solution is then allowed to stir for 3 h to ensure complete diazotization before the excess of nitrous acid is removed by the addition of 10% w/v sulfamic acid.

This diazo solution is then added over 30 min to a solution of *N,N*-diethylaniline (6.4 g, 0.03 *M*) in (70 ml) of 10% v/v hydrochloric acid below 5 °C. After stirring for 1.5 h the acidity to Congo Red paper is removed by the addition of sodium acetate crystals. The dyestuff is extracted into chloroform and purified by column chromatography using alumina-type "0" and chloroform as eluent, to yield 1.8 g of pure dye.

c. Quaternization. 3-Amino-1,2,4-triazole → *N,N*-diethylaniline (1.1 g, 0.0033 *M*) is dissolved in glacial acetic acid (20 ml) containing (0.35 g) of magnesium oxide at ambient temperature. The temperature is then raised to 70 °C and dimethyl sulfate (2 ml; 0.0165 *M*) is added dropwise over 15 min. After 1 h at 80–90 °C, the quaternization is judged to be complete by TLC,* and the reaction mixture is then poured into water (300 ml) and filtered through hyflo-supercel. The solution is salted to 15% w/v and the dye precipitated by the addition of 100% w/v zinc chloride solution (3 ml) and is filtered off and dried at 60 °C to give 0.9 g (59% of theory).

6. Oxazine Blue

a. Preparation of CI Basic Blue 3.[31] *N,N*-Diethyl-3-anisidine is nitrosated to give the 4-nitroso derivative. This is condensed with *N,N*-diethyl-3-aminophenol and the resulting CI Basic Blue 3 **(43)** is isolated as the zinc chloride double salt; see Scheme 10.

b. Nitrozation. N,N-Diethyl-3-anisidine (17.9 g; 0.1 *M*) is added to a stirred solution of water (120 ml) and 10 *M* hydrochloric acid (22 ml; 0.22 *M*) and the solution stirred and cooled to 0 °C by the addition of ice (*ca* 30 g). Sodium nitrite (7.1 g; 0.103 *M*) is dissolved in water (20 ml) and the solution added dropwise over about 1 h to the stirred *N,N*-diethyl-3-anisidine solution, keeping the temperature at 0–5 °C. On completion of the addition, an excess of nitrous acid should be present (detectable by an instant blue coloration of starch iodide paper)

* A major bluish-red and a minor orange-red spot, both of lower R_f value than the original yellow starting material, on silica gel plates with chloroform : methanol, 4 : 1, as eluent.

and the mixture should be strongly acid (blue color on Congo Red paper). Water (*ca* 100 ml) is then added and the mixture stirred for 2 h. Salt (*ca* 63 g) is added over 30 min and the mixture stirred for 4 h. The precipitated nitroso derivative is filtered off and the strength estimated by a titanous chloride titration. The obtained yield is normally about 85% theory.

c. Condensation. *N,N*-Diethyl-3-aminophenol (16.8 g; 0.102 *M*) is dissolved in ethanol (100 ml) by stirring under reflux. A suspension of the above nitroso paste in ethanol (50 ml) is added portionwise to the refluxing solution over about 1 h. The reaction mixture is then stirred at the reflux for a further 1 h and then allowed to cool to ambient temperature. A solution of anhydrous zinc chloride (8.0 g) in water (10 ml) is added and after stirring for 1 h the precipitated zinc chloride double salt of CI Basic Blue 3 is filtered off and dried at 60 °C.

7. Triphenylmethane Green[20]

Benzaldehyde is condensed with two equivalents of dialkylaniline in the presence of acid to form a leuco triphenylmethane. A sulfonamide group is introduced by chlorosulfonation and amination and the leuco dyebase is oxidized to the cationic triphenylmethane dye.

a. Condensation. A mixture of benzaldehyde (10.6 g), *N,N*-(bis-cyanoethyl)-*m*-toluidine (42.6 g), isopropyl alcohol (100 ml), 37% sulfuric acid (20 ml), and urea (5 g) is stirred for 3 h at 70 °C under nitrogen. The leuco triphenylmethane base crystallizes out and, after cooling to ambient temperature, is filtered off. The base thus obtained has a melting point of 168 °C.

b. Chlorosulfonation/Amination. The above leuco base (51.4 g) is added, over 2 h, to a stirred mixture of chlorosulfonic acid (160 g) and thionyl chloride (24 g) at 0–5 °C. After stirring for 4–5 h at 0–5 °C, the reaction mixture is drowned out into a stirred mixture of ice (2200 g) and concentrated ammonia (575 ml). The product is filtered off, washed with water, and dried.

c. Oxidation. The sulfonamido leuco base (59.2 g) is dissolved in 37% sulfuric acid solution (400 ml) and to this solution is added a solution of sodium bichromate (10 g) in water (30 ml) at 10–15 °C with stirring. The resulting cationic triphenylmethane dye is filtered off and dried.

8. Naphtholactam Blue

a. Preparation of the Naphtholactam Blue **(50)**. 4-Chloronaphtholactam is condensed with *N,N*-diethylaniline and the resulting dyebase quaternized with dimethyl sulfate to give **(50)**; see Scheme 12.

b. Condensation.[21] 4-Chloronaphtholactam (20.4 g) and *N,N*-diethylaniline (15 g) are triturated with chlorobenzene (120 ml) and phosphorus oxychloride

(34 g) is then added dropwise at 100–110 °C. The mixture is heated at the boil for a further 2 h and then cooled to 0 °C. The precipitated reaction product is filtered off and dissolved in hot ethanol (350 ml). Ethanolic potassium hydroxide (20% strength) is then added dropwise to the solution until the color changes from blue to red. The red solution is then diluted with water (300–500 ml) and the precipitate filtered off and dried. After recrystallization from cyclohexane (500 ml), the dye base has a melting point of 134 °C.

c. Quaternization.[21] The above dye base (38.5 g) is dissolved in toluene (500 ml) and the solution stirred at 100–105 °C. Methyl chloride is introduced into the solution until the reaction mixture is a pure blue color and a drop of the toluene solution on paper no longer spreads with a reddish color. The reaction mixture is then allowed to cool to ambient temperature and the dye **(50)** filtered off, washed with a little toluene, and dried.

9. Diazahemicyanine Blue

a. Preparation of CI Basic Blue 41. CI Basic Blue 41 **(51)** is prepared by the diazotization of 2-amino-6-methoxybenzothiazole **(52)** followed by coupling to *N*-ethyl-*N*-2-hydroxyethylaniline to give the dye base **(53)**. Quaternization of **53** with dimethyl sulfate produces **(51)**; see Scheme 13.

b. Preparation of the Dye Base.[22] 2-Amino-6-methoxybenzothiazole (1.8 g; 0.01 *M*) is dissolved in a stirred mixture of formic acid (7.5 ml) and 55% v/v sulfuric acid (11 ml) at ambient temperature. The solution is cooled to 0–5 °C (the amine separates as a fine white suspension) and diazotization is effected by the dropwise addition of nitrosyl sulfuric acid (1.5 ml; 0.01 *M*). The solution is then stirred at 0–5 °C for 1 h and the excess of nitrous acid destroyed by the addition of sulfamic acid.

The above diazo solution is added dropwise to a stirred solution of *N*-ethyl-*N*-2-hydroxyethylaniline (1.65 g; 0.01 *M*) dissolved in water (100 ml) and 2 *M* hydrochloric acid (10 ml) at 0–5 °C. The coupling mixture is then stirred at 0–5 °C for 1 h and sufficient sodium acetate solution added to remove the acidity to Congo Red paper. The precipitated dye base is then filtered off, washed with water, and dried at 40 °C to yield 2.9 g (81.5% of theory).

c. Quaternization.[22] The dye base (2.9 g) is dissolved in glacial acetic acid (30 ml) and magnesium oxide (0.5 g) added. Dimethyl sulfate (3 ml) is then added and the mixture stirred and heated at 70 °C for 30 min; TLC indicates the reaction is then complete. After cooling to room temperature, the blue solution is poured into water (*ca* 100 ml), filtered through "Hyflo," and the filtrate treated with a saturated solution of zinc chloride (2 ml). The precipitated CI Basic Blue 41 is then filtered off and dried.

V. SUMMARY

Cationic dyes for polyacrylonitrile fall into two main categories: the pendant cationic dyes in which the positive charge is localized on one atom, usually nitrogen, and the delocalized cationic dyes where the charge is delocalized over the whole molecule. The delocalized cationic dyes are by far the most important, being brighter, stronger, and more cost effective than the pendant cationic dyes. Indeed, the market-leading dyes in the three primary color areas of yellow, red, and blue are all delocalized cationic dyes. The leading yellow dye is CI Basic Yellow 28 **(29)**, a diazacarbocyanine, the leading red dye is **39** (R = Me, R′ = CH_2Ph), a triazolium diazahemicyanine, and the leading blue dye is CI Basic Blue 41 **(51)**, a benzothiazolium diazahemicyanine.

The current cationic dyes for polyacrylonitrile fibers constitute the most technically excellent dyes available today.

REFERENCES

1. P. L. Meunier, J. F. Lancius, J. A. Brooks, and R. J. Thomas, *Am. Dyest. Rep. 42,* 470 (1953).
2. D. R. Baer, Cationic Dyes for Synthetic Fibers, in: *The Chemistry of Synthetic Dyes* (K. Venkataraman, ed), Vol. IV, pp. 160–210, Academic Press, New York and London (1971).
3. B. Parton (ICI), British Patent 1,548,907 (1979).
4. W. Steinemann (Sandoz), British Patent 1,332,377 (1973).
5. S. N. Boyd (du Pont), U.S. Patent 2,821,526 (1958).
6. P. Gregory (ICI), British Patent 1,591,040 (1981).
7. R. Entschel, C. Muller, and W. Wehrli (Sandoz), British Patent 1,004,284 (1962); R. Entschel, C. Muller, and W. Wehrli (Sandoz), British Patent 1,008,636 (1962); J. Carbonell, V. Sanahuja, and H. Seigrist (Sandoz), British Patent 1,048,482 (1966).
8. P. F. Gordon and P. Gregory, *Organic Chemistry in Colour,* pp. 152–158, Springer-Verlag, Berlin, Heidelberg, New York, London, Paris, Tokyo (1987).
9. Ref. 8, pp. 124–126, and references cited therein.
10. Ref. 8, p. 236, and references cited therein.
11. Ref. 8, pp. 112–115.
12. P. Gregory and D. Thorp, *J. Chem. Soc., Perkin Trans. 1,* 1990 (1979).
13. Ref. 8, p. 228 and Ref. 2, p. 183.
14. Ref. 8, p. 229.
15. Bayer, British Patent 875,995 (1961), and references cited therein.
16. J. Voltz and W. Bossard, U.S. Patent 3,055,881 (1962).
17. D. Brierley, P. Gregory, and B. Parton, *J. Chem. Res.* (S), 174 (1980).
18. BASF British Patent 1,410,562 (1975).
19. P. F. Gordon and P. Gregory, Non-textile Uses of Dyes, in: *Critical Reports on Applied Chemistry,* Vol. 7, pp. 66–110, Developments in the Chemistry and Technology of Dyes (J. Griffiths, ed.), Blackwell, Oxford (1984).
20. Sandoz, British Patent 1,169,600 (1967).
21. Bayer, British Patent 973,259 (1964).
22. Geigy, British Patent 787,369 (1955); Geigy, British Patent 789,263 (1955); Ref. 8, pp. 46 and 50.
23. J. E. Fierz-David and L. Blangey, *Fundamental Processes of Dye Chemistry,* Interscience, New York and London (1949).
24. D. Haigh, *Int. Dyer, Text Printer, Bleacher Finish. 132.* 111 (1964).
25. Bayer, British Patent 807,241 (1955).
26. IG, British Patent 462,238 (1937).

27. A. S. Fern, C. A. Pulley, and S. M. Todd, British Patent 616,385 (1949).
28. Bayer, U.S. Patent 3,345,355 (1967).
29. IG, British Patent 463,042 (1935).
30. D. Brierley, P. Gregory, and B. Parton, *J. Chem. Res.* (M), 2476 (1980).
31. M. S. Moores, W. J. Balon, and C. W. Maynard, *J. Heterocycl. Chem. 6,* 755 (1969); H. Psaar and H. Heitzer, *Chem. Ber. 102,* 3603 (1969).

6

Dyes for Polyamide Fibers

F. WALKER

I. INTRODUCTION

The polyamides discussed with regard to their dyeing behavior in this chapter are of two types, the natural and the synthetic. The natural polyamides consist mainly of wool and other animal fibers such as those of the Angora goat (Mohair), the Cashmere goat, and, to a lesser extent, alpaca, llama, vicuna, guanaco, camel hair, and the hair of the Angora rabbit. The man-made fibers consist of a group under the generic name nylon. Although some six or seven nylons have found commercial value, only the two main types will be discussed.

II. DYES FOR WOOL AND OTHER ANIMAL FIBERS

A. Historical

The coloration of wool goes back to prehistoric times and probably involved immersing the animal fleece in aqueous extracts of plants and the bark of trees. The discovery that loose wool could be spun and then woven into fabrics was a major step forward such that the process of coloration, although still with natural products and by trial and error methods, achieved some sort of consistency.

It has only been in the last century that the coloring matters of the various plants and trees have been isolated and their structures determined. Thus, weld **(1)**, the coloring material extracted from the stem and leaves of *Reseda luteola,* contains luteolin as the active compound.[1] Madder, obtained from *Rubia tinctorum,* gave reds with an aluminum mordant and contained 1,2-dihydroxyanthraquinone **(2)** in the form of a glycoside.[2] The well-known

F. WALKER • 43 Child Lane, Roberttown, Liversedge, West Yorkshire WF15 7QN, England. Retired, formerly of L. B. Holliday & Co. Ltd., Leeds Road, Huddersfield HD2 1UH, England.

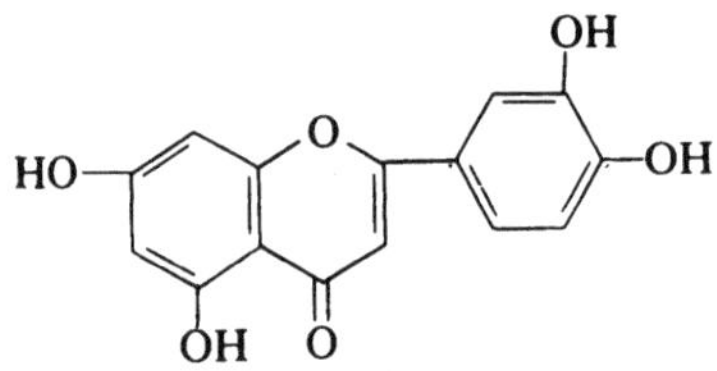

1

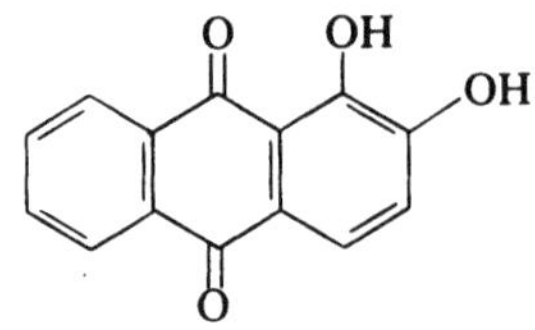

2

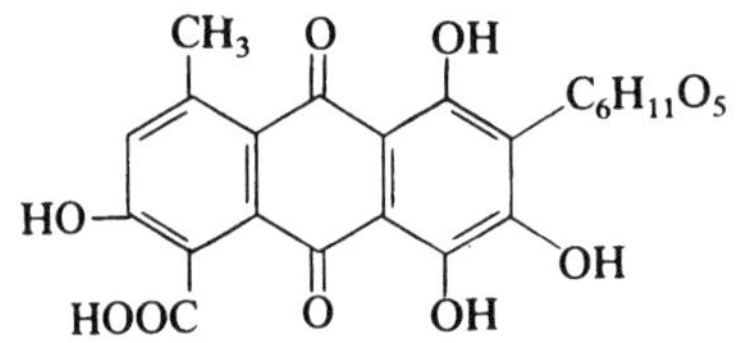

3

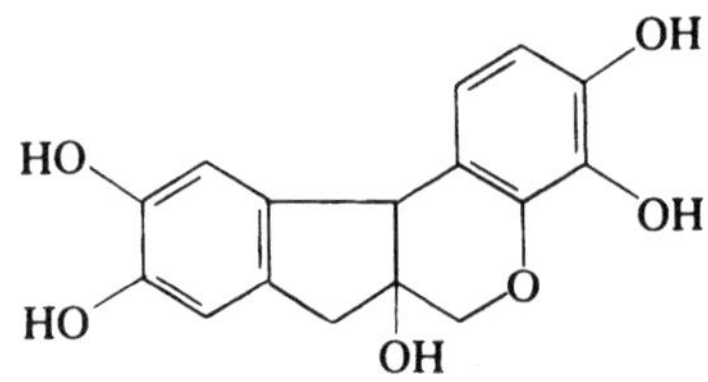

4

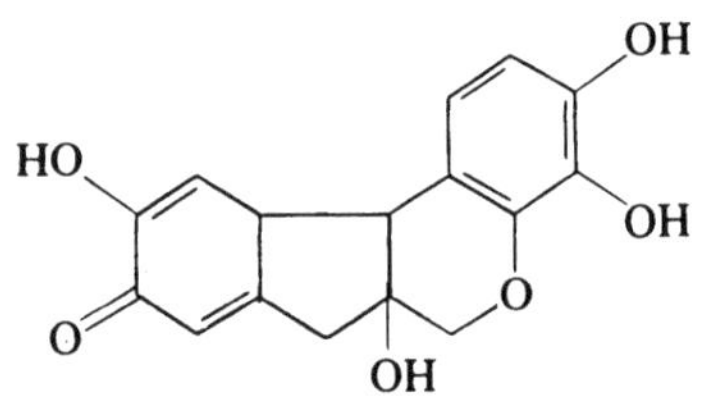

5

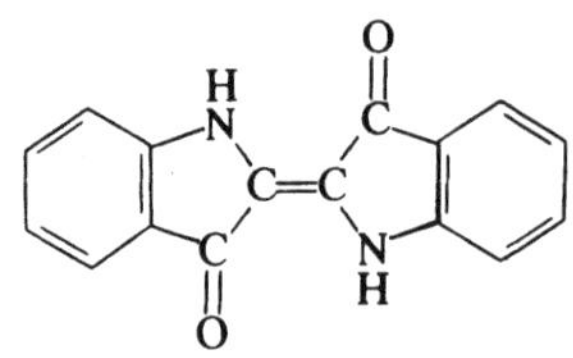

6

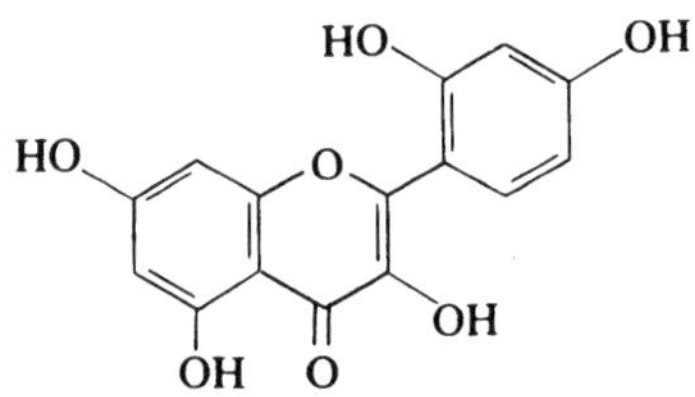

7

Cochineal[3] was obtained from the bodies of the female of the insect *Coccus cacti,* which was cultivated in Mexico on the Prickly Pear cactus, an *Opuntia* species; the pure dye has the structure **3**. Wool and silk are dyed crimson with an aluminum mordant and a bright scarlet from tin-mordanted material. Blacks were obtained from extracts of chips of the heartwood of the tree campeachy, *Haematoxylon campechianum,* which contains the coloring matter as hematoxylin **(4)**, readily oxidized during isolation to hematein **(5)**. The crude extract was referred to as Logwood. No doubt the most widely used blue was Indigo **(6)**, obtained from plants of the genus *Indigofera,* mainly *I. tinctora,* and grown widely in India and the Far East. This product soon surpassed the woad extracted from *Isatis tinctoria* which was widely grown in Europe.[6] Old Fustic, from *Chlorophora tinctoria,* contained Morin,[7] now identified as 3,5,7,2′,4′-pentahydroxyflavone **(7)**, and was used to improve the slightly bluish-black of Logwood on wool.

Lichens have been used until quite recently to dye wool and, in the 18th century, were the basis of an important industry in Scotland for dyeing the pleated skirts now known as kilts. Kilts, which provided a daytime covering and a blanket at night, were dyed with lichen products after the spinning stage and this was supposed to explain their aroma and moth-proofing property. The demand for the famous Harris Tweed soon necessitated the use of artificial dyes and it is now rare to find naturally-dyed material. There is a voluminous literature on the use of plant and lichen extracts for the coloration of wool from the earliest times to the present day and the reader is referred to some of these for a more complete account.[8–13] Although dyeings with some of these natural products approach many of the synthetic dyes in fastness properties, with the exception of Indigo and Alizarin, the light fastness is below that of the best currently available synthetic dyes. It is, however, in the field of their application, often very long and tedious, that the natural colors compare unfavorably with modern dyes. In nearly all cases, the use of plant products for the coloration of wool involved metal salts to fix the dyes onto the wool. Salts of aluminum, iron, tin, and chromium were commonly used, different metals often giving different hues.

Toward the end of the 19th century, the rise of the chemical industry became responsible for the availability, over a period of time, of dyes to give wool brighter, more lightfast, and more easily applied colorants. Stemming from the discovery by W. H. Perkin in 1856 of a crude dye which colored silk a rich purple, an industry arose, the ramifications of which spread rapidly in Europe, then America, and finally worldwide. Perkin's discovery sparked off a vigorous study of benzene, naphthalene, and anthracene, obtained by distillation of coal tar, which itself was then an unwanted byproduct of the infant coal-gas industry. By the end of the Great War, however, the initiative had passed to Germany which emerged as the principal dyestuff producer even to the end of World War II. It should not be forgotten, however, that the climate was favorable to the intellectual curiosity of a group of remarkable chemists in Europe from 1850 onward which resulted in the determination of the structures of many of the then useful natural dyes such as indigo, alizarin, and many others, and followed by technical production.

Thus, the indigo industry in the Far East began to collapse when the great German chemist Bayer, in 1878, first prepared indigo synthetically and this led to production on a large scale, although by a different route, which rapidly brought down the cost. The two decades from 1950 saw a remarkable surge in research work devoted to improving the fastness properties of existing dyes for wool and also to providing the dyer with materials for the coloration of the new fibers being produced commercially. Although these aims may now be approaching a plateau, it is true that research will not cease but may alter its direction somewhat. Economics will continue to assume a vital part of the dyer's concern. The need for dyes to exhibit fastnesses approaching the lifetime of the fiber is not so important as was thought in the 1950s; adequate fastness properties are required only in regard to the time scale of the usual life of a garment or furnishing.

Dye producers must also work within the constraints of environmental factors. For example, many afterchrome dyes have been replaced by the more expensive premetallized dyes, resulting in a switch from the dyer being held responsible for canal or river pollution with chromium salts to the manufacturer who must treat his own effluent. Because of improvements in dyeing techniques, this change has not been as drastic as was at first thought. Of the enormous number of acid dyes marketed at the turn of the century, not many remain due either to the improved fastness properties of later discoveries or to the banning of several widely-used intermediates, especially benzidine and 2-naphthylamine which were found to be carcinogenic.

B. Acid Dyes

1. Azo Dyes

The two main classes of acid azo dyes for the dyeing of wool are referred to as leveling dyes and milling dyes, but practically the line of demarcation is somewhat blurred. It is not the intention of this chapter to discuss the complex mechanisms by which the coloring matter diffuses into the fiber and is retained there by forces such as ionic, nonpolar van der Waals' forces or covalent bonds. It should be realized that temperature, pH of the dyebath, and the presence or absence of ionic or nonionic auxiliary compounds all play a part in giving a satisfactory level dyeing. The dyeing of wool is, of course, a cosmetic process applied at some stage in the processing of raw stock to saleable product. The stage of the dyeing process determines the properties that the dye must possess if it is to be satisfactory to the dyer and to the customer. Thus, the dyes used for yarn dyeing must migrate to the center or core of the fiber, particularly in the case of carpet yarns which are viewed end-on, and, at the same time, have adequate fastness to water, shampooing, and light. The dyes for piece-dyeing must have adequate fastness to milling and cross-staining of other fibers such as cellulose. After treatments, such as shrink resistance, can also determine which dye can or cannot be used. In the dyeing of loose stock or slubbing the dyes must be fast to any or all of the succeeding treatments such as milling.

Leveling dyes for wool are defined as those which require strong acid to exhaust the dyebath to a practical extent and which possess the characteristics of low relative molecular mass, good solubility, and molecular solutions and whose affinity for anions is low. Their rate of dyeing is rapid and, although they may be unlevel at the start of the dyeing process due to a rapid initial strike, they readily migrate from the deeper dyed parts of the wool to the paler parts via the dyebath solution. These dyes are much influenced by the presence of salts, such as sodium sulfate, in the dyebath which move the equilibrium between the dyebath and the fiber in favor of the wool.

Structures **8–13** show examples of currently used leveling dyes. It will be seen that yellow leveling dyes often make use of pyrazolone compounds as coupling components as in, for example, CI Acid Yellow 17 **(8)** and CI Acid Yellow 23 **(13)**. CI Acid Yellow 17 is prepared by diazotizing sulfanilic acid and coupling the resulting diazonium salt with 1-(2′,5′-dichloro-4′-sulfophenyl)-3-methyl-5-pyrazolone in alkaline solution. The basic diazotization reaction was

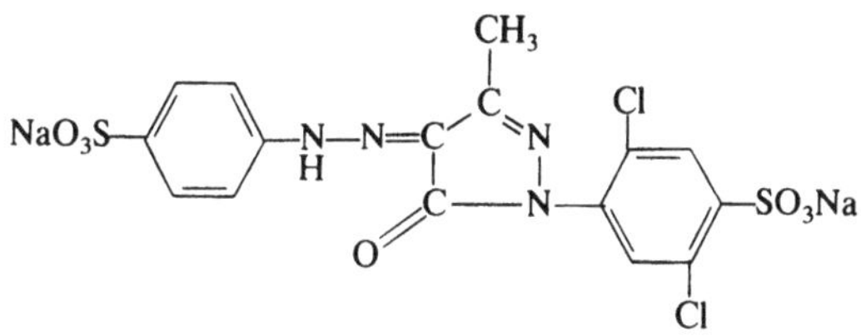

8

discovered by a chemist, Peter Greiss, who was working in a brewery in Burton on Trent in 1858, by reacting a primary aromatic amine with nitrous acid, formed *in situ* from sodium nitrite and hydrochloric acid. Since that time, many thousands of azo compounds have been prepared. It should be noted that, in this example, the presence of the two chlorine atoms confers an increase in light fastness and a desirable greening of the hue as compared with the dyestuff without these atoms, which, with the exchange of the 3-methyl group in the pyrazolone ring for a carboxylic acid group, is the well-known food dye Tartrazine and which can be prepared in a similar manner to **(8)**. CI Acid Yellow 23 was formerly prepared by the condensation of two molecules of phenylhydrazine-4-sulfonic acid and one molecule of dieketosuccinic acid as in Scheme 1.

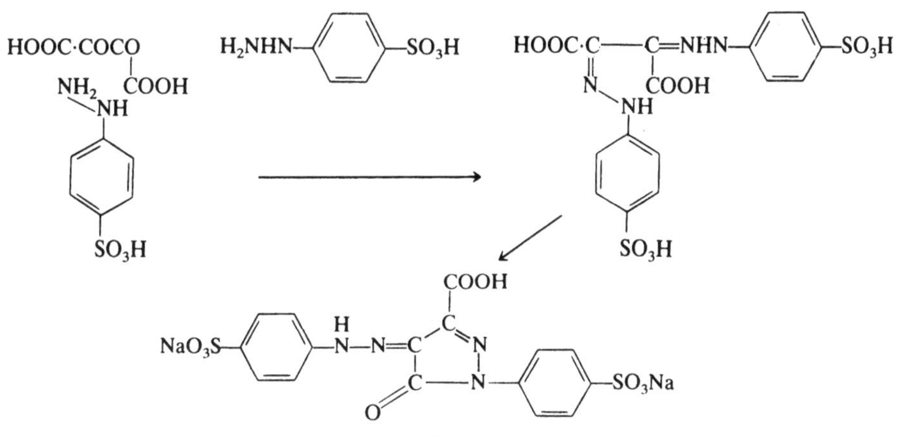

13

Scheme 1

Red dyes of this class frequently use a naphthol or an amino naphthol coupling component as in CI Acid Red 1 **(9)** derived from benzenediazonium chloride and *N*-acetyl-H-acid. The acetylation of the amino group of the H-acid moiety considerably improves the level dyeing properties as compared with a free amino group and also shifts the hue from a dull bluish-red to a bright red. However, the dye is easily hydrolzyed in acid media. Of interest here is that the replacement of a hydrogen atom in the benzene ring by a long alkyl chain, a shorter more bulky alkyl group such as a di-*t*-octyl group, or another weighting group converts this dye into one of the milling class. CI Acid Red 37 **(11)** uses an alternative aminonaphthol coupling component, Gamma acid (7-amino-1-naphthol-3-sulfonic acid), and 4-acetylaminoalinine-2-sulfonic acid as the diazonium component. In CI Acid Violet 7 **(12)**, a bathochromic shift results from the introduction of the 4-acetylamino group in to **(9)**. CI Acid Blue 92 **(10)** uses H-acid as the diazonium component which, in this case, is *N*-phenyl-1-naphthylamine-8-sulfonic acid and this dye is an example containing two naphthalene nulcei in the molecule.

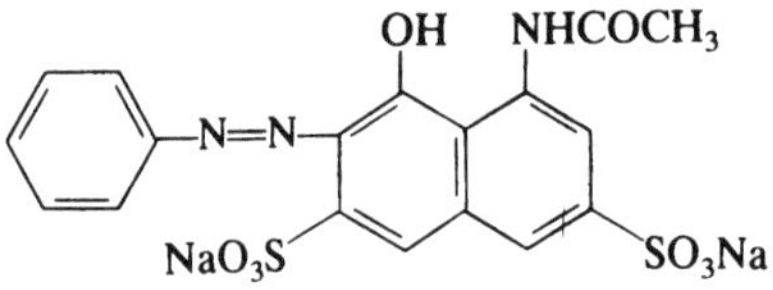

9

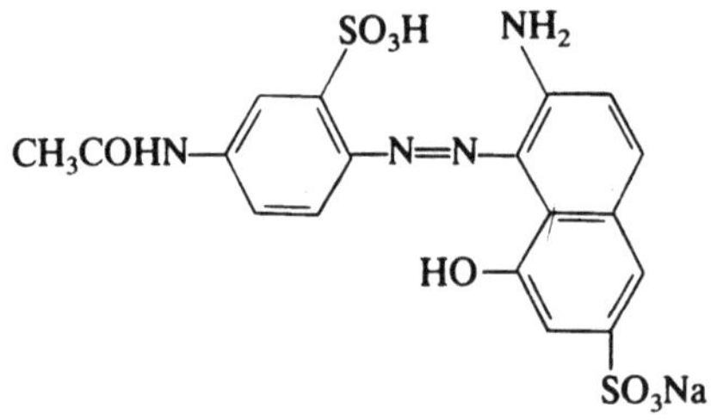

11

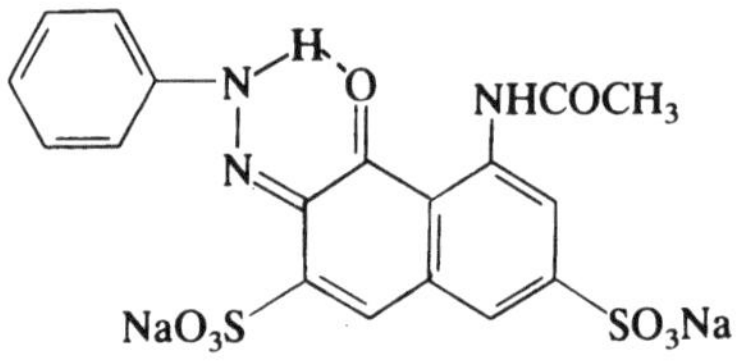

9a

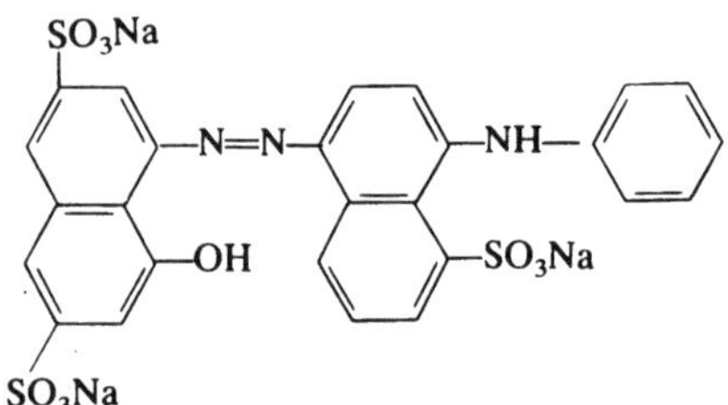

10

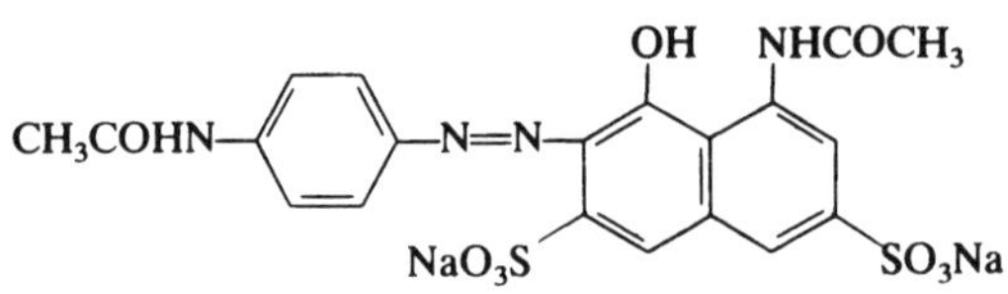

12

Azo dyes coupled in the *ortho* position to a hydroxy group are here formulated in the hydroxy-azo form for clarity rather than the predominant keto-hydrazone form (see Chapter 2), although the latter accounts better for the high tinctoral value of these dyes and their stability to alkali treatment. Compare **9** and **9a**. A useful triad for knitted ware, which demands good level dyeing properties, comprises CI Acid Yellow 17, CI Acid Red 37, and CI Acid Blue 92.

Azo milling dyes are characterized by a colloidal state in aqueous conditions, medium ion affinity, medium rate of dyeing, and slow migration. These dyes are, however, much faster to wet treatments, perspiration, and milling than leveling dyes. A weak acid, usually formic or acetic, is used in the dyebath together, in some cases, with an auxiliary to control the rate of dyeing. Some examples of this

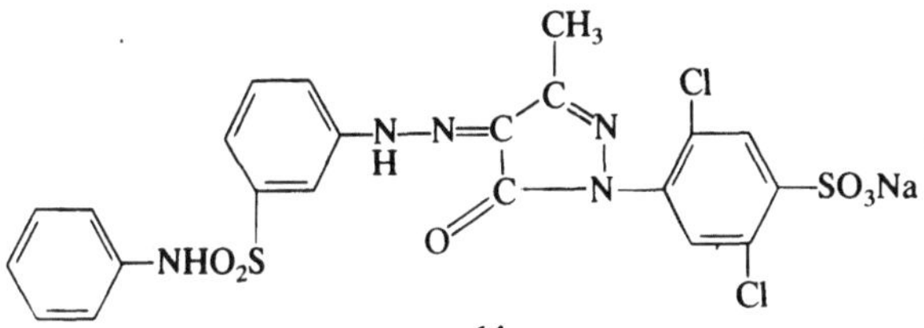

14

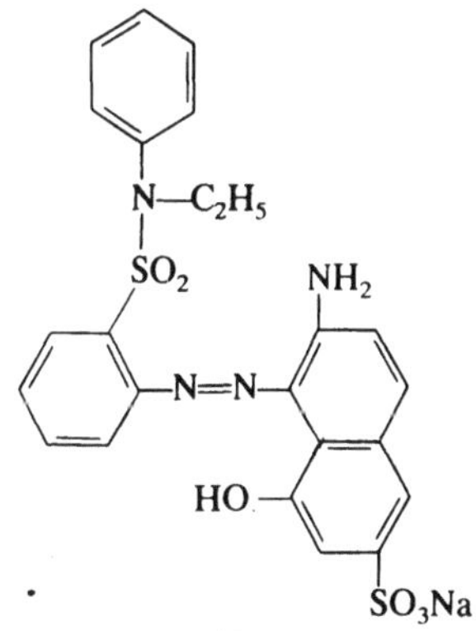

15

class are shown in structures **14–22**. Pyrazolone compounds again find use for yellow dyes in this class and CI Acid Yellow 29 **(14)** can be prepared by diazotizing 3-aminobenzenesulfanilide and coupling with the same pyrazolone derivative as in CI Acid Yellow 17 **(8)**. CI Acid Red 57 **(15)** is prepared by coupling 2-aminobenzenesulfon-*N*-ethylanilide in weakly acid solution with 7-amino-1-naphthol-3-sulfonic acid. This compound exemplifies a large class of dyes which is characterized by reasonably good leveling properties, good perspiration fastness, and very good fastness to light. The colors range from reddish-orange through red to bordeaux; the desirable application properties can be associated with the grouping **22a**. Compound **23** is scarlet and dye **24** is a bordeaux. CI Acid Orange 67 **(16)**, a useful component of mixture shades, is made by reacting diazotized 4-amino-4′-nitro-2′-sulfodiphenylamine with *m*-cresol and esterifying the free hydroxy group with toluene-*p*-sulfonyl chloride (tosyl chloride). CI Acid Blue 113 **(17)** is a bis-azo dye which is prepared by coupling diazotized metanilic

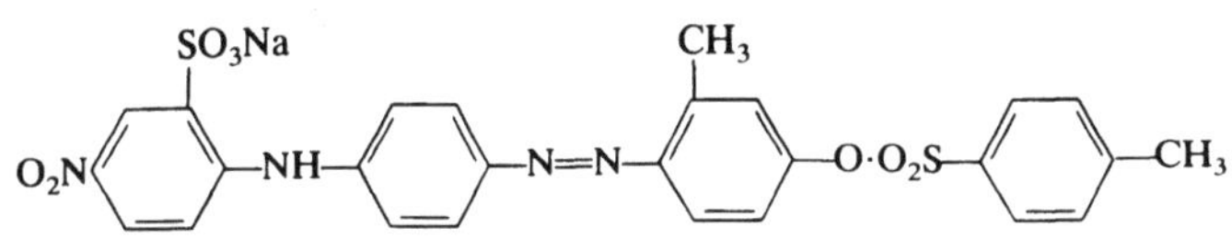

16

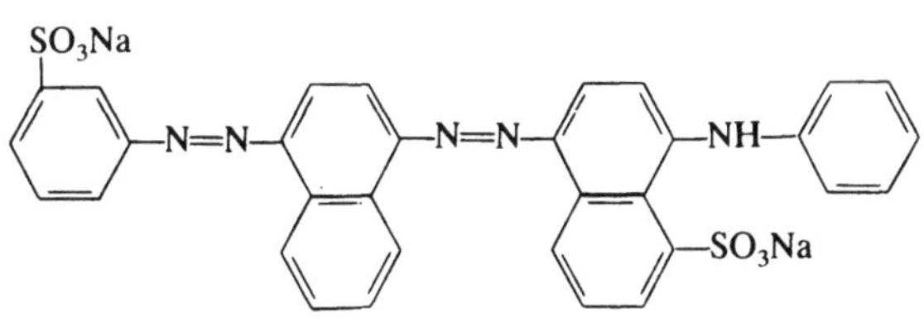

17

acid with 1-naphthylamine, rediazotizing, and then coupling the isolated diazonium compound with 1-(*N*-phenylaminonaphthalene)-8-sulfonic acid in weak acid solution. The product is widely used in producing deep navy-blue hues on wool.

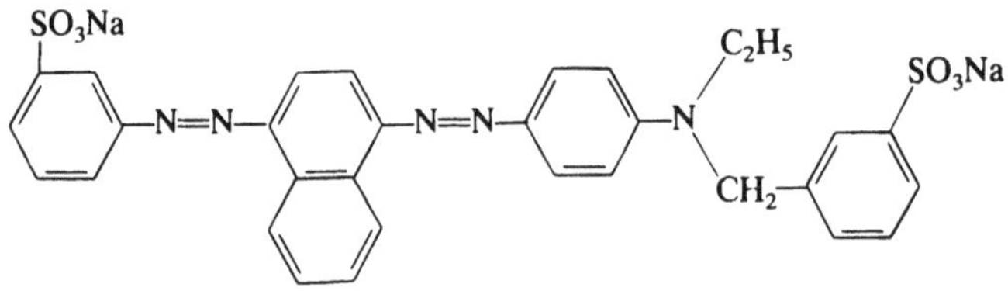

18

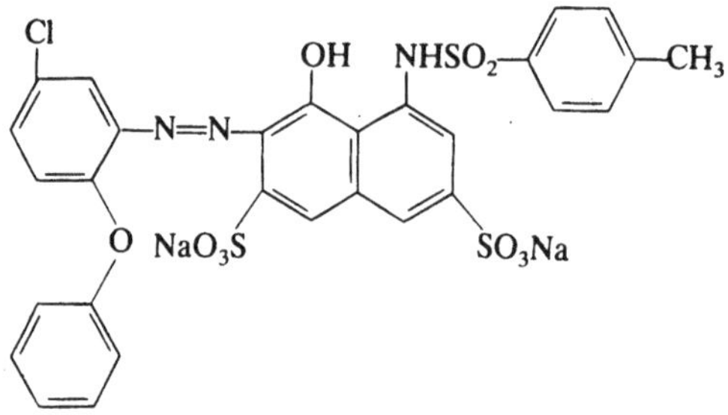

19

CI Acid Red 119 **(18)** is related to CI Acid Blue 113 except that the end component is *N*-benzyl-*N*-ethylaniline-3′-sulfonic acid. The maroon shade builds up well and the dye is useful in the production of brown hues. CI Acid Red 249 **(19)** is a bright red dye obtained from 2-amino-4-chlorodiphenyl ether and *N*-(*p*-toluenesulfonyl)-H-acid as the second component. The *N*-tosyl-H-acid group is much faster to hydrolysis than the *N*-acetyl group as in CI Acid Red 1. CI Acid Orange 19 **(20)** gives reddish-orange hues on wool from a weakly acid bath and is prepared from diazotized 4-methyl-3-aminobenzenesulfanilide and coupling to 1-naphthol-4-sulfonic acid. CI Acid Red 114 **(21)**, a bis-azo dye, is

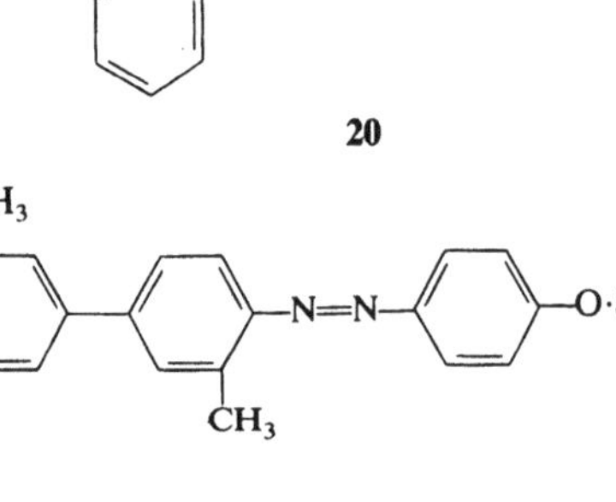

20

21

prepared by coupling tetrazotized *o*-tolidine first with 2-naphthol-6,8-disulfonic acid and then with phenol; the free hydroxy group of the phenol is then tosylated. This type of dye was discovered by chemists working for the Swiss firm Geigy who, having accumulated stocks of toluene-*p*-sulfonyl chloride from the preparation of the *ortho* isomer (used in the manufacture of saccharin), found that blocking the free hydroxy group in the above dye with tosyl chloride gave the resulting product neutral-dyeing properties and an enhanced fastness to alkaline perspiration.

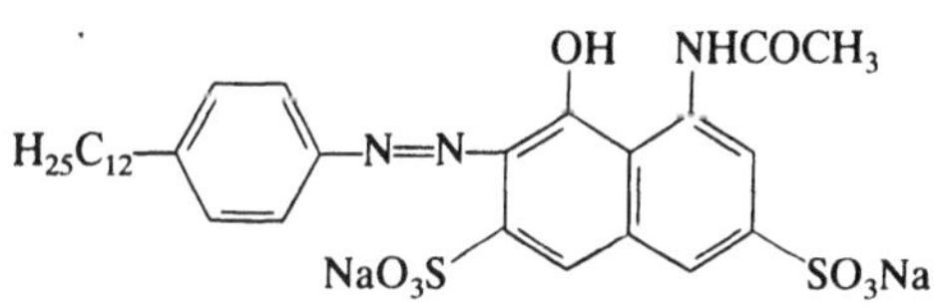

22

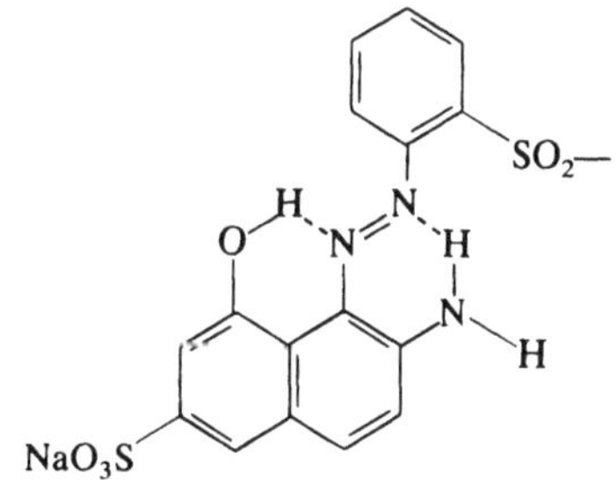

22a

CI Acid Red 138 **(22)** exemplifies the effect of introducing a long alkyl chain into an otherwise simple azo compound. The dye can be prepared by diazotizing *p*-dodecylaniline and coupling to *N*-acetyl-*H*-acid. The product dyes wool in bright red hues which, due to the hydrophobic nature of the first component, have very good fastness properties. Level dyeing can only be achieved in the presence of auxiliaries in the dyebath and the dye finds its main application in loose stock dyeing. A generally applicable triad of dyes having good wet fastness on wool together with acceptable light fastness consists of CI Acid Yellow 29, CI Acid Red 57, and a CI Acid Blue of the anthraquinone series of similar fastness properties.

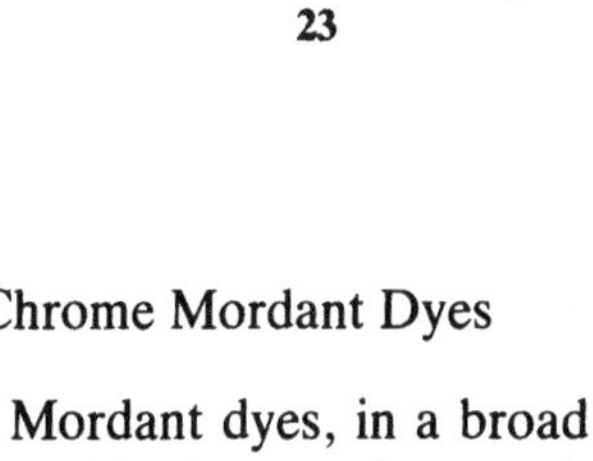

23

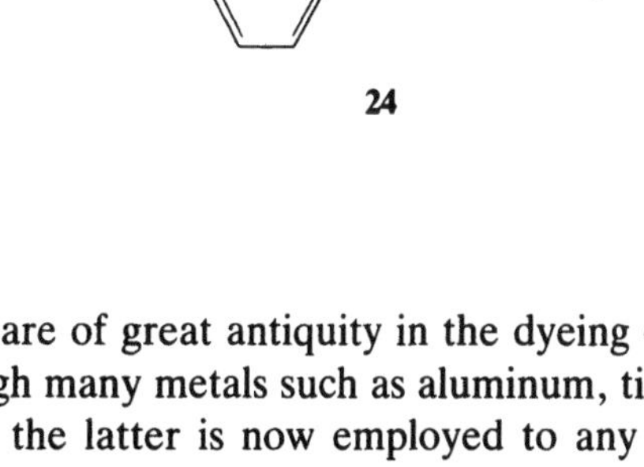

24

2. Chrome Mordant Dyes

Mordant dyes, in a broad sense, are of great antiquity in the dyeing of wool fibers with plant products and, although many metals such as aluminum, tin, iron, and chromium have been used, only the latter is now employed to any extent.

Logwood, however, which gives a good black on tin-weighted silk, still finds a use in dyeing nylon a deep black. Mordant dyes as a class still represent some 30% of all dyes for wool on a worldwide basis and can be used at all stages in wool processing. Three application techniques are used: the on-chrome method, the afterchrome process, and the single bath or metachrome method. In the first case, the wool is treated initially with a dichromate solution and then with a reducing agent such as formic acid or cream of tartar at the boil. The mordanted material is then treated in a fresh bath with the dye. The technique of afterchrome dyeing is self-explanatory, the fiber being dyed in an acetic or formic acid bath until exhausted, due regard being taken for the possible need to include auxiliaries to control the rate of dyeing to give levelness. After a slight reduction in temperature, dichromate is added and the temperature raised to the boil until a full shade is developed. The third technique is only applicable with relatively few dyes and proceeds by dyeing in a bath of dichromate and ammonium sulfate or ammonia together with the dye.

Nietzki,[14] in 1889, is credited with the discovery that the azo dye obtained from diazotized *m*-nitroaniline and salicyclic acid gave dyeings that could be rendered faster to washing by treatment with a chromium compound. It was not until the beginning of the 20th century that any real progress was made, particularly in the case of dyes containing *o,o'*-dihydroxy and *o*-hydroxy-*o'*-carboxylic acid groups. Mordant dyes find their use in the dyeing of deep shades for suitings and related applications where fastness to perspiration, milling, and light make them particularly valuable. The formation of the chrome complex during the dyeing process results in a considerable bathochromic shift except in those cases where the groups chelating with the chromium atom are spatially separated from the azo group and are present as a pendant grouping. This latter type of chromable dye was devised as a means of improving the west fastness properties and also light fastness without much change in the original hue. There is no space in this chapter to discuss the structures of the complexes as they exist in the wool fiber in detail. Typical mordant dyes are illustrated in compounds **25–33**. Salicyclic acid is a frequent end component used for yellow to orange mordant dyes, as, for example, in Mordant Yellow 3 **(25)** where 2-naphthylamine-6-sulfonic acid is the first component and in Mordant Orange 6 **(27)** where the first component is 4-aminoazobenzene-4-sulfonic acid. CI Mordant Yellow 8 **(26)** is obtained from diazotized anthranilic acid and 1-(*p*-sulfophenyl)-3-methyl-5-pyrazolone. CI Mordant Red 5 **(28)** results from the coupling of diazotized 2-aminophenol-4-sulfonic acid and resorcinol, giving a brownish-red hue. An important bluish-red dye is CI Mordant Red 7 **(29)** which is prepared by

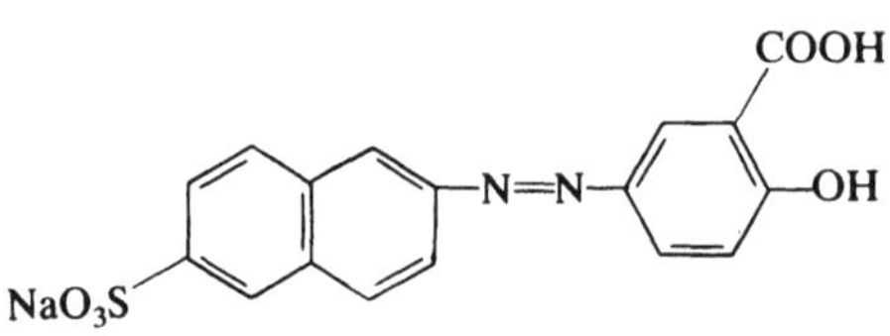

25

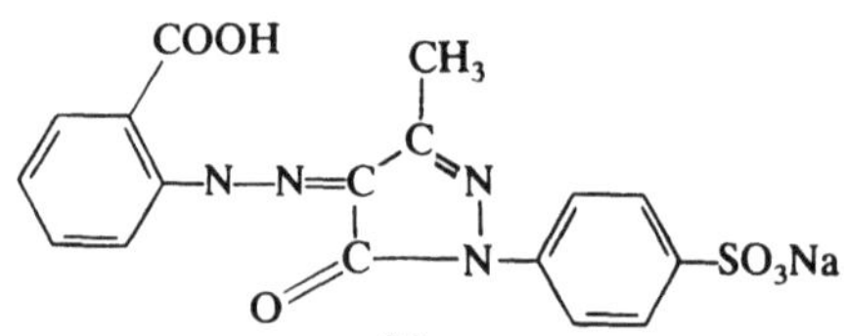

26

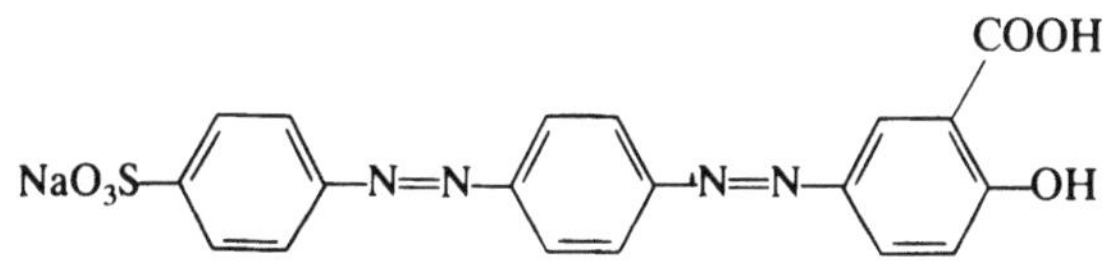

27

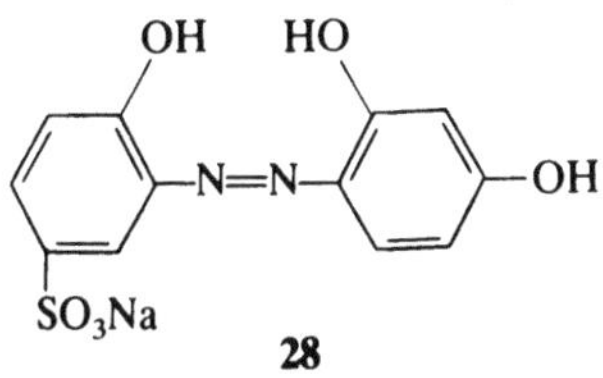

28

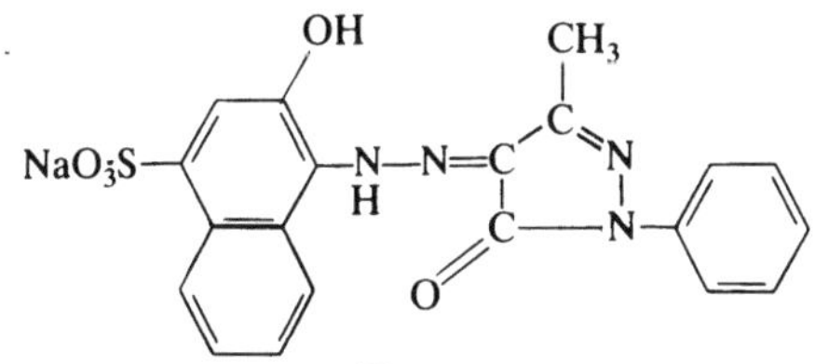

29

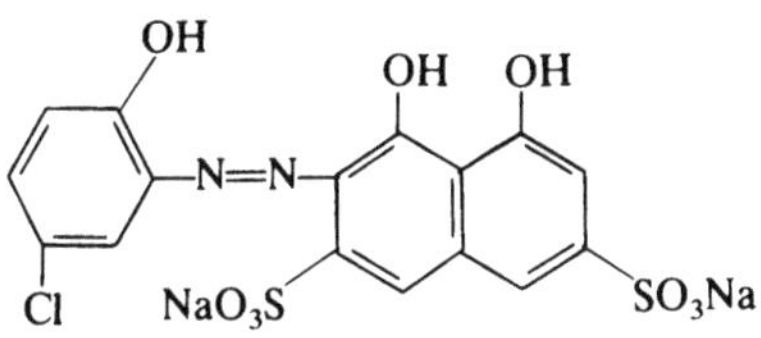

30

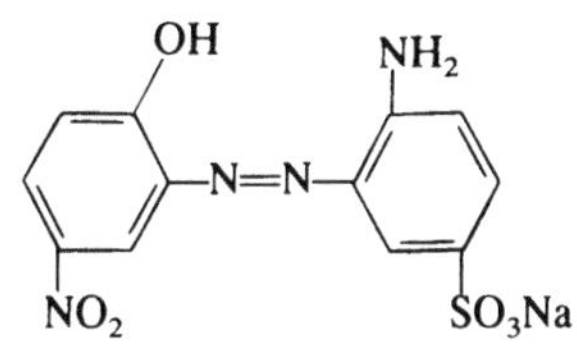

31

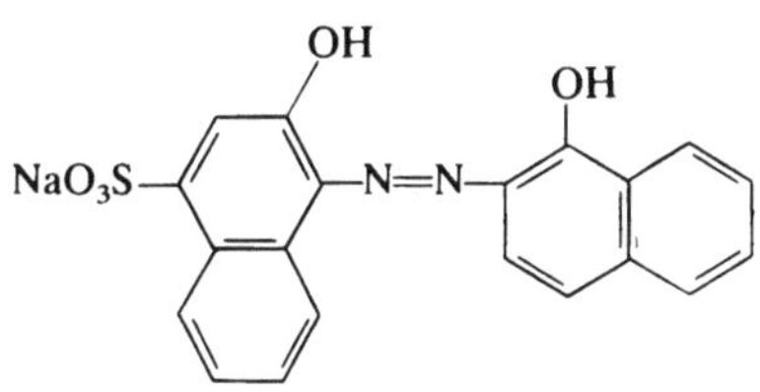

32

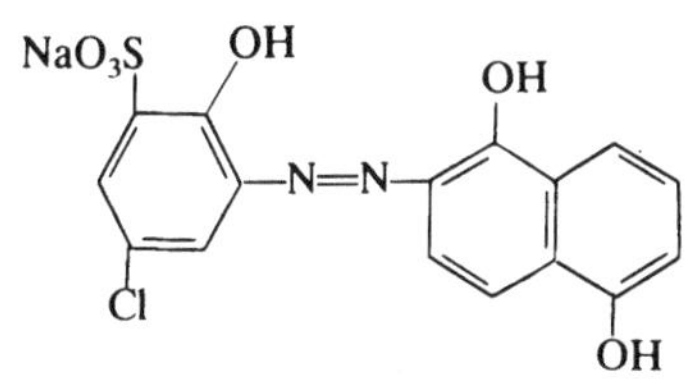

33

coupling the diazo-oxide of 1-amino-2-naphthol-4-sulfonic acid with 1-phenyl-3-methyl-5-pyrazolone. CI Mordant Blue 13 **(30)** results from the coupling of 2-amino-4-chlorophenol with 1,8-dihydroxynaphthalene-3,6-disulfonic acid, while CI Mordant Brown 33 **(31)** is obtained when 2-amino-4-nitrophenol is diazotized and coupled to 1,3-diaminobenzene-4-sulfonic acid. CI Mordant Black 11 **(32)**, a very important dye, is made by coupling the diazo-oxide of 1-amino-6-nitro-2-naphthol-4-sulfonic acid with 1-naphthol and is often marketed as a dispersion of its almost insoluble acid form. CI Mordant Black 7 **(33)** is produced by coupling diazotized 2-amino-4-chlorophenol-6-sulfonic acid with 1,5-dihydroxynaphthalene; the slightly bluish-black is fast to potting.

3. Metallized Dyes

a. 1:1 Complexes. It would have seemed a logical step from the chroming of the mordant dye on the fiber to prechroming in substance but, although successful attempts were made in 1912, lack of adequate acid-proof materials of construction for the prechroming reaction hindered the earlier attempts. The Swiss firm C.I.B.A. and the German company B.A.S.F. were primarily responsible for this innovation, marketing the ranges of Neolan and Palatine Fast dyes, respectively, about 1916. Having one or more sulfonic acid groups, these dyes are soluble in water but need to be applied from a 5–8% sulfuric acid bath with 10% sodium sulfate in order to achieve level results.[15] The wet fastness may, in some cases, be below that of the corresponding afterchrome dyes, but the brightness may be improved.

Production of the dyes entails heating an *o,o'*-dihydroxy or an *o*-hydroxy-*o'*-carboxylic acid azo derivative in an acid medium with chromium(III) compounds such as chromium fluoride or formate, often under pressure. The reader is referred to B.I.O.S. 1548 for details used by the I.G. prior to and during World War II. Structures **34–39** provide some examples of 1:1 complexes which have been used commercially. CI Acid blue 159 **(34)** is prepared by coupling the diazo-oxide of 1-amino-2-naphthol-4-sulfonic acid with 1-naphthol-8-sulfonic acid and heating the resulting azo dye at 115 °C with chromium sulfate solution. CI Acid Yellow 54 **(35)** results on chroming with chromium formate the azo dye obtained from diazotized 2-amino-4-sulfobenzoic acid and 1-(2'-methyl-4'-sulfophenyl)-3-methyl-5-pyrazolone. 2-Amino-4-chloroanisole is used as the first component in CI Acid Violet 58 **(36)**, the end component being 2-naphthol-6,8-disulfonic acid, and the resulting azo derivative is chromed at 120–135 °C with

34

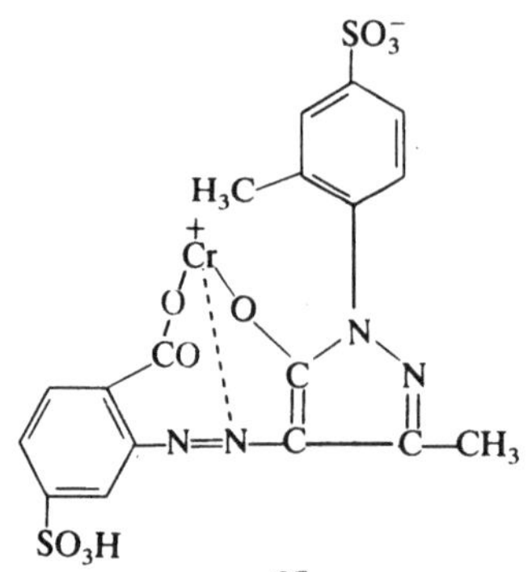

35

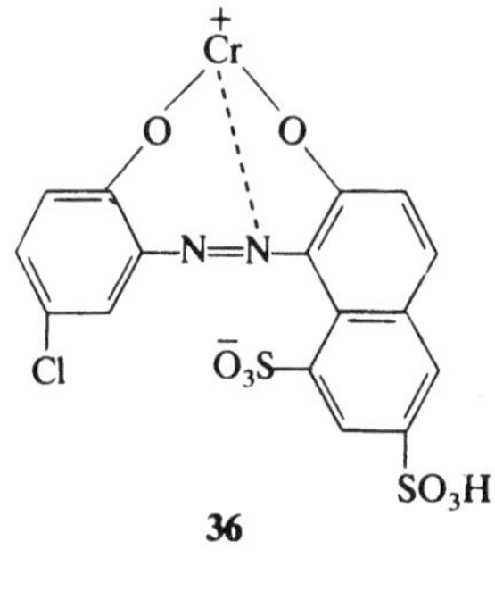

36

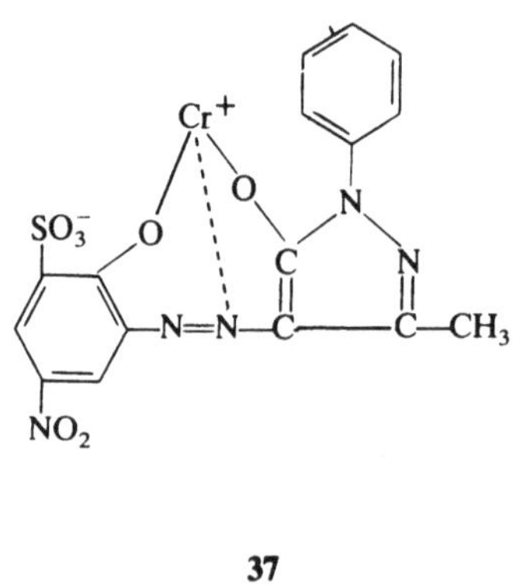

37

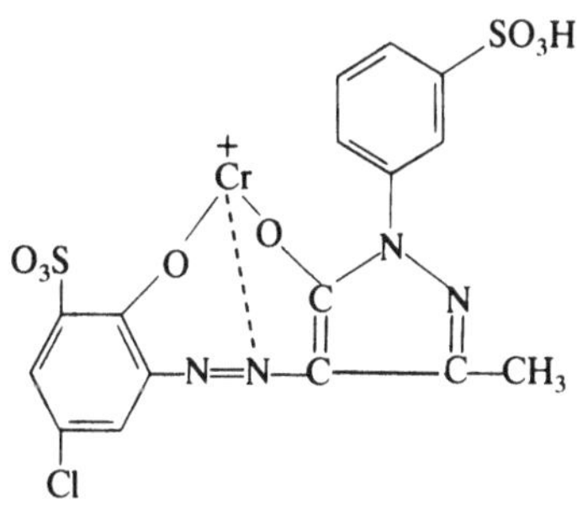

38

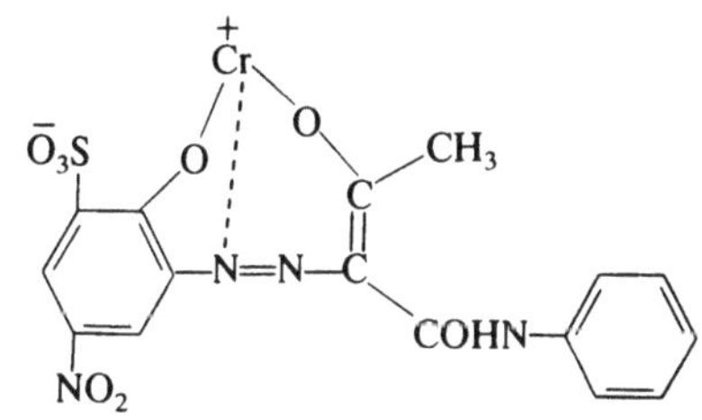

39

chromium formate, the methyl group of the anisole being displaced in the complex formation. CI Acid Orange 74 **(37)** is obtained by coupling diazotized 2-amino-4-nitrophenol-6-sulfonic acid with 1-phenyl-3-methyl-5-pyrazolone and chroming the resultant product with chromium formate at 130 °C. CI Acid Red 183 **(38)** is formed when diazotized 2-amino-4-chlorophenol-6-sulfonic acid is coupled with 1-(3′-sulfophenyl)-3-methyl-5-pyrazolone and the resulting azo derivative chromed with chromium formate. When diazotized 2-amino-4-nitrophenol-6-sulfonic acid is coupled to acetoacetanilide and the product chromed at just below the boiling point, CI Acid Yellow 99 **(39)** is produced.

b. 2:1 Complexes. In the early 1940s, several firms in Germany, Switzerland, and the U.S.A. turned their attention to neutral dyeing products for wool (and later for the synthetic polyamides), which contained in their structures one atom of chromium combined with two molecules of an *o,o′*-dihydroxyazo or azomethine compound containing no sulfonic acid groups.

Early products were found to have insufficient solubility to be commercially acceptable other than in a dispersed form. The firm Geigy were the first to market a technically superior product, Polar Grey BL (later renamed Irgalan Grey BL), and this compound was the forerunner of a range of neutral dyeing products characterized by their having an SO_2CH_3 group in the molecule.

The classical paper by Schetty[16] gives a complete review of the state of the art at that time. The commercial success of these dyes resulted in an explosion of research work by the major dye producers and some hundreds of patents were published in an effort to market ranges of these useful dyes for wool and nylon. It was soon found by C.I.B.A. and others that the methylsulfonyl group could advantageously be replaced by a sulfonamide or a substituted sulfonamide group and also, of equal importance, that metals other than chromium could provide useful products. Thus, cobalt complexes were found to have superior light fastness as compared with their chromium counterparts but were, in the main, hypsochromic in hue. The field was, of course, very extensive because any appropriate *o*-aminophenol devoid of sulfonic acid or carboxylic acid groups could be diazotized and coupled to a large number of phenols, naphthols, or acyclic, or heterocyclic compounds capable of keto–enol tautomerism.

The originally difficult process of introducing a chromium atom into the appropriate azo compound in the neutral or alkaline media required to produce

2:1 complexes was solved, first, by the use of hexavalent chromium complexing agents such as sodium chromosalicylate solutions and, second, by the elegant method of generating a chrome complex of, for example, a gluconic acid, by slowly adding a solution of a mixture of sodium dichromate and a reducing sugar, such as glucose, to the hot alkaline solution of the *o,o'*-dihydroxyazo compound. The cobalt complexes can be prepared similarly by reacting a solution of sodium cabaltitartrate or cobalticitrate with an alkaline solution of the appropriate dye. The structures of very few of the commercially available premetallized dyes have been revealed, despite their acceptance by the dyer and the very large numbers made. However, the examples **40–45** give an indication of the range of possibilities. The 2:1 premetallized dyes find a use at all stages in wool dyeing and in the dyeing of wool and nylon mixtures, sometimes with the aid of auxiliaries to restrain the nylon component. The dyes cover tippy wool well and are not so selective as milling dyes in dyeing mixtures of different types of wool; they are much used in dyeing carpet yarns particularly with wool and nylon blends, knitting yarns, and upholstery fabrics.

CI Acid Violet 78 **(40)** is prepared by chroming the azo dye obtained by coupling diazotised 2-amino-4-hydroxyphenylmethylsulfone with 2-naphthol. CI Acid Black 60 **(41)** is obtained by chroming the azo dye from 2-amino-4-methylsulfonamidophenol and 1-acetylamino-7-naphthol. CI Acid Red 308 **(42)** can be prepared by diazotizing 2-amino-4-(2'-carboxybenzoyl)phenol, coupling with 2-naphthol, and then converting into the cobalt complex.

The dye is an example in which the 2'-carboxybenzoyl group confers adequate solubility but requires a weak acid system to exhaust the dyebath. The same very high level of fastness properties as those dyes having sulfonamide groups are reached. CI Acid Orange 148 **(43)** containing sulfonic acid groups is prepared by the chroming of a mixture of dyes, one from diazotized 2-amino-4-nitrophenol-6-sulfonic acid and 1-phenyl-3-methyl-5-pyrazolone and the other

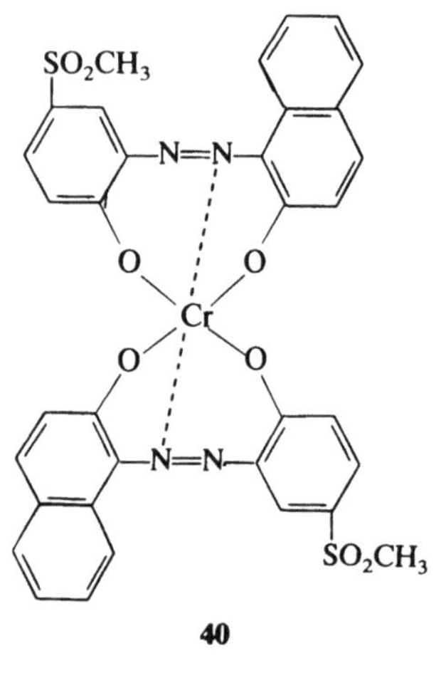

40

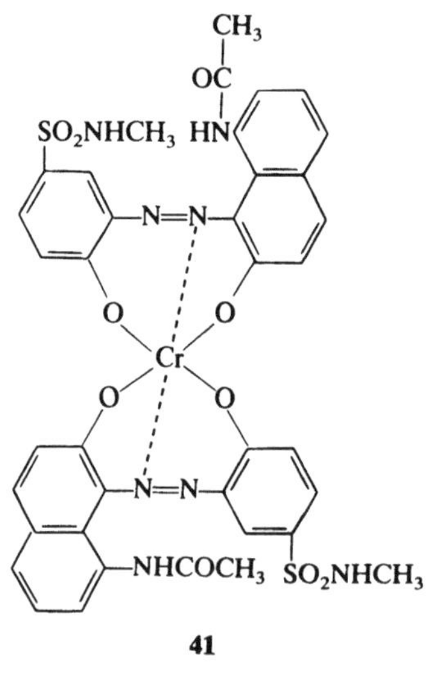

41

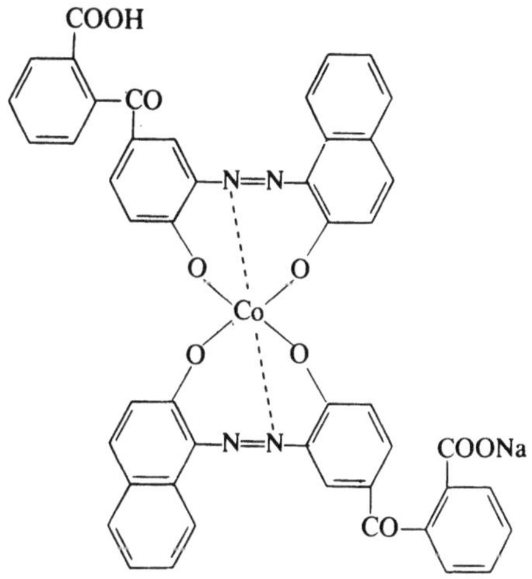

42

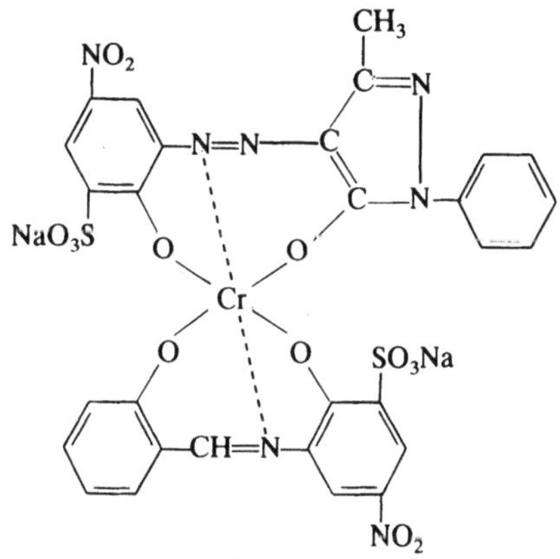

43

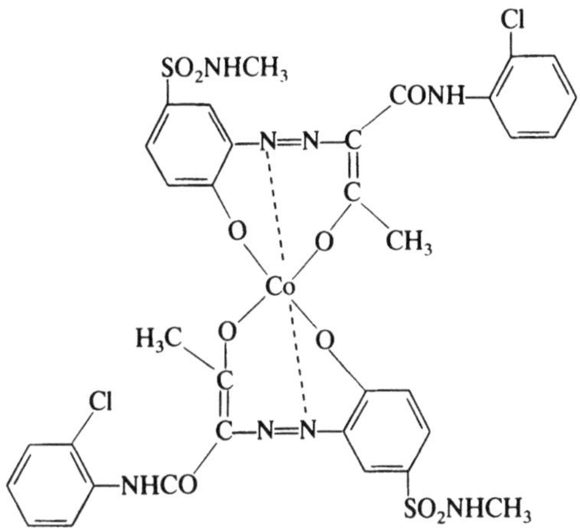

44

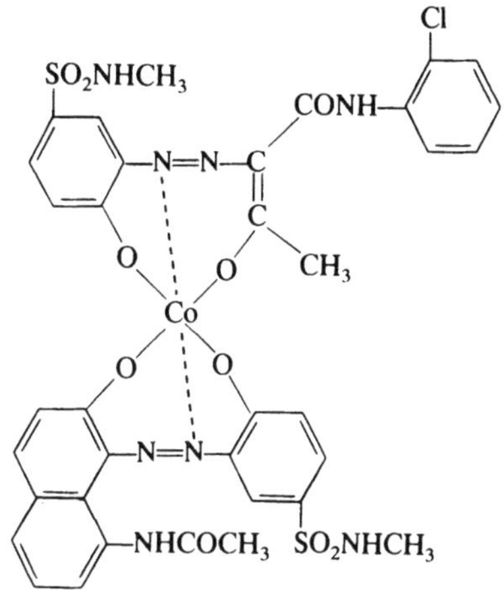

45

being the azomethine from salicylaldehyde and the same amino phenol. Compound **44** is an example of a fast mustard-yellow dye, which is made by cobalting the azo dye obtained by coupling 2-amino-4-methylsulfonamidophenol with 2-chloroacetoacetanilide. Dye **45** provides an example of a mixed azo type in which one molecule of the azo component of **44** is cobalted in the presence of the product obtained from diazotized 2-amino-4-methylsulfonamido phenol coupled with 1-acetylamino-7-naphthol and gives olive-brown hues on wool and nylon.

4. Anthraquinone Acid Dyes

With the exception of a few reds, the majority of anthraquinone acid dyes for wool range from violet through blue to green. The very large numbers that have been described indicate that none is perfect from the point of view of fastness properties and dyeing characteristics. However, many quite old products are still

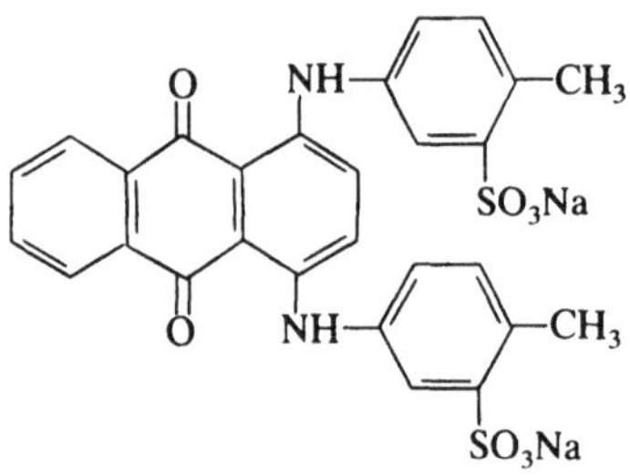

46

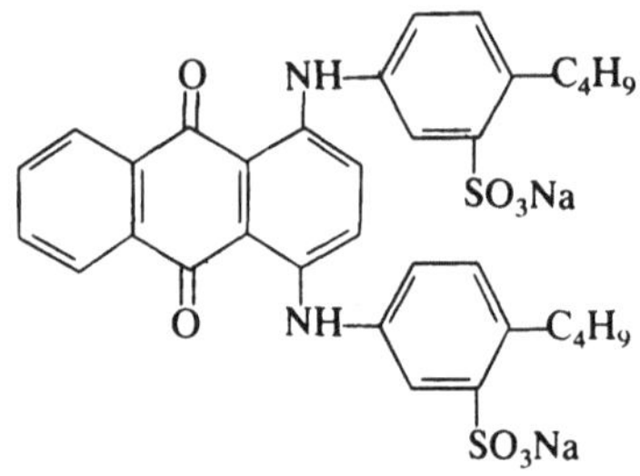

47

in use because of hue, price, or dyeing behavior. These dyes are invariably brighter than their azo counterparts and are used extensively as the blue components in triads for leveling dyes, while some have acceptable milling fastness. Structures **46–53** illustrate the range of compounds encountered in this class of dyes which are dominated by 1,4-disubstituted products.

CI Acid Green 25 **(46),** one of the oldest acid greens, is prepared by heating quinizarin with an excess of *p*-toluidine, boric acid, hydrochloric acid, and zinc dust, or by using leucoquinizarin and then sulfonating the purified base with oleum. CI Acid Green 27 **(47)** has considerably improved milling-fastness properties due to the butyl chains on the phenyl radicals but lower level-dyeing properties in the absence of an appropriate auxiliary. CI Acid Blue 80 **(48)** is made by substituting mesidine for *p*-toluidine in the CI Acid Green 25 synthesis; the change in color can be attributed to steric factors. This dye has very good light and good wet fastness properties. CI Acid Violet 41 **(49)** is prepared by heating 1,4-diamino-2,3-dichloroanthraquinone with a mixture of phenol and aqueous sodium sulfite under pressure. This dye has good leveling properties. CI Acid Violet 80 **(50)** is a bright purple dye which is prepared by acetylation of 4-bromo-1-methylaminoanthraquinone, followed by ring closure under alkaline conditions, condensation with *p*-toluidine, and sulfonation. Other amines such as *m*-chloroaniline can be used in place of *p*-toluidine to bring about slight changes in hue or application properties. CI Acid Blue 45 **(51)**, although a long-established dye, is still used to a considerable extent. The dye is prepared by first sulfonating 1,5-dihydroxyanthraquinone (anthrarufin), nitrating the resulting disulfonic acid, and then reducing the isolated dinitro derivative with sodium

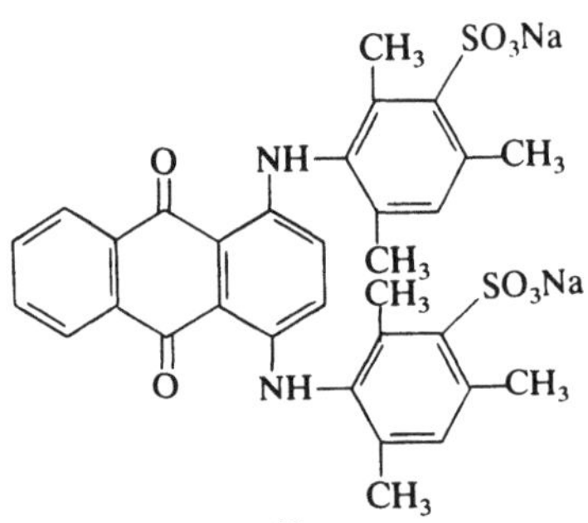

48

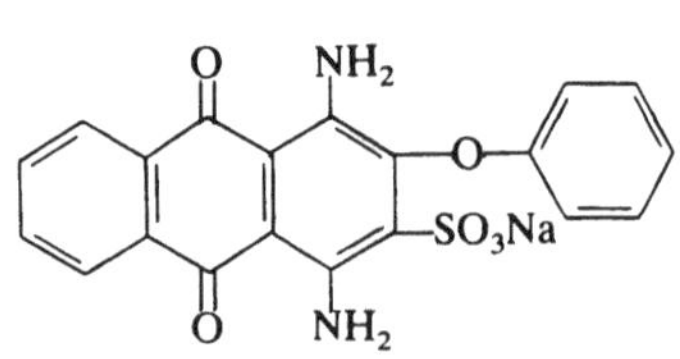

49

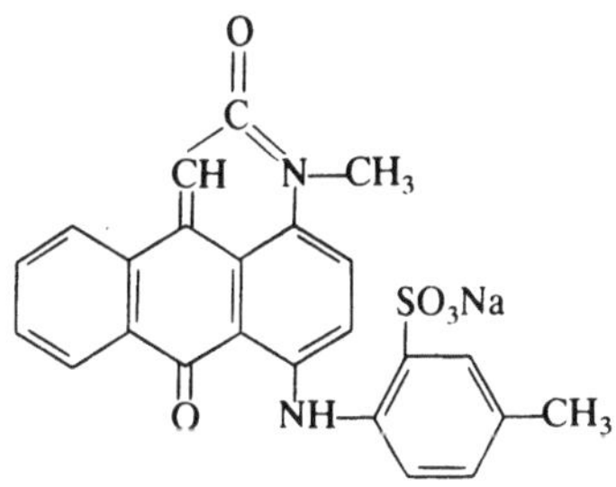

50

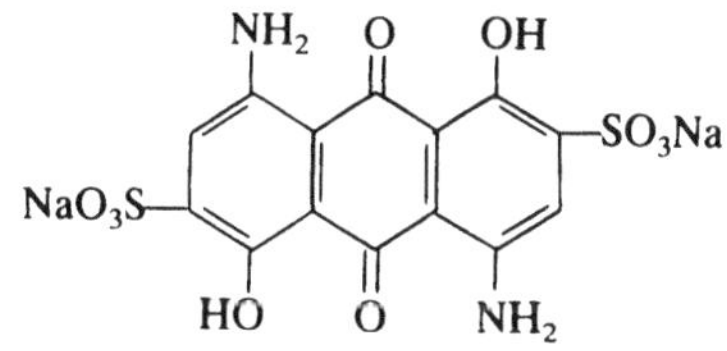

51

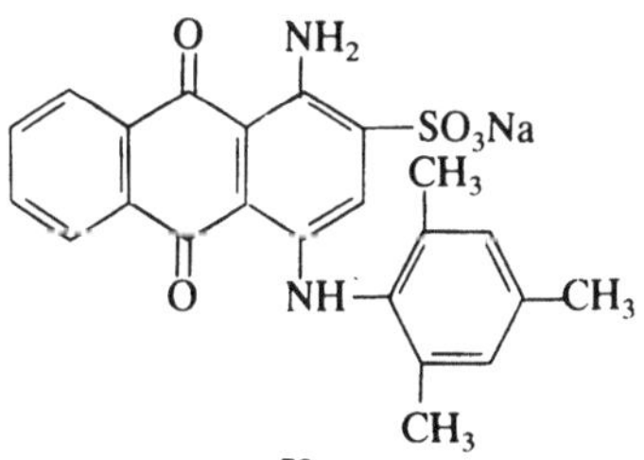

52

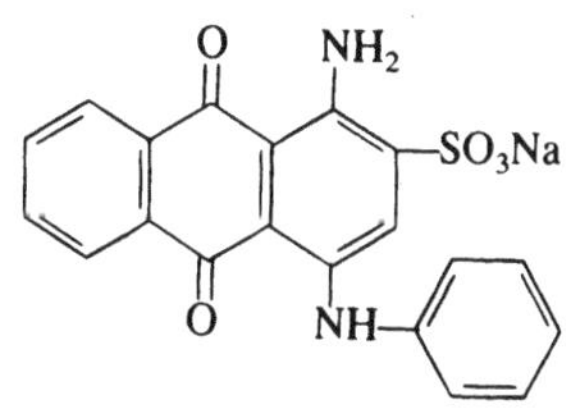

53

sulfide. CI Acid Blue 129 **(52)** can be prepared by replacing the bromine atom in 1-amino-4-bromoanthraquinone-2-sulfonic acid using mesidine. The dye gives bright reddish blues on wool and nylon. CI Acid Blue 25 **(53)**, an example of an early wool blue, is made by condensing 1-amino-4-bromoanthraquinone-2-sulfonic acid with aniline using a copper salt as catalyst.

As can be seen, the potential for variation in the 4-substituent is enormous and many hundreds or even thousands of aromatic or hydrogenated aromatic nuclei have been examined, some very complex. No doubt the fine tuning of this type of acid blue will continue until a product having the correct balance of all the desirable properties is obtained.

5. Miscellaneous Dyes

a. Vat Dyes. Indigo **(6)** is the only vat dye used to any extent for wool dyeing, particularly in conjunction with certain chrome-mordant dyes. Natural indigo is not a pure compound and contains Indirubin **(54)** and a variable brown contaminant so that the synthetic material is now always used. Until synthetic indigo became available and before the means of application became well understood so as to provide easy use by the dyer, many strange methods were devised to obtain satisfactory results. These stem from the need to keep the indigo, which has a very low solubility in water, in solution, so that penetration can take place to the core of the fiber before oxidation. Thus, a reducing bath of woad, bran, and lime was popular. Nowadays a bath is prepared which contains sodium hydrosulfite, ammonia, and a colloid such as glue; the indigo paste or powder is added at about 50 °C until a clear greenish-yellow solution is obtained, sometimes with the aid of a little sodium hydroxide, containing the leuco form **(56)**; see Scheme 2. Although the benzene rings of indigo have been substituted

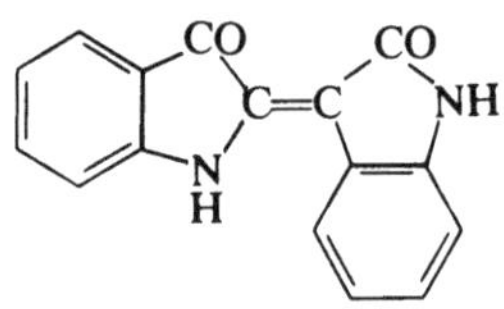

54

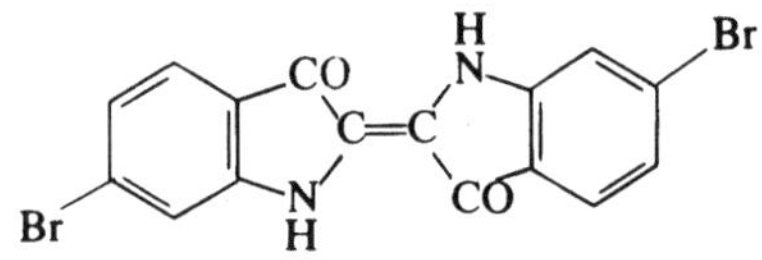

55

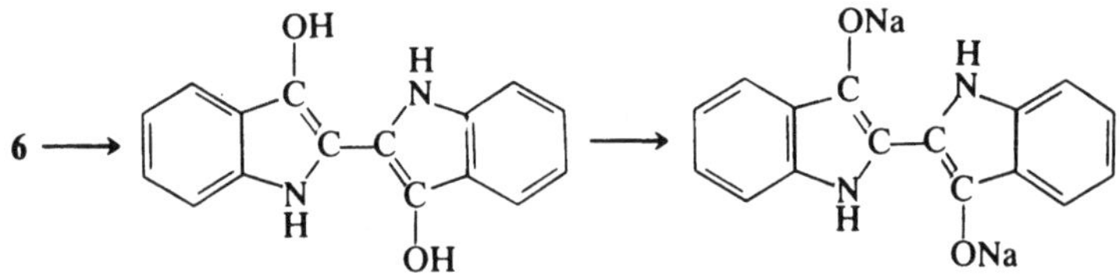

56

Scheme 2

with halogen, nitro, and other groups, it is doubtful if any of these derivatives are used on wool. A stabilized form of the leuco compound has been marketed since the 1940s to free the dyer from the need to vat the indigo, but this is not used to any great extent. Indigo-dyed cloth is frequently treated with afterchrome or milling dyes to deepen or modify the hue; examples of such dyes are CI Mordant Red 7 (**29**) and CI Acid Blue 113 (**17**). Indigo has a very high light fastness as well as very good wet fastness properties when applied to wool. However, there can be loose indigo on the surface of the fiber and this is noticeable when the dyed cloth is rubbed with a white material.

It is of interest here to note that the violet dye 6,6′-dibromoindigo (**55**), extracted from shell fish of the genus *Murex* in the Eastern Mediterranean regions and known for a very long time as Tyrian Purple, differs from indigo in that its formation from the precurser present in the body fluids of the shell fish requires the action of sunlight. The mechanism of the formation of the dye has been largely elucidated recently.[17]

b. Xanthene Dyes. The choice of bright pink dyes for wool, possibly for knitware, is somewhat restricted. However, the xanthene class offers a few useful products. For brilliance of shade but only moderate light fastness, compounds **57** and **58** are good examples. These sulforhodamine dyes are produced by

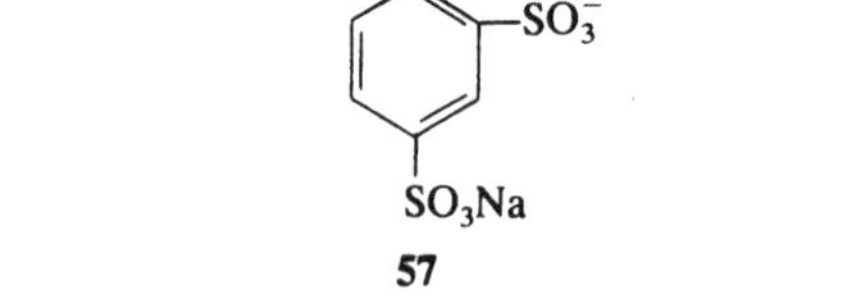

57

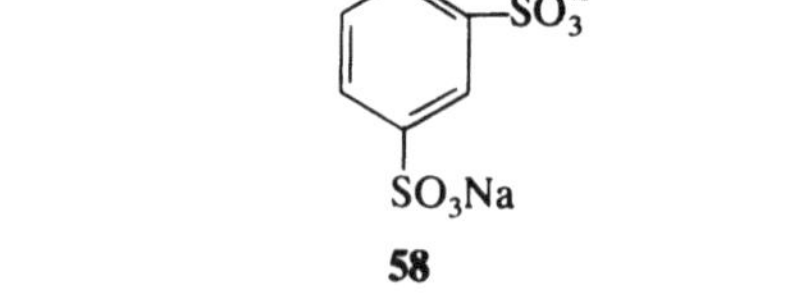

58

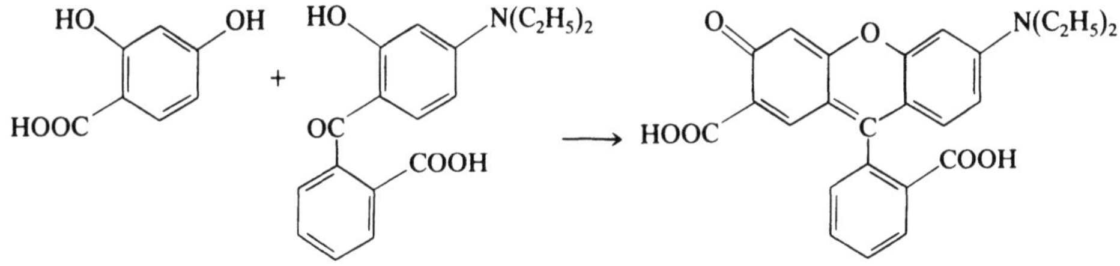

59

Scheme 3

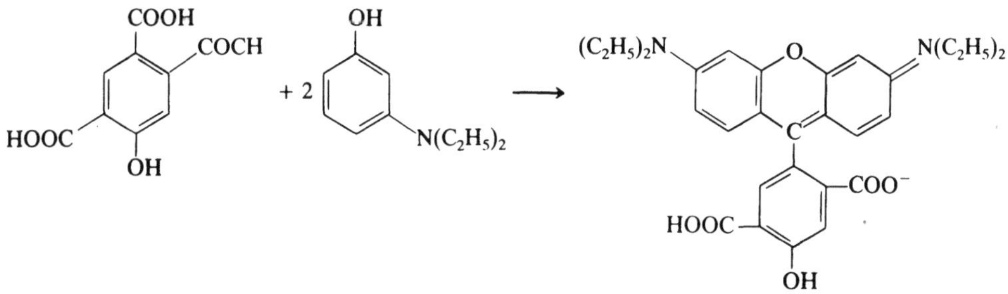

60

Scheme 4

condensation of *m*-diethylaminophenol or 2-ethylamino-*p*-cresol, respectively, with 2,4-disulfobenzaldehyde, ring closure by dehydration, and then oxidation using, for example, iron(III) chloride. Rather better light fastness is achieved with afterchrome dyes such as **59** and **60**. CI Mordant Red 15 **(59)** is prepared by condensing the intermediate stage involved in the manufacture of Rhodamine B (CI Basic Violet 10) with 2,4-dihydroxybenzoic acid as shown in Scheme 3. CI Mordant Red 27 **(60)** is made by condensing 5-hydroxytrimellitic acid with *m*-diethylaminophenol in a solvent (Scheme 4).

c. Triphenylmethane Dyes. Although of poor light fastness as a class, selected members of this type of dye find a use for knitted and hosiery ware. Examples of these blue to green dyes include compounds **61** and **62**. CI Acid Blue 9 **(61)** is prepared by condensing *o*-sulfobenzaldehyde with *N*-benzyl-*N*-ethylaniline-3′-sulfonic acid and then oxidizing the resulting leuco compound. CI Acid Green 22 **(62)** is similarly obtained by condensing *o*-chlorobenzaldehyde with *N*-benzyl-*N*-ethyl-*m*-toluidine-3′-sulfonic acid followed by oxidizing the leuco compound with lead dioxide. A somewhat more complex type with considerably higher light fastness is exemplified by Acid Blue 213 **(63)**, which can be produced by the condensation of 1-methyl-2-phenylindole with 4,4′-dichlorobenzophenone dichloride in the presence of zinc chloride, and the

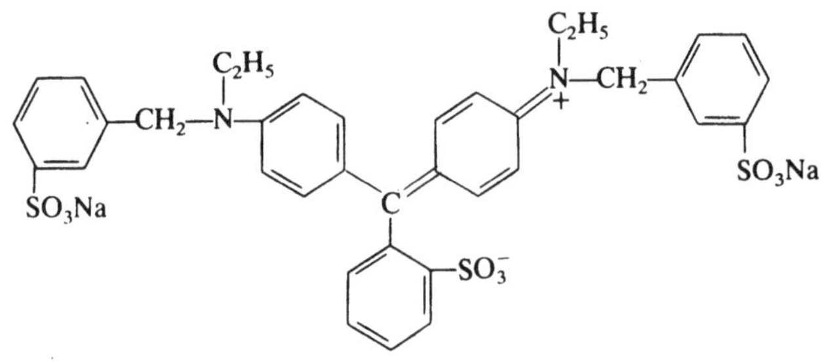

61

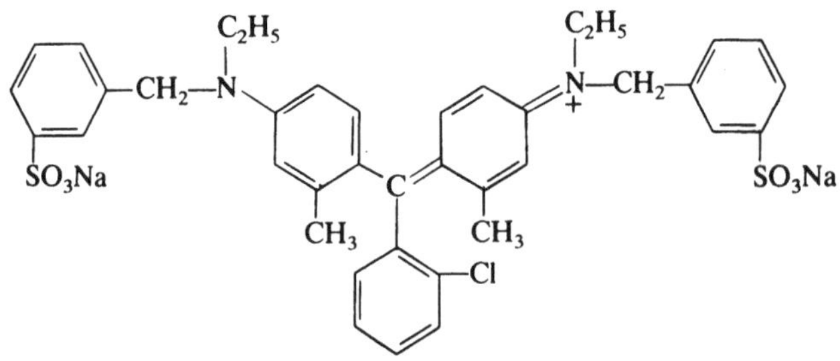

62

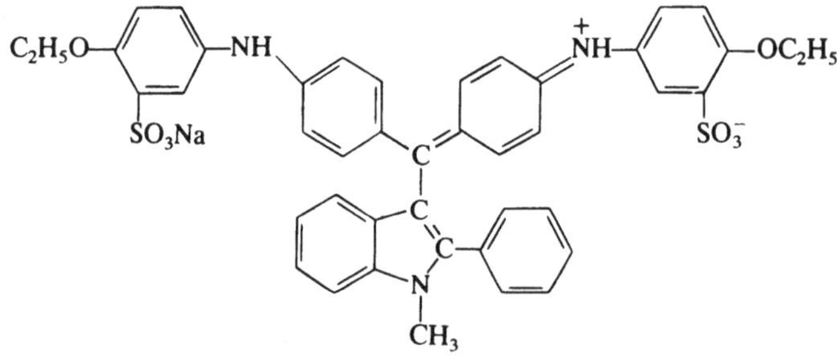

63

resulting product reacted with *p*-phenetidine to replace the halogen atoms before sulfonation.

d. Nitro Dyes. These dyes provide a small group of products for wool and nylon: although the wet fastness is only moderate, the light fastness is quite good. The most well-known example is **64** and it has been suggested that the satisfactory light fastness can be associated with hydrogen bonding as shown in **64a**. CI Acid Orange 3 **(64)** is prepared by condensing 4-amino-2-sulfodiphenylamine with 2,4-dinitrochlorobenzene in the presence of an acid-binding agent such as sodium hydrogencarbonate or chalk. Structure CI 10405 **(65)** is obtained by the reaction of 4′-acetylamino-4-amino-2-sulfodiphenylamine with 2-nitrochlorobenzene-4-sulfon-*N*-diethylamide in the presence of an acid-binding agent. CI Acid Brown 13 **(66)** is a more elaborate dye which illustrates well how increasing the size of the molecule improves the wet fastness. It is prepared by heating 3,3′-dinitro-4,4′-dichlorodiphenylsulfone with 4-aminodiphenylamine-2-sulfonic acid under the same conditions as previously described. CI Acid Brown 248 **(67)** is made by

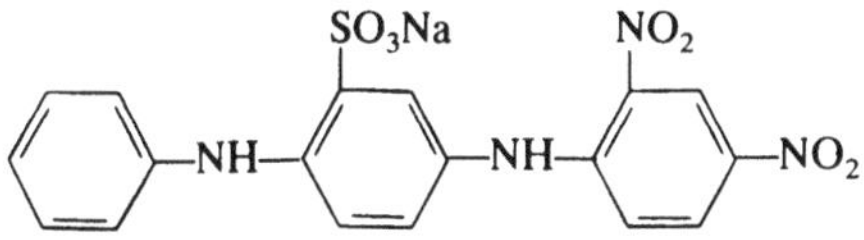

64

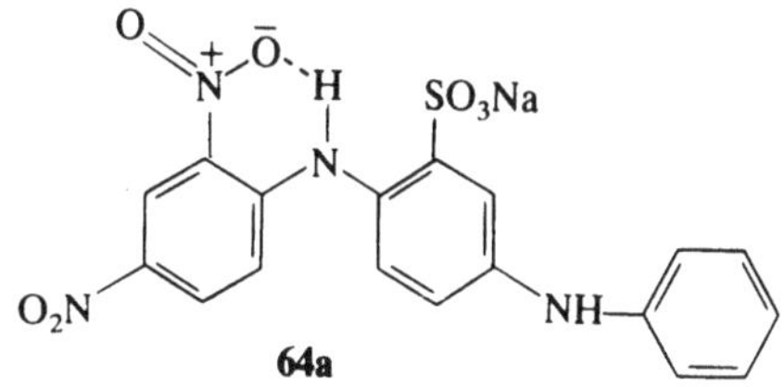

64a

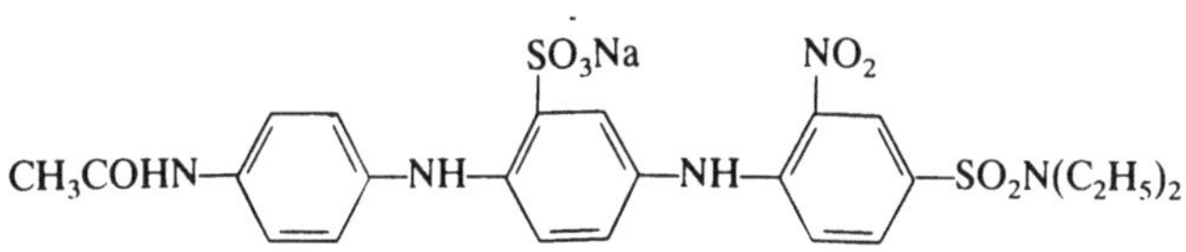

65

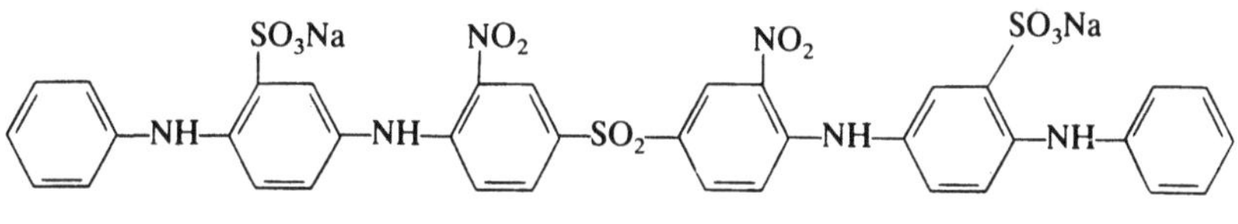

66

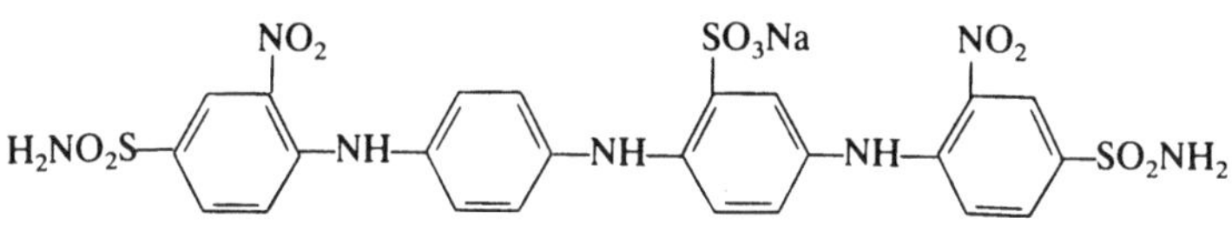

67

reducing 4-amino-4′-nitro-2′-sulfodiphenylamine and then condensing with 2-nitrochlorobenzene-4-sulfonamide.

e. Reactive Dyes. Many of the original monochlorotriazinyl dyes of the Procion or Cibacron ranges have been and still are used for wool dyeing, but the present tendency is to use dyes specially made for this purpose. Although the class as a whole provides only a small proportion of the total group for wool dyes, their use is increasing because very good wet fastness is combined with good light fastness and brightness. Relatively few constitutions have been disclosed. Any acid azo dye or indeed any acid dye with a free amino group which can be substituted with a reactive moiety is a potential reactive dye. Typical reactive systems for wool are shown below:

Procilan (ICI)	$D\cdot NHCOCH{=}CH_2$
Lanasol (Ciba-Geigy)	$D\cdot NHCOCBr{=}CH_2$
Hostalan (Hoechst)	$D\cdot NHSO_2CH_2CH_2N(CH_3)CH_2CH_2SO_3Na$
Remazol	$D\cdot NHSO_2CH{=}CH_2$
Remalan	$D\cdot NHSO_2CH_2CH_2OSO_3Na$

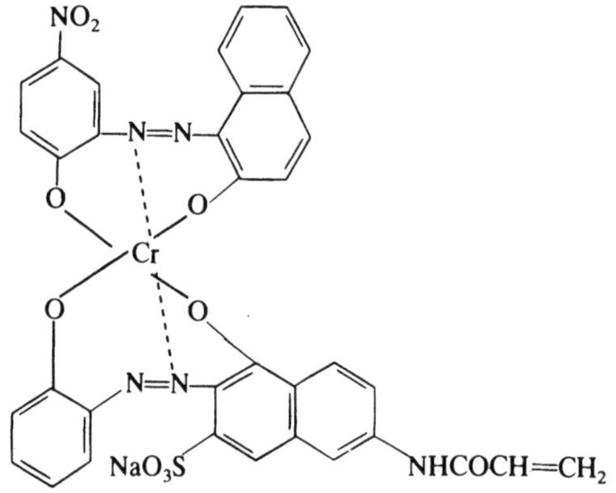

68

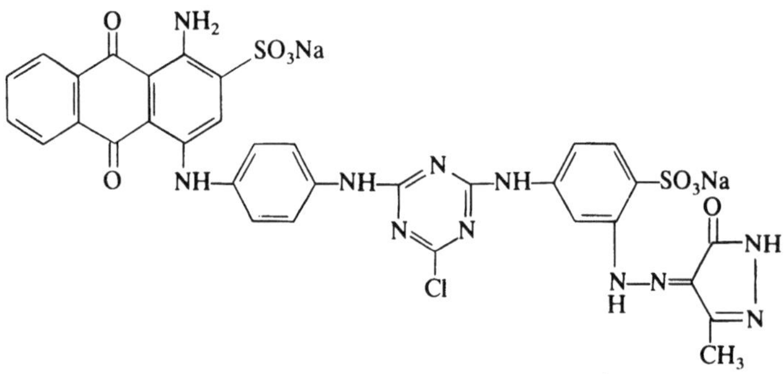

69

An example of a Procion type is structure **68** in which the combination of metal complex and reactive groups give the product extremely good wet and light fastness properties.[18] An example of a Procion type for wool is compound **69.**[19]

f. Phthalocyanine dyes. Although the phthalocyanines are mainly used for pigments and solvent dyes, controlled sulfonation give products suitable for wool, cotton, and nylon. Thus, for example, Coomassie Turquoise 3G (ICI) **(70)** gives a very bright greenish-blue hue on wool with a higher light fastness than

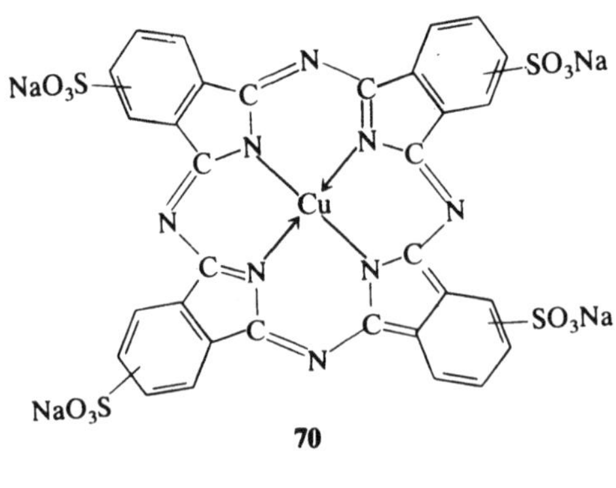

70

similarly-colored triphenylmethane dyes, but requires auxiliaries to ensure reasonably level-dyeing behavior. The dye is prepared by heating the pigment with 10–20% oleum at temperatures of 70–90 °C.

III. DYES FOR NYLON

In 1928, W. H. Carothers, working for the firm of E. I. du Pont de Nemours in Delaware, U.S.A. discovered the means by which long chain polymers could be produced. After a considerable amount of work in the U.S.A. and in Europe, the technical production of nylon was achieved. The main type of nylon is prepared by the condensation of adipic acid and hexamethylenediamine to give a nylon salt, $HOOC(CH_2)_4CONH(CH_2)_6NH_2$, which is heated in an autoclave in the absence of air to a prescribed temperature so that polymerization takes place with the elimination of water. After the desired degree of polymerization has been achieved, the mass is extruded in a ribbon form and chilled rapidly. The chipped ribbon is the base material from which nylon fiber is produced by melting and extrusion through fine spinnerets. Subsequent cold stretching of the filament orientates the fiber and makes it lustrous, stronger, and elastic. This produce is the Nylon 66 of commerce. Nylon 6 is prepared similarly by the polymerization of caprolactam, prepared by several routes on a technical scale.

Possibly the main physical difference between wool and nylon is that the latter absorbs only a fraction of the moisture absorbed by the former and this hydrophobic nature is responsible for most of the properties of nylon. Thus, as regards carpets, wool has a high resistance to soiling and is more readily cleaned by shampooing compared with nylon, which has also a higher flammability. Wool also acts as a moisture buffer in rooms, thereby dispelling electrostatic charges which can be troublesome with nylon carpets. Further developments in nylon manufacture have resulted in many variants which differ in their appearance and dyeing properties. Noteworthy are the types designated as deep-dye and basic-dyeable nylons, the latter usually being prepared by incorporating 5-sulfoisophthalic acid in the polymerization process.

Nylon can be dyed by many classes of dyes and by different techniques, but a comprehensive account is beyond the scope of this chapter. However, an outline of different dyeing processes used for nylon can be given according to end use. Hosiery ware, the original major use for nylon (and the source of the name nylons), is dyed on preformed blanks or cards by a triad of disperse dyes, such as compounds **71–73**. These dyes in admixture produce practically all the hues used in ladies' hosiery ware, but other disperse dyes can be used for some fashion shades.

Nylon yarns, both staple and continuous filament, are dyed in package form with selected acid dyes and also with the 2:1 premetallized dyes which are eminently suitable in order to obtain the required degree of fastness for weaving, knitted ware, and for carpets. Coloration can also be achieved by printing and by the dry technique of transfer printing. Thus, after much trial and research, it was established that azo dyes suitable for dyeing nylon required good aqueous

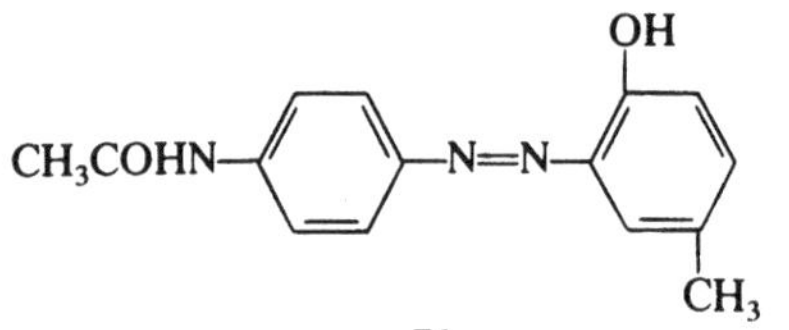

71

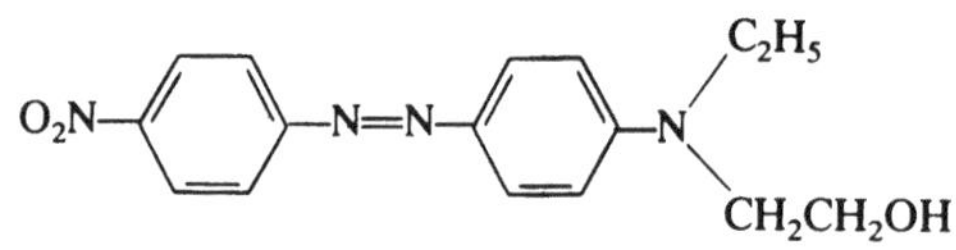

72

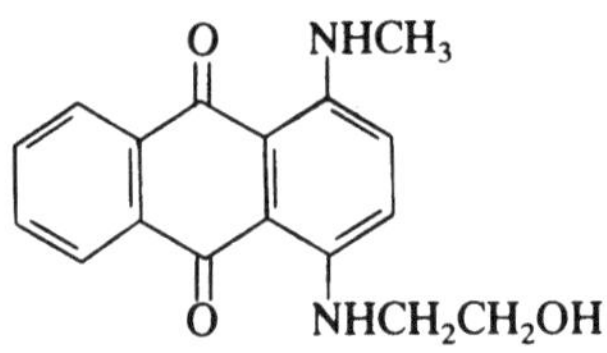

73

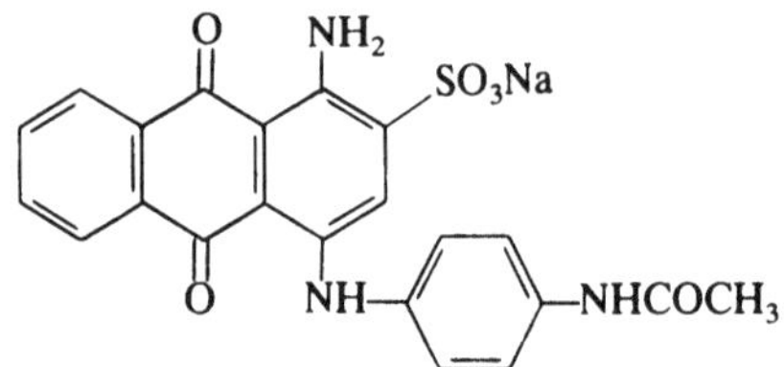

74

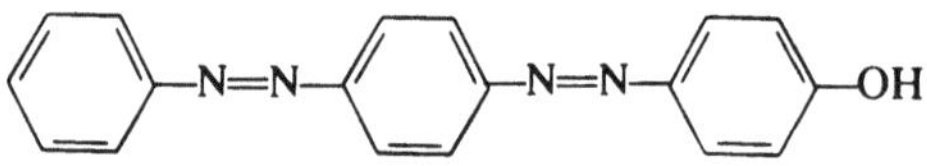

75

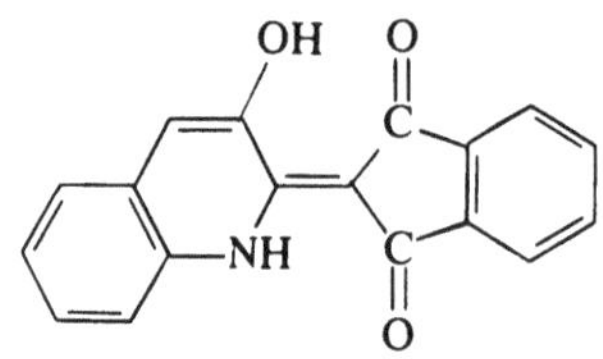

76

solubility, usually provided by one sulfonic acid group, and a relative molecular mass of about 450. Examples of dyes possessing these characteristics are CI Acid Yellow 29 **(14)**, CI Acid Red 57 **(15)**, and CI Acid Blue 40 **(74)** in the acid anthraquinone class.

As well as the three disperse dyes **71–73**, many other disperse dyes are widely used. Thus, for example, CI Disperse Yellow 3 **(71)** is probably the most widely-used yellow disperse dye; better fastness to sublimation, as required in ironing, is provided by CI Disperse Yellow 23 **(75)**; higher light fastness is given by CI Disperse Yellow 54 **(76)**. CI Disperse Yellow 3 **(71)** is prepared by coupling diazotized *p*-aminoacetanilide with *p*-cresol in alkaline solution. CI Disperse Yellow 23 **(75)** is manufactured by diazotizing *p*-aminoazobenzene and coupling with phenol in aqueous alkali. CI Disperse Yellow 54 **(76)** is prepared by heating phthalic anhydride and 3-hydroxy-2-methylquinoline-4-carboxylic acid with zinc chloride as catalyst, the carboxylic acid group being eliminated during the reaction. The hydroxyl group in the 3-position confers an increase in light fastness, associated with hydrogen bonding to the keto group (Chapter 4). An example of an orange dye for nylon is CI Disperse Orange 25 **(77)**, which is produced by coupling diazotized *p*-nitroaniline with *N*-ethyl-*N*-2-cyanoethylaniline. The cyanoethyl group confers improved light fastness when compared with the corresponding hydroxyethyl dye CI Disperse Red 1 **(72)** and also has a hypsochromic effect.

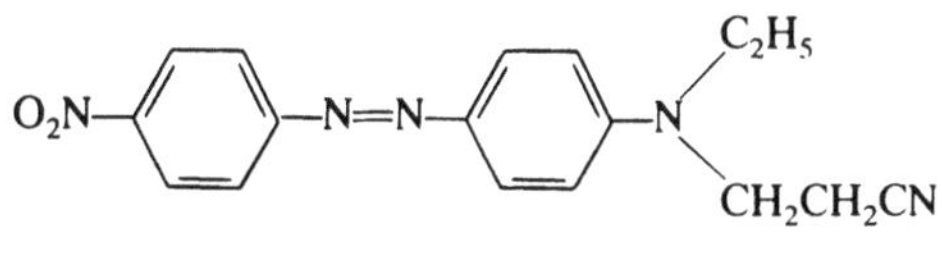

77

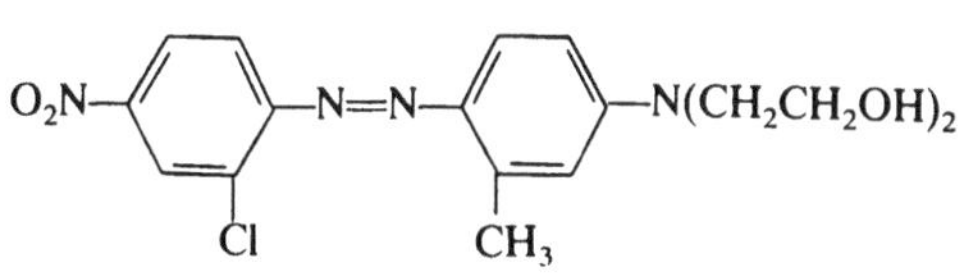

78

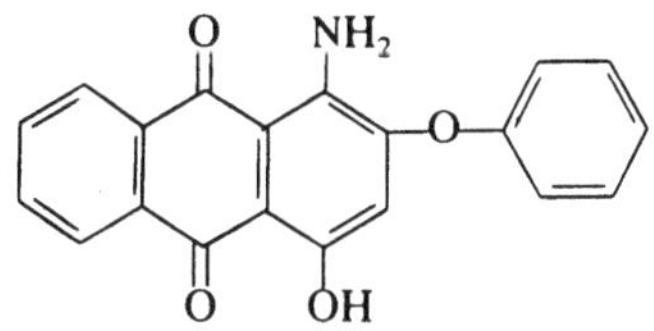

79

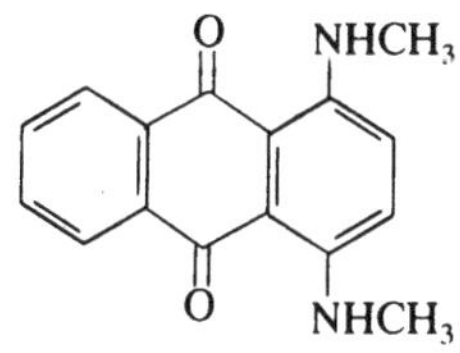

80

Disperse reds for nylon are typified by dyes **72**, **78**, and **79**. CI Disperse Red 1 **(72)** is made by coupling diazotized *p*-nitroaniline with *N*-2-hydroxyethyl-*N*-ethylaniline, CI Disperse Red 5 **(78)** by coupling diazotized 2-chloro-4-nitroaniline with *N,N*-bis-(2-hydroxyethyl)-*m*-toluidine, and CI Disperse Red 60 **(79)** by displacing the 4-bromine atom in 1-amino-2,4-dibromoanthraquinone by hydroxy using sulfuric acid followed by condensation of the resulting red dye with phenol in the presence of potassium hydroxide. The main disperse blue for nylon is CI Disperse Blue 3 **(73)** although CI Disperse Blue 14 **(80)** can also be used. These dyes are prepared by condensation of leuco quinizarin with a mixture of methylamine and ethanolamine to give CI Disperse Blue 3 **(73)** and with methylamine alone to give CI Disperse Blue 14 **(80)**. It is noteworthy that the former dye contains amounts of the symmetrical compounds 1,4-di(methylamino)anthraquinone and 1,4-di(2-hydroxyethylamino)anthraquinone; this situation leads, in many cases, to an improvement in dyeing properties.

The nylons are dyed to a high degree of fastness by the 2:1 premetallized dyes which can be used for the coloration of tufted carpets. The rise of continuous dyeing of tufted carpets by the print-steam technique has necessitated very careful choice of dyes: very good solubility or, in the case of disperse dyes, good stability and a high affinity for nylon. As well as the soluble 2:1 premetallized dyes, the insoluble analogs in a dispersed form have been used for some time as in the Perlon range developed by the I.G. and in patents of du Pont in America for the dyeing of nylon, and later revived in the Amichrome range of Francolor. Typical examples of these dyes are shown by structures **81–83**. CI Acid Orange 92 **(81)** is prepared by coupling diazotized 2-amino-4-nitrophenol with 3-methyl-1-phenyl-5-pyrazolone under alkaline conditions followed by chroming the azo dye in a solvent such as butanol. CI Acid Black 63 **(82)** can be prepared similarly by coupling diazotized 2-amino-5-nitrophenol with 2-naphthol and chroming the product in a solvent. This procedure gives a bluish-black which, on dilution, is a good gray; a more neutral black tone can be obtained by using a mixture of 2-amino-4- and -5-nitrophenols. CI Solvent Yellow 21 **(83)** is obtained by chroming the azo dye prepared by coupling diazotized anthranilic acid with

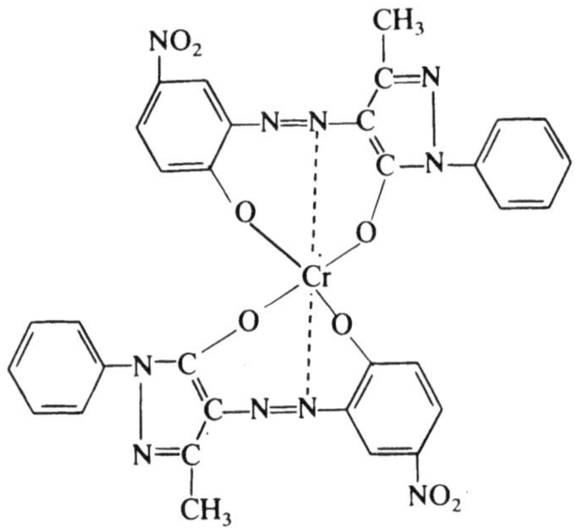

81

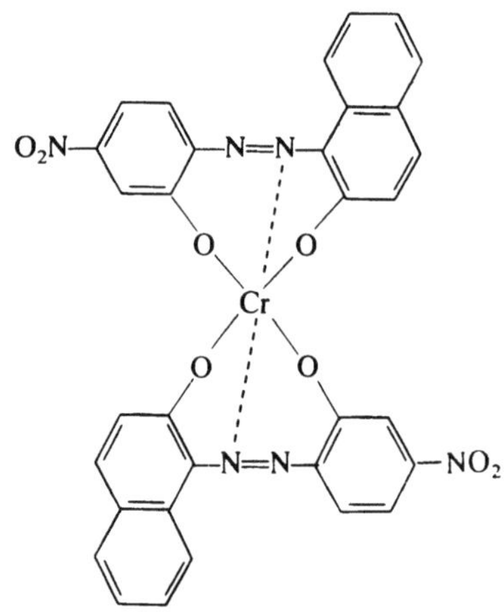

82

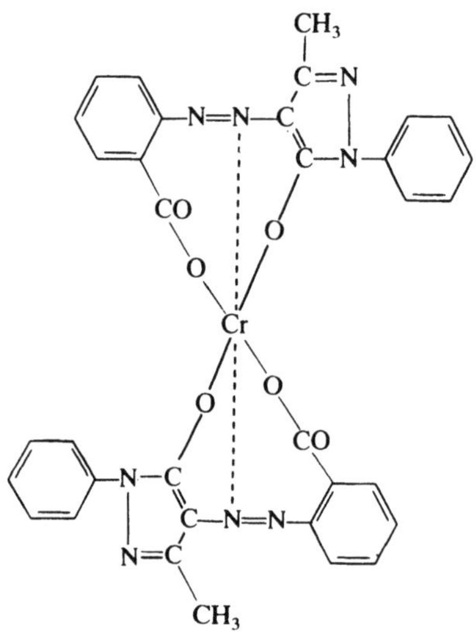

83

3-methyl-1-phenyl-5-pyrazolone. Many old well-established direct dyes have also found a use in dyeing nylon, such as CI Direct Yellow 12 and CI Direct Red 1. Unfortunately, the structures of many recent dyes marketed especially for nylon have not been disclosed.

In a great many cases, wool and nylon blends are used, especially for carpets (often 80% wool and 20% nylon), hosiery, and knitted ware; the proportion of nylon here is often higher because improved resistance to abrasion is desirable. Nylon has a high affinity for acid dyes but also a lower capacity than wool, so that with leveling dyes the nylon is dyed heavier in pale shades while the reverse is true in heavy shades. The effect is not so pronounced with milling dyes. Thus, care has to be exercised in the choice of dye in order to obtain even, solid shades. At the present time, many auxiliaries are used to correct this tendency to dye unevenly on the two fibers. It should be noted that the use of Nylon 6 in place of

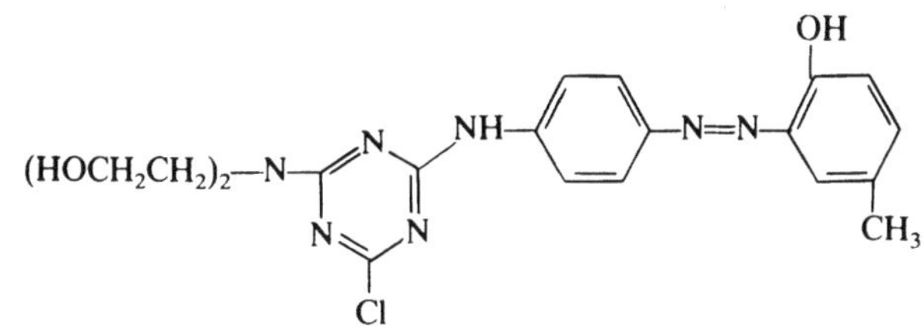

84

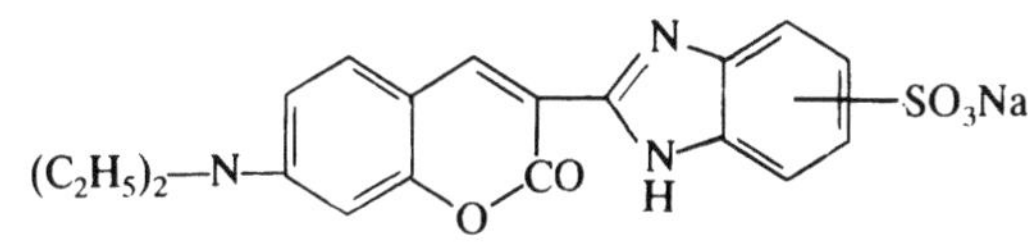

85

nylon 66 can affect the amount of auxiliary used, nylon 6 being easier to dye than nylon 66. Among the leveling dyes available are CI Acid Yellow 17, CI Acid Red 37, and CI Acid Blue 45; dyes with better milling fastness include the group CI Acid Yellow 29, CI Acid Red 57, CI Acid Red 80, CI Acid Violet 41, CI Acid Blue 40, and CI Acid Blue 47. For carpet yarn in wool/nylon blends the 2:1 premetallized dyes are eminently suitable with selected milling dyes such as Polar Brilliant Red B and 10B, Polar Brilliant Blue RWL, and Irganol Yellow 4GLS as bright colors. An improvement on CI Disperse Yellow 3 **(71)** is the reactive dye Procinyl Yellow GS **(84)** which can be made by sequential condensation of 4-amino-2′-hydroxy-5′-methylazobenzene with cyanuric chloride and then diethanolamine. Nylon cloth used, for example, for tennis ball coverings is dyed in very bright fluorescent yellows with 3-arylcoumarins of the type **(85)**.[20]

IV. SYNTHESIS OF DYES

A. CI Acid Yellow 19 (CI 18967)

2-Naphthylamine-1-sulfonic acid (22.3 g, 100%) is dissolved in water (250 ml) and aqueous sodium hydroxide (10 ml, 40%) to give a clear solution, warming if required. Crushed ice is added to bring the temperature down to 5 °C and hydrochloric acid (35 ml, 28%) is added with good agitation to precipitate the sulfonic acid in a fine form. Sodium nitrite (7 g) in water (25 ml) is added dropwise over 1 h with constant stirring and keeping the temperature at 5–10 °C. After stirring for a further hour, the suspension should be acid to Congo Red and show a slight positive reaction on starch-potassium iodide paper. Meanwhile, 1-(2′,5′-dichloro-4′-sulfophenyl)-3-methyl-5-pyrazolone (33 g, 100%) is dissolved in water (330 ml) by the addition of aqueous sodium hydroxide (10 ml, 40%) and sodium carbonate (30 g) is then added. The clear solution is cooled with crushed ice to 0 °C. The diazonium salt suspension is added to the pyrazolone solution with constant stirring. Reaction is very rapid and, after a further 2 h stirring, the dye is completely precipitated by adding salt (5% of the total volume). Stirring is continued for a further hour after which the dye is filtered on a Buchner funnel, sucked as dry as possible, and the product dried at about 80 °C to give a yield of about 65 g.

The resultant compound dyes wool and nylon in bright yellow hues of good wet fastness and very good fastness to light.

B. CI Acid Red 57 (CI 17053)

2-Aminobenzenesulfon-*N*-ethylanilide (27.6 g, 100%), finely ground, is stirred in water (150 ml) containing crushed ice to lower the temperature to 0 °C before hydrochloric acid (10 ml, 28%) is added. While stirring, a solution of sodium nitrite (7 g) in water (15 ml) is added slowly over 2 h, concurrently with a further 20 ml of hydrochloric acid, keeping the reaction mixture at 0–5 °C. After a further 30 min stirring, a turbid solution should be obtained, giving a positive reaction on starch-potassium iodide paper and an acid reaction on Congo Red paper. In the meantime a coupling solution is prepared of 7-amino-1-naphthol-3-sulfonic acid (24 g, 100%) in water (300 ml), aqueous sodium hydroxide (10 ml, 40%), and hydrated sodium acetate (15 g); crushed ice is added to bring the temperature down to 0 °C. The pH is now adjusted to 6.5–6.8 with dilute acetic acid if required. The diazonium solution is rapidly filtered and then added, over 5 min, to the stirred second component solution and the deep red mixture stirred for 4 h or overnight if convenient, before making slightly alkaline to glazed litmus paper with sodium hydroxide solution (40%); 10% of the total volume of salt is then added. The mixture, in a somewhat gelatinous form, is slowly heated with stirring to 50–55 °C whereupon it suddenly thins down and becomes crystalline. The precipitated material is then filtered on a Buchner funnel, sucked as dry as possible, and dried in an oven at 80 °C to give about 60 g of a dark red powder having a comparative strength of twice that of the standard material. The product dyes wool and nylon in red hues of good levelness and a high degree of fastness to light and washing.

C. CI Acid Orange 67 (CI 14172)

4-Amino-4′-nitro-2′-sulfodiphenylamine (31 g, 100%) is dissolved in water in water (250 ml) and sodium hydroxide solution (10 ml, 40%) by gentle warming. To the deep brown solution is added sodium nitrite (7 g) with stirring until dissolved and the solution is then added slowly to a constantly stirred solution of hydrochloric acid (40 ml, 100%) and water (600 ml) at room temperature and the diazonium suspension stirred overnight at 20 °C. To the greenish-gray suspension, which should be just acid to Congo Red paper and only faintly positive on starch-iodide paper, ice is added to bring the temperature to 0 °C and a little sodium hydrogencarbonate added to neutralize the free acid to Congo Red paper. A solution of *m*-cresol (11 g) in water (100 ml) and sodium hydroxide solution (10 ml, 40%) is added rapidly to the well-stirred diazonium suspension whereupon a deep brownish-yellow solution results. After 2 h stirring, salt to 5% of the total volume is added and, after stirring for a further hour to complete the precipitation, the azo compound is filtered off and washed with a little 5% salt solution and sucked dry. The filter cake is dissolved in water (500 ml) and sodium hydroxide solution (10 ml, 40%) and diluted to 900 ml before warming to 70–75 °C. With good stirring, toluene-*p*-sulfonyl chloride (30 g) is added over a period of 1 h (Hood), keeping the mixture slightly alkaline to glazed litmus paper. A spot of the solution on filter paper, originally brownish-orange, slowly becomes reddish-yellow as the reaction progresses. After heating for a little longer until the odor of tosyl chloride can no longer be detected, 5% of the total volume of potassium chloride is slowly added to precipitate the dye which is filtered off and sucked dry. After drying in an oven at 80 °C, 50 g of a brown powder is obtained which has a strength of about 180% of the standard. The product dyes wool and nylon in sandy-yellow hues of good light and wet fastness.

D. CI Acid Yellow 76 (CI 18850)

p-Aminophenol (11 g, 100%) is dissolved in water (150 ml) and hydrochloric acid (30 ml, 28%) by gentle warming, cooled to 0 °C with ice and diazotized by the dropwise addition of sodium nitrite (7 g) in water (15 ml) over 30 min at 0 °C. The clear solution is added to a solution prepared from 1-(4′-sulfophenyl)-3-methyl-5-pyrazolone (26 g, 100%), aqueous sodium hydroxide (10 ml, 40%), hydrated sodium acetate (15 g), and water (250 ml), and the mixture is stirred or allowed to stand overnight. The mixture is warmed to 70–75 °C before aqueous sodium hydroxide solution (15 ml, 40%) is added and then tosyl chloride (30 g) is slowly added with stirring, keeping the reaction mixture alkaline to glazed litmus paper by the addition, if required, of sodium hydroxide solution (Hood). After a further 30 min heating, salt to 5% of the total volume is added and the mixture stirred for a further 30 min. The precipitated crystalline dye is filtered off, washed with a little 5% salt solution, sucked dry, and then dried in an oven at 80 °C to give 45 g of a yellow powder used for dyeing wool and nylon in reddish-yellow hues from a neutral bath.

E. CI Acid Orange 3 (CI 10385)

4-Aminodiphenylamine-2-sulfonic acid (26.4 g, 100%) is dissolved in water (400 ml) and aqueous sodium hydroxide (10 ml, 40%) and the solution heated to 90 °C. With good stirring, under an efficient hood, sodium hydrogencarbonate (10 g) is first added and then, keeping the temperature at 90–95 °C, 2,4-dinitrochlorobenzene (22 g) (care: can be dermatitic) is added portionwise. After heating for a further 1 h, the hot solution is filtered through a layer of filter aid on a preheated Buchner funnel and the funnel washed with hot water (100 ml). To the hot liquor is added salt to 2.5% of the total volume and the mixture cooled to 50 °C, with stirring, then filtered, washed with 5% salt (100 ml), and dried at about 60 °C to give 40 g of a tan-colored product for the dyeing of wool and nylon in dull orange-yellow hues of good light fastness but moderate wet fastness.

F. CI Mordant Yellow 8 (CI 18821)

Anthranilic acid (13.7 g) is stirred with water (150 ml) and hydrochloric acid (25 ml, 28%) and the mixture warmed to dissolve the acid before cooling with crushed ice to 0 °C and diazotizing by the addition of sodium nitrite (7 g) in water (15 ml) over a period of 15 min. The diazonium solution should be clear, acid to Congo Red paper, and give a slightly positive reaction on starch-potassium iodide paper. 1-(4′-Sulfophenyl)-3-methyl-5-pyrazolone (26 g) and sodium carbonate (35 g) are dissolved in water (400 ml; care: foaming) and the solution cooled to 10 °C with crushed ice. The diazonium solution is then added. Coupling takes place rapidly and, after 1 h, the mixture is heated to 50 °C. Salt to 10% of the volume is added over a period to precipitate the dye in a good form for filtration. After stirring for a further 1 h, the product is filtered off, sucked dry, and dried at 80 °C to give about 50 g of a light brown powder suitable for dyeing wool by the afterchrome method in mustard-yellow hues of very good light and wet fastness.

G. CI Mordant Red 7 (CI 18760)

To a suspension of the diazo-oxide of 1-amino-2-naphthol-4-sulfonic acid (25 g, 100%) in water (300 ml) is added a solution of 1-phenyl-3-methyl-5-pyrazolone (18 g) in sodium hydroxide solution (10 ml, 40%) and water (200 ml); the mixture is then stirred overnight at 20 °C. If coupling is not complete, as shown by the lack of a purple spot when the reaction mixture is spotted at the edge of filter paper with a freshly made solution of resorcinol in a few drops of aqueous ammonia, a few grams of sodium carbonate are added and the mixture warmed to 30 °C for about 1 h. Hydrochloric acid (28%) is added until the brownish-red mixture changes to a bright red and the dye is completely precipitated. After filtration, the dye is dried at 70 °C to give 40–45 g of a bright red powder which dyes wool by the afterchrome method in bluish-red hues of very good light and wet fastness.

H. CI Mordant Orange 6 (CI 26520)

4-Aminoazobenzene-4′-sulfonic acid (28 g, 100%) as sodium salt is dissolved in water (500 ml) and hydrochloric acid (30 ml, 28%) is added slowly with good stirring to precipitate the free acid as a voluminous purple-gray precipitate. At 20 °C, with good stirring, sodium nitrite (7 g) in water (15 ml) is added over 1 h and the suspension stirred overnight at room temperature. The clay-colored diazonium salt should be acid to Congo Red and show a slight excess of nitrous acid. Ice and salt (100 g) are now added to bring the temperature below 0 °C, and a solution of salicyclic acid (15 g) in sodium hydroxide solution (11 ml, 40%) and water is added rapidly. With good stirring, sodium hydroxide solution (40%) is added over about 1 h until the mixture is slightly alkaline to litmus, keeping the temperature at 0 °C. The mass is then stirred, allowing the temperature to rise slowly and maintaining an alkaline medium. After a further 2 h, the mixture is warmed to 50–60 °C, maintained for 1 h, then filtered, sucked dry, and dried at 70 °C to give a yield of 50 g.

The product dyes wool a deep orange hue by the afterchrome method. Nylon is dyed as with an acid dye in bright orange hues, fast to light, and wet treatments.

I. CI Acid Red 249 (CI 18134)

2-Amino-4-chlorodiphenyl ether (22 g) is dissolved in hydrochloric acid (30 ml, 28%) and glacial acetic acid (30 ml) and the solution cooled externally to 0 °C. At 0–5 °C, sodium nitrite (7 g) in water (15 ml) is slowly added. After stirring for a further 30 min, the diazonium solution, which is somewhat turbid, is added in a thin stream to a previously prepared mixture of *N*-benzenesulfonyl-H-acid (47 g, 100%) as a neutral solution, together with calcium hydroxide (43 g) and ice and water to a volume of 500 ml at 0 °C. The mixture is stirred overnight, warmed to 50–60 °C, and the calcium precipitated by slowly adding sodium carbonate (60 g). The hot solution is filtered from the calcium carbonate which is washed with hot water (200 ml) and the liquor salted to 10% of the total volume. The precipitated dye is filtered off on a Buchner funnel and dried at 80 °C to give 100 g of standard strength product. The dark red powder dyes wool and nylon from a weak acid bath in bright red hues of good light and wet fastness properties. Use of the *p*-tosyl or *p*-chlorobenzenesulfonyl derivatives of H-acid gives dyes having rather better fastness to perspiration.

J. CI Acid Black 60 (CI 18165)

2-Aminophenol-4-sulfon-*N*-methylamide (20.2 g, 100%) is dissolved in hydrochloric acid (20 ml, 28%) and water (200 ml) by gentle warming, then cooled to 0 °C and diazotized by the slow addition of sodium nitrite (7 g) in water (15 ml), the diazo-oxide precipitating in a bright yellow form. A little sodium hydrogencarbonate is added to make the mixture neutral to Congo Red, and a solution of 1-acetylamino-7-naphthol (20.1 g, 100%) in sodium hydroxide solution (10 ml, 40%) and water (100 ml) to which sodium carbonate (11 g) has been added is then poured in with good stirring. The mixture is stirred overnight at about 20 °C. Salt to 10% of the volume is now added and the precipitated azo compound filtered off after stirring for a further hour. The filter cake is pasted with water, diluted to 1 liter, and dissolved by the addition of sodium hydroxide solution (20 ml, 40%); the temperature is raised to 90 °C. At this temperature, a solution containing sodium dichromate (8 g) and glucose (20 g) in water (80 ml) is added slowly over a period of 20–30 min. After stirring for a further hour at 90 °C, the mixture is filtered through a Buchner funnel with the aid of a pad of filter-aid, the liquors cooled to 50 °C and, after adding sodium hydrogencarbonate (11 g), salt to 10% of the volume is added, precipitating the chromium complex in a form easily filtered. The filter cake is dried at 80 °C to give 100 g of a product which dyes wool and nylon from a neutral bath in bluish-gray hues of excellent light and wet fastness.

K. CI Acid Blue 129 (CI 62058)

1-Amino-4-bromoanthraquinone-2-sulfonic acid (40 g, 100%) as sodium salt is stirred into water (250 ml); ethanol (100 ml), mesidine (16 ml), sodium hydrogencarbonate (30 g), and glucose (6 g) are then added to the suspension, which is stirred and heated to 40 °C. Copper(I) chloride (1 g) is added and the temperature raised to 70 °C and maintained at 70–75 °C for 2 h. On cooling to 55 °C, the precipitated dye is filtered, washed with sodium hydrogencarbonate solution (400 ml, 5%) at 50 °C, and then dried at 80 °C to give a yield of 40 g. The product dyes wool and nylon in bright reddish-blue hues of good light and wet fastness.

L. CI Acid Green 25 (CI 61570)

Quinizarin (24 g), *p*-toluidine (100 g), hydrochloric acid (10 ml, 28%), and boric acid (3 g) are mixed together in a 500 ml flask immersed in an oil or other bath and zinc dust (2 g) added with stirring and the mixture heated to 95 °C (Hood) for 4–5 h. To the dark-colored melt, after cooling to 70 °C, is added methanol or ethanol (100 ml), before cooling to 20 °C. After standing for 1 h, the precipitated material is filtered, washed with four portions of methanol (50 ml), then once with water, and dried to give a yield of 38 g of base.

To a mixture of oleum (50 ml, 20%) and sulfuric acid (16 ml, 98%) at 20 °C is added with stirring finely-powdered base at below 30 °C. A further 22 ml of oleum are added at 20 °C and the mixture stirred until a sample is completely soluble in aqueous ammonia. The sulfonation mixture is poured into a solution of salt (60 g) in water (700 ml). The suspension is stirred until in a granular form, then filtered, and washed with a solution of salt (45 g) in water (700 ml) and sucked dry.

The filter cake is placed in a dish and made slightly alkaline with sodium hydroxide solution (40%) and dried at 80–90 °C to give a yield of 55 g, equivalent to about 70 g of standard. The dark greenish powder dyes wool and nylon from a weak acid bath in green hues of good fastness to light and wet treatments. The fastness to wet treatments on nylon depends to some extent on the exact degree of sulfonation, and optimum conditions must be found by trial and error then rigidly adhered to. In a similar manner, CI Acid Green 27 (CI 61580) can be obtained by using *p*-butylaniline instead of *p*-toluidine; this dye has much improved milling fastness properties.

M. Coumarin Acid Dye

3-(Benzimidazol-2′-yl)-7-diethylaminocoumarin (33 g), itself a fluorescent pigment prepared by the condensation of 4-diethylamino-2-hydroxybenzaldehyde with 2-cyanomethylbenzimidazole, is added with stirring over 30 min to oleum (125 g, 10%) at room temperature. The temperature is raised, over 30 min, to 50 °C and held at this value for 3 h until a sample, diluted with water and made alkaline with ammonia, gives a clear solution. The sulfonation mixture is poured onto ice and water (1 liter) and stirred for 30 min, then filtered, and washed with a little cold water and sucked as dry as possible. The filter cake is added to cold water (250 ml) and sodium carbonate added to give a clear solution (care: foaming). After warming to 70 °C, any impurities are filtered off and salt to 10% of the total volume is added before stirring for 30 min. Filtration and drying at 70 °C gives a yield of 50 g.

The product dyes nylon from a weakly acid bath in very bright fluorescent greenish-yellow hues of good light and wet fastness properties.

N. Phthalocyanine Acid Dye (71)

Copper phthalocyanine (20 g), finely powdered, is added to oleum (145 g, 20%) and the temperature raised to 80 °C and held at this for 12 h. The solution is run slowly into a well-stirred mixture of salt solution (1 liter, 30%) and ice (1 kg), the final temperature being 20 °C. The fine precipitate is filtered by gravity, then transferred to a Buchner funnel, sucked as dry as possible, and washed with a small quantity of water and dried at 80 °C to yield 40 g of greenish-blue product. The free acid is ground with sodium carbonate (10 g) to give a readily-soluble product which dyes wool and nylon from a weakly acid bath, with the aid of an appropriate auxiliary to promote level dyeing, in bright greenish-blue hues.

REFERENCES

1. W. H. Perkin, *J. Chem. Soc. 69,* 800 (1896); W. A. Hutchins and T. S. Wheeler, *J. Chem. Soc., 91* (1939).
2. W. H. Perkin, *Chem. Ber. 9,* 281 (1876).
3. M. M. Pelletier and Caventou, *Ann. Chim. Phys. 8,* 250 (1818).
4. W. H. Perkin and R. Robinson, *J. Chem. Soc. 93,* 489 (1908); O. Dann and H. Hofmann, *Angew. Chem. 75,* 1125 (1963).
5. P. Engels, W. H. Perkin, and R. Robinson, *J. Chem. Soc. 93,* 1115 (1908).
6. F. Bayer, *Chem. Ber. 11,* 1296 (1878); J. Harley-Mason, *J. Chem. Soc.,* 2907 (1950).
7. W. H. Perkin and L. Pate, *J. Chem. Soc. 67,* 644 (1895); see also *The Colour Index,* 3rd edn., Vol. 3, p. 3588 (CI 75660).
8. K. L. Casselman, *Craft of the Dyer,* University of Toronto Press (1980).
9. W. Partridge, *A Practical Treatise on Dyeing of Wool, Cotton and Silk,* H. Walker & Co. (1823); reprinted by S. Maney & Sons Ltd. for Pasold Research Fund Ltd. (1973).
10. E. M. Bolton, *Lichens for Vegetable Dyeing,* Studio Press, London (1960).
11. S. Grierson, *J. Soc. Dyers Colour. 100,* 209 (1984).
12. S. Grierson, D. G. Duff, and R. S. Sinclair, *J. Soc. Dyers Colour. 101,* 220 (1985).
13. D. H. S. Richardson, *The Vanishing Lichens,* David and Charles, London (1975).
14. R. Nietzki, *J. Soc. Dyers Colour. 5,* 175 (1889).
15. I. D. Rattee, *J. Soc. Dyers Colour. 69,* 288 (1953).
16. G. Schetty, *J. Soc. Dyers Colour. 71,* 705 (1955).
17. I. Ziderman, *Rev. Prog. Colour. Relat. Top. 16,* 46 (1986).
18. A. N. Derbyshire and G. R. Tristam, *J. Soc. Dyers Colour. 81,* 584 (1965).
19. K. Venkataraman, *The Chemistry of Synthetic Dyes,* Vol. VI, p. 227, John Wiley & Sons, New York (1978).
20. B.A.S.F., B.P. 1275778 (10.10.69).

7

Application of Dyes

S. M. BURKINSHAW

I. INTRODUCTION

Dyeing is believed to have originated in Neolithic times, some 4000 to 9000 years ago.[1] The coloring matters used were obtained from natural sources such as plants, insects, and shells,[2,3] often by means of complex and time-consuming operations. The practice of pretreating fabrics with mordants to give improved dyeings and sometimes new hues was reported by Pliny in the first century AD.[4]

The recipes and techniques used by the early dyers were often closely guarded secrets and, as a consequence, fame and prosperity attended certain regions. This secrecy persisted for many centuries and, in the thirteenth and fourteenth centuries, led to the formation, throughout Europe, of the Dyers' Guilds which exercised control of trading practices. Gradually, the Guilds became more open associations and the development of book printing and availability of paper enabled dissemination of the vast amounts of accumulated empirical knowledge in the form of recipe books such as that by Rosetti published in 1548.[5] Further rationalization of the dyeing process gained impetus in the mid-seventeenth century from the development of systematic quality control in the French dyeing industry initiated by Colbert.[6] During the eighteenth and early nineteenth centuries, a more scientific approach to dyeing was adopted which paralleled the advancements in general scientific enquiry. Such investigations gave rise to detailed accounts of dyeing processes and to theories of dyeing.[1,6,7] The quest for scientific understanding emerged during the late eighteenth and early nineteenth centuries in the founding of centers of technical education and training in chemistry in France, Germany, and Britain.[8] In 1856, while at The Royal College of Chemistry in London, W. H. Perkin isolated Mauve, the first synthetic dye to be commercially produced and used.[9] Perkin's successful

S. M. BURKINSHAW • Department of Colour Chemistry and Dyeing, The University of Leeds, Leeds, West Yorkshire LS2 9JT, England.

application of Mauve to silk, wool and mordanted cotton, together with the discovery of the diazo reaction shortly afterward by Greiss, stimulated enormous interest in the preparation of synthetic dyes or "coal tar colors," which in turn gave rise to the emergence of dyestuff manufacturers such as Geigy, Sandoz, Bayer, and others, which now constitute the major part of the world chemical industry.[10]

The search for new and improved dyes was accompanied by investigations into the nature of dye–fiber interactions, an area which became of great importance with the introduction of regenerated and, especially, synthetic fibers.

The following discussion concerns the dyeing of the major fibers. For each dye–fiber system considered the relevant mechanism of interaction is also discussed. The dyeing of fiber blends is not covered and readers are directed elsewhere.[11,12] While textile printing can, in essence, be regarded as "localized dyeing" this large and important application technique, which has received much attention,[13–23] is not discussed.

A. Classification of Dyes and Pigments

Until the latter half of the nineteenth century, relatively few publications concerned the classification of the natural coloring matters available. The introduction of synthetic dyes, however, augmented the already large number of naturally-occurring coloring matters and several publications appeared in which coloring matters were classified according to name, hue, and chemical structure.[24]

Each of these classifications, although useful, was superseded by the Colour Index (CI), which rapidly became the internationally accepted catalog of dyes and pigments. The first edition of the *Colour Index* was published in 1924 by the Society of Dyers and Colourists (S.D.C.) and used chemical constitution as the basis for classification. A second (1956), third (1971), and the current revised third editions (1976, 1981, and 1987), which are jointly published by the S.D.C. and the American Association of Textile Chemists and Colorists (AATCC), classify colorants according to:

1. *Constitution,* for the chemist requiring information on chemical structure, preparative methods, etc.
2. *Application,* for the user seeking information on commercial products, dyeing properties, etc.

The revised third edition consists of several volumes:

Volumes 1 to 3. In these, colorants are classified according to application type. Each coloring matter within each of the nineteen classes is ascribed a Colour Index Generic Name according to hue, e.g., CI Acid Red 11, CI Reactive Yellow 16, etc. The different classes of colorant are arranged alphabetically (Table 1); the information that Volumes 1 to 3 provide for each coloring matter includes chemical type, CI Constitution Number and colorant constitution (if declared), dyeing behavior, etc.

Volume 4. This divides colorants on the basis of their chemical structure; the different classes recognized are shown in Table 1.

Table 1. CI Classification of Colorants

Application	Constitution
Acid dyes	Nitroso
Azoic coloring matters	Nitro
Basic dyes	Azo
Developers	Azoic
Direct dyes	Stilbene
Disperse dyes	Carotenoid
Fluorescent brightening agents	Diphenylmethane
Food dyes	Triarylmethane
Ingrain dyes	Xanthene
Leather dyes	Acridine
Mordant dyes	Quinoline
Natural dyes	Methine
Oxidation bases	Thiazole
Pigments	Indamine
Reactive dyes	Indophenol
Reducing agents	Azine
Solvent dyes	Oxazine
Sulfur dyes	Thiazine
Vat dyes	Sulfur
	Lactone
	Aminoketone
	Hydroxyketone
	Anthraquinone
	Indigoid
	Phthalocyanine
	Natural organic coloring matters
	Oxidation bases
	Inorganic coloring matters

Volume 5. This includes:

(a) An index of commercial names in alphabetical order. Since colorants are given a variety of commercial names which may disguise the fact that one colorant is sold under different names by different manufacturers, for each commercial product the CI Generic Name and CI Constitution Number (if the colorant structure has been declared) is provided so as to permit equivalent products to be identified.

(b) An index of CI Generic Names, each entry including the corresponding commercial name to again allow the identification of equivalent products.

(c) An index of manufacturers of colorants.

Volume 6. This is the first (1976) supplement to Volumes 1 to 4.

Volume 7. This is the second (1982) supplement to Volumes 1 to 4 and a supplement to Volume 6.

Volume 8. This is the third (1987) supplement to Volumes 1 to 4 and a supplement to Volume 7.

For the purposes of this discussion, the classification of colorants according

Table 2. Usage of the Major Dye Classes

Dye class	Fiber
Acid[a]	Wool, polyamide
Azoic	Cotton, viscose rayon, secondary acetate, triacetate, polyamide,[b] polyester[b]
Basic	Acrylic
Direct	Cotton, viscose rayon, polyamide[b]
Disperse	Secondary acetate, triacetate, polyester, polyamide, acrylic[b]
Reactive	Wool, cotton, viscose rayon, polyamide[b]
Sulfur	Cotton, viscose rayon
Vat	Cotton, viscose rayon, polyester,[b] polyamide[b]

[a] Including premetallized.
[b] Minor use.

to application type will be employed. Some details of the important characteristics of each dye class studied will be found within those sections that deal with their use.

Table 2 summarizes the usage of the most widely used application classes of dye.

B. Textile Fibers

Textile fibers can be classified in several ways; the system shown in Table 3 is useful from the viewpoint of dyeing.

Until the latter half of the nineteenth century, all textile fibers were obtained from natural sources. The regenerated fibers, although essentially man-made, are simply modifications of these natural polymers, being prepared by the regeneration of one fibrous material from another. Synthetic fibers, however, are derived

Table 3. Classification of Textile Fibers

Hydrophilic
(1) *Cellulosic fibers*
    (a) Natural-cotton, linen
    (b) Regenerated-viscose rayon
(2) *Protein fibers*
    (a) Natural-wool, cashmere
    (b) Regenerated-casein

Hydrophobic
(1) *Regenerated fibers*
    Cellulose acetate, cellulose triacetate
(2) *Synthetic fibers*
    Polyamide, polyester, polyacrylonitrile

Table 4. Stages of Processing at Which Coloration May Be Carried Out

Fiber	Mass coloration	Gel dyeing	Loose stock	Sliver	Top	Tow	Yarn	Piece	Garment
Wool	–	–	+	+	+	–	+	+	+
Cotton	–	–	+	–	–	–	+	+	+
Viscose	+	–	+	–	–	–	+	+	+
Acetates	+	–	–	–	–	–	+	+	+
Polyester	+	–	+	+	+	+	+	+	+
Polyamide	+	–	+	+	+	–	+	+	+
Acrylic	+	+	+	+	+	+	+	+	+

from relatively simple, small molecules; for example, Nylon 6.6, the first commercially-available wholly-synthetic fiber, is prepared by condensation polymerization of hexamethylene diamine and adipic acid. Inorganic fibers such as glass fiber enjoy only relatively limited, mostly speciality usage. Furthermore, owing to their inert nature, such fibers are rarely dyed and as a consequence will not be considered herein.

In the following discussion of the dyeing of the major fiber types, the essential details of the physical and chemical properties of the various fibers in relation to dyeing are given. More detailed accounts of fiber structure in relation to dyeing are available.[25–29]

Textile materials can be dyed at various stages of their manufacture (Table 4). The particular stage at which coloration is affected depends on many factors, such as the manufacturing process involved, processing costs, end use, and fastness requirements. Current trends are away from dyeing fiber in the loose state to dyeing fiber in yarn, piece, and garment forms so as to defer coloration until the last possible stage of textile manufacture. In this way the quantity of colored goods held in stock is reduced and the dyer has greater flexibility in response to fashion trends.

The following discussion serves only as an indication of some of the features of coloration at the various stages of manufacture.

1. Mass Pigmentation

The pigmentation of the polymer mass prior to extrusion yields colored regenerated and synthetic fiber of excellent all-round fastness properties with low coloration costs, but the color ranges are limited and there is poor flexibility in production.[30]

2. Gel Dyeing[30,31]

Acrylic fibers are wet and dry spun from solution in a strongly polar solvent. In wet spinning, residual solvent is removed from the extruded filaments by a washing treatment. Gel dyeing consists of passing the washed filaments, in the form of continuous tow, through an aqueous basic (cationic) dyebath, followed by washing and rinsing to remove surplus dye. The tow is in the gel state (i.e., prior to drawing and stretching which develop the fiber's final crystallinity and orientation) and therefore dye molecules can readily penetrate the fiber, even at the relatively low temperatures commonly used in gel dyeing, giving rapid and uniform coloration. The dyes used must be capable of withstanding the subsequent processing that the acrylic fiber undergoes.

3. Loose State (Loose Stock, Top, Tow, and Sliver) Dyeing[31]

Although level dyeing is not essential, since this can be disguised by subsequent processing (e.g., blending), dye selection is of vital importance with regard to end-use requirements and especially further processing (e.g., weaving, knitting) that the dyed fiber will undergo. This latter consideration is of particular importance in the case of loose stock, which generally is exposed to more severe processing. Thus, dyes of high fastness can be used, any unlevelness being corrected by later processing.

During dyeing, the loose fiber can undergo mechanical damage owing to the often high rate of liquor flow necessary to achieve penetration of densely-packed material. Such damage impairs the subsequent processing ability of the fiber thereby increasing waste, processing time, and costs. This problem can be reduced by careful control of liquor flow during dyeing. In wool dyeing, fiber damage can also be reduced by the addition of fiber protective agents to the dyebath.

4. Yarn Dyeing[32]

While yarn dyeing is generally more costly than dyeing in the loose state, it is nevertheless more flexible in response to changes in fashion.

Hank dyeing is nowadays mostly confined to high bulk yarns, hand-knitting yarns, and coarser carpet yarns owing to a lower loss of handle and bulk in comparison to dyeing yarn in package form. The sometimes inefficient liquor circulation can result in unlevelness with certain fibers and also limit the selection of dyes; the poor liquor flow characteristics may also necessitate the use of prolonged dyeing times to achieve adequate penetration, giving rise to high production costs and the risk of fiber damage. Air-channeling is a major cause of damage with some fibers (such as wool) although this problem can be reduced by the use of suitable deaerating agents.

Package dyeing can be used, for example, for texturized yarns, sewing thread, carpet, and weaving yarns. A higher degree of levelness is obtainable and consequently dyes of higher fastness can be used. Good package preparation is crucial for successful dyeing.[33] The relatively recent introduction of parallel-sided packages eliminates the need for rewinding before knitting or weaving. Such package centers permit press-packing to be achieved so as to produce large dyepacks of uniform column density which results in improved levelness and reduced fiber damage through improved liquor flow characteristics.

5. Fabric Dyeing[34–38]

This includes woven and knitted fabrics for which level dyeing is essential. Both fabric and garment dyeing can in many cases be integrated with finishing and, since dyeing in both fabric and garment forms is closest to the point of delivery, these methods are the most flexible in response to fashion trends.

6. Garment Dyeing[34,39]

Leveling and, especially, adequate seam penetration are of vital importance in garment dyeing. With some fibers, however, adequate dye exhaustion may require prolonged dyeing times which can cause fiber damage.

C. Dyeing Methods

These can conveniently be divided into two groups depending on the general application procedure employed.

1. Immersion (Exhaustion) Methods

The whole of the textile material is accessible to all the dye liquor throughout the dyeing process. The degree of levelness and penetration achieved is mainly determined by the rate of dyeing which is, in turn, controlled by the rate of dye adsorption at the fiber surface and rate of dye diffusion within the fiber. These two latter factors are, in turn, determined by dye–fiber substantivity so that the establishment of appropriate and uniform conditions of pH, temperature, electrolyte concentration, etc., throughout the material is thus of vital importance; an adequate and uniform rate of dye liquor–fiber agitation is also of great importance in this context.

Details of dyeing machinery and of various techniques employed for this batchwise processing are described elsewhere.[31–45]

2. Impregnation–Fixation Methods

The material is first impregnated with a solution or dispersion of the dye which, ideally, remains deposited mechanically within the material until the dye is fixed within the fiber during the subsequent fixation stage.

Usually, fixation only permits dye diffusion to occur at the initial point of deposition and thus uniform distribution of dye during the impregnation process is in most cases essential for level dyeing.

This dyeing method is utilized in semicontinuous and continuous application processes.[46,47]

In the sections dealing with the dyeing of the different fiber types, commercial names of dyes, auxiliaries, and fibers are mentioned; this does not infer superiority of product but instead serves only as a guide.

D. Fastness of Dyed Textiles[48]

The dyed textile must be sufficiently permanent or fast to conditions encountered during processing (e.g., weaving, knitting) and use (e.g., washing, ironing). Color fastness refers to a dye on a given fiber and not to the dye in isolation. Furthermore, fastness is a function of the particular dye–fiber system so that a given dye will often possess different fastness properties on different fibers. There is no permanently-fast dye in as much as no dye can indefinitely withstand all treatments or agencies to which it is exposed and consequently "fastness" is a relative term.

The current wide range of fastness tests has been developed over many years and in various countries.[49] The current internationally accepted color fastness testing procedures are contained in the International Organization for Standardization (ISO) publication.[50]

Because of the continual changes occurring in the area of textile coloration and in the use of colored textiles (and therefore in the conditions that dyed textiles encounter), fastness testing is a dynamic field and new and revised test procedures are constantly under consideration.

In the following sections that deal with the various dye–fiber systems, an attempt has been made to give an indication of the fastness properties of the dyeings to some agencies. This account is by no means comprehensive and the fastness ratings quoted, which relate to change in shade of the dyeing and are based on dye manufacturers' literature, are intended only as a general guide. Detailed information of the fastness characteristics of specific dyes on particular substrates is available in the relevant dye-makers' literature.

II. DYEING OF WOOL

Wool is a staple fiber obtained from the fleeces of various breeds of sheep. Wools vary markedly in strength, lustre, crimp, etc. and are classified as Fine, Medium, Long, Cross-bred, or Mixed according to the breed of sheep from which they are obtained and are graded as 32s, 50s, 70s, etc. according to the finest count (the number of 560 yard hanks per lb) possible to which the wool can be spun. Other less common wools such as Mohair and Cashmere, which are derived from the hair of different breeds of goat, are much prized for their fineness and delicate nature.

While the following account concerns the dyeing of wool obtained from sheep, the principles outlined nevertheless apply in general terms to other wools such as those mentioned above.

Many processes are involved in the conversion of raw wool into a useful textile fiber, accounts of which can be found elsewhere.[34,51,52] Yarn is produced either on the woollen or worsted system and fabric is obtained by weaving and/or knitting.

Briefly, wool as obtained from the sheep contains dirt and vegetable matter and is coated with wool grease and sweat. These impurities, which can be present by up to 50% by mass of the fiber, are removed by processes such as carbonizing, scouring and bleaching and the wool is then subjected to various mechanical and chemical operations such as carding, combing, and chlorination.

Comprehensive accounts of the structure and composition of wool can be found in other texts.[53–56] Although wool fibers are complex structures, they consist essentially of two main parts, namely the cuticle and the cortex.

a. Cuticle. The overlapping scaler structure of the cuticle is of prime importance in dyeing. The epicuticle or outer layer of these scales which covers the whole of the fiber, except the tip where it has been removed by weathering, is thin and hydrophobic and therefore provides a barrier to ingress of water and dye molecules. The epicuticle is easily removed by chemical treatment, usually chlorination, thereby facilitating ease of dyeing.

b. Cortex. The spindle-shaped cortical cells constitute more than 90% of the whole fiber and are easily penetrated by dye molecules.

Chemically, wool consists of the protein keratin which is made up of the imino acid proline and eighteen L-α-amino acids of general formula $NH_2CHRCOOH$, joined by peptide links to give polypeptide chains. The relative proportions of the constituent amino acids vary between samples of wool even within one breed of sheep. The side groups R vary markedly in size and chemical nature and are mainly responsible for the dyeing and other characteristics (e.g., water absorption and swelling) of the fiber. From the standpoint of dyeing these side groups can be divided according to chemical nature. The compositions of some of the constituent L-α-amino acids in wool are given in Table 5.

The bulky nature of many of these side groups does not facilitate the close approach of adjacent polypeptide chains and, as a consequence, the fiber is predominantly amorphous (of the order of 80%) and is therefore relatively easily

Table 5. Some of the L-α-Amino Acids in Wool

Chemical nature	Amino acid	Side group (R)
Hydroxy	Serine	$-CH_2OH$
	Tyrosine	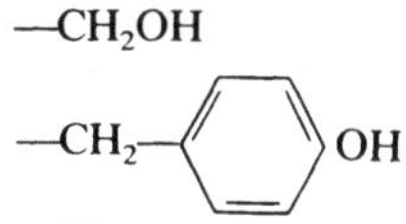
Hydrocarbon	Glycine	$-H$
	Alanine	$-CH_3$
Acidic	Aspartic acid	$-CH_2COOH$
	Glutamic acid	$-(CH_2)_2COOH$
Basic	Lysine	$-(CH_2)_4NH_2$
	Arginine	$-(CH_2)_3NH-C(=NH)-NH_2$
	Histidine	$-CH_2$
Sulfur	Cystine	$-CH_2 \cdot S \cdot S \cdot CH_2-$
	Methionine	$-(CH_2) \cdot S \cdot CH_3$

penetrated by water and dye molecules. The acidic, and especially the basic, side groups are by far the most important in relation to dyeing. The arginine, lysine, and histidine residues provide sites for attachment of anionic dyes; the latter two side groups are also important sites for covalent linkage of reactive dyes. Both *N*-terminal and *C*-terminal amino acid residues are present, but these are in relatively small amounts and contribute little to the acidic and basic character of wool[55] although the *N*-terminal residues provide sites for the attachment of reactive dyes.

The keratin fiber contains several types of cross-link,[54] namely hydrogen bonds, hydrophobic bonds, electrostatic links, and disulfide linkages. The relatively high sulfur content of wool fibers is due primarily to the disulfide group of the amino acid cystine. Cystine residues **(1)**, by virtue of their disulfide groups, are the most abundant and important crosslink in wool. Cysteine residues **(2)** provide sites for the attachment of reactive dyes.

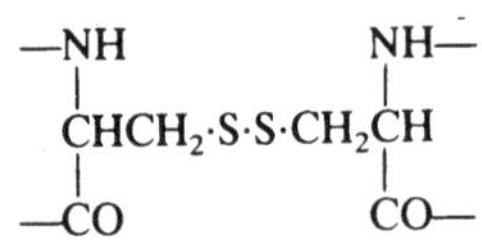

1

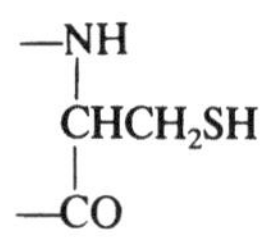

2

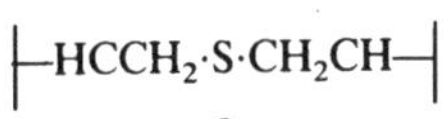

3

These cross-linkages contribute to the physical and mechanical properties of the wool fiber and are greatly affected by the various wet treatments, including dyeing, that wool encounters during preparation and use.

When dry wool is stretched by up to about 30% and the tension then quickly released, the fiber contracts to its original length. This inherent resistance to creasing accrues from the presence of the various interchain cross-linkages. Wool, which has had the epicuticle removed, is relatively easily wetted by and swells appreciably in water. Swelling, which increases with increase in temperature, disrupts the interchain bonds thereby increasing penetration of the fiber by dye molecules. However, when wool is subjected to prolonged boiling in water, some of the disulfide links are hydrolyzed. If the wool is subjected to tension during boiling, the fiber will not contract after release of the tension owing to the formation of new cystine links in the extended form; the fiber has acquired a permanent set. Because, as will be discussed later, wool is commonly dyed at or about the boil for long periods, care must be exercised so as to avoid the introduction of permanent set, which is often seen as crease marks in the dyed fibre.

Hydrolysis of the cystine links occurs more readily under alkaline conditions especially at high temperatures. Alkaline hydrolysis also results in the formation of lanthionine cross-linkages **(3)** which, although quite stable, cause undesirable yellowing of the wool.

The polypeptide chain is hydrolyzed by prolonged boiling in water. This rupture is increased in acidic conditions such as those used in dyeing and results in a decrease in strength and abrasion resistance of the wool.

The different stages of manufacture at which wool is dyed are in the loose state, yarn, and piece (i.e., fabric and garment) forms. The estimated worldwide usage[57] of the different dyeing routes is as follows: loose stock dyeing, 18%; top dyeing, 18%; yarn dyeing, 38%; piece dyeing, 26%.

c. Dye Types Used on Wool. Wool is dyed using the following classes of dye:[58] acid (including premetallized), mordant and reactive.

Acid dyes are water-soluble, anionic dyes that are primarily used to dye wool, nylon, and silk from acidic or neutral dyebaths. The name acid dye is derived from the use of the acidic dyebath conditions that were originally necessary for their application to wool and silk. In the Colour Index, metal-complex or premetallized acid dyes are included within the class of acid dyes but, for the purposes of the present discussion, the two dye types will be considered separately.

A. Nonmetallized Acid Dyes

In the majority of cases, water-solubility is conferred upon the dye by sulfonic acid groups, commonly as the sodium salt (—SO_3Na).

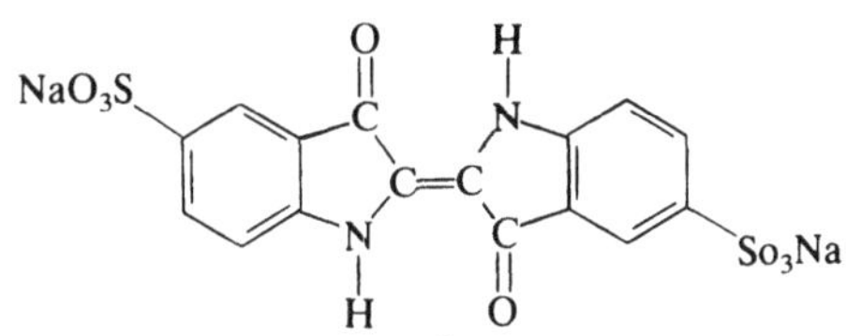

4

The first member of this dye class, CI Acid Blue 74 (**4**), was prepared by Barth in 1740 by the sulfonation of indigo. Picric acid, the first synthetic organic dye, obtained by Woulfe in 1771, was the second member of the class and gained some success in the dyeing of silk during the latter half of the last century. However, it is perhaps to Nicholson that the preparation of the first synthetic, nonmetallized acid dye might be attributed since, in 1862, he sulfonated the sparingly water-soluble and cationic Aniline Blue (CI Solvent Blue 3) to yield the water-soluble, predictably named, Soluble Blue (CI Acid Blue 22, **5**) which could dye silk and wool from an acidic dyebath.

There are nine chemical classes of nonmetallized acid dye of which three are the most important.

1. Azo

This class constitutes by far the largest number of acid dyes, and includes the majority of red, yellow, orange, brown, and black hues. The monoazo acid dyes

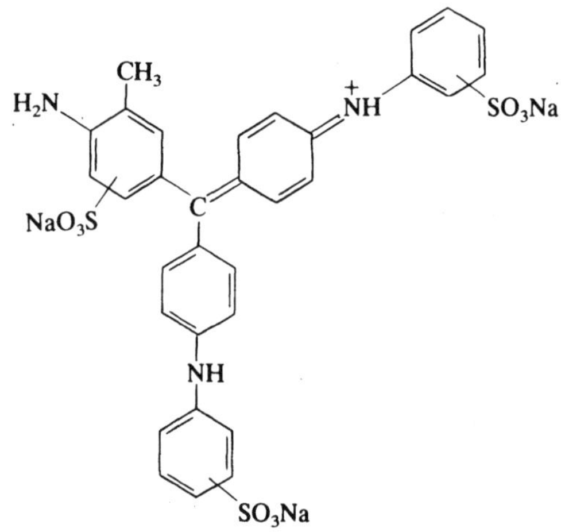

5

range from relatively simple, low molecular weight dyes such as CI Acid Orange 7 (**6**), the first azo acid dye, that generally possess low affinity for wool and nylon, to the larger molecular weight dyes typified by CI Acid Red 138 (**7**), on which higher affinity is imparted by the presence of alkyl chains. A fairly small number of mostly yellow hues is provided by the monopyrazolone dyes, such as CI Acid Yellow 11 (**8**). Disazo types, such as CI Acid Orange 51 (**9**), tend in general to possess high affinity owing to their complexity and large size.

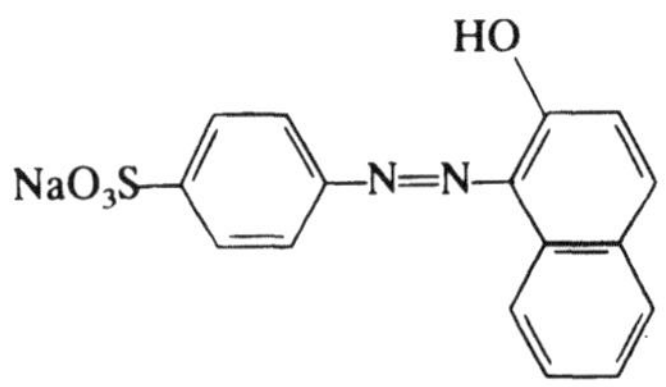

6

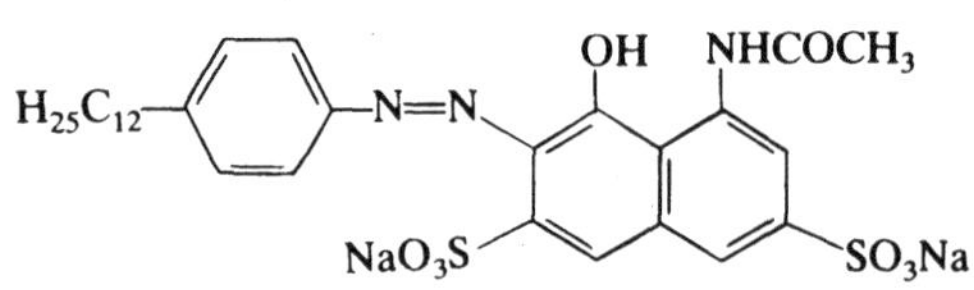

7

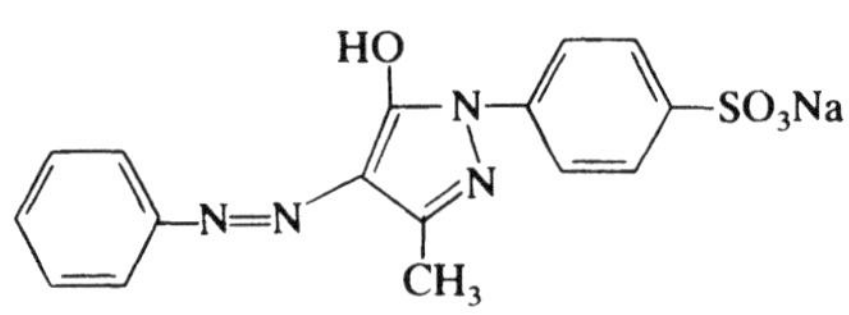

8

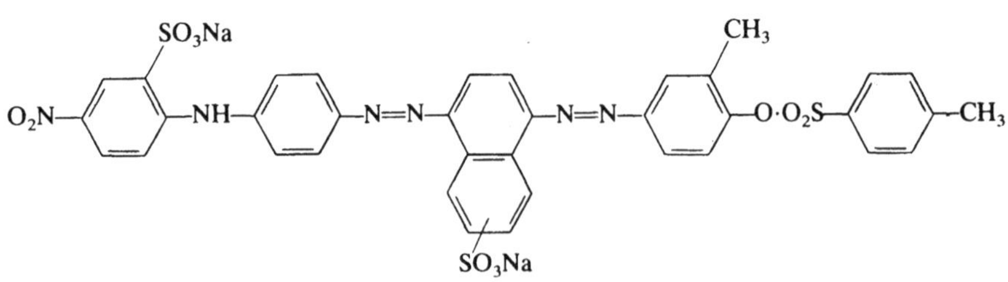

9

2. Anthraquinoid (AQ)

Currently, AQ acid dyes provide a range of mostly bright blues, violets, and greens of generally high fastness to light on wool and nylon, such as CI Acid Blue 25 **(10)**. Although, in general, dyeings have only low to moderate fastness to wet treatments, certain members such as CI Acid Green 27 **(11)** have good all-round fastness properties on wool and nylon fibers.

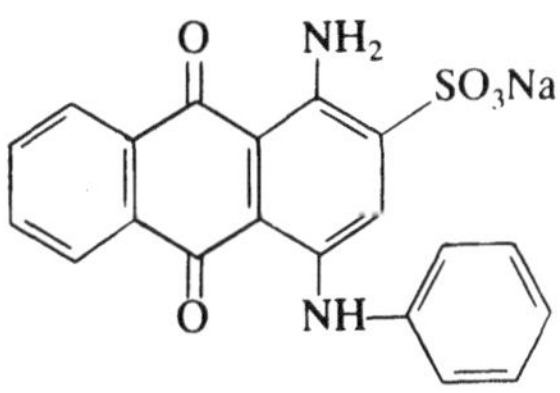

10

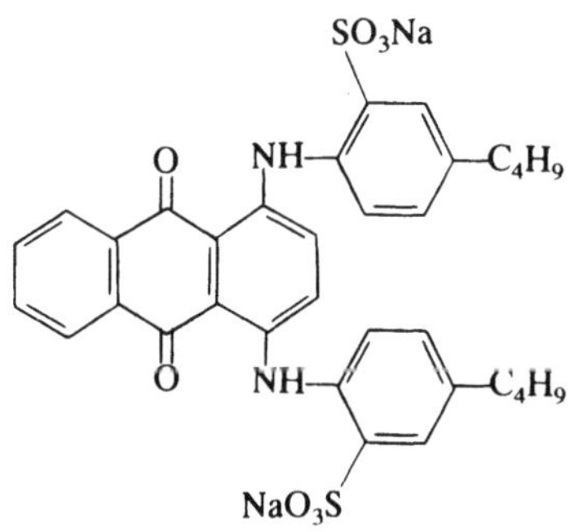

11

3. Triphenylmethane

Characteristically, triphenylmethane acid dyes, as typified by CI Acid Violet 19 **(12)**, provide brilliant violets, blues, and greens of relatively low all-round fastness on wool and nylon. The remaining six chemical classes of nonmetallized acid dye, namely Xanthene, Azine, Nitro, Indigoid, Quinoline, and Phthalocyanine, provide only a relatively small number of commercial acid dyes which enjoy limited use that generally accrues owing to matters of cheapness or hue. Details of these latter six dye types can be found elsewhere.[59,60]

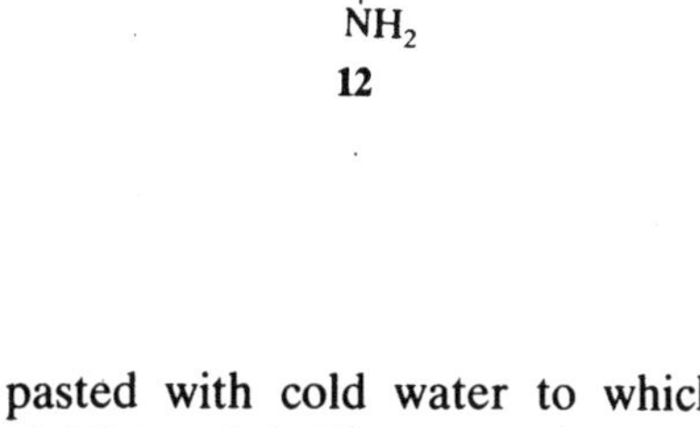

12

4. Dissolving the Dyes

Powder forms are pasted with cold water to which is then added, with stirring, boiling water. Additional boiling may be necessary in the case of concentrated solutions. Grain forms are stirred into about ten times their volume of cold or warm (60–70 °C) water.

The ensuing solutions are added to the dyebath through a sieve.

5. Dyeing Behavior

In general, small molecular size, relatively simple acid dyes require strongly acidic conditions for exhaustion onto wool and nylon and exhibit only low to moderate fastness to washing and to wet treatments on such substrates owing to the dye's low affinity for the fibers. Generally, an increase in complexity or molecular weight of acid dyes leads to increased affinity which, in turn, results in less acidic application conditions being required for exhaustion and higher fastness of the dyeings to washing and wet treatments. As a consequence, acid dyes for wool and nylon are subdivided according to their dyeing behavior. The following subgroups are realized for wool.

a. Acid Leveling (Equalizing Acid: Level Dyeing Acid) Dyes. These dyes, as represented by the Supracen (Bayer), Neolan E (Ciba–Geigy), Lissamine (ICI), and Sandolan E (Sandoz) ranges, are of small to moderate molecular size and are commonly applied to wool from a strongly acidic dyebath (pH 2.5–4.0) containing sodium sulfate (Glauber's salt). A typical exhaustion application procedure is shown in Scheme 1. The mechanism of adsorption of the dyes can be represented as follows.

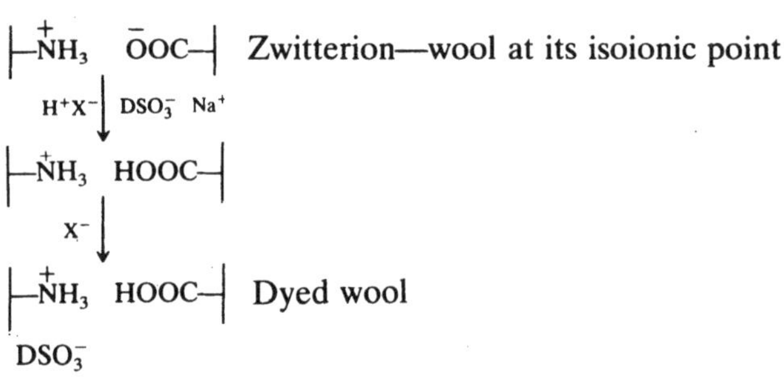

On immersion in the acidic dyebath, the rapid diffusing protons are adsorbed onto the carboxyl groups (the carboxyl groups are "back-titrated") and the small molecular size acid anions (X^-) are adsorbed onto the protonated amino groups. Owing to their greater affinity, the larger molecular size, slower diffusing dye anions displace the adsorbed acid anions from the protonated amino groups.

While electrostatic interaction is the predominant contributor to dye adsorption, it may, depending on dye, be augmented by van der Waals forces.

The low-pH conditions make the initial strike very rapid and consequently unlevel. These dyes, however, as their name implies, level out on boiling, i.e.,

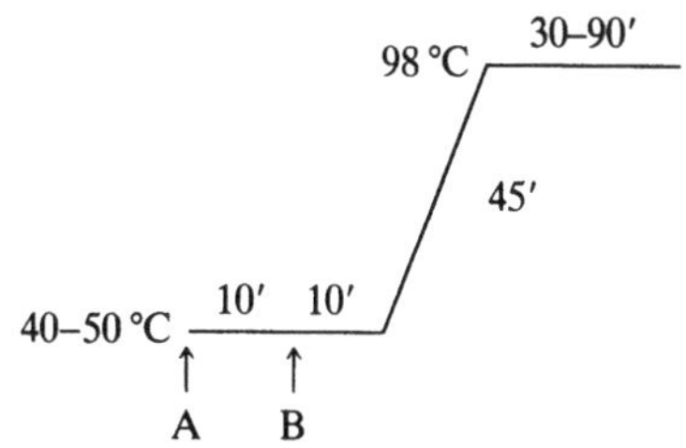

A. H_2SO_4 (96%)	2–4% owf (pH 2.5–3.5)
Sodium sulfate (anhydrous)	5–10% owf
B. Dye	x% owf

Shading. Cool to 80 °C prior to addition of dye.

Scheme 1

there is a redistribution of dye within the substrate with time. Because of the relatively low affinity which these small molecular size dyes have for wool, coupled with the fact that weak, electrostatic forces are mainly responsible for their attachment to the fiber, prolonged boiling causes the attached dye molecules to desorb into the dyebath. These desorbed dye molecules can then become readsorbed onto different sites within the substrate. This migration of dye molecules from one site to another within the substrate on boiling results in leveling of an initially unlevel dyeing.

In some cases the substantivity of the dye may be such that the initial strike is extremely rapid with the result that self-leveling is not entirely effective. Consequently, a low affinity, fast diffusing anion is added to the dyebath which can compete with the dye anions during the initial stage of dyeing; this is the function of the sodium sulfate. It also may be necessary to further reduce dye substantivity by using less acidic conditions in the initial dyeing stage, further acid being added to complete exhaustion. This is illustrated in Scheme 2. The light fastness of the dyeings varies from moderate to very good depending on the dye, being in the range (ISO BO1) 3 to 5–6 1/1 SD; 2 to 4–5 1/12 SD. The wash fastness properties of acid leveling dyes on wool generally leave much to be desired, being of the order (1/1 SD) 1 to 2 (ISO CO2) and 4 to 4–5 (ISO CO1). Fastness to wet treatments is also generally low, for example, alkaline perspiration 3 to 4–5, alkaline milling (severe) 1 to 2. Consequently, the dyes are used for those applications where a high degree of levelness is required but high fastness to aqueous agencies is not necessary, for example, knitting yarns and some garments.

b. Intermediate (Weakly Acid) Dyeing Dyes. As their name suggests, such dyes, as represented by the Supramin (Bayer), Coomassie P (ICI), and Sandolan P (Sandoz) ranges, have slightly greater affinity for wool and consequently are

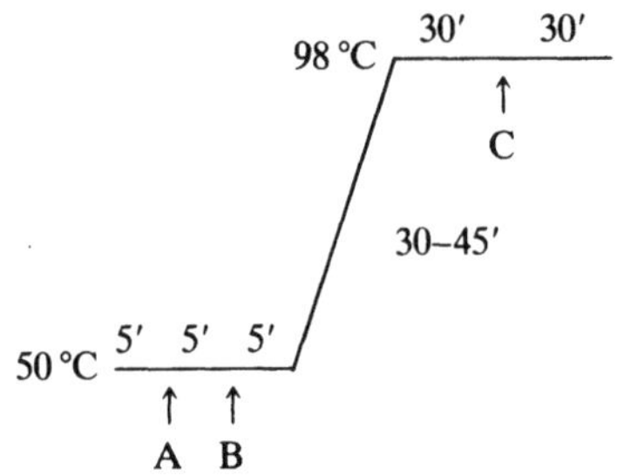

A. HCOOH (85%) 2% owf (pH 3.5–4.5)
   Sodium sulfate (anhydrous) 5–10% owf
B. Dye *x*% owf
C. Extra acid (2% owf) [H_2SO_4 (96%) or HCOOH (85%)] added to promote exhaustion

Scheme 2

applied from weakly acidic (pH 3.5 to 5) dyebaths containing sodium sulfate. The migrating power of the dyes is lower than that of acid leveling dyes and, as a consequence, the time period over which the boil is reached is generally longer. A typical dyeing process is shown in Scheme 3. The wash and wet fastness properties of the dyeings vary from moderate to good, for example, (1/1 SD) 3 to 4 (ISO CO3) and 4–5 to 5 (ISO CO1); alkaline perspiration 3 to 3–4; alkaline milling (severe) 2 to 3. Light fastness varies widely, for example, (ISO BO1) 3 to 6–7 1/1 SD; 2–3 to 4–5 1/12 SD.

c. *Acid Milling Dyes.* These dyes derive their name from the high fastness of their dyeings on wool to the milling process, being of the order (1/1 SD) alkaline milling (severe) 4 to 4–5. Fastness of the dyeings to both washing and wet treatments is very good, typically (1/1 SD) 4–5 to 5 (ISO CO6/B2), 4–5 to 5 (ISO CO3), alkaline perspiration 4–5. Light fastness of the dyes on wool ranges from good to very good (with the exception of certain blue dyes), typically (ISO BO2) 5–6 to 6 1/1 SD; 4–5 1/12 SD.

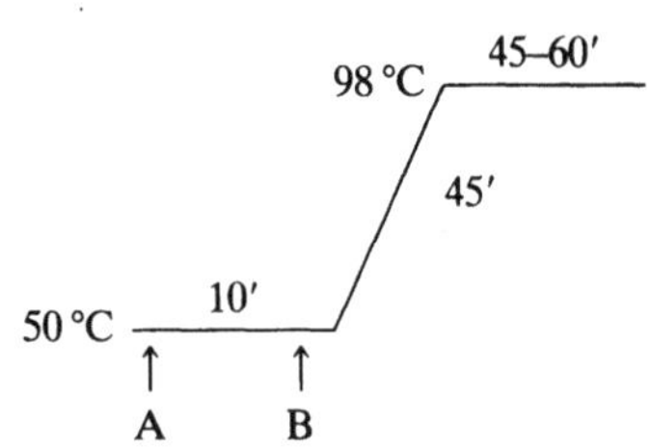

A. Sodium sulfate (anhydrous) 5–10% owf
   Acetic acid (85%) 1–3% owf
B. Dye *x*% owf

Scheme 3

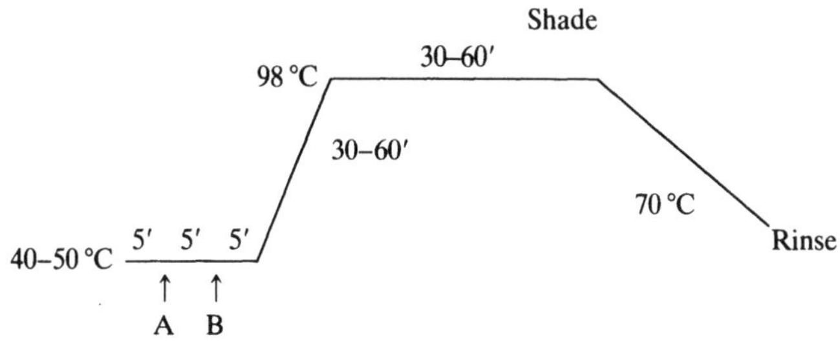

A. Ammonium sulfate	2–3% owf (pH 5 to 6)
Unisol C (ICI)	1–2% owf (leveling agent)
B. Carbolan dye (ICI)	x% owf

Shading. The dyebath is cooled to 70–75 °C and the dye is added. The bath is then slowly raised to the boil and dyeing continued at this temperature for 15 min.

Scheme 4

The dyes, as exemplified by the Carbolan and Coomassie (ICI), Polar and Erionyl (Ciba–Geigy), Sandolan N (Sandoz), and Supranol and Alizarin (Bayer) ranges, have very high affinity for the fiber which, together with their generally large molecular size, leads to low rates of diffusion within the substrate and to virtually no self-leveling ability. Therefore, dyes of this type are applied from neutral or weakly acidic (pH 4.5–6.5) dyebaths. The sulfate anion (from Glauber's salt) cannot compete with the very high affinity of the dye anion and the dyes are therefore applied in the presence of a proprietary leveling agent such as Lyogen WD (Sandoz) or Unisol WL (ICI), sometimes in conjunction with Glauber's salt. Careful control of the rate of heating of the dyebath is also used to promote level dyeing.

The dyes are widely used in the dyeing of carpet yarns, machine knitting yarns, and yarns for weaving, where their characteristically high fastness to aqueous agencies is required. The application procedure shown in Scheme 4 is typical. Because of the dye's high affinity and the neutral or slightly acidic dyeing conditions used, the dyeing mechanism involved differs from that of acid leveling dyes. Although electrostatic interaction will contribute to dye adsorption, van der Waals forces are the primary contributor.

The brilliance of pale and medium depth dyeings can be increased by the use of dyebath bleaching aids, such as Lanalbin B (Sandoz), which remove the yellowing of wool that accompanies long-liquor dyeing.[61] Pale depths are preferably dyed at low temperatures (80–90 °C) to reduce yellowing; these temperatures also reduce felting and creasing of delicate materials and improve the subsequent processing ability of loose stock and slubbing. The use of a leveling agent is required to promote dye leveling and penetration and helps prevent skittery dyeings.[62] High temperatures can also be used to expedite dyeing as shown in Scheme 5.

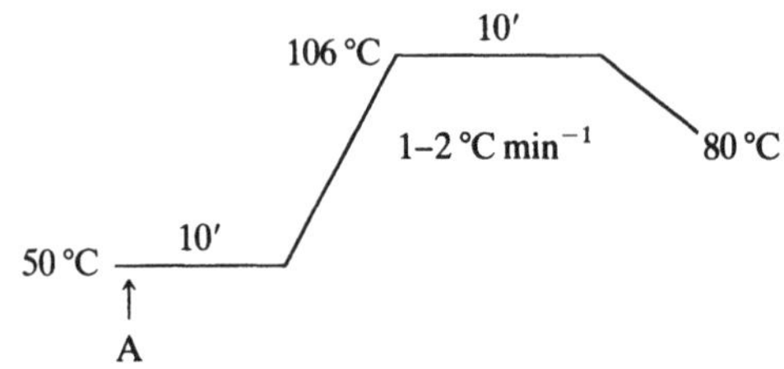

A. Supramin (Bayer) dye	x% owf
Sodium sulfate (anhydrous)	1–2 g l^{-1}
Sulfuric acid (96%)	2–4 owf
(pH 2–3.5)	

Scheme 5

B. Mordant Dyes

Many natural dyes have little, if any, substantivity and affinity for textile fibers and consequently their use often necessitated pretreatment of the textile with a mordant (tannins or metallic salts) by means of which the dye's substantivity and affinity was increased. In this manner, mordanting enabled deeper dyeings of improved fastness to be attained.

The term "Mordant Dye" refers to a dye which is applied to a fiber in conjunction with a metallic mordant; because of the predominance of chromium salts as mordants, the term "Chrome Dye" is often popularly used.

The name "mordant" stems from the French word "mordre," to bite or corrode, and the belief of early French dyers that the metallic salt "corroded" and opened-up the porous textile substrate making it more permeable to dyes.

Characteristically, mordant dyes contain groups (e.g., —OH, —COOH) which are capable of forming a stable coordination complex with the chromium ion inside the fiber. The formation of this large molecular size complex results in an often dramatic increase in fastness of the dyeing to light and to wet treatments and is usually accompanied by a considerable change in hue of the dyeing.

The vast majority of mordant dyes closely resemble acid dyes in their dyeing behavior (hence their older name "acid mordant dyes") but the generally very poor fastness properties of the unmordanted dyeings negates their use as acid dyes *per se*. As a class, mordant dyes give fast, full, but generally dull shades on wool and nylon fibers.

There are several chemical classes to which mordant dyes belong.

1. Anthraquinone

This group provides mostly blues, reds, and browns. The parent dye of this group is Alizarin (CI Mordant Red 11; **13**), the major ligand in Madder, a natural dye derived from the root of the Madder plant, which gained fame because of the fast, bright red shades it gave on cotton mordanted with alumina and calcium.[63]

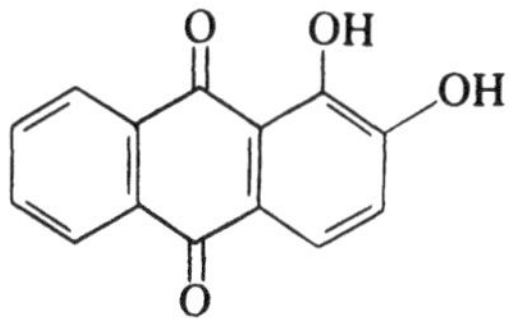

13

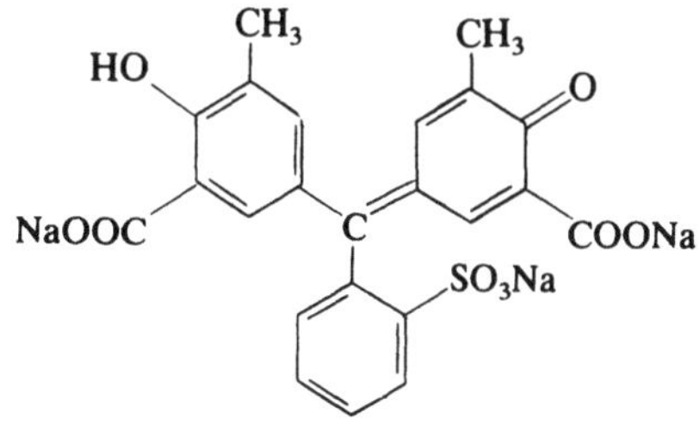

14

Alizarin, like many mordant dyes, is polygenetic, in that the dye yields a variety of hues with different mordants; for example, Alizarin gives a red-violet on wool mordanted with tin and a rose-red on an alumina mordant. The elucidation of the structure of Alizarin and its synthesis (which was in fact the first synthesis of a natural dye) by Graebe and Liebermann in 1868 led to the commercial introduction of synthetic Alizarin in the following year.[62] The sulfonation of Alizarin by Graebe and Liebermann in 1871 to yield Alizarine Red 8 (CI Mordant Red 3; **14**), which gives a bordeaux shade on chrome mordanted wool, stimulated the search for other mordant anthraquinone dyes.

An important representative of this group is CI Mordant Black 13 **(15)**.

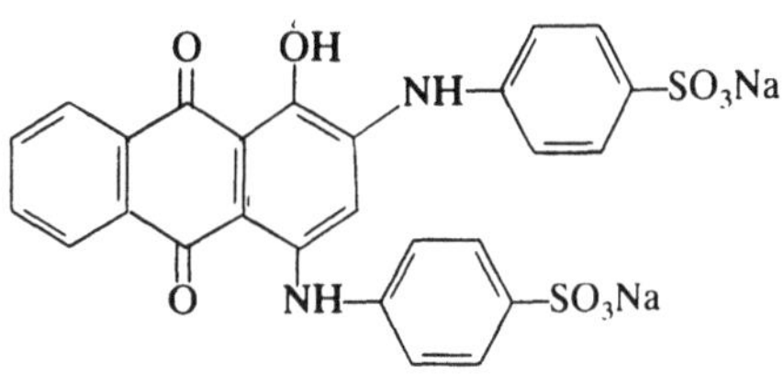

15

2. Azo

This group, consisting mostly of monoazo dyes, comprises the majority of mordant dyes and, with the exception of bright greens, blues, and violets, provides a comprehensive range of shades on wool and nylon characterized by very high fastness to light and to wet treatments. The azo mordant dyes can be subdivided as follows.

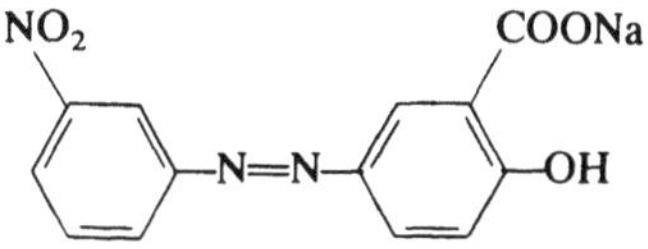

16

a. o-Hydroxy Carboxyazo Dyes. Most yellow and orange dyes are of this type, as exemplified by the first azo mordant dye, CI Mordant Yellow 1 **(16)**, produced by Nietzke in 1887. A few disazo types are important, such as CI Mordant Yellow 26 **(17)**.

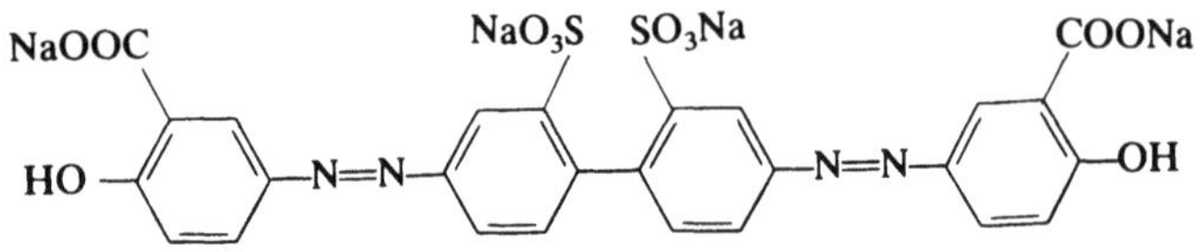

17

b. o,o'-Dihydroxyazo Dyes. This most important subgroup provides the majority of brown, black, blue, and red dyes, an important representative being CI Mordant Blue 13 **(18)**. This group also contains pyrazolone types, such as CI Mordant Red 19 **(19)**.

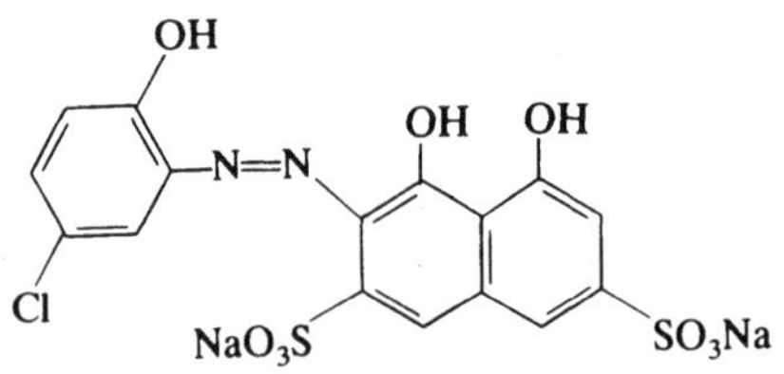

18

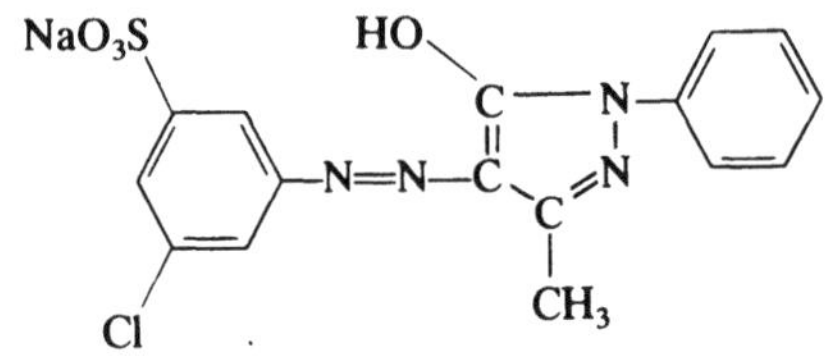

19

c. o-Amino,o'-Hydroxyazo Dyes. Many brown and a few black mordant dyes are of this type, for example, CI Mordant Brown 13 **(20)**.

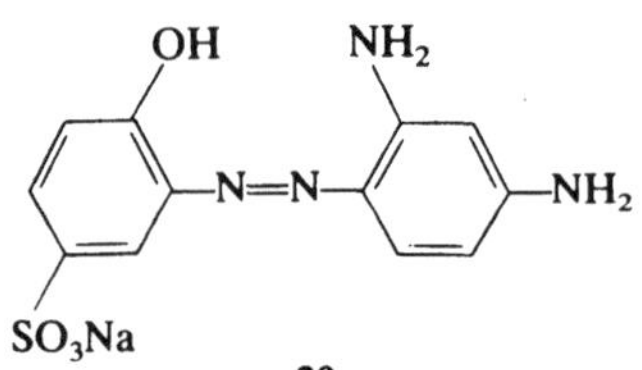

20

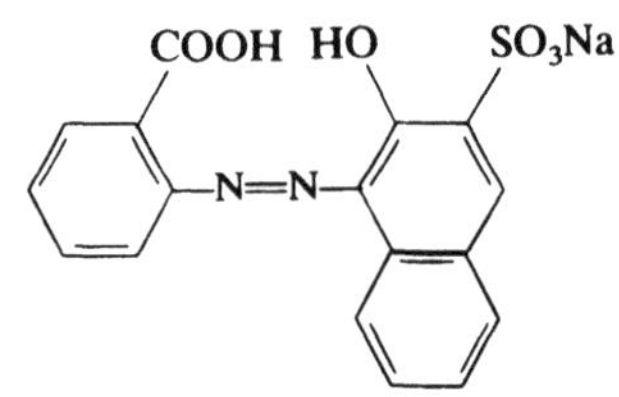

21

d. o-Hydroxy,o'-Carboxyazo Dyes. There are few dyes of this type, an example being CI Mordant Red 9 **(21)**.

In 1890 Meister, Lucius, and Bruning introduced the Chromotrope range of mordant dyes, which were mordanted subsequent to their exhaustion on wool. Such dyes, as typified by CI Mordant Blue 79 **(22)**, require oxidation prior to their metallization.

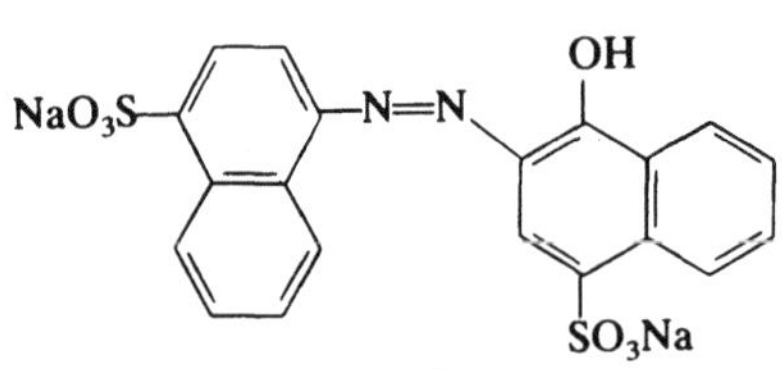

22

3. Triphenylmethane

Members of this group, which consist mostly of bright violets and blues, are characterized by only moderate fastness to light on wool and nylon. An example is CI Mordant Blue 3 **(23)**.

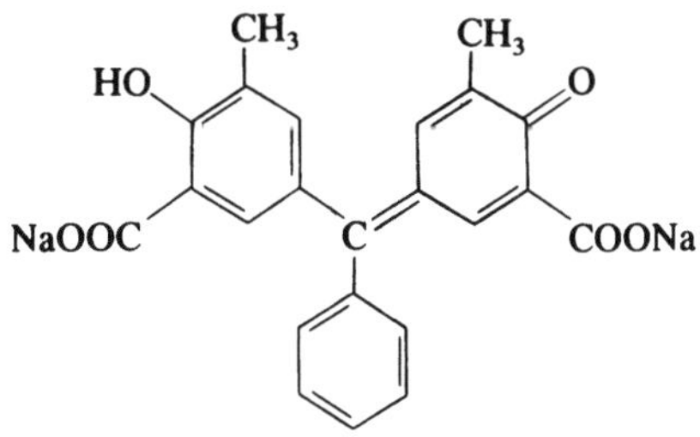

23

4. Xanthene

This group contributes only a very small number of brilliant, mostly red dyes, of which CI Mordant Red 27 **(24)** is an example.

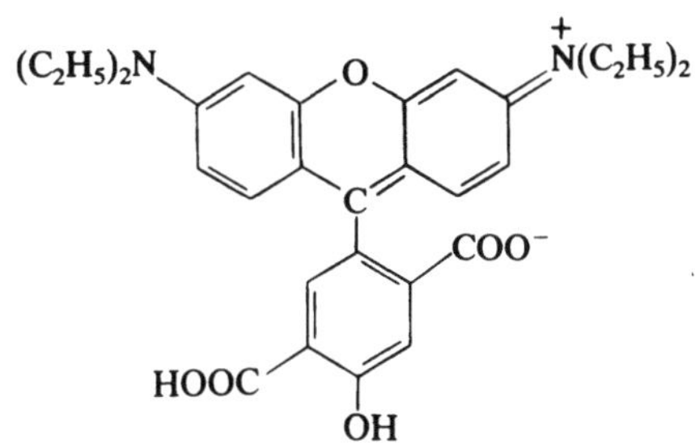

24

5. Dissolving the Dyes

The dyes are pasted with cold water to which is added boiling water; additional boiling may be necessary with concentrated solutions. The dye solution is sieved as it is added to the dyebath.

6. Dyeing Behavior

Mordant dyes yield deep, generally dull shades of good to very good fastness to light, with the exception of some blue dyes, typically 5 to 6 1/1 SD; 4 to 5–6 1/3 SD (ISO BO2). Fastness of dyeings to washing and to wet treatments is excellent, for example, (1/1 SD), ISO CO6/B2 4–5 to 5; alkaline perspiration 5, alkaline milling (severe) 4–5.

The dyes are suitable for applications that require generally excellent all-round fastness and are applied to wool in loose stock, slubbing yarn, and fabric forms. However, usage of the dyes tends to be restricted to those materials that are not adversely damaged by the relatively long dyeing times involved.

Numerous chromium salts have been utilized as mordants[64] but sodium and potassium dichromate are the most important on economic and technical grounds; sodium and potassium chromate are also used. There is little difference between chromate and dichromate, one changing into the other at an appropriate pH.[65]

The chromium can be applied to the fiber: (a) before dyeing (chrome-mordant or prechrome method), (b) together with the mordant dye (metachrome method), and (c) after dyeing (afterchrome method). In all cases, the conditions under which mordant dyes are applied to wool must enable adequate exhaustion and fixation of the uncomplexed dye to be achieved and complex formation to occur within the fiber.

The hexavalent chromium anions, Cr(VI), either as chromate or dichromate, are adsorbed onto the protonated amino groups in wool, this adsorption increasing with decrease in pH of application as with other anions. The Cr(VI) anions are then reduced within the fiber to the chromic cation Cr^{3+}. Reduction involves the oxidation of cystine groups[65,66,118] and results in loss of wet strength. Reduction results in liberation of alkali and the pH of the bath therefore increases during chroming.[52,65,67,118]

The preferential use of Cr(VI) in wool mordanting is due to its rapid adsorption and desorption. The chromic cation (Cr^{3+}) in contrast is adsorbed more slowly and, once adsorbed, is unable to migrate and become evenly distributed throughout the fiber by virtue of chelation with carboxyl groups.

Race *et al.*[68] gained evidence of the formation of a 2:1 dye–chromium Cr(III) complex in the case of all three application methods. It is considered that this large-size complex is attached to the fiber primarily by van der Waals' forces, electrostatic forces being of secondary importance. The characteristically high wet and washing fastness of mordant dyes on wool can be attributed to both van der Waals forces of interaction and to the very low diffusional behavior of the large-size 2:1 complex within the fiber. 1:1 complexes, however, may be also present.[67] in which case, besides van der Waals forces, the 1:1 dye–chromium complex may be chelated with appropriate groups (e.g., carboxyl) in the fiber, which would also account for the characteristically high fastness of mordant (chrome) dyes on wool to aqueous agencies. The precise nature of the dye–chromium(III) complex remains a matter of debate.

a. Chrome-Mordant or Prechrome Method. This two-stage process is the oldest application method, but is the least popular owing to its greater length and expense. Furthermore, both mordanting and dyeing must be level and the material undergoes prolonged boiling, which can impair fiber quality. A typical process is shown in Scheme 6. The acetic acid is used to neutralize alkali liberated during chroming and any that is carried over from scouring. The addition of chelating agents, such as formic or oxalic acid (1–3% owf), to the chrome bath

Chroming

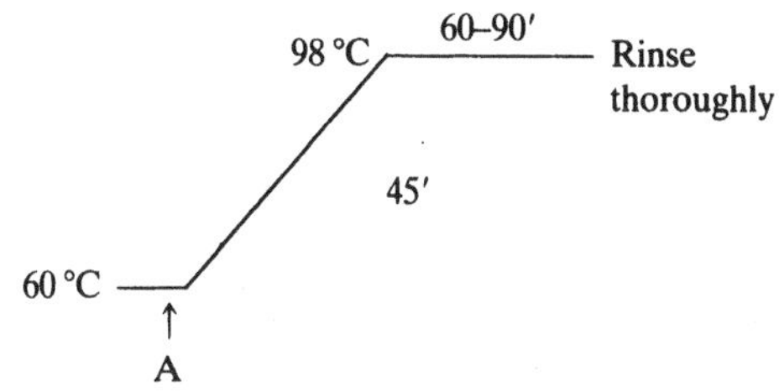

A. Sodium dichromate 2–4% owf
   Acetic acid (30%) 0.5% owf

Dyeing

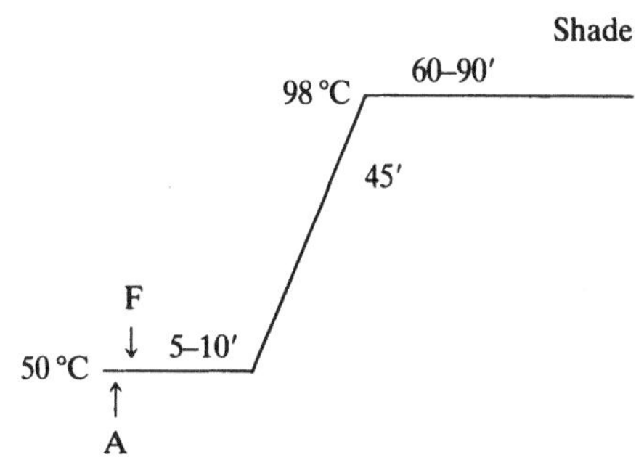

A. Acetic acid (30%) 1–5% owf
   Dye *x*% owf
F. Fiber added

Scheme 6

facilitate migration of the Cr(III) and promote its even uptake within the fiber.[67] Modifications of this method are used.[52]

b. Metachrome Method. This method, first introduced in 1900,[64] used potassium chromate and ammonium sulfate as mordant. Several variations of the original method have been developed.[52,64]

Since dye and mordant are applied together, the process is consequently much shorter than that of prechroming but is not significantly shorter than the afterchrome method. However, the number of mordant dyes that are suitable for application by this method is relatively small, since they must have adequate substantivity in the pH region 5–7[68] and must not reduce the Cr(VI) in the bath,[52] which limits the shade range available with this methods.

The process depicted in Scheme 7 is intended for the application of the Metomega (Sandoz) range of dyes to wool.

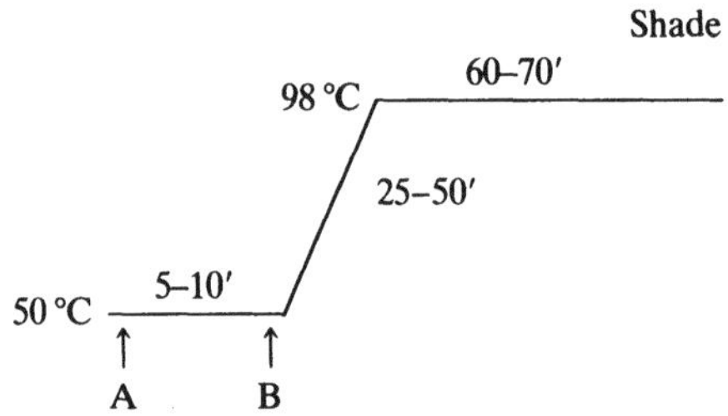

A. Ammonium sulfate 2–5% owf pH 6.2–6.5
Metomega chrome mordant (potassium chromate)[a] 1–5% owf
Sodium sulfate (anhydrous)[b] 10% owf
B. Megomega dye (Sandoz) 1–5% owf

[a] Amount of mordant applied is usually equal to that of the dye.
[b] For yarn and piece goods only.

Shading. In the case of loose wool and slubbing the dye is added directly to the dyebath. For yarn, the bath is cooled to 70 °C before addition of the dye.

Scheme 7

c. Afterchrome Method. This is the most popular application method and a wide range of dyes is available, such as the Solochrome (ICI), Eriochrome (Ciba-Geigy), Diamond (Bayer), and Omega Chrome (Sandoz) ranges. The method yields dyeings of superior levelness and penetration. The dyes are most commonly applied using the minimum of dichromate so as to produce effluent of

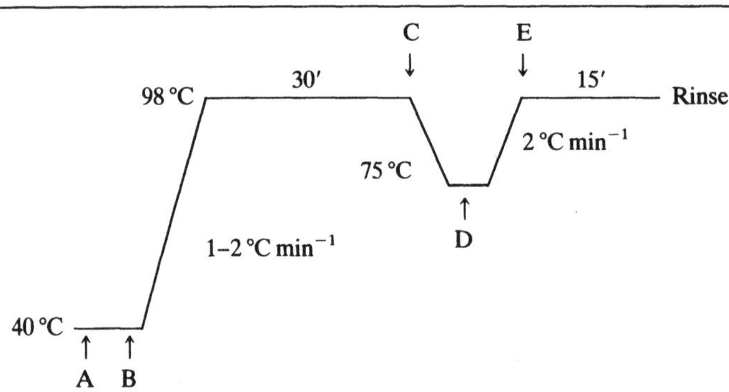

A.	Acetic acid (60%)	3% owf
	Alvolan SCN (Bayer) 150%	0–1.0% owf
B.	Diamond dye (Bayer)	*x*% owf
C.	Formic acid (85%) to pH 4–4.5	1–2% owf
D.	Potassium dichromate	*y*% owf
E.	Sodium sulfate (anhydrous)	7.5% owf

Scheme 8

low chrome content, as exemplified by the Bayer low-chrome method[69] (Scheme 8). The chrome requirement of the dyes in the Diamond range has been determined and is expressed as the "GCr" factor. The amount of each dye used (%) is multiplied by its particular GCr factor and the quantities are totaled to arrive at the amount (%) of potassium dichromate required (0.25% minimum).

As previously mentioned, low dyeing temperatures reduce processing costs and minimize fiber yellowing and damage. The latter consideration is of great significance in afterchrome dyeing, since the fiber undergoes prolonged boiling; chroming at the boil is a major cause of fiber damage. Although adequate dyebath exhaustion can be achieved at low temperatures using techniques similar to those employed for the application of nonmetallized acid dyes, inadequate chroming of the dyed fiber may occur at low temperatures.

Typical of a low-temperature afterchrome method is that depicted in Scheme 9, developed by Ciba-Geigy for their Eriochrome dye range. With selected afterchrome mordant dyes, adequate chroming can be achieved at 80 °C by the addition of sodium thiosulfate to the chrome bath. The function of the thiosulfate is to promote reduction of Cr(VI) to Cr(III) at the relatively low chroming temperature. This can be compared to the role of thiosulfate in the application of mordant dyes to polyamide fibers (see Section VI.C below). An afterchrome dyeing method based on the use of sodium thiosulfate is being developed by the IWS.[57]

The addition of selected fiber protective agents such as Sustilan N (Bayer)

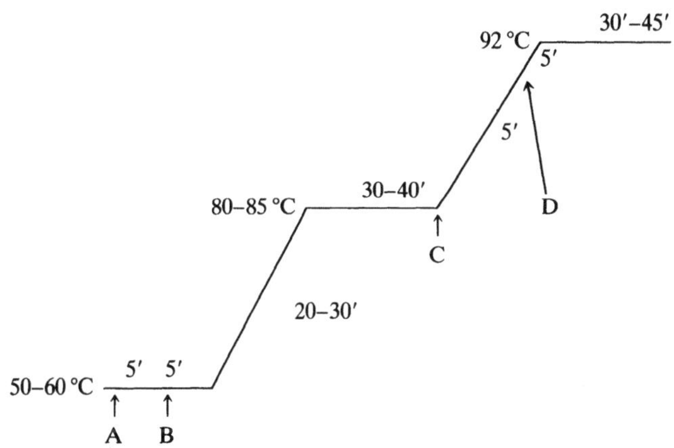

A.	Acetic acid (80)%	1–2% owf to pH 4 to 4.5
	Albegal LS (Ciba-Geigy)	0.5–0.7% owf if necessary to remove air, also
	Albegal FFD (Ciba-Geigy)	0.2–0.4% owf
B.	Eriochrome dye (Ciba-Geigy)	*x*% owf
C.	Sodium dichromate (15% of the added amount of dye)	0.2% owf
D.	Formic acid (85%)	2–3% owf to pH 3.5–3.8

Scheme 9

and Egalsil (Grunan) at a concentration of 1–2% owf to the chroming bath minimizes chemical damage during conventional afterchrome dyeing.

C. METAL-COMPLEX OR PREMETALLIZED DYES

Dyes of this type yield shades that generally are duller than nonmetallized and reactive dyes but which are brighter than mordant dyes.

The use of mordant dyes in conjunction with chromium mordants (chrome salts) gained widespread commercial importance during the latter half of the last century and the early part of the present century, owing to the excellent all-round fastness of these dyeings on wool. However, as described in the previous section, the application of these "chrome" dyes is often complicated and time-consuming and, as a consequence, efforts were directed at simplifying and economizing their application.

These investigations led to the introduction of the metal-complex or premetallized acid dyes in which one metal atom, commonly chromium, is complexed with either one or two molecules of a monoazo dye containing groups (—OH, —NH_2, —COOH), located *o,o′* to the azo linkage, that are capable of chelating the metal.

In the 1:1 metal-complex dyes the trivalent, six-coordinate chromium ion is coordinated with the tridentate *o,o′*-monoazo dye ligand and three monodentate ligands which are commonly water molecules, although other colorless ligands are employed.

The metal ion in 2:1 metal-complexes is coordinated with two tridentate *o,o′*-monoazo dye ligands which may either be identical (symmetrical 2:1 metal-complex dye, **25**) or different (unsymmetrical 2:1 metal-complex dye, **26**).

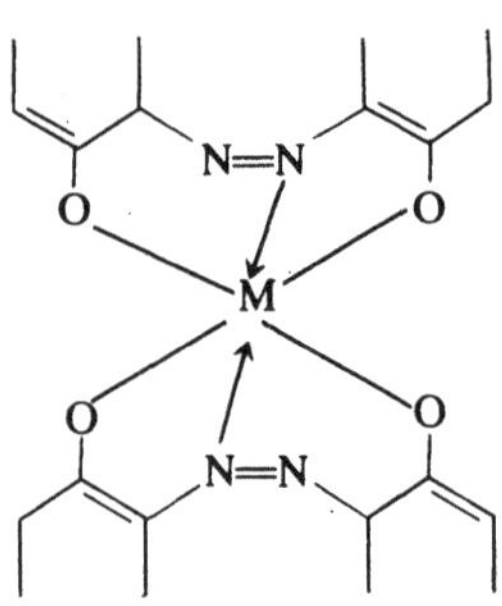

25

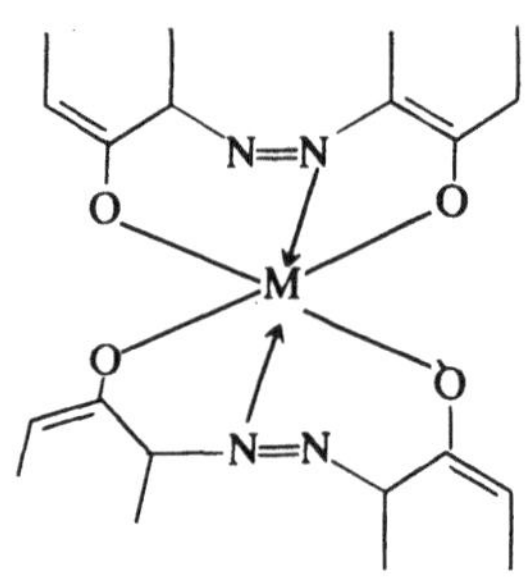

26

1. Dissolving the Dyes

The dye is pasted with cold water to which is added boiling water. Boiling may be necessary in the case of concentrated solutions. The dissolved dye is sieved during addition to the dyebath.

2. 1:1 Metal-Complex (Acid Dyeing Premetallized) Dyes

Although the main use of this dye type is on wool, selected members are employed for dyeing polyamide fibers.

The preparation of water-soluble 1:1 chromium complexes of sulfonated hydroxyanthraquinone dyes by Bohn in 1912, and similar complexes of azo mordant dyes in the same year,[70] led to the introduction of the Palatin Fast (BASF) and Neolan (Geigy) ranges of 1:1 premetallized dyes for wool in 1919. The majority of these dyes, as exemplified by CI Acid Red 214 **(27)**, are almost exclusively 1:1 chromium complexes of *o,o'*-dihydroxy azo or *o,o'*-hydroxyamino azo dyes,[71] water solubility being conferred by the presence of one or more sulfonic acid groups. In 1954, BASF introduced the now-withdrawn Neopalatin range of 1:1 metal-complex dyes in which the three molecules of water that normally served as the colorless monodentate ligands in conventional 1:1 complexes were replaced by other colorless bi- and tridentate coordinating agents such as salicyclic acid. Although the Neopalatin dyes possessed good leveling properties and high all-round fastness on wool, their tinctorial strength was much less than that of the parent 1:1 metal-complex dyes.

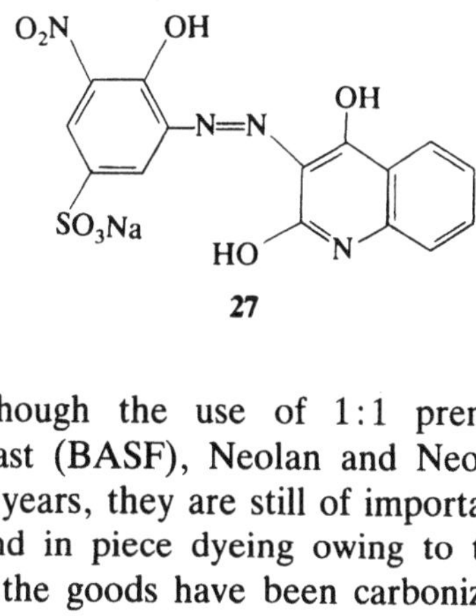

27

a. Dyeing Behavior. Although the use of 1:1 premetallized dyes, as represented by the Palatin Fast (BASF), Neolan and Neolan P (Ciba-Geigy) ranges, has declined in recent years, they are still of importance in dyeing loose stock and yarn for carpets and in piece dyeing owing to their excellent level dyeing behavior, especially if the goods have been carbonized prior to dyeing. They yield dyeings of good to very good light fastness, typically in the range 5 to 7 1/1 SD (ISO BO2). The fastness of dyeings to washing and to wet treatments is generally good, for example, 3–4 to 4 (ISO CO3) and 4 to 4–5 alkaline perspiration. The range of shades is restricted to fairly dull hues.

The dyes are applied to wool from a strongly acidic (sulfuric acid) dyebath (approximately pH 2) under which conditions they migrate very well at the boil and so give level dyeings. A typical application procedure is shown in Scheme 10. Prolonged boiling at such low pH has an adverse effect upon the fiber's physical properties and consequently the use of reduced amounts of sulfuric acid, or the use of other acids, such as 8–10% owf formic acid (85%), in conjunction with a leveling agent, is recommended in order to reduce fiber damage (Scheme 11). BASF have developed a dyeing method for their Palatin Fast dyes in which the sulfuric acid is replaced with sulfamic acid[57] (Scheme 12). At the start of dyeing

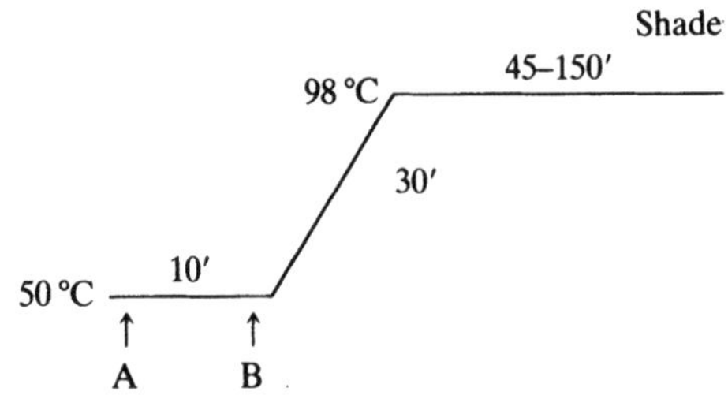

A. Sulfuric acid (96%)	9–10% owf (50:1 LR)
B. Palatin Fast (BASF) dye	x% owf

Shading. Heating is stopped, the dye added, and dyeing continued for 10 min. The bath is then raised to 98 °C and dyeing continued at this temperature for 30 min.

Scheme 10

The dyeing cycle of Scheme 10 is used with the following recipe:

A. Sulfuric acid (96)%	6% owf (50:1 LR)
Uniperol O (BASF)	3% owf
B. Palatin Fast (BASF) dye	x% owf

Shading is carried out as described in Scheme 10.

Scheme 11

the pH is 2, but as the temperature approaches the boil the sulfamic acid hydrolyzes to give ammonium hydrogensulfate, causing the dyebath pH to rise to 3–3.5 (equation 1).

$$H_2O + NH_2SO_3H \rightarrow NH_4HSO_4 \qquad (1)$$

As a consequence, fiber damage is lower than when using sulfuric acid.

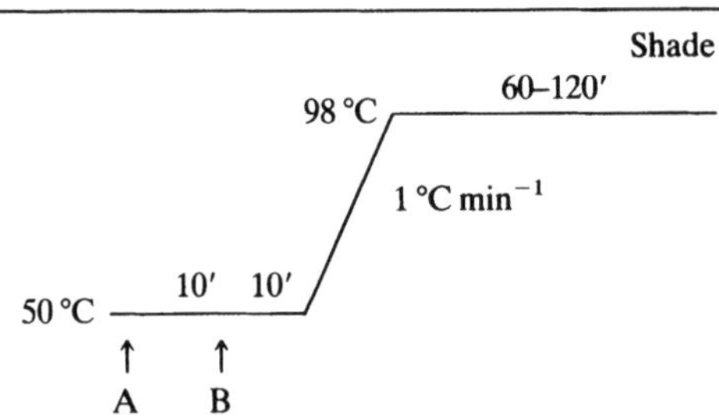

A. Sulfamic acid	6% owf
Glauber's salt	5% owf
Uniperol 5700 (BASF)	2% owf
B. Palatin Fast (BASF) dye	x% owf

Scheme 12

Pad	Acetic acid (60)%	15–20 $g\,l^{-1}$
	Primasol KW (BASF)	25–40 $g\,l^{-1}$
	Thickener	4–8 $g\,l^{-1}$
↓	Palatin Fast (BASF) dye	$x\,g\,l^{-1}$
Steam	45–60′; 100–102 °C	
↓		
Rinse	Acetic acid to pH 5–6; 40 °C	

Scheme 13

The process shown in Scheme 13 outlines the continuous dyeing of slubbing. The mechanism of interaction of 1:1 metal-complex dyes with wool has been studied by various workers but remains a matter of debate. The dyes exhibit maximum exhaustion in the range pH 3–5, depending on dye,[72] but such conditions result in tippy dyeings of poor wet fastness. Under these conditions, the dye can be expected to interact with the fiber by virtue of electrostatic forces operating between dye and protonated amino groups in the fiber, complex formation between chromium in the dye and appropriate groups (such as imino, —NH—) in some of the side chains in the fiber, and also van der Waals' forces. Dye exhaustion increases with reduction in pH of the dyebath below 3[72] and, as previously mentioned, under these conditions the dye is able to migrate well at the boil. At these pH values (i.e., below 3) the imino groups will be protonated (i.e., N^+H_2) and coordination with the chromium in the dye is no longer possible. However, electrostatic interaction between the dye and these and other positive amine sites in the substrate can be expected. Thus, at pH 2, the conditions commonly employed for their application, the 1:1 premetallized dyes will be adsorbed onto wool by virtue of electrostatic and van der Waals' forces of interaction, the latter being less important owing to the relatively small size of the metal-complex dye molecule. This would explain the observed migration of the dyes when applied at pH 2, the dyes in effect behaving as acid leveling dyes. The reduction in exhaustion with decrease in application pH below 3 can be attributed to a corresponding absence of dye–fiber coordination due to protonation of the appropriate ligands (e.g., imino groups) in the fiber.

Hartley has demonstrated that 1:1 premetallized dyes coordinate with carboxylate groups in wool.[67,73] If this coordination were to occur to a significant extent, then the dye could be expected to be very strongly attached to the fiber and consequently exhibit correspondingly high fastness to wet treatments. This is not, however, observed in practice, the fastness of 1:1 metal-complex dyes on wool to wet and washing treatments being only moderate to good. Thus, the contribution that dye–fiber coordination makes toward attachment of these dyes to wool can, at most, be relatively minor.

Giles *et al.* propose that 1:1 premetallized dyes behave as simple acid dyes, coordination with carboxylate groups not playing a part in their adsorption.[74] These workers propose that the higher fastness of 1:1 metal-complex dyes as compared with their nonmetallized counterparts is solely due to the former's increased tendency to aggregate in the fiber.

Peters[65] suggests that, once adsorbed, the 1:1 metal-complex dye may revert to a 2:1 complex but, were this the case, then the fastness of 1:1 premetallized dyes on wool might be expected to closely approach those of their 2:1 counterparts.

3. 2:1 Metal-Complex (Neutral Dyeing Premetallized) Dyes

In 1951 Geigy introduced the Irgalan range of 2:1 premetallized dyes for wool. Typically, the dyes are 2:1 symmetrical chromium or cobalt complexes of, commonly, *o,o′*-dihydroxy azo dyes, and are solubilized by the presence of nonionic substituents such as methyl sulphonyl (—SO_2CH_3). The weakly acidic solubilizing group confers low, cold aqueous solubility on the dyes but the ensuing dispersion passes into solution with increase in temperature. The dyes are applied to wool from a neutral or weakly acid dyebath (pH 5.5 to 7) and are also applicable to nylon.

Ranges of 2:1 dyes that contained various nonionic solubilizing substituents, such as alkyl-amino sulphonyl (—SO_2NHR) and cyclic sulphonyl groups, were soon introduced by other manufacturers,[71] for example, CI Acid Black 60.

In 1951 BASF introduced the Vialon Fast range of 2:1 metal-complex dyes for nylon. Members of this dye type are notable for their complete lack of solubilizing groups, dyeing taking place from dispersion to give dyeings of very good fastness to light and wet treatments.

In the 1960s monosulfonated, asymmetrical 2:1 chromium complexes, for example, the Lanacron (Ciba) and Irgaren (Geigy) ranges, and the Elbilan (LBH) range of symmetrical 2:1 complexes solubilized by two carboxyl groups, were marketed. In 1970 the Acidol M range of disulfonated, symmetrical 2:1 chromium complexes was marketed by BASF. Each of these dye types was applicable to both wool and nylon from weakly acidic dyebaths in the presence of an auxiliary leveling agent, yielding dyeings of good all-round fastness.

a. Dyeing Behavior. 2:1 premetallized dyes, as represented by the ranges shown below, possess very high affinity for wool and are applied under neutral or slightly acidic conditions in the presence of a proprietary leveling agent. In general their dyeing behavior resembles that of acid milling dyes.

1. Weakly polar dyes

Isolan K	(Bayer)
Ortolan	(BASF)
Irgalan	(Ciba-Geigy)
Lanasyn	(Sandoz)

2. Strongly polar dyes

Monosulfonated

Isolan S	(Bayer)
Lanacron S	(Ciba-Geigy)
Neutrichrome S	(ICI)
Lanasyn S	(Sandoz)

Disulfonated

Acidol M	(BASF)
Azarin	(Hoechst)
Neutrichrome M	(ICI)

The weakly polar types display very good light fastness on wool, for example (ISO BO2), 6 to 6–7, 1/1 SD. The fastness of dyeings to washing and to wet treatments is generally good, typically (1/1 SD) 4 to 4–5 (ISO CO6/B2), 3 to 3–4 alkaline milling, and 4 to 4–5 alkaline perspiration. In deep shades, however, their wash and wet fastness properties are usually inferior to those of mordant dyes. The dyes are principally used on loose stock, slubbing, yarn, and knitted fabric. They are less widely used on woven pieces owing to their tendency to highlight irregularities in the structure and quality of the material. A typical application procedure is shown in Scheme 14.

The strongly polar 2:1 metal-complex dyes possess generally higher fastness to wet and washing treatments on wool than their weakly polar counterparts, being of the order (1/1 SD) alkaline perspiration 5, alkaline milling 4–5 to 5, ISO CO6/B2 5. Light fastness (ISO BO2) of dyeings is in the range 6 to 7 1/1 SD and 5–6 to 6 1/12 SD. The disulfonated dyes possess the highest all-round fastness properties. The dyes exhibit slightly lower-level dyeing behavior than weakly polar types, the disulfonated variants possessing the lowest leveling characteristics and are therefore rarely used on woven fabrics. Strongly polar 2:1 premetallized dyes are therefore suitable for those applications in which mordant dyes are used, as well as in cases where the latter dye class cannot be employed owing to the prolonged dyeing times involved.

Schemes 15 and 16 are typical exhaustion dyeing processes. In contrast to the 1:1 premetallized dyes, the metal atom of 2:1 metal-complexes is fully chelated with the dye and there is no possibility of coordination with appropriate ligands in the fiber. Because of the large size of these dyes, van der Waals' forces can be considered to play a major role in their attachment to wool. The generally high wet and washing fastness of 2:1 premetallized dyes on wool can be attributed to

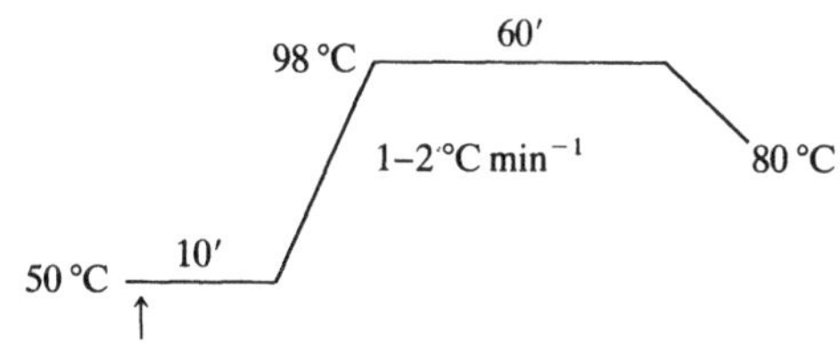

A. Isolan K (Bayer) dye	x% owf
Sodium sulfate (anhydrous)	1–2 $g\,l^{-1}$
Acetic acid (60%) (pH 5–6)	0.5–1.0% owf
Avolan IL (Bayer)	1.5–3% owf

Scheme 14

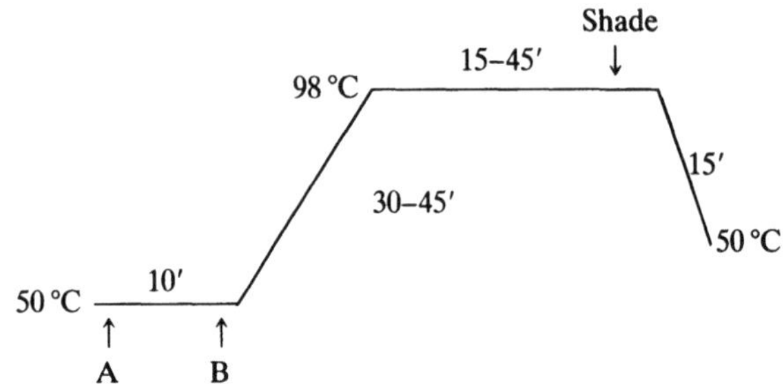

A. Ammonium sulfate 3% owf
pH 6 to 6.5 (adjust with acetic acid if required)
Lyogen SU (Sandoz) 2% owf
B. Lanasyn S (Sandoz) dye x% owf

Shading. Cool the dyebath to 60–80 °C before adding the dye.

Scheme 15

The dyeing cycle of Scheme 4 is used with the following recipe:

A.	Acetic acid (30%)	pH 4–5
	Unisol WL (ICI)	0.5–1.5% owf
B.	Neutrichrome M (ICI) dye	x%owf

Shading is carried out as described in Scheme 4.

Scheme 16

the nature of this interaction and to the low diffusional behavior of the large dye molecules within the fiber.

Electrostatic forces can also be expected to contribute to adsorption. The weakly polar types carry an overall (i.e., nonlocalized) negative charge that is capable of interaction with basic groups in the substrate. This type of interaction can be expected to be more pronounced in the case of the strongly polar dyes which carry localized negative charges. The reduced-level dyeing characteristics of strongly polar 2:1 premetallized dyes as compared with their weakly polar counterparts may well be due to the correspondingly greater contribution of electrostatic interaction to dye–fiber substantivity. The relative importance of electrostatic forces in the uptake of sulfonated 2:1 dyes is shown by the use of generally lower pH dyebaths (i.e., pH 4–5) for the application of disulfonated dyes as compared with higher pH values (i.e., pH 5–6) employed for applying the monosulfonated variants.

Furthermore, Rexroth[75] has related the structure of 2:1 complexes, especially with respect to the solubilizing group present, and wet fastness properties on wool, disulfonated 2:1 metal-complexes having overall higher fastness.

2:1 premetallized dyes can also be applied to slubbing using a Pad–Steam

	25–30 °C	
	Lanasyn (Sandoz) dye	x g l^{-1}
	Lyogen CW (Sandoz)	10–20 g l^{-1}
Pad	Lyogen V (Sandoz)	10 g l^{-1}
↓	Thickener	4–8 g l^{-1}
	Formic acid (85%) pH 2.8–3.0	10–20 ml l^{-1}
	100–120% pick up	
Steam	15–40′; 100–102 °C	
↓		
Wash		

Scheme 17

process[62] as shown in Scheme 17. The semicontinuous IWS Lanapad process, which utilizes RF heating for dye fixation, has been described for the application of 2:1 premetallized dyes to wool tops.[76]

D. REACTIVE DYES

Reactive dyes are covalently bound to the fiber and, in this respect, therefore differ from all other classes of dye. As a consequence of their mode of attachment, the dyeings characteristically possess a very high degree of fastness to washing and wet treatments.

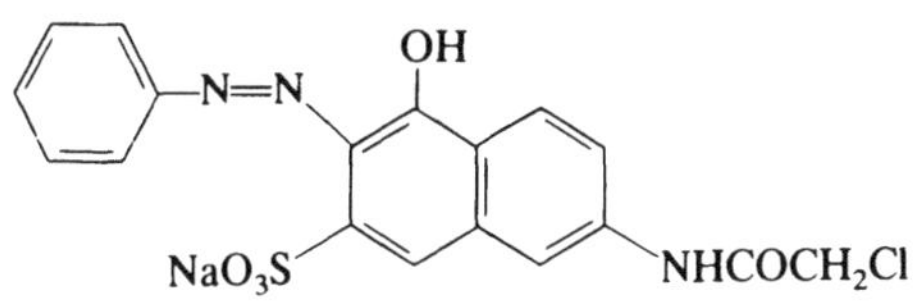

28

The first reactive dye, CI Acid Orange 30 **(28)**, was introduced in 1932 by IG Farben for dyeing wool, but the reactivity of the labile chlorine atom in the chloroacetylamino group was at the time not recognized.[77] In 1952 Hoechst introduced the Remalan range (β-sulfatoethylsulfone) and two years later Ciba marketed the Cibalan Brilliant range (α-chloroacetylamino) of dyes for wool, both of which had only limited success owing to the low rate of fixation of the dyes on the fiber.[78] The introduction by ICI in 1956 of the Procion (dichlorotriazine) range of reactive dyes for cotton, as typified by CI Reactive Red 1 **(29)**,

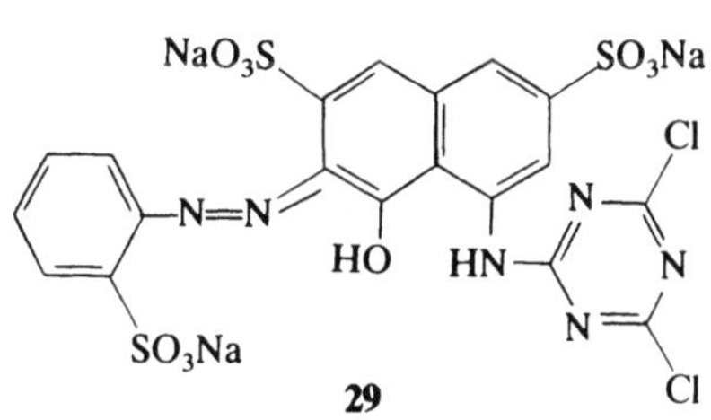

29

lead rapidly to many dye makers introducing ranges of reactive dyes for both cotton and wool based on a variety of reactive systems.

The development and chemistry of reactive dyes have received much attention.[10,65,77–82] The use of dyebath auxiliaries is of vital significance in the dyeing of wool with reactive dyes. Like all other water-soluble, anionic dyes, reactives exhibit selectivity in initial uptake, i.e., dye adsorption occurs more rapidly at the weathered fiber tip. Since a fairly significant degree of dye–fiber reaction will have occurred before the liquor has reached the boil, dye migration from tip to root will be severly impaired and be manifest as skittery dyeing. In contrast to the majority of dyebath auxiliaries employed to control the dyeing rate, those designed for use with reactive dyes increase the rate of dyeing,[77] possibly by disaggregating the dye in aqueous solution, although the precise nature of their action is as yet unclear.

Current reactive dyes for wool, as typified by the ranges shown below, are highly resistant to hydrolysis during immersion dyeing and, as a consequence, exhibit high fixation efficiency. In addition, the reactivity of the majority of dyes is such that, with the aid of a suitable auxiliary, satisfactory rates of dye–fiber reaction can be achieved so as to give level dyeing. As is the case of reactive dyes on cotton, the dyes react with wool either by nucleophilic substitution or by nucleophilic addition. The mechanism of interaction of the major reactive dyes for wool has been extensively reviewed.[77]

Drimalan F (Sandoz)
2,4-Difluorochloropyrimidine

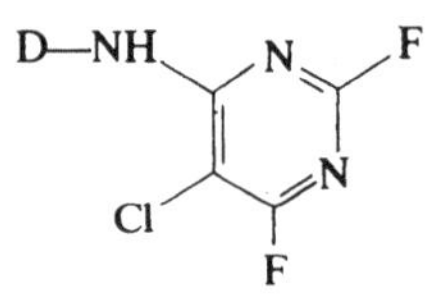

30

The high fixation yield exhibited by these dyes, which is often in excess of 95%, has been attributed to their high resistance to hydrolysis and the ability of both fluorine atoms to react with wool.

Lanasol (Ciba-Geigy)
α-Bromoacrylamido

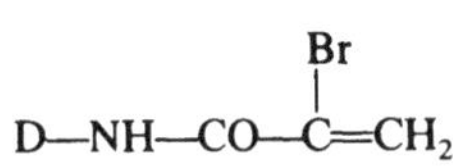

31

The dyes are essentially bifunctional, provided that sufficient accessible nucleophilic groups are available in the fiber.

Procilan E (ICI)
N-Methyltaurine-ethyl sulfone

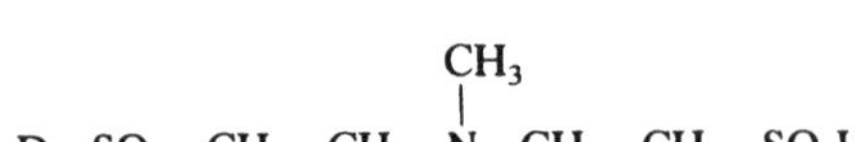

32

Under hot, aqueous acidic (pH 5) dyebath conditions, the vinyl sulfone derivative **(33)** is released gradually so that, in the early stages of dyeing at least, the dye behaves as an acid dye, being able to migrate and thus facilitate level dyeing. As a consequence, the dyes are especially suitable for hank and winch dyeing where mechanical conditions lend themselves to unlevel dyeing.

$$D{-}SO_2{-}CH{=}CH_2$$

33

1. Dissolving the Dyes

The powder is pasted with cold water to which is then added, with stirring, hot water. The resulting solution is added to the dyebath through a sieve.

2. Dyeing Behavior

Reactive dyes for wool are applied to loose stock, slubbing, yarn, and piece goods. Dyeings on wool possess good fastness to light, typically (ISO BO1) 5–6 to 6 1/1 SD, 4–5 to 5 1/12 SD, and excellent fastness to washing, typically (1/1 SD) 5 (ISO CO3 and ISO CO6/B2), and to wet treatments, for example, alkaline perspiration 5, alkaline milling 4–5.

a. Batchwise Dyeing Process. Reactive dyes are usually applied to wool by exhaustion methods at pH 4.5–6.5. At lower pH values, the dyeing is unlevel owing to a high rate of dye uptake while at higher pH exhaustion is low. The particular pH employed depends on depth of shade, being commonly in the range pH 5.5–6.0 for pale depths and pH 5.0–5.5 for full depths. The application of reactive dyes to wool generally involves two stages:

1. low temperature (40–50 °C) exhaustion at pH 5–6, followed by
2. fixation, most commonly at a high temperature (100 °C).

A typical application process is shown in Scheme 18.

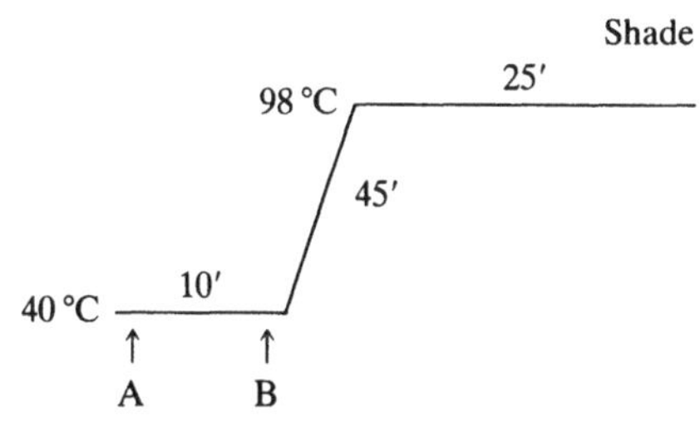

A.	Ammonium sulfate	4% owf
	Acetic acid (80%)	2% owf
	Drimagen F (Sandoz)	1% owf
	Sodium sulfate (anhydrous)	0–5% owf
B.	Drimalan F (Sandoz) dye	2% owf

Shading. Cool to 60 °C, add the dye, raise to the boil, and continue dyeing for 10–20 min.

Scheme 18

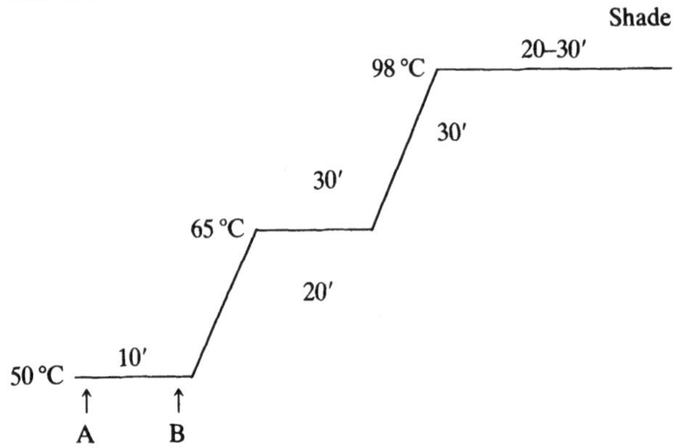

A. Ammonium sulfate (anhydrous)	4% owf
Acetic acid (80)%	1.5–2% owf (to pH 4.5 to 5)
Lyogen F (Sandoz)	$1\ \mathrm{g\,l^{-1}}$
B. Drimalan F (Sandoz) dye	2% owf

Shading. Cool to 60 °C, add the dye, raise to the boil, and continue dyeing for 10–20 min.

Scheme 19

Some dyeing methods include a hold period at 60–70 °C to promote dye migration (Scheme 19). The period of time at the boil depends on depth of shade, deep shades necessitating longer dyeing times to achieve maximum dye–fiber fixation.

To develop maximum wet and washing fastness properties the dyeings require an alkaline aftertreatment, commonly by using ammonia at pH 8–8.5 for 15–20 min at 80 °C, to remove unfixed and hydrolyzed dye. Alternative methods of aftertreatment include the use of hexamine[83] at the boil which enables more efficient removal of unfixed dye at pH 6.5 thereby reducing fiber damage, and also the use of sodium sulfite which was found to be more effective than ammonia.[84] Recent studies[57] have demonstrated that reactive dyes exert a fiber-protective effect during dyeing, possibly by alteration of the thiol/disulfide interchange process.

b. Semicontinuous Dyeing Processes. Reductions in dyeing costs and fiber damage can be secured by using the IWS cold pad-batch application method[85–87] for tops and woollen cloth (Scheme 20). The urea promotes dye uptake under the room temperature conditions.[88] The sodium bisulfite also promotes dye uptake, especially in the case of high molecular weight dyes. With some reactive dyes, the bisulfite increases the extent of fixation. The beneficial effects of bisulfite on dyeing are due to cleavage of disulfide cross links within the fiber. This sulfitolysis enables the dyes to more easily penetrate the fiber and also provides groups with which the dyes can interact.[55] If sodium bisulfite is added to the pad liquor, then dyes that are stable to this agent must be used. Reactive dyes recommended for

	25–30 °C	
	Lanasan (Sandoz) dye	$x\ g\,l^{-1}$
	Lyogen CW (Sandoz)	$2\ g\,l^{-1}$
Pad	Lyogen V (Sandoz)	$10\ g\,l^{-1}$
	Thickener	$8\ g\,l^{-1}$
	Acetic acid (80%) to pH 6	
	Urea	$350\ g\,l^{-1}$
	Sodium hydrogensulfite	$10\ g\,l^{-1}$
	100–130% pick up	
	20–30 °C	
Batch	24–48 hours pale shades	
	48–72 hours deep shades	
	Ammonia (880) 60 °C	$1\ ml\,l^{-1}$
Rinse	Water (twice) 60 °C	
	Acetic acid to pH 4.5 (cold)	
	Water (cold)	

Scheme 20

pad-batch application often include those used in cotton dyeing, such as the Drimarene K and Drimarene R (Sandoz) ranges. The pad-steam process can also be used for dyeing slubbing as shown in Scheme 21. RF heating has also been examined as a fixation medium.[89,90]

Wool is most stable, i.e., undergoes least chemical and physical damage, in the isoelectric region at pH 4.5–6. Recently, two dye ranges have been introduced, namely, the Lanaset (Ciba–Geigy) and Sandolan MF (Sandoz), which are intended for dyeing wool in the isoelectric region. Both dye ranges are designed to exhibit excellent migration and leveling when applied in the presence of specially developed dyeing auxiliaries, giving good coverage of tippy wool.

	25–30 °C	
	Drimalan F (Sandoz) dye	$x\ g\,l^{-1}$
	Lyogen CW (Sandoz)	$10\text{–}20\ g\,l^{-1}$
Pad	Lyogen V (Sandoz)	$10\ g\,l^{-1}$
	Thickener	$4\text{–}8\ g\,l^{-1}$
	Acetic acid (80%) to pH 6–6.5	
	100–120% pick up	
Steam	15–40′; 100 °C	
	Ammonia (880) 60 °C	$1\ ml\,l^{-1}$
Rinse	Water (twice) 60 °C	
	Acetic acid to pH 4.5 (cold)	
	Water (cold)	

Scheme 21

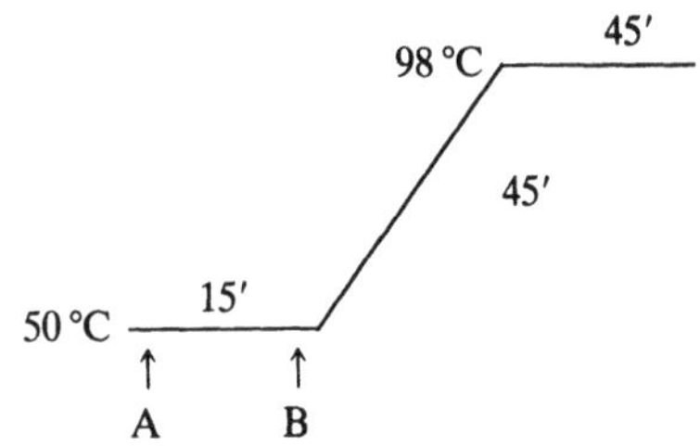

A.	Sodium acetate	$1\ g l^{-1}$
	Acetic acid	to pH 4.5–5
	Sodium sulfate (anhydrous)	5% owf
	Lyogen MF (Sandoz)	0.5–2% owf
B.	Sandolan MF dye (Sandoz))	x% owf

Scheme 22

The Lanaset[91] range comprises some 15 mutually compatible, modified, tinctorially strong, 2:1 reactive and premetallized dyes. They are applied to wool at pH 4.5–5 using an acetic acid/sodium acetate buffer in conjunction with an amphoteric auxiliary Albegal SET (Ciba-Geigy), that promotes dye disaggregation to facilitate leveling.

The Sandolan MF dyes occupy an intermediate position between acid leveling dyes and acid milling or 2:1 premetallized dyes[92] and are suitable for dyeing yarn, fabric, and garments. They are applied in the range pH 4.5–6, the higher pH being used for pale shades. The auxiliary Lyogen MF (Sandoz) promotes uniform initial dyeing and dye migration during fixation; Glauber's salt assists dye migration (Scheme 22). In the case of high twist yarns, for example, where dye penetration is difficult to achieve, the Sandacid V (acid donor) technique can be used (Scheme 23). Dyeings possess good fastness to light, of the order 5 to 5–6 1/1 SD, 4 to 5 1/12 SD (ISO BO1). Fastness to washing and wet

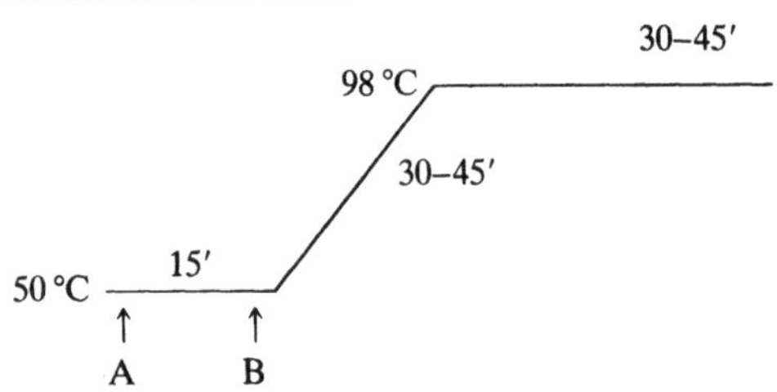

A.	pH 7–7.5 using ammonia if necessary	
	Sandacid V Liquid (Sandoz)	$0.5–1\ ml\,l^{-1}$
	Lyogen MF (Sandoz)	1% owf
	Sodium sulfate (anhydrous)	5% owf
B.	Sandolan MF (Sandoz) dye	x% owf

Scheme 23

treatments is very good, typically (1/1 SD) 5 (ISO CO6/A2E) and 4–5 to 5 (ISO EO4).

E. Dyeing Shrink-Resist Treated Wool

When a garment made from any fiber type is washed, the strains imposed on the material during manufacture are relaxed and the garment shrinks. In the case of some knitted materials, relaxation may be accompanied by extension. Such dimensional changes are reversible and are fairly easily rectified by, for instance, stretching.

With wool garments, felting shrinkage also occurs during washing and wear. The major cause of felting of wool is the preferential movement of the fibers toward their root ends that occurs as a consequence of directional friction between interlocking surface scales. The fibers are thereby released from the interwoven yarn structure and become entangled, resulting in irreversible shrinkage.

The traditional method of imparting shrink-resist to wool is by modifying the scaler structure of the fiber by oxidative attack. The most common oxidant employed for this purpose is chlorine and its derivatives.[51,55]

Chlorination is normally carried out under aqueous acidic (pH 2 to 6) conditions (acid chlorination) so as to minimize yellowing and fiber damage. The chlorinating species in this treatment may be hypochlorous acid, generated by the action of sulfuric acid on sodium hypochlorite, or chlorine gas dissolved in water. Dry chlorination using chlorine gas is also employed. Sodium hypochlorite and chlorine react very rapidly with wool and more control of the process can be obtained using DCCA or its sodium salt.[34]

Chlorination is considered to modify the hydrophobic nature of the epicuticle, increasing its permeability to water; cleavage both of the peptide chain at the tyrosine residues and of the disulfide cross-linkages occurs. The cuticle is substantially more modified than is the cortex. Excessive oxidation leads to substantial fiber damage. Such oxidative treatments produce water-soluble, high molecular weight peptides derived from the proteins within the cuticle which, upon immersion in water, absorb water and soften. As a result, the directional frictional effect is reduced and, with it, felting.

Chlorination is often followed by an "anti-chlor" treatment with sodium bisulfite. In many cases, maximum shrink-resist is achieved only after bisulfite treatment and it is considered that the bisulfite cleaves the partially oxidized disulfide groups.

In 1965 the IWS introduced "Superwash" wool, i.e., wool that can be washed confidently in the washing machine. This concept relies upon the deposition of a permanently bound, highly cross-linked, cationic resin on the surface of the fibers. Such resins restrict relative movement of the fibers, impart a smooth coating to the scaler fiber surface, and resist desorption of the sulfonated peptides from the cortex.

Machine washable wool exhibits substantially different dyeing behavior to both untreated and chlorinated wool. The cationic nature of the fiber surface results in anionic dyes having greater substantivity for superwash wool leading to high initial strike. To achieve level dyeing, strict control of the rate of dye uptake is required.

Since machine washable wool can withstand quite severe wet treatments, then equally high standards of wet fastness are required of the dyes employed in its coloration. With the exception of certain acid-milling dyes, which can be used only in pale to medium shades, dyeing of machine washable wool to meet "Superwash" label requirements is carried out using reactive, mordant, and mono- and disulfonated 2:1 metal-complex dyes. With each of the three latter dye types an appropriate aftertreatment is often necessary to develop maximum wet fastness.

1. Reactive Dyes

These are widely used for bright hues and are aftertreated with ammonia or other compounds (see above). High-temperature dyeing can be carried out as the Hercosett resin provides a measure of fiber protection.

2. Mordant Dyes

This dye class provides deep, economical shades on machine washable wool and, with the possible exception of the Omega Chrome FL (Sandoz) dye range, requires an ammonia (pH 8–8.5) aftertreatment to develop maximum fastness properties.

3. 2:1 Premetallized Dyes

A typical application procedure is shown in Scheme 24. While in pale depths dyeings are of adequate fastness, medium and deep depths usually require an aftertreatment with a cationic product, such as Sandofix L (Sandoz) or Basolan F (BASF), the latter being a highly cationic quaternary polyammonium compound, to achieve machine washability. A typical aftertreatment consists of treating the

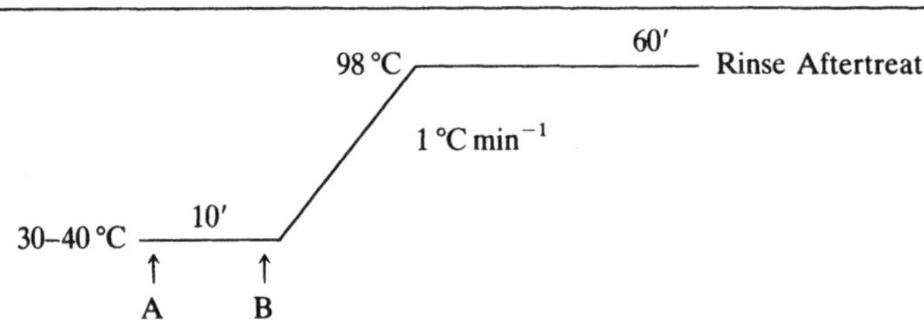

A. Ammonium sulfate 0–3% owf to pH 6.3
Lyogen SU (Sandoz) 1–3% owf
B. Lanasyn S (Sandoz) dye x% owf

Scheme 24

rinsed, dyed material in a fresh bath containing Basolan F (4% owf) at 40 °C for 15 min at pH 7.5, followed by rinsing.

III. DYEING OF CELLULOSIC FIBERS

This section considers the dyeing of the two most important cellulosic fibers, cotton and viscose rayon. Linen, obtained from flax, enjoys comparatively limited textile usage while the other major bast fibers, namely hemp and jute, find use as industrial textiles.

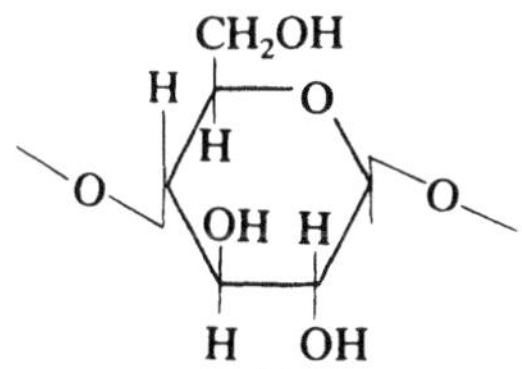

34

All cellulosic fibers are essentially condensation polymers of β-D-glucopyranose with 1,4-glycosidic linkages **(34)**. The polymeric chains are held together by extensive hydrogen bonding between hydroxyl groups. The large number of hydroxyl groups readily form hydrogen bonds with water and are therefore responsible for the hydrophilic nature of cellulosic fibers. Owing to degradation during processing, a C-6 primary hydroxyl group (—CH_2OH) is replaced by a carboxyl group (—COOH) periodically along the chain; because of more severe conditions of manufacture, these acidic groups are in greater concentration in viscose rayon than in normal cotton. Since the pK_a of these carboxyl groups is approximately 3.2, the fibers consequently carry a negative charge in aqueous conditions. Under strongly aqueous alkaline conditions, ionization of the hydroxyl groups confers a high negative charge on the fiber and, in addition, breaks down interchain hydrogen bonding, causing charge repulsion between the polymer chains and leading to swelling (the principle of mercerization).

Since the major classes of dye used for dyeing cellulosic fibers are applied as anions, exhaustion dyeing of such fibers is carried out in the presence of sodium chloride which reduces dye–fiber repulsion and thereby promotes dye uptake.

a. Cotton. This, perhaps most important of all textile fibers, is obtained from the seed hair of plants of the genus Gossypium. The mature staple fiber takes the form of a flat, convoluted ribbon, varying in length, fineness, color, etc., according to source. The cotton fiber consists of four main parts.[53,93] The cuticle comprises a very thin outer layer of wax, protein, pectin, and mineral matter and covers the primary wall which constitutes only a small proportion of the fiber and consists mainly of a network of cellulose fibrils together with pectin and fats. The secondary wall constitutes the bulk of the mature fiber and comprises cellulose fibrils arranged spirally around the longitudinal fiber axis laid down in successive layers, the direction of spiral reversing at intervals along the fiber length. The lumen or cavity is the remainder of the central canal from which these layers were produced and contains residual protein, mineral salts, and the natural coloring matter of the fiber.

Accounts of the numerous processes involved in the conversion of the fibers from the cotton seed pod into a useful textile fiber are given elsewhere.[53] The textile fibers undergo various processes to improve their quality and dyeability, a comprehensive account of which is provided by Dickinson.[94] The primary wall and cuticle are of great importance in this context, since they contain the greatest proportion of naturally occurring impurities.

b. Viscose Rayon. Accounts of the production of viscose rayon fibers are available elsewhere.[93,95] Owing to the low degree of orientation of the cellulose molecules, regular rayon filament has a lower tenacity, both wet and dry, than cotton. Modifications to the regeneration process result in the production of various types of viscose fiber, namely, high-tenacity polynosic, modal, and high wet modulus (HWM) fibers, of improved physical properties.[95] Briefly, high-tenacity viscose fibers enjoy mostly industrial applications and are rarely dyed. HWM and modal fibers have a higher degree of crystallinity and orientation than regular viscose fibers. As a consequence, the polynosic and HWM fibers imbibe water and swell to a lesser extent than fibers of regular viscose and possess different dyeability. Viscose fibers are more sensitive to degradation by aqueous alkalis and acids, regular viscose being most sensitive.

c. Dyes Used on Cellulosic Fibers. Cotton and staple viscose fiber is dyed as loose stock, yarn, and as knitted and woven fabrics. Continuous viscose filament is commonly dyed in the form of the cake that is produced in fiber production.

Cellulosic fibers are dyed using the following classes of dye: direct, sulfur, vat, azoic, and reactive.

A. DIRECT DYES

Until the latter part of the last century, cotton could be dyed directly without the aid of a mordant (metal salt or tannin) only with a few natural colorants such as Annatto, Safflower and Indigo; the early synthetic dyes also required the use of mordants. In 1884 Boettiger prepared a red disazo dye derived from benzidine which dyed cotton directly from a neutral dyebath containing salt (sodium chloride). The dye, CI Direct Red 8 **(35)**, was marketed in the same year by

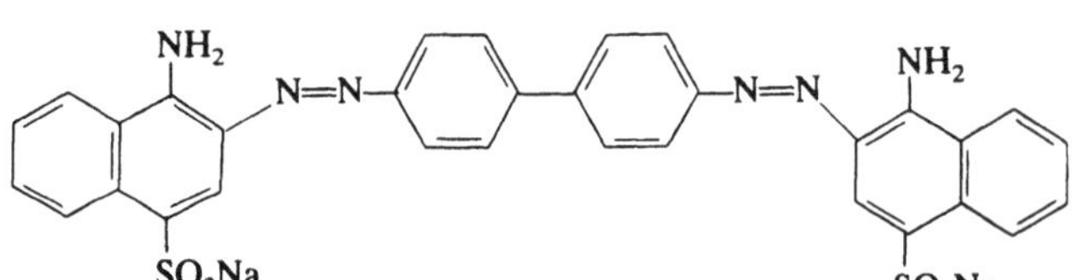

35

AGFA as Congo Red and became the first member of one of the most important classes of dye for cellulosic fibers.

Direct dyes are relatively inexpensive and provide an outstandingly wide range of hues. However, the fastness, especially to washing, of dyeings on cellulosic fibers generally leaves much to be desired. While this has led to their replacement with reactive dyes, which have much higher wet and washing fastness properties on such substrates, direct dyes nevertheless enjoy fairly widespread use where high standards of fastness are not required. Furthermore, a variety of methods of aftertreating dyeings is employed to improve the fastness properties of the dyes on cellulosic materials.

Direct dyes belong to several chemical classes.

1. Azo

The majority of direct dyes are chiefly of the disazo, trisazo, and polyazo types although there are a few monoazo representatives such as CI Direct Orange 75 **(36)**. Many of the earlier disazo dyes derived from benzidine and its derivatives are no longer manufactured in the UK and in other countries, owing to regulations covering the use of carcinogens.

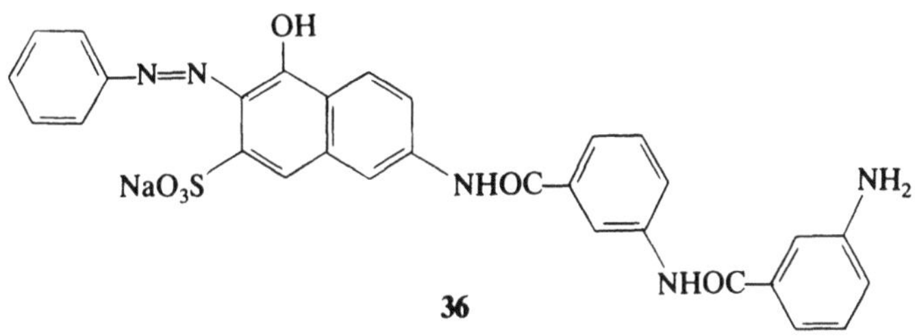

36

J-acid and its derivatives provide important coupling components for many dis- and higher azo dyes, such as CI Direct Red 81 **(37)** and CI Direct Red 4 **(38)**. The tris- and polyazo dyes are mostly browns and blacks as exemplified by CI Direct Black 9 **(39)** and CI Direct Black 19 **(40)**. The triazine ring is used as a bridge to link two separate chromophoric systems, as for example in CI Direct Green 26 **(41)**.

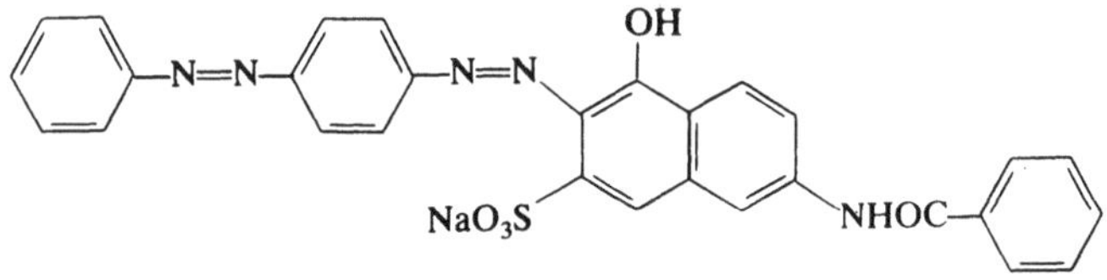

37

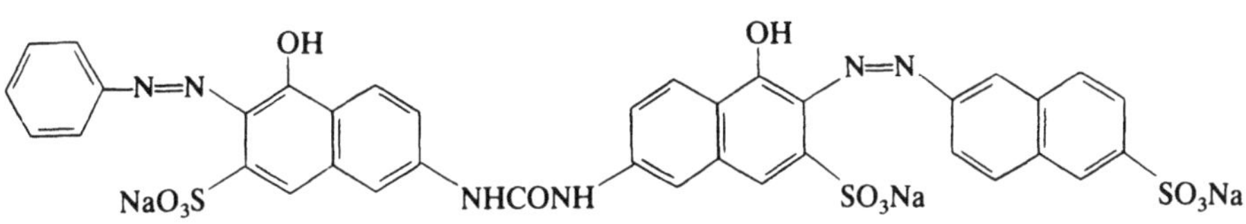

38

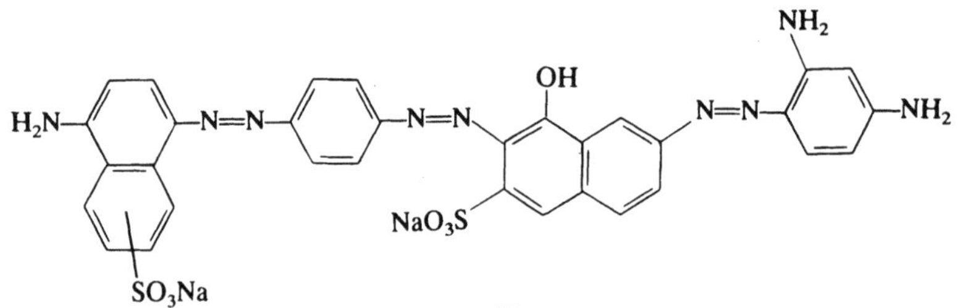

39

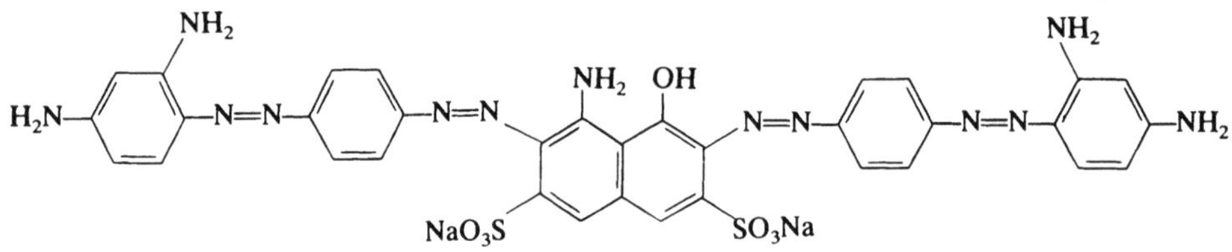

40

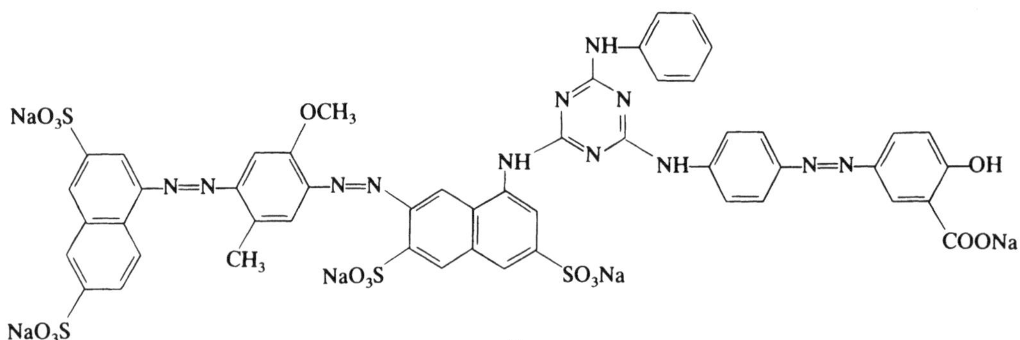

41

2. Stilbene

This is a relatively small but important group of mostly red, yellow, and orange dyes, such as CI Direct Yellow 12 **(42)**.

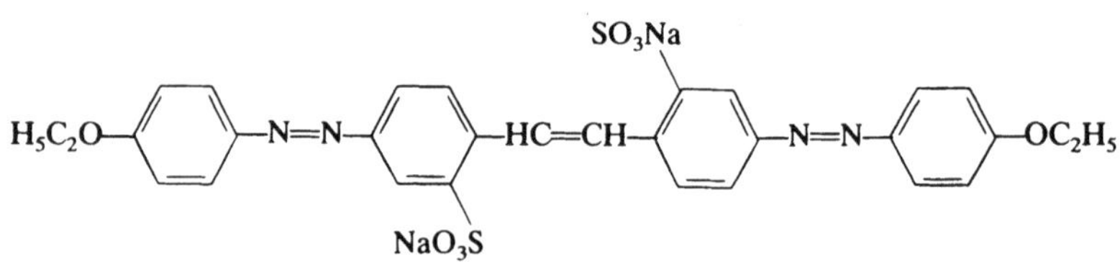

42

3. Phthalocyanine

Characteristically, these bright blue and turquoise-blue dyes, as typified by CI Direct Blue 86 **(43)**, possess high fastness to light but low fastness to wet treatments on cotton.

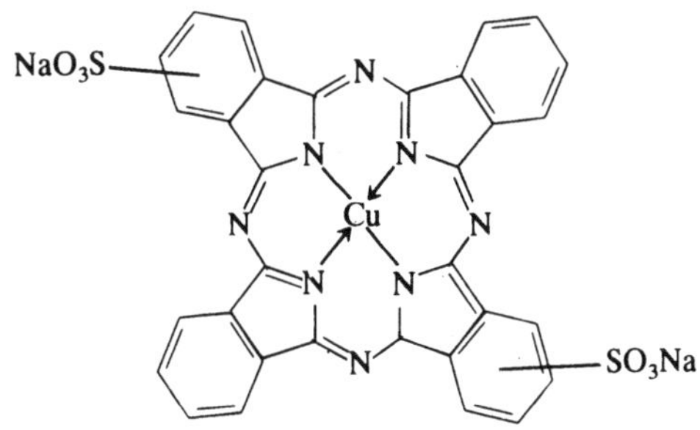

43

4. Dioxazine

This small group provides mostly blue dyes such as CI Direct Blue 106 **(44)**.

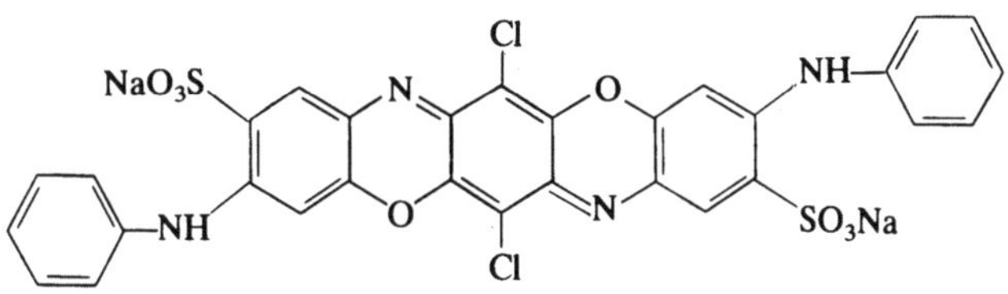

44

5. Miscellaneous Dyes

Other classes, namely AQ, quinoline, and thiazole, contribute only a small number of direct dyes.

Premetallized, water-soluble copper complexes of *o,o'*-dihydroxy, dis-, and trisazo direct dyes, as exemplified by CI Direct Blue 84 **(45)**, display very good wash and light fastness properties on cellulosics.

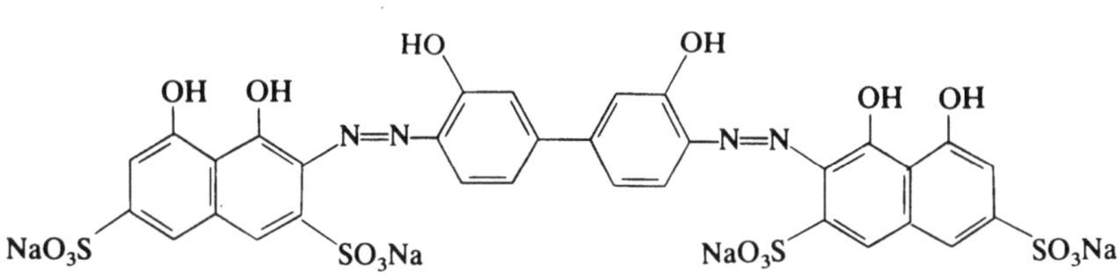

45

6. Dissolving the Dyes

The dye powder is pasted with water to which is then added boiling water. The ensuing solution is sieved during addition to the dyebath.

7. Dyeing Behavior

With the possible exceptions of diazotizable and premetallized types, direct dyes, as typified by the Solar (Sandoz), Duranol (ICI), and Sirius Supra (Bayer) ranges, generally give dyeings on cellulosic fibers of poor to moderate fastness to washing (1/1 SD) ISO CO1, cotton 2 to 3; viscose 3 to 4 and wet treatments (1/1 SD) ISO EO4, cotton 3 to 4, viscose 3–4 to 4. Light fastness of dyeings varies from moderate to good, in the range (ISO BO1): cotton 5 to 6 1/1 SD, 4–5 to 5 1/12 SD; viscose 6 to 6–7 1/1 SD, 4–5 to 5 1/12 SD. Nevertheless, dyes of this class are widely used where higher standards of wet and wash fastness are not required.

The forces of interaction operating between this class of dye and cellulosic fibers are predominantly van der Waals, arising from the highly conjugated nature of the large, linear, and coplanar dye molecules. Multilayer attachment also occurs as a result of the structural features of the dyes. The polyhydroxy (hydrophilic) character of the cellulosic substrate prevents hydrophobic interaction between dye and fiber, although hydrogen bonding between the cellulosic hydroxyl groups and appropriate groups in the dye may also contribute to dye–fiber attachment in the dry, dyed state. These relatively weak dye–fiber bonds, coupled with the fairly high water solubility of the dyes and the hydrophilic nature of the cellulosic fibers, are responsible for the generally poor to moderate wet fastness properties of dyeings on both cotton and viscose.

Direct dyes vary markedly in dyeing behavior which, obviously, is of great significance with respect to compatibility in mixtures. This variation predominates as sensitivity to sodium chloride. Since, as discussed previously, salt is used to suppress charge repulsion effects operating between dye and fiber, then salt sensitivity will depend upon the charge on the dye anion. It will also depend upon the affinity of the dye for the substrate. Because of the nature of direct dyes (i.e., their extensive conjugated system), they are prone to aggregate in solution. This aggregation is increased by an increase in electrolyte concentration and is more pronounced in the case of high affinity (i.e., large size), low solubility (low charge) dyes. The differences in dyeing behavior are covered by the following S.D.C. classification.

Class A. These dyes have little salt sensitivity and are self-leveling, i.e., have good migration properties. Generally, Class A dyes are relatively highly water-soluble monoazo and disazo dyes which consequently exhibit a small degree of aggregation in solution and little sensitivity to electrolyte, Their low substantivity results in good migration characteristics.

Class B. These are not self-leveling but dye uptake is controlled by careful salt addition, i.e., they are salt-sensitive. These mostly disazo and trisazo dyes are of relatively high ionic charge and thus have low to moderate substantivity. Their exhaustion behavior is markedly dependent upon electrolyte concentration because of their high ionic charge.

Class C. These are not self-leveling and are highly sensitive to salt. Exhaustion cannot be adequately controlled solely by salt addition and additional control of temperature is required for level dyeing, i.e., they are temperature-sensitive. The low ionic charge of these disazo and trisazo dyes results in high substantivity and the dyes therefore exhibit rapid exhaustion. Level dyeing cannot be achieved solely by electrolyte addition because, owing to their small anionic charge, the dyes are very prone to aggregation in solution and this is accentuated by electrolyte addition. Thus, level dyeing is controlled by careful and gradual increase in dyeing temperature, with addition of electrolyte when required to increase exhaustion once level dyeing has been achieved, i.e., at high temperature.

Generally, fastness of dyeings to wet treatments follows the dye classification in that Class C dyes show higher fastness to wet treatments than Class A dyes.

The ABC classification is of great practical use in selecting dyes for use in admixture; many dye makers include the classification in their pattern cards. As a general rule, dyes belonging to different classes should not be applied together although overlap between the different classes is to be expected. The different dyeing behavior of direct dyes, namely substantitivy, leveling, salt- and temperature-sensitivity, is often reflected in the different dyeing conditions recommended by dye manufacturers.

Temperature greatly affects strike, leveling, and fiber penetration; a high dyeing temperature (98 °C), which facilitates leveling, penetration, and expedites dyeing, results in low exhaustion of many direct dyes in conventional dyeing times (45–60 min). Consequently, many dye manufacturers give details of the optimum dyeing temperature for the various dyes in their ranges. The salt requirements of the dyes are also usually reflected in the dye makers' literature. Liquor to goods ratio (LR) greatly affects exhaustion and leveling properties. Dye aggregation in solution increases with increase in dye concentration, i.e., with decrease in LR. The above S.D.C. classification is based on a 30:1 LR, thus the use of lower liquor to goods ratios (e.g., 3–5:1 in Jig dyeing) can greatly affect this classification; for instance, Class B dyes when applied at low values of LR may require additional temperature control for level dyeing to be achieved.

The more soluble Class A dyes are preferred for package dyeing owing to their superior migration characteristics. However, Class B and Class C dyes may also be used for package dyeing yarn but high temperatures (120 °C) are generally required to achieve adequate penetration and leveling. Careful choice of dye and strict control of dyebath pH must be exercised when dyeing at such high temperatures since many direct dyes are reduced at high temperatures, especially under alkaline conditions. This particular problem is of great significance in high-temperature dyeing of viscose fibers. The use of ammonium sulfate as acid donor and a mild oxidant overcomes the problem.

Scheme 25 shows typical application conditions used for the Diphenyl (Ciba-Geigy) dye range.

Direct dyes have greater substantivity for viscose than for cotton owing to the generally lower degree of crystallinity and orientation of the former fiber. However, viscose swells appreciably under aqueous conditions which leads to difficulties in dye penetration, especially in the case of dyeing continuous viscose cakes and staple viscose loose stock. Care must therefore be taken to avoid dense packing of dyeing machines.

Impregnation-fixation dyeing methods are also used. However, such applica-

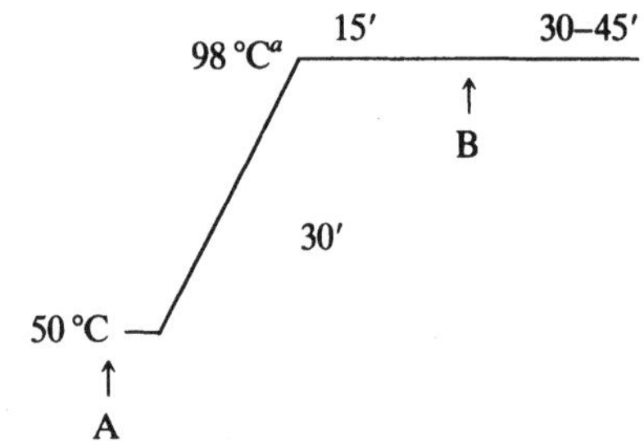

A.	Albatex PON conc. (Ciba–Geigy)	0–0.1 $g l^{-1}$
	Diphenyl (Ciba–Geigy) dye	x% owf
B.	Sodium chloride[b]	
	or	
	Sodium sulfate (anhydrous)[b]	

[a] 85 °C for viscose rayon.
[b] 0–20 $g l^{-1}$ in portions.

Scheme 25

Pad	30–80 °C	
	Dipheyl (Ciba–Geigy) dye	x $g l^{-1}$
	Irgapadol FFU (Ciba–Geigy)	0–3 $g l^{-1}$
	60–80% pick-up[a]	
	70–90% pick-up[b]	
↓ Steam	2–3 min	
	102–105 °C	
↓ Rinse	Hot and cold	

[a] Cotton.
[b] Viscose rayon.

Scheme 26

tion techniques, e.g., pad–steam (Scheme 26), pad–roll, pad–jig, are usually restricted to selected dyes.

8. Aftertreatments

A variety of aftertreatments has been devised to improve the fastness to wet treatments of direct dyes on cellulosic fibers; certain aftertreatments also improve the light fastness of some direct dyeings. In each method, fastness to wet treatments is improved by reducing the diffusional behavior of the dye within the fiber by means of an increase in molecular size of the dye.

a. Diazotization and Coupling. This technique, first employed with Primuline, is little used nowadays. It is applicable to many direct dyes that contain a primary amino group. The dyed material is diazotized using nitrous acid produced *in situ* by the action of, for example, cold hydrochloric acid on sodium nitrite. The material is then well rinsed and passed into a cold solution of the developer (e.g., 2-naphthol, *m*-phenylenediamine) and coupling occurs to yield a larger dye molecule of improved fastness to aqueous treatments. Coupling is accompanied by an often dramatic change in hue of the original dyeing.

b. Metal Salts. Aftertreatment of dyeings of certain direct dyes with copper salts, usually copper sulfate, is mostly commonly used and results in an often marked improvement in light fastness together with some increase in wash fastness of the dyeings. Such dyes are typically *o,o'*-dihydroxyazo, *o*-hydroxy, *o'*-amino, and salicyclic acid derivatives, such as CI Direct Blue 1 **(46)**.

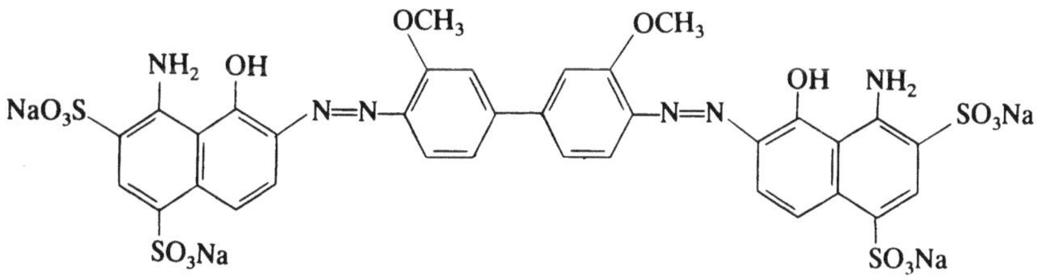

46

The following aftertreatment is suggested for the Benzanil Fast Copper (YCL) dyes: The rinsed dyeing is treated in a fresh bath at 30 °C prepared by first adding acetic acid (0.5–1.5% owf) followed by copper sulfate (1–3% owf; predissolved). The bath is raised to 70–80 °C and aftertreatment continued at this temperature for 20 min.

The hue of the dyeing may be reduced as a result of the treatment. Demetallization (loss of copper) can occur during subsequent wet treatments, causing a drop in fastness and a change in hue.

c. Formaldehyde. This is used with certain, mostly black, dyes. The mechanism of reaction is unclear but it is suggested that dye molecules become linked through methylene bridges. A typical treatment consists of: formaldehyde (30%), 2–3%; acetic acid (30%), 1%; 70–80 °C; 30 min.

Although both fastness to washing and to water is improved, this aftertreatment can result in a reduction of light fastness of the dyeing.

d. Cationic Fixing Agents. These compounds, as typified by Fixogene CD and Matexil FC-PN (ICI), form a large-size complex with the anionic dye within the fiber. For example, the following aftertreatment of rinsed material; Matexil FC-PN, 0.5–3% owf; 20–40 °C; 20 min, increases the fastness of dyeings to water, perspiration, and mild washing, although[96] a small shade change may occur.

e. Resin Treatment and Cross-Linking Agents. Resin finishing is widely used to impart dimensional stability to cotton and viscose fabrics. This treatment also serves to improve the wet and washing fastness of direct dyes. The light fastness of the dyeing may, however, be reduced and the hue also affected by such treatments.

The Indosol SF (Sandoz) range of water-soluble, reactant-fixable (Classes B and B/C) direct dyes is fixed after dyeing using one or a combination of reactant fixatives such as Indosol CR liquid (Sandoz), which also cross-links the fiber and therefore imparts dimensional stability and crease recovery. The dyeings exhibit very good fastness to washing [typically ISO CO3 (1/1 SD) 4 to 4–5 on both cotton and viscose] and wet treatments [alkaline perspiration (1/1 SD) 4–5 to 5 cotton and 4 to 4–5 viscose] and satisfactory light fastness [e.g., ISO BO2 (1/1 SD) 5 to 5–6 cotton and 5 to 6 viscose].

The dyes are pasted with hot (50 °C) water to which is then added boiling water; the solution may require brief boliing. Scheme 27 shows a typical exhaustion (e.g., winch) application procedure.

Pad–batch, pad–stream, and pad–jig processes may also be employed.

Dyeing

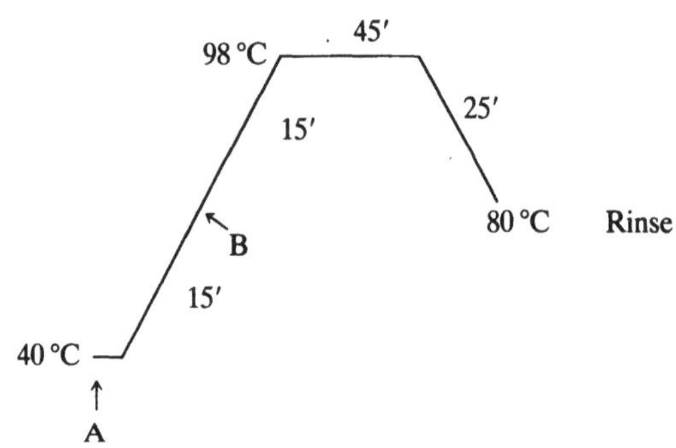

A. Indosol SF (Sandoz) dye	x% owf
B. Sodium chloride or Sodium sulfate (anhydrous)	0–5 g l^{-1} in several portions

Aftertreatment

Pad	25 °C 70–100% pick-up Indosol CR (Sandoz) Acetic acid to pH 4–4.5 Sandozin NIT (Sandoz) (wetting atent)	 60–13 g l^{-1} 1 g l^{-1}
↓ Dry	to 3–6% moisture content	
↓ Fix	30 s 180 °C	

An exhaust method can also be used for aftertreatment of pale depth dyeings.

Scheme 27

B. VAT DYES

These are water-insoluble dyes which contain at least two conjugated carbonyl groups (>C=O) that enable the dyes to be converted, by means of reduction under alkaline conditions, into the corresponding, water-soluble, ionized, "leuco compound;" it is in this form that the dye is adsorbed by the substrate. Subsequent oxidation of the leuco compound *in situ* regenerates the parent, insoluble vat dye within the fiber. This can be represented simply as follows:

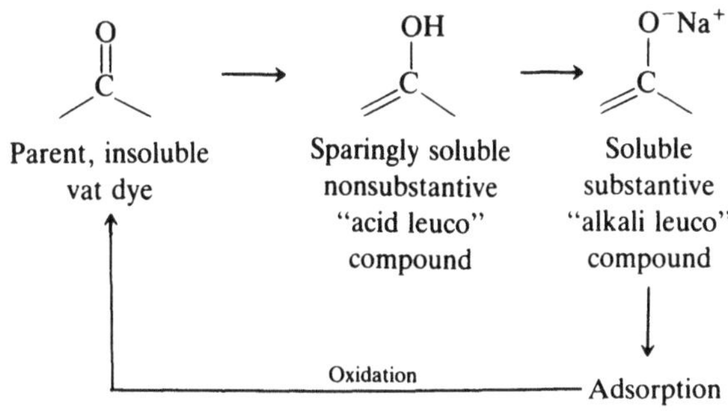

The dyed material is then "soaped" in order to develop the true hue and optimum fastness properties of the dyeing.

The principal use of vat dyes, which are sold as powders, granules, and liquids, is in the dyeing of cellulosic fibers on which they provide a very wide range of hues that, generally, are of outstanding all-round fastness. Selected vat dyes also yield pale to medium depths on polyester fibers.

There are several chemical classes of vat dyes.

1. Indigoid and Thioindigoid

Vat dyeing has been practiced for many centuries using natural coloring matters of both vegetable (e.g., Indigo, Woad) and animal (e.g., Tyrian Purple) origin. The most important of these natural vat dyes is Indigo or Indigotin **(47)** found as the glucoside Indican in various species of the Indigo plant (Indigofera) and obtained by hydrolysis of the glucoside and subsequent oxidation.[97] The use of natural Indigo, which consists of Indigotin together with small amounts of related substances, on cotton, linen, and wool, gained considerable impetus from the commercial introduction of the reducing agent sodium hydrosulfite (dithionite) discovered by Schutenberger in 1871.[98] Following Baeyer's work on the

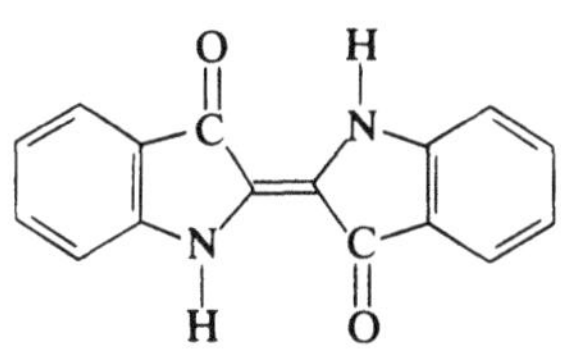

Indigotin

47

structure and synthesis of Indigo BASF introduced synthetic Indigo (CI Vat Blue 1) in 1897 which quickly replaced the natural product.

Derivatives of Indigo, mostly halogenated (especially bromo substituents), provide other vat dyes of this class, for example, CI Vat Blue 5 **(48)**. The introduction of CI Vat Red 41 in 1905 led to production of a large number of thioindigoid vat dyes which provide mostly red and orange hues, a popular example being CI Vat Orange 5 **(49)**.

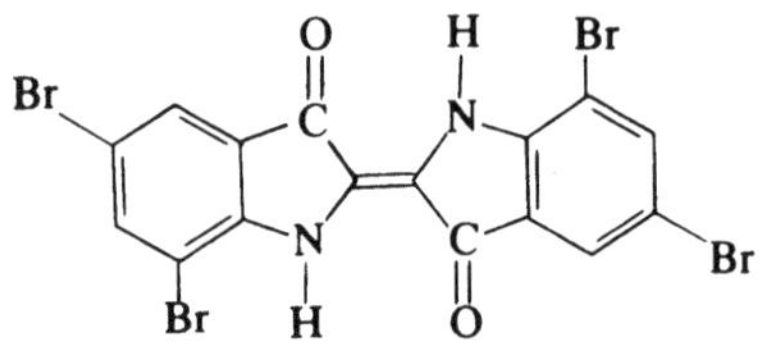

48

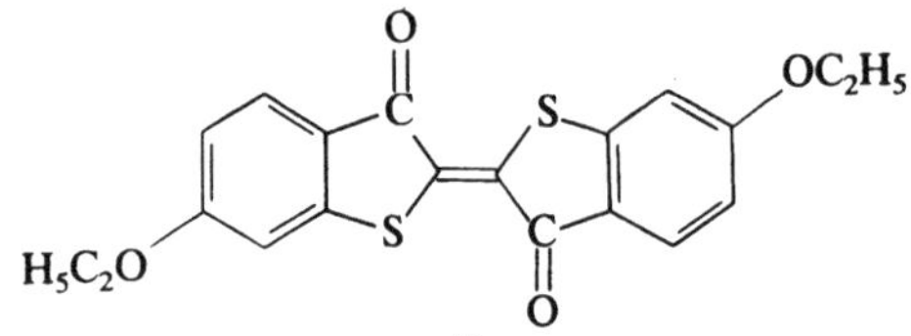

49

2. Anthraquinone

This largest and most important class of vat dyes can be further subdivided.

a. Indanthrone. This group of blue dyes, typified by CI Vat Blue 6 **(50)**, is of historical importance since it contains the first synthetic AQ vat dye CI Vat Blue 4, discovered by Bohn in 1901. Characteristically, these dyes display excellent fastness to light and wet treatments on cotton but are prone to over-reduction and over-oxidation.

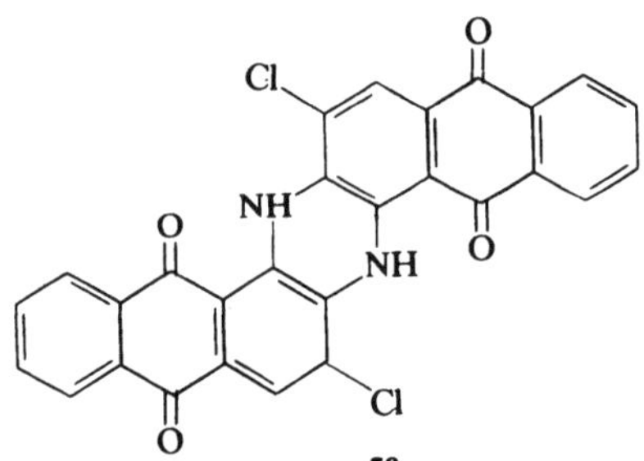

50

b. Flavanthrones. There are few dyes of this type, the most important being CI Vat Yellow 1 **(51)** (Flavanthrone itself). The dyes are characterized by slow oxidation rates and lower wash fastness.

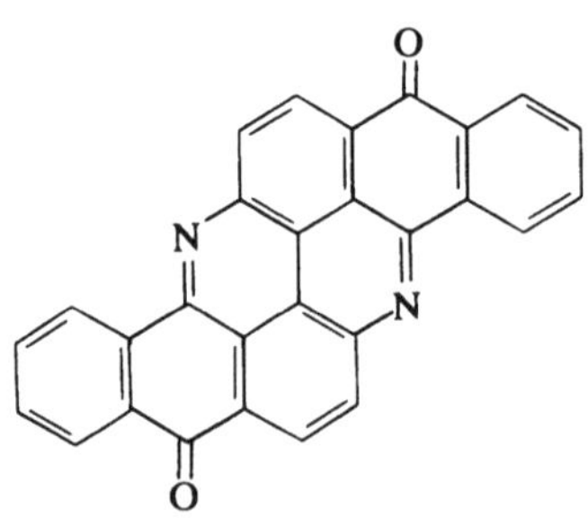

51

c. Pyranthrones. On cotton, dyeings of these mostly orange dyes such as CI Vat Orange 7 **(52)** possess poor light fastness and are prone to phototendering. Also, the dyes tend to cause fiber damage during dyeing.

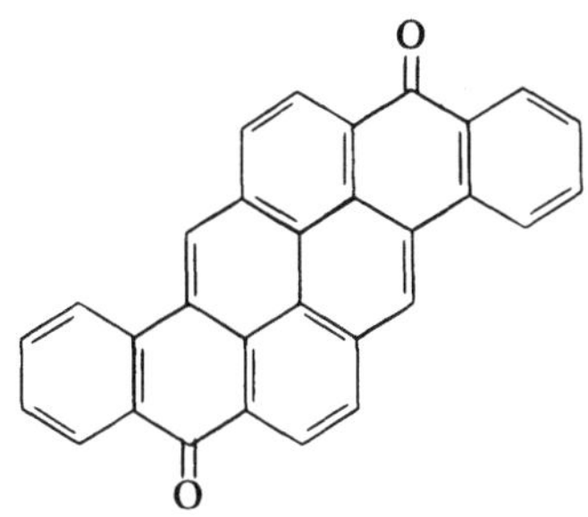

52

d. Acylaminoanthraquinones. This fairly small group contains mostly yellow and orange dyes, such as CI Vat Yellow 26 **(53)**.

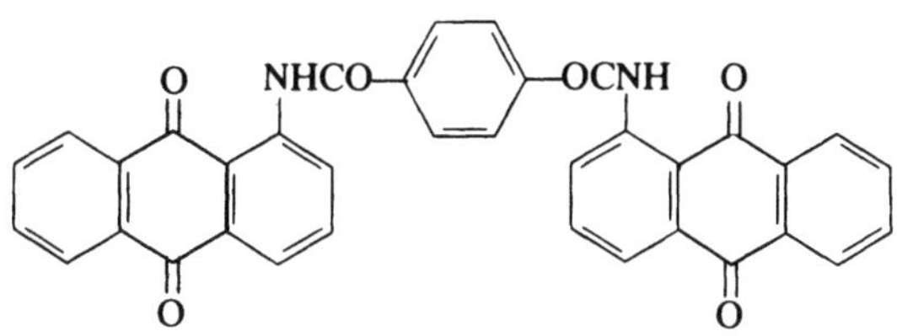

53

e. Carbazoles. These dyes, which provide a wide range of hues, have outstanding all-round fastness on cotton fibers, an example being CI Vat Green 8 **(54)**.

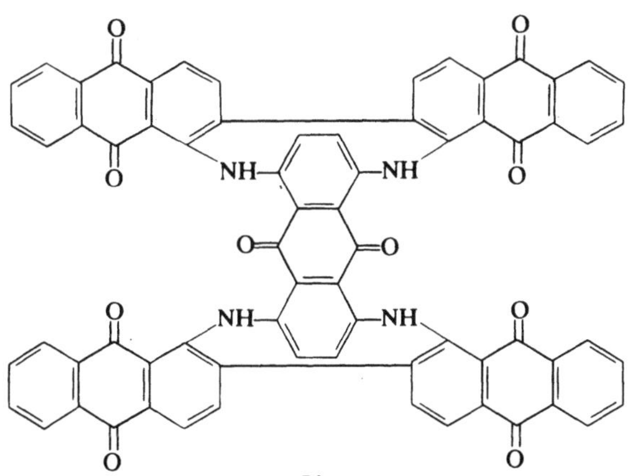

54

f. Anthrimides. These mostly yellow and red dyes, as exemplified by CI Vat Red 48 **(55)**, possess good all-round fastness on cotton but are prone to phototendering.

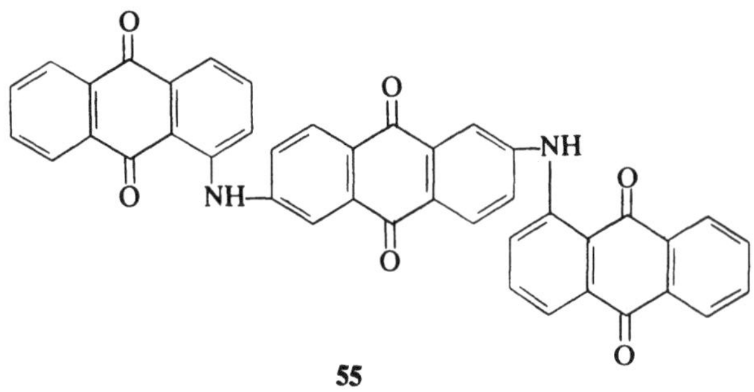

55

g. Dibenzanthrones. This large group of predominantly dark blue, green, and black dyes, such as CI Vat Green 1 (**56**), display very good light and wet fastness properties on cotton.

There are several other chemical classes which provide a relatively small number of commercial dyes.[97,99,100]

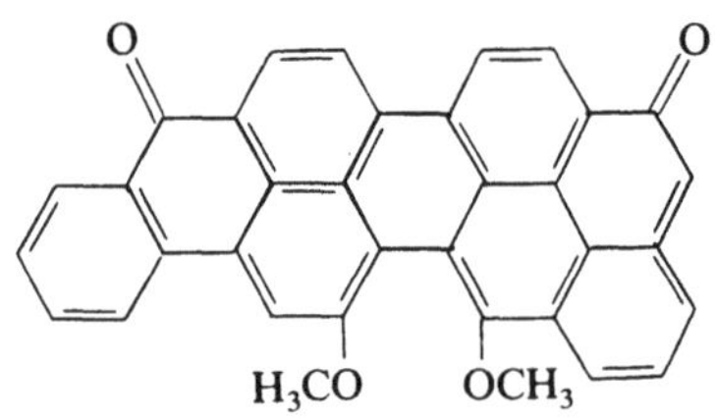

56

3. Dyeing Behavior

Vat dyes, such as the Solanthrene (ICI) and Indanthrene (BASF) ranges, yield dyeings on cellulosic fibers of an overall higher standard of fastness than any other dye class. Typically, fastness to washing on both cotton and viscose (1/1 SD) is 5 (ISO CO5) and light fastness (ISO BO2) cotton 6–7 1/1 and 1/10 SD; viscose 7–8 1/1 SD, 6–7 1/10 SD.

Vat dyeing of cotton can be carried out at various stages of processing, namely, as loose stock, sliver, yarn, and woven and knitted fabrics. Because of the high alkali-sensitivity of viscose, care must be exercised to minimize fiber damage. Consequently, high dyeing temperatures (i.e., >80 °C) should whenever possible be avoided. Furthermore, owing to the fiber's strong swelling under aqueous alkaline conditions, there is a high risk of obtaining poor dye penetration and thus unlevel dyeing in tightly packed materials because of reduced liquor flow. As a result, viscose is rarely dyed in loose form and the fiber is most commonly dyed as yarn, in hank and package forms, and as woven fabric.

The dyes can be applied to cellulosic materials using batchwise, semicontinuous, and continuous processes. Several dyeing methods are employed.

a. Leuco Dyeing. The material is dyed using a solution of the leuco compound. This traditional method is suitable for virtually all forms of material.

b. Pigmentation.

Prepigmentation. The unreduced dye is distributed throughout the material and subsequently converted (reduced) into the leuco compound within the fiber. This technique is used for dyeing sliver, yarn in package form, and for woven and knitted goods.

Semipigmentation. In this method, employed for dyeing yarn in both hank and package forms, together with woven and knitted goods, the unreduced dye is gradually reduced to the leuco compound during the course of dyeing.

The different dyeing methods arise from the versatility of vat dyes and the difficulties often encountered in securing adequate penetration and a high standard of levelness on some materials.

The physical characteristics of vat dyes are of considerable importance with regard to their application. To facilitate ease of dispersion, the hydrophobic dye particles must readily disperse and aggregates of particles break down to give the desired degree of dispersion. Dispersion stability is also important, the degree of stability required depending on the particular dyeing process used. Particle size and particle size distribution are of significance in this context. The dyes are commonly milled to a particle size of well below 1 μm in the presence of a dispersing agent[101,102] and the dyes, which are available in powder, liquid, and granular forms to suit particular methods of application, usually have a high standard of particle size distribution. Dispersing agents such as Setamol WS (BASF), added to the liquor prior to addition of the dye, maintain undissolved dye particles produced by oxidation or by lack of reducing agent in fine dispersion.

c. Vatting. This is the process of converting the water-insoluble vat dye into the water-soluble leuco compound by the action of a reducing agent, commonly sodium dithionite, in the presence of alkali, usually sodium hydroxide. The leuco compound is usually of a different hue from the oxidized form: indigoid dyes normally yield pale yellow or brown leuco solutions while leuco solutions of anthraquinonoid dyes are usually of an intense and different hue from that of the oxidized form.

The effective reducing agent is most probably the dithionite radical $\cdot SO_2^{2-}$. Anhydrous sodium dithionite, commonly referred to as hydrosulfite or "hydros," is stable but, on contact with water, is readily decomposed to sodium thiosulfate and sodium bisulfite (equation 2).

$$2Na_2S_2O_4 + H_2O \rightarrow Na_2S_2O_3 + 2NaHSO_3 \qquad (2)$$

These acidic products accelerate this exothermic hydrolytic reaction which can result in spontaneous ignition.

In the absence of air, alkaline solutions of dithionite are very stable. Dithionite is very sensitive to oxygen, however, and oxidative decomposition occurs readily in alkaline solution (equation 3).

$$Na_2S_2O_4 + 2NaOH + O_2 \rightarrow Na_2SO_3 + Na_2SO_4 + H_2O \qquad (3)$$

The quantity of dithionite necessary for vatting is therefore determined by that required for the particular dyes, together with an excess to compensate for oxidation of the reagent, the amount of which depends on the application conditions in use, such as surface area of the liquor, extent of exposure of the impregnated fiber to atmospheric oxygen, and initial dithionite concentration. In some instances the reducing agent is consumed by the fiber.

An excess of sodium hydroxide must also be used since alkali is consumed both in the vatting process and in the oxidation of dithionite. The amount of sodium hydroxide required therefore depends on the dye used and the extent of oxidation. A pH of between 12 and 14 is normally employed to prevent precipitation of the sparingly-soluble acid leuco compound or other products which can give off-shade dyeings and reduced rubbing fastness.

The rate of vatting is very important especially in continuous dyeing (e.g., pad–steam) processes where the time available for reduction is small. Vatting rate depends on: crystal habit, particle size, particle size distribution of the dye, and the temperature and concentration of both dye and dithionite.[101]

Different rates of reduction are reflected in dye makers' recommendations for vatting different dyes and are important in determining the suitability of a given dye for a particular application method.

Some leuco compounds are prone to over-reduction, which results in low color yield and in off-shade dyeings. The problem can be prevented by the use of reduction inhibitors or redox buffers such as sodium nitrite (10–15% on weight of dithionite) for temperatures of up to 80 °C. In leuco dyeing, the nitrite is added when vatting is complete, while in prepigmentation dyeing it is added immediately after the alkali and dithionite. At dyeing temperatures in excess of 90 °C, glucose (80%; $2\,g\,l^{-1}$) can be used.[103]

d. Dyeing. The substantivity of leuco compounds toward cellulosic fibers can be attributed to ion–dipole and dispersion forces operating between the dye anion and the substrate. Owing to the very high electrolyte (reducing agent and alkali) content of the dyeing system, the substantivity of leuco dyes is very high and consequently the dye anions exhaust very rapidly even at relatively low temperatures. In solution, leuco compounds are present in monomolecular form or as dimers or, to a lesser extent, as higher aggregates. The rate of diffusion of the dye anions within the substrate is characteristically very low because of their large molecular size; this is most pronounced in the case of aggregates and conditions of high substantivity. The combination of these factors results in poor dye penetration and unlevel dyeing.

Retarding or leveling agents are often used to reduce the rate of exhaustion of the dye anion and thereby promote level dyeing. The retarding agent, such as Matexil DN-VL 200 (ICI) and Peregal P (BASF), forms a complex with the leuco compound that is less substantive toward the fiber than is the original dye anion. The quantity of retarding agent required depends primarily on the substantivity of the particular leuco compound and is specified within the dye manufacturer's literature. An excess of retarding agent can result in weak dyeings and, in some cases, aggregation of the dye anion.

Leuco compounds have a higher substantivity for viscose fibers than for cotton and there is therefore a greater tendency for unlevel dyeing. This can be compensated for by employing a larger amount of retarding agent. Sodium sulfate is used to increase the substantivity of leuco compound toward cellulosic fibers; sodium chloride is rarely used since the impurities normally present (i.e., Mg^{2+} salts) cause formation of salts of leuco compounds of low water-solubility.

At high temperatures (85–90 °C) the dye anions are able to migrate so that level dyeing and penetration are improved. Certain leuco compounds, however, are unstable at such high temperatures, especially with prolonged dyeing times, and yield off-shade dyeings. With some temperature-sensitive dyes this over-reduction can be offset by the use of glucose.

Rongal HT (BASF) is used as reducing agent in leuco dyeing methods at temperatures above 90 °C; up to 115 °C is possible. This agent causes some fiber bleaching, especially at elevated temperatures, which is advantageous for brightening shades. The technique is suitable for those dyes that are stable to the high temperatures used. A reduction inhibitor (e.g., glucose) may be required for dyes that are prone to over-reduction. Dyes initially applied using the prepigmentation process generally exhibit greater stability to the high temperatures used in the reduction stage, although some dyes require the addition of a redox buffer to the dye liquor.

4. Dyeing Auxiliaries

Sequestering agents such as Trilon TB (BASF) combine with the alkaline-earth (Ca^{2+}, Mg^{2+}) and heavy metal (Fe^{2+}, Fe^{3+}) ions present in the water and leached from the cotton, thus preventing the formation of sparingly-soluble salts of the leuco compounds which otherwise result in weak shades of low fastness to washing and rubbing.

Protective colloids are also employed, such as Dekol 8 (BASF), which binds calcium ions and prevents the formation of sparingly-soluble salts of leuco compounds.

Wetting agents such as Primasol FP (BASF) are used to enhance wetting of the material.

5. Oxidation

After rinsing to remove loosely adherent dye and most of the residual alkali and reducing agent, the dyeing is oxidized to regenerate the parent, insoluble vat dye *in situ* in the fiber. Although air is the cheapest and most plentiful oxidant, it is slow in action and unreliable because side reactions can occur leading to off-shade dyeings. More rapid and reproducible chemical oxidation is therefore preferred by means of dilute aqueous solutions of hydrogen peroxide, sodium perborate, or ammonium persulfate. Typical batchwise conditions are shown below:

Hydrogen peroxide (130 vol.)	0.5–1.0 $g l^{-1}$
Acetic acid (80%)	0.5–1.0 $g l^{-1}$
50 °C	
15–20 min	
Sodium perborate	1–2 $g l^{-1}$
50 °C	
15–20 min	

6. Soaping

Following oxidation, the dyeing is "soaped" in a dilute aqueous detergent solution (e.g., 2 $g l^{-1}$) for 15–20 min at the boil and then finally rinsed. The soaping treatment has a threefold function: to remove any surface deposited dye which would otherwise impair the rubbing fastness of the dyeing; to produce the permanent, true hue of the dyeing—were this not done then subsequent laundering would produce a modification of shade; and to develop maximum light and washing fastness properties of the dyeing in some cases. Thorough soaping is therefore an essential stage in vat dyeing.

The precise mechanism of soaping remains unclear. Leuco dye anions are considered to be adsorbed in monomolecular form oriented mainly parallel to the longitudinal fiber axis. After oxidation the insoluble dye will be similarly disposed. It is suggested[65,104,105] that, upon soaping, the dye molecules associate and form crystals within which the dye molecules are oriented mainly perpendicular to the fiber axis. An alternative view[106,107] is that the changes which accompany soaping are due to stereoisomerism effects; during oxidation the metastable form of the dye crystal is formed, but this is then converted into the stable form during soaping.

Some vat dyes show only a small change in shade upon soaping and are therefore most suitable for continuous dyeing in which only a small soaping time is available.

7. Acidification

After soaping, a treatment with dilute acetic acid solution (e.g., 30%; 1–2 $ml l^{-1}$; 5–10 min at 20–25 °C) may be given to neutralize alkali residues in the dyed material.

8. Dispersing the Dyes

Powder and grain forms are sprinkled into water (e.g., approximately 10–20 times their volume) at 30–50 °C with stirring. Concentrated powder forms (e.g., Indanthren N dyes), which are not suitable for pigmentation processes, are pasted with water at about 50 °C with the addition of a wetting agent and the paste then diluted with water at 50–60 °C to the desired volume. These dyes are usually vatted using the stock vat method described below. Liquid dyes do not require dispersing and are poured with stirring into cold or warm water.

9. Leuco Dyeing

This is a two-stage process in which the dye is first vatted (reduced) to produce the leuco compound and the material then dyed using this alkaline leuco solution.

The rate of dyeing in this exhaustion process increases with increasing temperature, and with decreasing liquor ratio and dye concentration; dithionite or alkali have little effect upon dyeing rate.

In the *Colour Index*, vat dyes are classified into three groups according to the method of their application. Many dye manufacturers have developed their own system of classification. BASF, for example, divide their Indanthren dyes into five groups with similar dyeing behavior.

- *IK* dyes have low substantivity for cellulosic materials and are applied at low temperature (20–25 °C) using a small amount of sodium hydroxide but a large amount of electrolyte to promote substantivity.
- *IW* dyes have higher substantivity and are applied at 45–50 °C using more alkali but less electrolyte. Viscose fibers are dyed without electrolyte owing to the higher substantivity of the dyes toward these fibers.
- *IN* dyes are highly substantive and, although requiring greater amounts of alkali, are applied at 60 °C without electrolyte.
- *IN Special* dyes require more sodium hydroxide than the IN dyes, but are dyed at 60 °C without salt.
- Dyes that require a special dyeing method.

Although the Indanthren range contains no IK dyes, some IW dyes can be applied using this method.

The quantities of dithionite, alkali, and sodium sulfate required for dyeing are provided in tabular or graphical forms by the dye maker, and vary according to the type and amount of dye applied and liquor ratio used. Typical values for the Indanthren IW dyes are given in Table 6.

a. Vatting. Two methods are used.

Vatting in Long Liquor. This preferred method entails vatting the dye in the bath in which subsequent dyeing is carried out, commonly at the temperature used for dyeing. The dispersed dye is sieved into the bath which contains the

Table 6. Additions of Chemicals for IW Dyeing Method[103]

	1–3% dye		3–5% dye	
Sodium dithionite ($g\,l^{-1}$)	3	3–5	3–4	5–7
Sodium hydroxide (30%) ($ml\,l^{-1}$)	6–7	9–12	7–9	12–15
Glauber's salt ($g\,l^{-1}$)	10–15	10–15	15–25	15–20

Table 7. BASF Stock Vat Recipes[103]

	I	II	II
Sodium dithionite (g)[a]	0.5	0.75	1.5
Sodium hydroxide (30%) (ml)[a]	1.5	3.0	6.0
Water (ml)[a]	50	50	100
Vatting temperature (°C)	50	60	60
Vatting time (min)	10–15	10–15	10–15

[a] Per gram dye used.

appropriate amounts of dispersing agent, protective colloid, electrolyte, sequestering agents, and leveling agents set at the correct temperature for dyeing. The amounts of alkali and reducing agent required for dyeing are then added and the liquor left to stand for 5 to 10 min to effect vatting.

Vatting in the Stock Vat. A concentrated (i.e., stock) solution of the leuco compound is prepared by dissolving the vat dye in a proportion of the total amount of alkali and reducing agent used for dyeing. The quantity of dithionite and alkali used for preparation of the stock vat is specified by the dye manufacturer. BASF have three recipes for stock vatting according to the solubility of their Indanthren dyes. Table 7 shows the chemical requirements of the Indanthren Colloisal form dyes. After vatting, the stock vat is added to the dyebath, often in portions, that is set with alkali and reducing agent as recommended for the particular dyes.

b. Dyeing. The dye liquor must contain sufficient dithionite throughout the dyeing process. This is easily determined by using Caledon or Indanthren Yellow paper which turns blue within three seconds if sufficient reducing agent is present. Phenolphthalein paper can be used to check whether sufficient sodium hydroxide is present, the paper turning red immediately if there is a sufficiency of alkali.

The temperature at which dyeing is carried out depends mainly on the form of material being dyed and the dyes used. For example, loose stock and sliver can be dyed at 50–60 °C and yarn in package form a 80 °C to promote levelness. The procedure shown in Scheme 28 is for the dyeing of cotton in package form using a liquor ratio of 10:1.

Dyeing can be carried out at temperatures of up to 115 °C in which case Rongal HT (BASF) should be used as reducing agent; for dyes that are sensitive to overreduction, glucose is added to the dyebath.

10. Pigmentation Processes

These techniques utilize the low substantivity of the unreduced or vat dye to achieve uniform distribution throughout the material, thereby overcoming the problems of poor penetration and unlevel dyeing which often arise as a result of the very high substantivity of leuco compounds for the cellulosic substrate. A very high standard of particle size distribution is necessary for these processes.

	10′	30′	15–30′	
50 °C				Rinse
	↑ A	↑ B	↑ C	Aftertreat

Stock vat preparation

Solanthrene (ICI) dye (1% owf) dispersed in water at 50 °C to which is added:

NaOH (38 Be)	6.25 $ml\,l^{-1}$
$Na_2S_2O_4$	2.5 $g\,l^{-1}$

Vatting is continued for 10 min at 50–60 °C.

Dyeing (liquor ratio 10:1)

A. NaOH (38 Be) 3.75 $ml\,l^{-1}$
   $Na_2S_2O_4$ 1.5 $g\,l^{-1}$

B. Leuco dye added in 2–4 portions

C. Sodium sulfate (anhydrous) 8 $g\,l^{-1}$ in portions (predissolved).

Scheme 28

a. Prepigmentation Processes. An aqueous dispersion of the unreduced vat dye is evenly distributed within the material and is subsequently reduced *in situ.*

Batchwise Processes. A typical process for dyeing fabric is described in Scheme 29. The quantities of chemicals and auxiliaries are obtained from the dye manufacturers literature. Aftertreatment is carried out as for leuco dyeing using the batchwise method.

The bath is filled to approximately 70% of the final volume required. Half the quantity of Solanthrene (ICI) dye to be used, dispersed in water, is added and the bath heated to 80 °C. Dyeing is carried out for one end. The remaining quantity of Solanthrene dye, dispersed in water, is added and the fabric run for one end at 80 °C. The bath is made up to the final volume required using cold water, the temperature adjusted to 60 °C, and two-thirds of the required[a] amounts of sodium dithionite and sodium hydroxide added. The fabric is run for one end at this temperature. The remaining quantities[a] of sodium hydroxide and sodium dithionite are added, and the fabric run for one end at 60 °C. Further ends are run to obtain the required depth of shade; the dyed fabric is then rinsed and aftertreated.

[a] Obtained from dye makers' literature.

Scheme 29

Semicontinuous Processes. Pad–Jig Develop. This method permits a higher standard of levelness and penetration to be obtained than is achieved using the traditional leuco dyeing method. A typical process is shown in Scheme 30. Insoluble dye is removed mechanically from the padded material during the initial stages of jig development. A small quantity of pad liquor is therefore added to the development liquor and allowed to "vat" prior to the start of development. In

Pad	20–30 °C	
	Indanthren (BASF) dye	x g l^{-1}
	Primasol FP (BASF)	1–10 g l^{-1}
	Primasol AMK (BASF)[a]	1–10 g l^{-1}
	60–80% pick up	
↓		
Dry (optional)		
↓		
Jig Develop	50–80 °C	
	Setamol WS (BASF)	1–2 g l^{-1}
	Dekol S (BASF)	0.5–1.5 g l^{-1}
	Pad liquor	Y ml l^{-1}
	NaOH (38 Be)[b]	A ml l^{-1}
	$Na_2S_2O_4$[b]	B g l^{-1}
	Peregal P (BASF)	1–2 g l^{-1}
↓		
Rinse		
↓		
Aftertreat		

[a] If intermediate drying is carried out.
[b] Calculated from tables.

Scheme 30

Pad	20–30 °C	
	Indanthren (BASF) dye	x g l^{-1}
	Primasol FP (BASF)	1–3 g l^{-1}
	Primasol AMK (BASF)	10 g l^{-1}
	Trilon TB powder (BASF)	1–2 g l^{-1}
	60–80% pick-up	
↓		
Dry (optional)	80–120 °C	
↓		
Pad	20–30 °C	
	NaOH (38° Be)[a]	45–75 ml l^{-1}
	$Na_2S_2O_4$[a]	30–50 g l^{-1}
	Pad liquor	20–30 ml l^{-1}
	100% pick-up	
↓		
Batch	2–4 h	
↓		
Rinse		
↓		
Aftertreat		

[a] From tables.

Scheme 31

this manner, an equilibrium is then established between the insoluble dye removed from the material and the leuco dye being adsorbed from the development liquor, so preventing the incidence of "ending" or "tailing." The cold Pad-Batch method is suitable for certain speciality goods. Scheme 31 shows the dyeing of cotton woven fabrics.

Continuous Processes. Pad–Steam. This is the most important method for dyeing cotton and viscose rayon fabrics and consists of the stages shown in Scheme 32. Liquid forms of vat dyes are preferred because of their low tendency to migrate during drying and, since they can be metered, provide a convenient way of preparing the large volume of dye liquor required. Dye migration during drying is also reduced by the incorporation of an inhibitor in the pad liquor. After drying, the fabric is cooled prior to chemical padding. The quantities of dithionite and caustic used depend on the amount of dye applied and on the processing conditions, since some dithionite will be consumed during the development (steaming) stage. These amounts are specified within the relevant literature. Steaming, using dry saturated steam, is carried out under "air-free" conditions to avoid premature oxidation of the reduced dye. The steamer is usually conditioned to minimize condensation marks on the material. Following steaming, the fabric is sprayed with water to reduce the alkali and reducing agent content.

Wet-Steam. The drying stage that follows impregnation of the fabric with the unreduced dye can be omitted to reduce energy costs and dye migration.

	20–30 °C	
	Solanthrene (ICI) dye	x g l^{-1}
Pad	Matexil WA-HS (ICI)	2–5 g l^{-1}
↓	Matexil FA-MIV (ICI)	10–20 g l^{-1}
	60–90% pick up	
Dry	Hot Flue	
↓		
Cool		
↓	20–30 °C	
	80–110% pick-up	
Pad	Pad liquor	10–30 ml l^{-1}
↓	NaOH (38 Be)[a]	A ml l^{-1}
	$Na_2S_2O_4$[a]	B g l^{-1}
Steam	102–104 °C; 30–40 s	
↓		
Rinse		

Aftertreat continuously

[a] Calculated from dye makers' literature.

Scheme 32

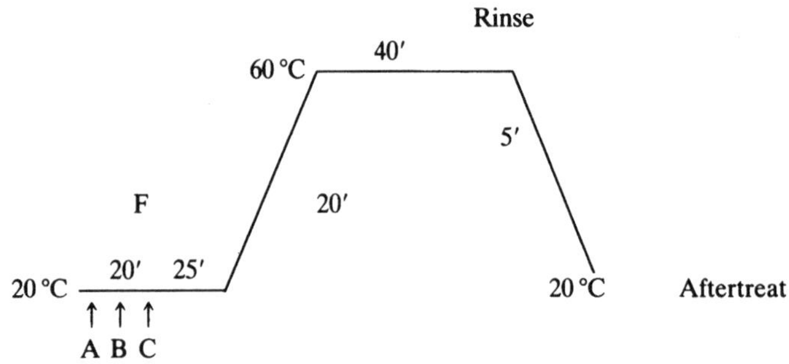

A.	Dekol S (BASF)	$2\ g l^{-1}$
	Peregal P (BASF)	$0.5\ g l^{-1}$
	Sodium sulfate (anhydrous)	$15\ g l^{-1}$
	NaOH (38° Bé)	$7\ ml l^{-1}$
B.	Indanthren (BASF) dye	3% owf
C.	NaOH (38° Be)	$5\ m l^{-1}$
	$Na_2S_2O_4$	$5\ g l^{-1}$

Liquor ratio 10 : 1

Scheme 33

b. Semipigmentation Process. This batchwise method utilizes the slow vatting rate of the dyes at low temperatures. The dispersed, unreduced dye, in the presence of reducing agent and alkali, is gradually converted into the leuco compound by means of temperature control. This technique gives better penetration and therefore more uniform dyeings than the leuco dyeing method, especially in the case of, for instance, tightly twisted yarns. A typical procedure is as shown in Scheme 33. The amounts of reducing agent and alkali necessary for vatting are the same as those employed in leuco dyeing.

Dyeing is commenced immediately at low temperature under which conditions the rate of vatting is very slow, and substantially unreduced dye becomes distributed throughout the material. As the temperature is carefully and slowly raised, the vatting rate increases and the resultant leuco compound is adsorbed onto the fiber.

The previously mentioned Rongal HT (BASF) method can also be used.

11. Solubilized Vat Dyes

In 1912 Bader and Sunder prepared CI Solubilized Vat Blue 1 **(57)**, a stable, water-soluble sulfuric acid ester derivative of leuco indigo, which could be applied to wool and cotton fibers and the parent insoluble indigo then regenerated by subsequent aqueous acidic oxidation.[97]

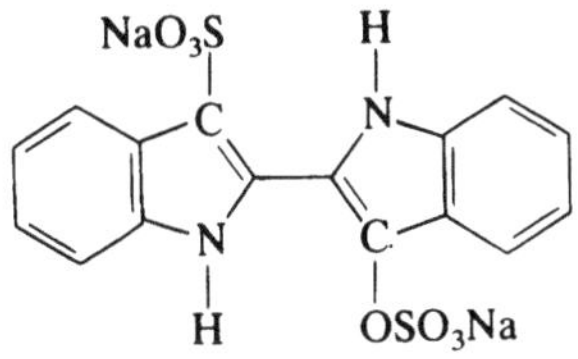

57

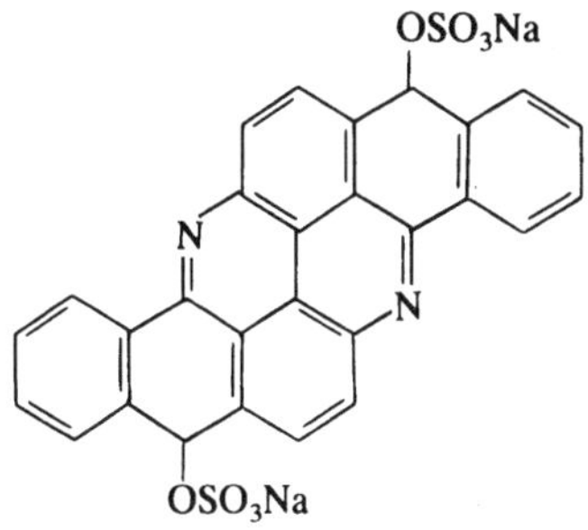

58

Stable, soluble sodium salts of the sulfuric ester of the leuco compounds of other vat dyes were subsequently introduced by various manufacturers. These highly water-soluble vat dyes, such as CI Solubilized Vat Yellow 1 **(58)**, obviate the need for reduction by the dyer.

The dyes, as represented by the Indigosol (Sandoz) and Anthrasol (H) ranges, are normally marketed in powder form and have high storage stability. They are mostly used for dyeing piece goods in pale shades, but are also employed for dyeing yarn in both hank and package forms. Dyeings exhibit very good all-round fastness properties.

Solubilized vat dyes are applied to cellulosic materials by batchwise and continuous processes. In essence, their application involves even distribution throughout the material followed by oxidation *in situ* to regenerate the parent, insoluble vat dye within the material. Subsequent soaping produces the permanent hue and maximum fastness properties of the dyeings. The dyes are applied from neutral or from slightly alkaline solution and possess relatively low substantivity toward cellulosic fibers. Consequently, the dyes exhibit a low rate of exhaustion that is controllable by electrolyte addition and therefore this type of dye offers advantages over conventional vat dyes with regard to uniform dyeing and penetration.

Oxidation is most commonly carried out using sulfuric acid and sodium nitrite. Following neutralization with aqueous sodium carbonate, the dyeing is soaped and finally rinsed.

The dyes vary markedly in dyeing behavior and in the conditions required for oxidation (development). Thus, the particular conditions employed for dyeing and development vary according to both the dye and the application process used. The following batchwise application method (Scheme 34) shows typical ranges of temperature, electrolyte concentration, etc., that may be used, detailed information being available in the dye makers' literature.

Unlike conventional vat dyes, the oxidation of solubilized vat dyes involve a chain reaction[118] that requires an excess of nitrous acid, formed by the action of sulfuric acid on sodium nitrite (equation 4).

$$NaNO_2 + H_2SO_4 \rightarrow HNO_2 + NaHSO_4 \quad (4)$$

The nitrous acid slowly hydrolyzes the leuco ester producing a small amount of

	10′	10′	10′	15′	15′	
30° C	↑ A	↑ B	↑ C	↑ D	↑ E	Rinse

A	Anthrasol (Hoechst) dye	x g l^{-1}
	Na_2CO_3	1 g l^{-1}
	Hostapal CV highly conc. (Hoechst)	0.25 g l^{-1}
B	Leonil DB (Hoechst)	1 g l^{-1}
C	Leonil DB (Hoechst)	2 g l^{-1}
D	Leonil DB (Hoechst)	3 g l^{-1}
	NaCl	25 g l^{-1}
E	$NaNO_2$	1 g l^{-1}
	H_2SO_4 (96%)	5 ml l^{-1}

Scheme 34

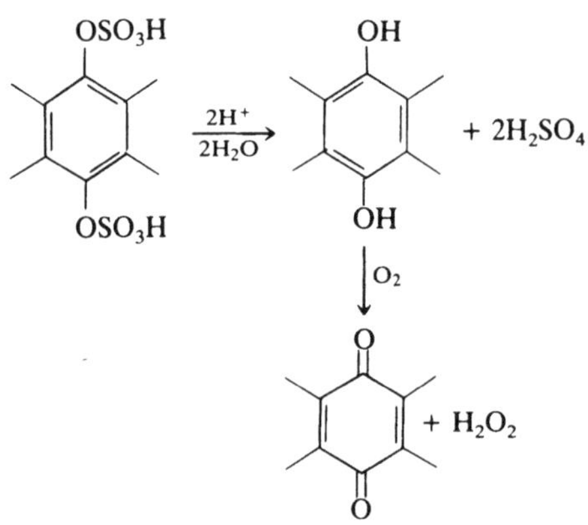

(5)

Scheme 35

the corresponding acid leuco compound which, in turn, is oxidized by air to the insoluble vat dye (equation 5; Scheme 35). The hydrogen peroxide produced by this reaction then reacts with nitrous acid initiating a chain reaction.

Pernitrous acid is formed (equation 6);

$$H_2O_2 + HO{\cdot}NO \rightarrow H_2O + HO{\cdot}ONO \qquad (6)$$

this dissociates to yield hydroxyl and nitrogen dioxide radicals (equation 7),

$$HO{\cdot}ONO \rightarrow HO{\cdot} + {\cdot}ONO \qquad (7)$$

the latter reacting with excess nitrous acid to form further pernitrous acid (equation 8)

$$HO{\cdot}NO + {\cdot}ONO \rightarrow HO{\cdot}ONO + NO \tag{8}$$

which, in turn, dissociates to yield more hydroxyl and nitrogen dioxide radicals, thus propagating the chain reaction. As a consequence, a large amount of hydroxyl radical are produced which oxidize the majority of the leuco ester (equation 9; Scheme 36).

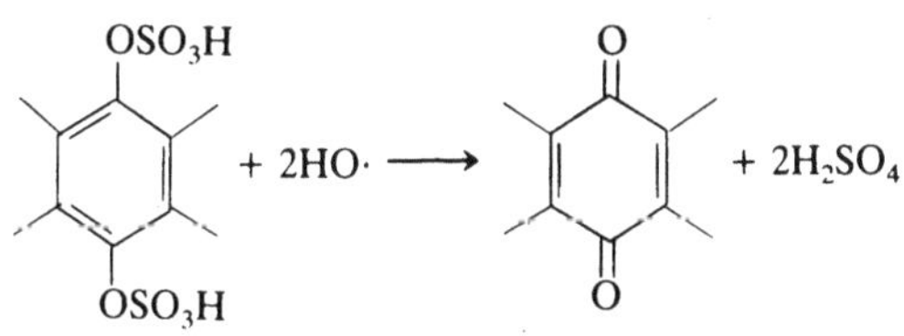

(9)

Scheme 36

12. Sulfurized Vat Dyes

This small, but nevertheless important, group of vat dyes resembles sulfur dyes in the method of their preparation (i.e., thionation), although the method of their application is that of vat dyes, namely, vatting using alkaline sodium dithionite. The first of these dyes, CI Vat Blue 43 **(59)**, was introduced in 1909 by Casella under the trade name Hydron Blue R; another popular example is CI Vat Blue 42 **(60)**.

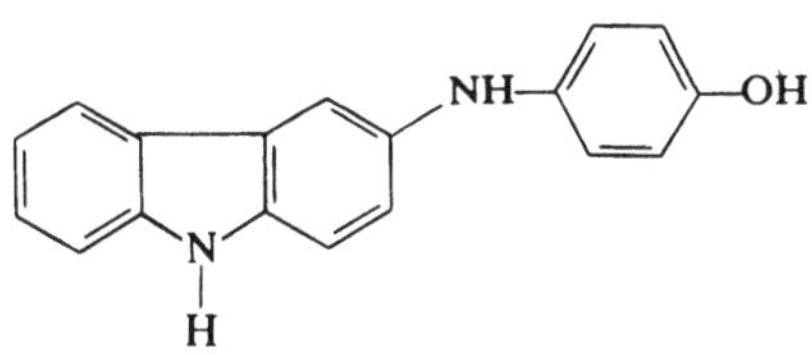

59

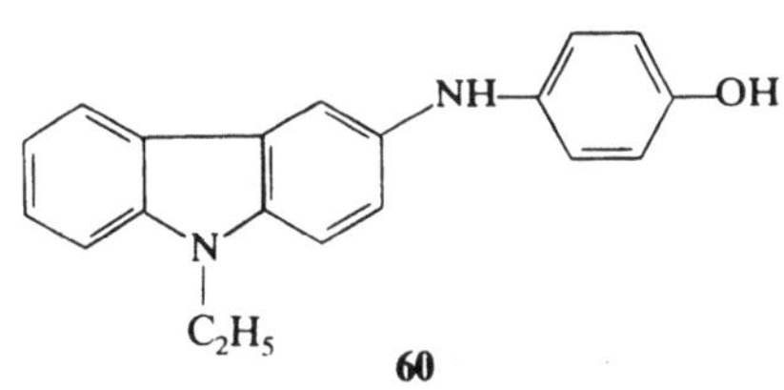

60

13. Dyeing with Indigo

Leuco indigo has only low substantivity for cellulosic fibers and consequently is applied, usually from a dithionite/sodium hydroxide vat, by a series of short "dips" with intermediate squeezing and air oxidation, so building up the desired depth of shade. Details of the dyeing of cotton in the form of warp yarn or woven fabric, which is more commonly carried out continuously, are given in the BASF Manual.[103]

C. SULFUR DYES

Sulfur dyes are used for the dyeing of cellulosic fibers in medium to deep shades of generally dull brown, black, olive, blue, green, maroon, and khaki hues.

Sulfur dyes are chemically complex and, for the main part, are of unknown structure, the majority of which are prepared by thionation of various aromatic intermediates.[108]

The first commercial sulfur dye, CI Sulfur Brown 1, marketed as Cachou de Laval, was prepared by Croissant and Bretonnière in 1873 by heating organic refuse with sodium sulfide or polysulfide.[109] However, Vidal in 1893 obtained the first dyes of this class from intermediates of known structure.[110] The first of these dyes was Vidal Black (CI Sulfur Black 3) (1893), prepared by heating *p*-phenylenediamine (and later *p*-aminophenol) with sulfur or with sodium polysulfide, which could be applied to cotton from a sodium sulfide bath and, subsequently, oxidized on the fiber using dichromate. The use of various intermediates.[108,111] such as indophenols, substituted naphthalenes, and phenols, has resulted in a large number of dyes.

In the third edition of the *Colour Index* the following subgroups are listed.[112]

1. CI Sulfur Dyes

These are water-insoluble dyes of low substantivity towards cellulosic fibers that are normally applied in the leuco (alkaline-reduced) form from a sodium sulfide bath and subsequently oxidized to the insoluble parent dye *in situ* in the fiber:

$$\text{D—S—S—D}' \xrightarrow[\text{NaOH}]{\text{Na}_2\text{S}} \text{D—S Na} + \text{Na S—D}'$$

Insoluble, low substantivity parent dye → Soluble and substantive leuco form

$$\downarrow$$

$$\text{D—S—S—D}'$$

In situ

The dyes are commonly sold as powders and also as dispersed powders or pastes containing dispersing agents. An important example of this subgroup is Sulfur Black T (CI Sulfur Black 1) first prepared in 1896.

2. CI Leuco Sulfur Dyes

These nonsubstantive dyes, as represented by the Sulfol (JR) range, are CI Sulfur dyes in soluble, prereduced (leuco) form suitable for application either directly or with the addition of extra reducing agent. The dyes are produced almost exclusively as liquid mixtures of the leuco dye and reducing agent, the latter commonly being sodium sulfide or sodium formaldehyde sulfoxylate in the case of powders and grains or sodium sulfide and/or sodium hydrosulfide in liquid brands.

3. CI Solubilized Sulfur Dyes

These, as typified by the Hydrosol (Hoechst) or Hydrosol (H) range, are highly water-soluble thiosulfonic acid derivatives of the parent sulfur dye ($DS{\cdot}SO_3Na$), and are sold in either powder or liquid form. They are applied to cellulosic fibers in the presence of alkali and reducing agent, during which process the nonsubstantive solubilized dye is converted into the substantive, alkali-soluble thiol form:

$$\underset{\text{Solubilized dye}}{DS{\cdot}SO_3} \rightarrow \underset{\text{Thiol}}{DS^-} \rightarrow \underset{\text{Parent dye}}{DSSO}$$

In each of the above three subgroups, the dye is applied to the fiber as the alkali-soluble leuco compound which is subsequently oxidized to the parent sulfur dye *in situ* in the fiber.

4. CI Condense Sulfur Dyes

These dyes are no longer manufactured[112] and consequently their application to cellulosic fibers will not be considered. The general characteristics of the dyes are discussed elsewhere.[65,113,114,118]

5. Dyeing Behavior

As mentioned above, this account considers only CI Sulfur, CI Leuco Sulfur, and CI Solubilized Sulfur dyes. For each of these three types of dye, it is in the soluble leuco form that the dye is adsorbed by the substrate. This leuco compound is produced by reduction under alkaline conditions and is possibly the sodium salt of the thiol derivative of the sulfur dye. Subsequent oxidation *in situ* regenerates the parent, insoluble dye within the fiber. Heavy depth dyeings are usually soaped to improve brightness of shade and fastness to washing.

Irrespective of subgroup, at the end of the dyeing process the sulfur dye is present in the fiber in the form of large, water-insoluble particles and, as a consequence, on cotton, sulfur dyeings that have been aftertreated with an acidified oxidant in general exhibit good fastness to washing in the presence of soap, typically 4–5 (ISO CO3), but are less resistant to domestic laundering, for example, 3–4 to 4–5 (ISO CO6/B2). Fastness of such dyeings to wet treatments is usually very good, for example, 4–5 to 5 (ISO EO4), and light fastness varies from moderate to good, typical values being (ISO BO2) 3 to 5–6 1/1 SD, 4 to 4–5 1/3 SD, some dyes having excellent fastness (i.e., 7 1/1 SD), although some exhibit poor light fastness in pale depths.

a. Reduction. Reduction of the water-insoluble sulfur dye to the leuco compound involves cleavage of disulfide (or disulfoxide) bonds within the sulfur dye molecule. The resulting alkali-soluble thiols are therefore of much lower molecular weight than the original sulfur dye.

Sodium sulfide (Na_2S) is the traditional reducing agent used in sulfur dyeing. Sodium hydrosulfide (NaHS) is widely used as an alternative to sulfide but must be used in conjunction with alkali, commonly sodium hydroxide or sodium carbonate, at a concentration of 5 g and 10 g, respectively, per 7 g of hydrosulfide.[115]

Sodium dithionite, when used together with sodium carbonate, is suitable for CI Solubilized Sulfur dyes and, in conjunction with sodium hydroxide, can be used for certain CI Sulfur dyes. Each of these dithionite reducing systems can in some cases give rise to over-reduction, resulting in low color yields or off-shades.

A major problem accompanying the use of sulfide-based reducing systems is that the discharge of dyehouse effluent containing sulfides is restricted on environmental grounds; oxidation by aeration is an effective treatment for such waste dye liquors.[109,115]

The use of glucose, in conjunction with sodium carbonate, sodium hydroxide, or both, as reducing agent is of increasing popularity because of environmental considerations. In the case of CI Solubilized Sulfur dyes the glucose reducing system may be used as the sole reducing agent. With liquid forms of CI Leuco Sulfur dyes the system may be employed as the additional reducing agent and with insoluble dyes in combination with sodium sulfide, thus effectively lowering the amount of sulfide employed for dyeing. In batchwise application a temperature of 90–95 °C must be maintained when using glucose as reducing agent.[115]

The quantity of reducing agent required for dyeing depends on the particular dye and amount used, liquor ratio, and application conditions; detailed information is provided in the manufacturer's literature. The agent must be in sufficient amount so as to reduce the dye and compensate for air present in the dyeing system. Sodium polysulfide inhibits the premature oxidation of sodium sulfide, sodium hydrosulfide and glucose during dyeing thereby lowering the risk of obtaining bronzy dyeings of low wet rubbing fastness. The amount of this antioxidant required varies according to dye, liquor ratio, and particular application conditions, typically of the order of 2–5 $g l^{-1}$.

b. Dyeing. The substantivity of leuco sulfur dyes toward cellulosic fibers, which is attributable to ion–dipole and dispersion forces operating between the dye anion and the substrate, is relatively low but is increased by the addition of sodium chloride or sodium sulfate. Substantivity varies markedly between dyes and differs according to substrate; the dye anions have greater substantivity for viscose rayon fibers than cotton.

The often slow rate of exhaustion of the leuco compound is also increased by electrolyte addition. As a consequence of the slow but controllable rate of exhaustion of the dye anions, satisfactory penetration and uniform dyeing can be achieved on most fiber constructions although precautions are necessary when dyeing yarn in package form.

c. Oxidation. This treatment regenerates the parent, insoluble sulfur dye within the fiber and, in pale to medium depths, produces the permanent hue of the dyeing. Chemical means of oxidation are normally employed although air is used in, for instance, the Pad–Sky dyeing process.

Loosely adhering dye must be removed by rinsing prior to oxidation to avoid bronziness and inferior wet rubbing fastness. The traditional oxidizing system is acidified sodium dichromate, typical conditions being[115]: Batchwise: $Na_2Cr_2O_7$, 1 gl^{-1}; CH_3COOH (80%), 0.8 gl^{-1}; 60 °C; 15–20 min. Continuous: $Na_2Cr_2O_7$, 5 gl^{-1}; CH_3COOH (80%), 6 gl^{-1}; 60–70 °C; 20–40 s.

The addition of copper sulfate (1 gl^{-1}) to the above batchwise oxidation liquor improves the light fastness of many sulfur dyeings, but the improvement diminishes as the metal is desorbed from the material during subsequent washing.

Since chromium is undesirable in effluent, several alternative oxidizing systems are in use.

Hydrogen peroxide under slightly acidic conditions is used, for example: H_2O_2 (35%) 1–2%; CH_3COOH (60%) 1–3%; pH 4–4.5; 70 °C; 15 min.

Other oxidizing agents have been proposed[109,115] including sodium bromate, the active constituent in several proprietary oxidizing agents such as Oxydant 584 (JR), potassium iodate/acetic acid, the sodium salt of *m*-nitrobenzenesulfonate and sodium chlorite, the active agent in Sandopur DSC (Sandoz).

The fastness of sulfur dyeings to washing in the presence of peroxy compounds can be markedly improved by alkylating the unoxidized dyeing with epichlorohydrin-based agents under alkaline conditions; subsequent oxidation of the alkylated material is often not required.[109,115]

d. Soaping. This stage is not essential for pale to medium depths of shade, in which cases thorough rinsing of the oxidized dyeing is normally sufficient. With heavy depths, soaping brightens and improves the wash fastness of the dyeing. Suitable conditions are[115]: Batchwise: detergent, 1–3 gl^{-1}; Na_2CO_3, 2 gl^{-1}; 90–95 °C; 15–20 min. Continuous: detergent, 5–10 gl^{-1}; Na_2CO_3, 2–5 gl^{-1}; 98 °C; 30–60 s.

e. Dissolving the Dyes. Liquid dyes are poured into about 50–70% of the final volume of water required for dyeing, which can be set at the dyeing temperature, to which has been previously added any alkali and reducing agent as required. Wetting agent is then added and the pad or dye bath then adjusted to full volume.

Insoluble dyes are pasted with water and wetting agent, and the desired amount of reducing agent and water to give a dye concentration of 20–30 gl^{-1} added. The stirred liquor is boiled and allowed to simmer for 2–5 min with occasional stirring and then diluted to the required volume for dyeing. Alternatively, the required amount of sodium hydroxide is added to the concentrated (i.e., 20–30 gl^{-1}) dye liquor which is then heated to 60–70 °C and the required amount of sodium dithionite added. Vatting is carried out for 10–15 min at this temperature and the ensuing leuco solution added to the "sharpened" dye bath, i.e., set with a small amount of dithionite and caustic.

CI Solubilized Sulfur dyes are sprinkled onto warm water containing wetting agent, and the liquor boiled and then simmered for 2–3 min before being diluted to full volume.

Dispersed powders or pastes are sprinkled, with stirring, into warm water which may contain a dispersing agent.

The application of sulfur dyes to cellulosic fibers has been discussed comprehensively by Senior and Clarke.[115]

6. Batchwise Dyeing Processes

Cotton yarn is dyed in package form. CI Solubilized Sulfur dyes are often applied in their nonsubstantive, unreduced form to achieve even penetration of the material, followed by reduction *in situ*. Liquid forms of CI Leuco Sulfur dyes are also used, the dye being added in portions to the circulating bath. Electrolyte addition is made gradually at the appropriate dyeing temperature (75–90 °C) to facilitate level dyeing.

Batchwise dyeing of woven fabrics may be carried out on the jig using liquid forms of prereduced dyes. Knitted fabrics are dyed using jig, winch, and jet dyeing machines, CI Leuco and CI Solubilized Sulfur dyes being preferred. The following procedure (Scheme 37) is typical for winch dyeing.

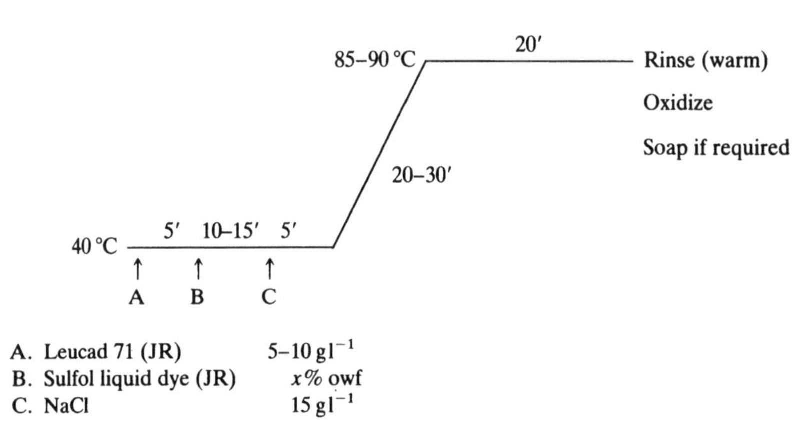

Scheme 37

7. Semicontinuous Processes

A common method for dyeing woven fabrics is the Pad–Jig process using CI Solubilized or dispersed CI Sulfur dyes; alternative application methods are Pad–Batch and Pad–Roll.[115]

8. Continuous Processes

These are not only the most popular methods for dyeing woven fabric but are also the most widely used of all processes for the application of sulfur dyes.

Several variants are used, a common example being the Pad-Steam process; typical conditions for dyeing cotton are shown below.

Pad	30–40 °C 80% Pick-up Sulfol Liquid Dye (JR), *x*% owf Matexil WA-HS (ICI), 2 ml l^{-1} (wetting agent) Leucad 71 (JR), 5–10 ml l^{-1} (NaHS)
↓ Steam	102–104 °C 60 s Air-free steam
↓ Rinse	Water 40 → 50 → 60 °C C
↓ Oxidize	Oxydant 584 (JR), 4.5–5 ml l^{-1} Acetic acid (80%), 7–9 ml l^{-1} 60–70 °C; 15–20 s
↓ Rinse	Water, 60 °C
↓ Soap	90–95 °C
↓ Rinse	Water, 60 °C

D. AZOIC COLORANTS

An azoic coloring matter is a water-insoluble azo compound (predominantly monazo and a few disazo) produced *in situ* in textile fibers by the interaction of a diazo component with a coupling component. Azoics are principally used for dyeing cellulosic fibers on which they provide a wide range of bright hues, with the exception of greens and bright blues; navies and blacks are also given. The actual hue produced with these colorants is determined by the particular combination of diazo and coupling components used.

The use of azoic coloring matters on polyamide, polyester, and cellulose di- and triacetate fibers is mostly limited to the production of deep shades, especially blacks.

The concept of an azoic coloring matter dates back to 1880 when Read Holliday introduced Vacanceine Red (CI, 1st edn., No. 93) by the process of first impregnating cotton with a solution of sodium 2-naphtholate and treating the impregnated fabric with a solution of diazotized 2-naphthylamine. Replacement of the naphthylamine with *p*-nitroaniline yielded the more commercially important Para Red (CI Pigment Red 1) in 1889. Other "ice colors," so called because of the low temperatures required for preparation of the diazo solutions, using 2-naphthol as coupling component, followed.

However, the fastness to light, washing, and rubbing of these early dyeings

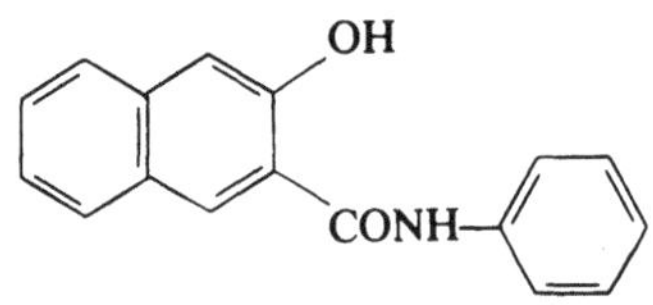

61

left much to be desired. Furthermore, 2-naphthol has negligible substantivity for cotton, so that the impregnated fabric had to be dried prior to development with the diazo component to minimize migration and consequent non-uniform dyeing. In 1912 the Griesheim Elektron Co. introduced Naphtol AS, the anilide of 2,3-hydroxynaphthoic acid (CI Azoic Coupling Component 2, **61**), the first of a large series of coupling components based on arylamides of 2,3-hydroxynaphthoic acid, and also proposed the name "azoic" to describe their combinations with diazotized amine. A major feature of such coupling components is their substantivity for cotton which, if required, enables them to be developed in the fiber without recourse to intermediate drying. In the *Colour Index,* three products which produce azoic coloring matters are listed.

1. Coupling Components

Of the more than thirty of these Naphtols, or naphthols as they are generically known, listed in the *Colour Index,* the majority are anilides of 2,3-hydroxynaphthoic acid. Arylamides of other 2-hydroxy-3-carboxylic acids, such as anthracene, for example CI Coupling Component 36 **(62)**, which gives yellower hues when coupled with a given diazo component and carbazole, for example, CI Coupling Component 5 **(63)**, to give components that yield dark browns and blacks, have been utilized as also have acylacetic arylamides which produce greenish-yellow to orange hues.[116,117] The coupling components vary considerably in their substantivity toward cotton; high substantivity facilitates high exhaustion of the naphthol and imparts high rubbing fastness to the dyeings produced. The water-insoluble coupling components are rendered soluble by conversion into their sodium salts, commonly by pasting with sodium hydroxide.

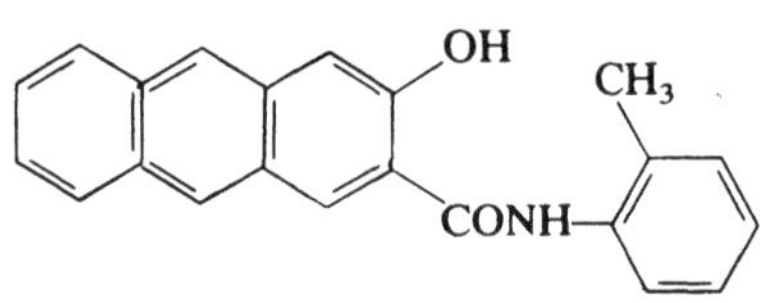

62

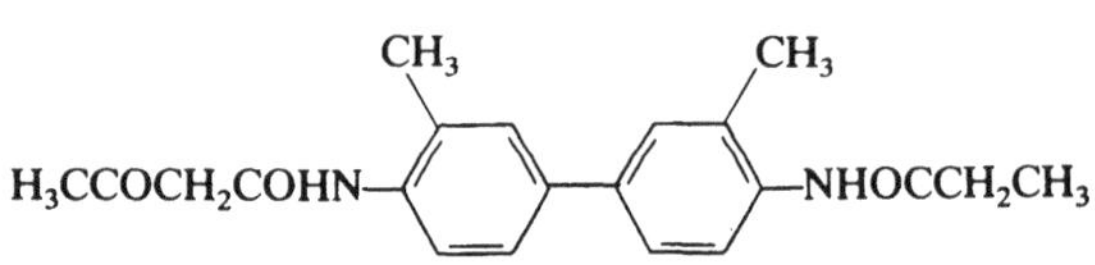

63

2. Diazo Components

The current wide range of diazotizable aromatic primary amines is marketed as "Fast Color Bases" in the case of easily diazotized amines in the form of the free base, hydrochloride or sulfate, and as stabilized diazonium compounds or "Fast Color Salts," such as CI Azoic Diazo Component 4 **(64)**. The color name of the diazo component (e.g., Fast Red B Salt) is derived from the hue of the azo compound obtained when it is coupled with CI Azoic Coupling Component 2.

The earlier diazo compounds were relatively simple aromatic primary amines. The introduction of alkoxy, phenoxy, halogen, or sulfone substituents into the diazotizable amine imparts higher light fastness to the azoic dyeings produced.[59,117] Other chemical classes of diazo components include aminoazobenzene, diphenylamine, and anthraquinone.

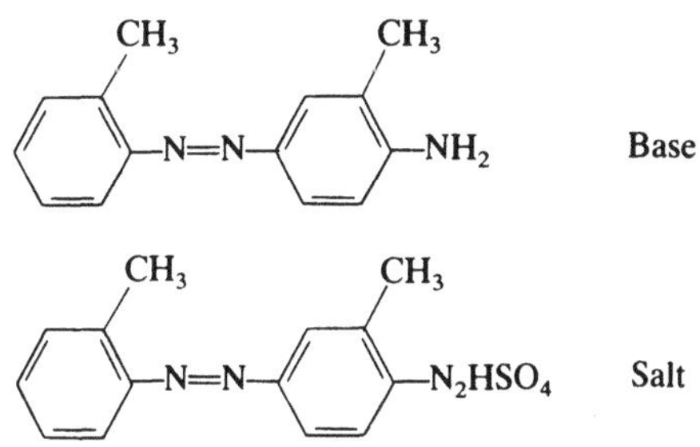

64

Stabilized diazo compounds, which overcome the often problematic diazotization of amines and the poor storage stability of their stock solutions, are readily soluble in water. Five types of stabilized compound are available.[118]

a. Anti-Diazotates. The early stabilized diazo compounds were of this type.

b. Diazonium Salts. The majority of fast color salts are sold as the zinc chloride double salts; cobalt chloride and borofluoride complexes, such as CI Azoic Diazo Component 2 **(65)**, are also marketed. In addition salts with naphthalene sulfonic acids **(66)** are available.

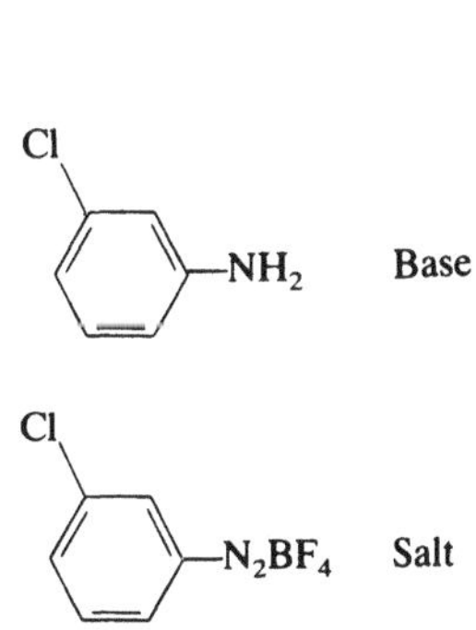

65

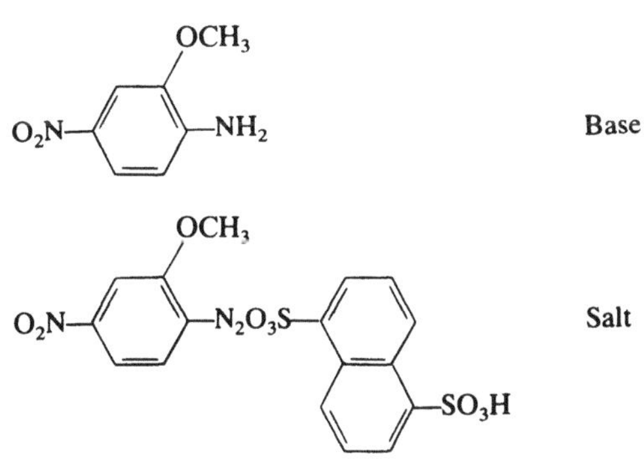

66

c. Diazosulfonates. These derivatives (i.e., $ArN{=}N{-}SO_3Na$) are stable under alkaline conditions and are coupled under neutral conditions.

d. Diazoamino Compounds. These are obtained by reacting an aromatic or aliphatic primary or secondary amine with a diazo compound under acidic conditions (equation 10).

$$RN_2Cl + R'NHR'' \rightarrow RN{=}N{-}NR'R'' \tag{10}$$

The corresponding diazoamino compound is readily decomposed under acid conditions to regenerate the parent diazonium compound.

e. Water-Soluble Diazonium Salts. The Azanil (Hoechst) range of soluble diazonium salts is applied in conjunction with a highly substantive Naphtol by exhaustion dyeing. When the two components have been absorbed by the fiber, coupling is effected by means of acidification of the dyebath.

The rate at which coupling occurs varies with different diazo components and is most commonly controlled by means of pH. Diazo components are classified into four groups according to both the rate at which coupling occurs and the method of their application. Groups I to III contain fast bases and fast salts which have high, medium, and low coupling energy, respectively, and which are suitable for both batchwise and impregnation application techniques, the optimum pH ranges for coupling in the case of groups I to III being,[116] respectively, 4–5.5, 5–6.5, and 6–7. Group IV contains Fast Color Salts which are best suited to continuous application and consists of diazo compounds of low coupling energy as well as compounds of very high coupling energy which are unsuitable for conventional batchwise application of azoic colorants.

3. Azoic Compositions

These mixtures of azoic coupling components and, usually, stabilized azoic diazo components[65] are used primarily for textile printing.

4. Dyeing Behavior

A major feature of this class of colorant is that following the development of the azoic compound in the fiber, dye which is deposited at the surface of the fiber must be removed (e.g., by "soaping") in order to secure maximum fastness of the dyeings. This is of special importance with regard to cellulosic fibers on which azoics exhibit very good fastness to washing. A characteristic feature of azoic colorants on cellulosic fibers is that the generally very good to excellent (5–6 to 7) light fastness of deep shades falls dramatically with increasing humidity; light fastness is also often very poor (2 to 2–3) in pale depths. Fastness to other agencies is generally good.[116]

Cotton is dyed most commonly in yarn form. Viscose staple is dyed as loose stock and continuous viscose filament in cake form. Cotton and viscose fabrics are commonly dyed using continuous methods.

The stages involved in azoic dyeing are:

- Application of the coupling component.
- Removal of surplus coupling component.
- Development (coupling).
- Aftertreatment.

A single-stage exhaustion application technique is available which comprises the simultaneous application of an Azanil (Hoechst) water-soluble diazoamino compound and a highly substantive coupling component. Coupling is achieved by acidification of the application bath.

5. Application of Coupling Components

The depth of shade of an azoic dyeing depends on the amount of coupling component applied to the substrate which, in turn, is determined by the following factors:

- The substantivity of the component.

For batchwise dyeing, coupling components of high substantivity are preferred since they yield dyeings of high rubbing fastness. Components of both low and moderate substantivity can also be used in the presence of, most commonly, sodium chloride to promote exhaustion. Coupling components of low substantivity are mostly used in continuous (padding) dyeing processes although moderate substantivity components may also be used, in which case substantivity is reduced by increasing the speed and the temperature of padding. Details regarding the suitability of particular coupling components are provided in the dye maker's literature.

- The concentration of the applied solution of coupling component
- The nature of the substrate
- The liquor ratio employed

Coupling components or "naphthols" are insoluble in water but their sodium salts or "naphtholates" are soluble. Naphtholation can be carried out by two methods.

Cold Dissolution. This is mostly used for batchwise application and commonly comprises pasting the naphthol with methylated spirit and warm (30 °C) water, to which is then added aqueous sodium hydroxide. The amount of sodium hydroxide is normally equivalent in mass to the amount of naphthol used.

Naphtholates are prone to hydrolysis with the formation of the corresponding free acid form of the naphthol, as a result of which subsequent coupling is impaired, giving rise to weak and often patchy dyeings. This effect is especially prevalent in batchwise application wherein an appreciable length of time can elapse between naphthol impregnation and subsequent development. The stability of many naphthol-impregnated materials can be improved considerably by the presence of formaldehyde in the naphtholate solution. The corresponding methylol derivative is less readily hydrolyzed than is the parent arylamide. During coupling, the hydroxymethyl group (—CH_2OH) is cleaved, yielding the parent arylamide. The formation of the above methylol derivative takes place more rapidly in concentrated solution so that addition of the formaldehyde is made to the concentrated naphtholate solution. However, formaldehyde cannot be used with certain coupling components and also at high application temperatures due to precipitation of the naphthol and consequent prevention of coupling.

The above concentrated naphtholate solution is then diluted using aqueous sodium hydroxide solution and a dispersing agent is added to maintain the naphtholate in solution.

Hot Dissolution. In this method, the naphthol is mixed with a hot solution of dispersing agent and, after boiling for a short time, the required quantity of hot, aqueous sodium hydroxide is added. Some naphtholates may require reboiling to yield clear solutions. If formaldehyde is to be added, the naphtholate solution must be cooled to a temperature specified in the literature supplied by the dye manufacturer to avoid precipitation of a methylene derivative.

The amount of sodium hydroxide required for naphtholation is specified in the dye maker's literature. The functions of the alkali are to convert the naphthol into the corresponding naphtholate and to swell the cellulosic fiber so as to facilitate penetration and thereby absorption of the naphtholate. Since viscose fibers have greater alkali absorbency and also swell to a much greater extent than cotton under alkaline conditions, the amount of sodium hydroxide required for treatment with the naphthol is less than for cotton materials.

For batchwise application, the quantity of sodium chloride used to increase the substantivity of the coupling component, especially in the case of low and moderate substantivity components, depends on the particular naphthol and again is specified within the manufacturers' literature. Sodium chloride may be replaced with anhydrous sodium sulfate or with twice the amount of Glauber's salt (hydrated sodium sulfate).

Batchwise application is usually carried out for 30–60 min at 20–30 °C (although higher temperatures of up to 50 °C may be used to facilitate greater penetration) while padding application is carried out at 80–95 °C.

6. Removal of Surplus Coupling Component

It is important that mechanically held, i.e., loosely bound, naphtholate liquor is removed from the material prior to development, otherwise azoic colorant is deposited on the surface of the fibers, leading to poor fastness to washing and especially rubbing. In batchwise methods of applying the naphthol to loose stock and yarn and in the application of fabrics using the winch, this is commonly achieved by hydroextraction.

Package dyed yarn and also fabrics subsequently developed on the winch and jig may be rinsed with a cold aqueous, alkaline brine solution to remove surplus naphtholate liquor; subsequent removal of liquid is often carried out. This process is especially suitable for high substantivity naphthols.

When development is carried out by padding, it is necessary to dry the naphtholated fabric after mangling. Drying, which is usually carried out at 90–110 °C, serves to fix the loosely-held naphtholate in the fabric thereby resulting in good fastness to rubbing of the dyeing. Drying is preferred in the case of low substantivity coupling components which have been applied to fabric by jig or pad techniques.

7. Development

Owing to the instability of diazo compounds, it is unwise to store development liquors.

a. Dissolution of Fast Salts. These stabilized, diazotized amines are readily dissolved in a cool (25–30 °C) aqueous solution of a nonionic dispersing agent. The development bath is prepared by diluting the concentrated development liquor to the required volume with water. Higher temperatures should not be used to dissolve the fast salt so as to avoid decomposition of the diazonium salt. The dispersing agent aids dissolution of the fast salt and also disperses the azoic colorant that is precipitated in the development bath, thereby facilitating removal of surface deposited azoic colorant during washing off.

Some fast salts require the addition of acetic acid to the development bath to neutralize residual alkali from the naphtholated material so as to achieve the correct pH for coupling to take place. However, many fast salts contain alkali-binding agents, usually aluminum sulfate in the case of moderate to high coupling salts, zinc sulfate for low-energy coupling salts or chromium acetate in the case of certain low coupling energy salts intended for continuous application, so that no further addition is normally required.

b. Diazotization of Fast Bases. Fast bases vary markedly in the ease with which they are diazotized and dye makers provide comprehensive details of the conditions required for each of the various bases.

In direct diazotization, the aromatic amine is dissolved in dilute hydrochloric acid and an aqueous solution of sodium nitrite is added slowly. In the case of amines that are insoluble in the acid, then either indirect diazotization can be carried out in which case the amine and sodium nitrite are pasted with water and the mixture is then added slowly to dilute hydrochloric acid, or the amine is pasted with the acid and aqueous sodium nitrite is slowly added. Diazotization is improved by the addition of a nonionic dispersing agent, the presence of which in the development bath also serves to improve removal of surface-deposited azoic colorant as discussed above.

Diazotization is carried out at the specified temperature, usually below 15 °C and in many cases with the aid of ice, to avoid decomposition of the diazonium salt. The excess of hydrochloric acid, used to ensure efficient and complete diazotization, must be eliminated to enable coupling with the naphthol to occur.

This is normally achieved by the addition of sodium acetate to the above diazotized base. The acetic acid thus produced may provide sufficient alkali-binding capacity and the correct pH for coupling in the development bath (prepared by diluting the concentrated diazonium solution with water) but, in the case of many diazotized bases, the appropriate coupling pH is obtained by the addition to the development bath of either further alkali-binding agent (e.g., acetic acid, zinc sulfate, etc.) or sodium acetate. Certain diazotized bases require near-neutral pH conditions for coupling which are achieved solely by the addition of either sodium hydrogencarbonate or phosphate buffer to the development bath.

The rate at which development (coupling) takes place depends on the pH and the temperature which, in turn, are determined by the particular coupling component and diazo component used. The optimum pH for coupling, as mentioned earlier, will depend on the coupling energy of the naphthol in use:

Group I, pH 4–5.5.
Group II, pH 5–6.5,
Group III, pH 6–7.
Group IV, pH varies.

Development is usually carried out at 15 °C, although higher temperatures (20–25 °C) are permissible in the case of low-energy coupling components.

In batchwise techniques, development is completed within 20–30 min; in padding processes the impregnated material is "skyed" for between 30 and 60 s for development to occur.

8. Aftertreatment

This stage is crucial for the production of a satisfactory dyeing. Following development, the material is usually rinsed with a cool (30–40 °C) aqueous solution of hydrochloric acid to remove the excess of diazonium compound and the residual sodium hydroxide produced during development. After rinsing in water, the material is then treated in an aqueous alkaline soap or detergent solution (i.e., "soaped") at between 65 and 100 °C depending on the particular azoic combination used. A final treatment in a bath containing containing a nonionic dispersing agent completes the aftertreatment.

The treatment with hot soap or detergent performs three important functions:

- It removes loosely bound azoic colorant which is essential for development of maximum rub fastness.
- The true hue of the dyeing is produced; many azoic combinations exhibit marked changes in hue upon soaping.
- Maximum degree of fastness, especially to light and chlorine bleach, is developed.

The latter two effects are attributable to aggregation of azoic colorant particles which assume a crystalline structure. In the case of dyeings on viscose rayon, the magnitude of this aggregation may be such that "blinding" or delustering of the fiber occurs; this effect is minimized by soaping at low temperatures (50–60 °C).

E. REACTIVE DYES

A general introduction to this dye class is given in Section II.D. The first reactive dyes for cellulosic fibers, the Procion M (later Procion MX) dichloro-*s*-triazine dyes, were introduced by ICI in 1956. Other dye ranges for these fibers, using a variety of reactive groups as shown below, were subsequently introduced by many dye manufacturers.[78–80,119,120]

Procion MX **(67)** (ICI)
Dichloro-*s*-triazine

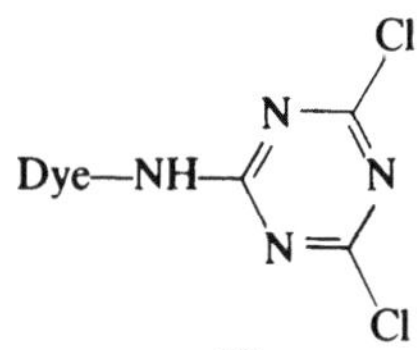

67

Procion H **(68)** and H-E **(69)** (ICI)
Monochloro-*s*-triazine

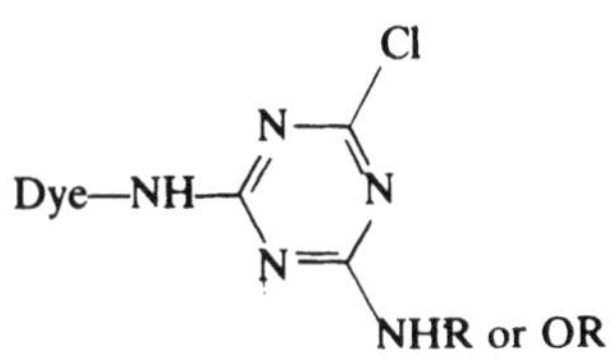

68

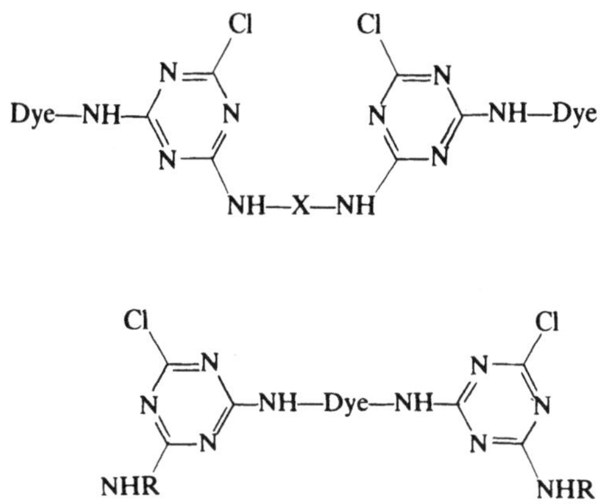

69

Cibacron F **(70)** (Ciba-Geigy)
Monofluoro-*s*-triazine

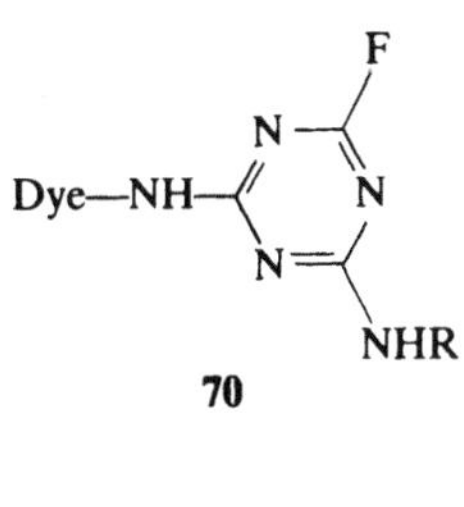

70

Drimarene R and K **(71)** (Sandoz)
Monochlorodifluoropyrimidine

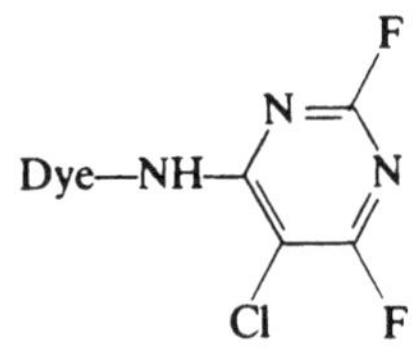

71

Drimarene X **(72)** (Sandoz)
Dichloropyrimidine

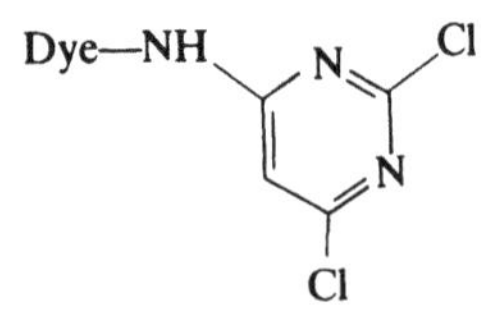

72

Remazol **(73)** (Hoechst)
Vinyl sulfone

$$D{-}SO_2{-}CH_2{-}CH_2{-}SO_3H$$

73

1. Dissolving the Dyes

The dyes are in most cases readily water-soluble. They are dissolved either by pasting with cold water to which is then added hot water, or by strewing the dye powder into hot powder which is stirred at high speed. Usually a temperature no greater than 80 °C is used for dissolution. In the case of highly reactive dyes, such as Procion MX (ICI), warm (50–60 °C) water is used. Remazol (Hoechst) dyes are dissolved by boiling in water.

Since the dyes are prone to hydrolysis, stock solutions should not be stored for long periods.

2. Dyeing Behavior

Reactive dyes, as typified by the ranges shown above, yield dyeings on cellulosic fibers of characteristically excellent fastness to washing and wet treatments, typically, on both cotton and viscose (1/1 SD) 4–5 to 5 ISO CO4, 5 ISO EO4 (alkaline) and good to very good light fastness, for example, (ISO BO2) 5–6 to 6–7 cotton and 6 to 6–7 viscose 1/1 SD, 5 to 5–6 cotton, and 5 to 6 viscose 1/12 SD.

The dyes are applied to cotton and viscose in loose fiber, yarn, and piece forms. Batchwise, semicontinuous and continuous processes are used.

The dyes react with ionized hydroxyl (i.e., —O^-) groups in the cellulosic substrate via a nucleophilic substitution or addition mechanism.[78–81] Because of the aqueous alkaline conditions required for ionization of the fiber, hydroxyl anions (i.e., OH^-) will always be present and therefore be in competition with the cellulosate anion as nucleophilic reagent; the dye is hydrolyzed as a result of reaction with the hydroxyl anions. With the possible exception of the Procion T (ICI) range of dyes,[119] all reactive dyes for cellulosic fibers undergo hydrolysis, the extent of which determines the efficiency of the dyeing process, which is of great practical importance, especially on economic grounds. 100% efficiency is, with few exceptions such as the Procion T (ICI) dyes and the Procion Resin Process (ICI).[119] never achieved in practice because of dye hydrolysis. The rate of dye–fiber reaction is also of major practical significance with regard to processing times. The reactivity of the different dye ranges varies from low through to high. The different ranges of dyes require different application conditions (e.g., concentration of electrolyte) so that dyes of different ranges are usually incompatible.

Electrolyte (sodium chloride or sulfate) increases the substantivity of the dyes toward the cellulosic fiber, by supressing dye–fiber repulsion and by increasing the activity of the anionic dyes in solution, thereby increasing the rate of dye exhaustion, rate of reaction, and efficiency of the dyeing process.[80]

Anhydrous sodium carbonate is the mostly widely used alkali; sodium bicarbonate and sodium hydroxide are also employed, enabling a pH range of 8–12 to be achieved. The alkali is added to the dyebath or pad liquor as a well-diluted solution. Electrolyte must be free of alkali so as to prevent premature dye hydrolysis or fixation and be added predissolved in water. Sequestering agents such as Calgon T (Albright and Wilson) are commonly used to soften process water and so avoid precipitation of alkaline earth metal salts during dye fixation and their deposition on the dyed material. An excess of inorganic sequesterant can, however, lead to reduced color yield. The use of organic sequestering agents such as EDTA is avoided, since they decopperize metal-complex reactive dyes to the detriment of light fastness and shade. Some reactive dyes are reduced, especially when dyeing is carried out in enclosed machines at high (i.e., >70 °C) temperatures, as a result of heat, alkali, and cellulose components and desizing residues. The addition of a mild oxidizing agent, based on sodium-*m*-nitrobenzene sulfonate, such as Matexil PA-L (ICI) or Revatol S (Sandoz) at the start of dyeing, protects the dyes against reduction. Lubricants such as Imacol J (Sandoz) prevent the formation of rope markes during winch dyeing of tubular-knitted fabrics, and antifoam agents such as Antimussol UP (Sandoz) are also used when required.

a. Batchwise Dyeing. This involves three stages:

- Exhaustion usually under neutral conditions in the presence of electrolyte.
- Addition of alkali to promote exhaustion and effect fixation (dye–fiber reaction).
- Rinsing and washing to remove unfixed dye, alkali, and electrolyte.

(Liquor ratio up to 7 : 1)

40 °C —— 45′ —— 5′ —— 40′ ——

↑ ↑ ↑ ↑

A B C D

A.	NaCl or Na_2SO_4 (anhydrous)	$40\ g l^{-1}$
B.	Cibacron (Ciba–Geigy) dye	2% owf
C.	Na_2CO_3	$1\ g l^{-1}$
D.	NaOH (30%)	$3\ ml l^{-1}$

Shading. After complete fixation, $\frac{1}{3}$ of the dyebath is dropped and the bath refilled using cold water. No electrolyte or alkali is added if shading additions are small. The predissolved dye is added and the dyebath raised to fixation temperature over 10–20 min and dyeing continued at this temperature for 10–20 min, depending on depth of shade difference.

Scheme 38

The temperature of dyeing ranges from ambient up to the boil, or even higher (e.g., in package dyeing), using selected dyes.

Dye exhaustion and fixation is generally higher on regular and HWM viscose than on cotton. In the case of polynosic fibers the substantivity of the dyes varies over a wide range.[80] Consequently, batchwise dyeing of viscose is usually carried out under different conditions of temperature and alkali than that of cotton so as to reduce dye–fiber substantivity and achieve adequate penetration and leveling.

A typical method for dyeing cotton using Cibacron F (Ciba-Geigy) dyes at short liquor ratio (up to 7 : 1) is shown in Scheme 38. For viscose rayon a dyeing temperature of 60 °C and the following recipe are used: A = Glauber's salt or sodium chloride, $40\ g l^{-1}$; B = Cibacron F dye, 2% owf; C = sodium carbonate, $10\ g l^{-1}$.

In addition to economies in consumption of dyes and chemicals, water, and heat, dyeing from short liquor ratio also increases the rate of reaction and efficiency of the dyeing process, primarily as a result of reduced dye hydrolysis.

For applying high reactivity dyes from longer liquor, a low alkali method is often used (Scheme 39). Higher dyeing temperatures are used for less reactive dyes as shown in Scheme 40. Alternatively, a simplified "all-in" procedure can be used (Scheme 41).

Washing-Off of Batchwise Dyeings. Unfixed dye can be removed by using the following typical procedure:

Rinse dyed material thoroughly.

"Soap" (should extended soaping be required, then a fresh soaping stage must be used) at long liquor, e.g., Lanapex R (ICI), $1\ g l^{-1}$; 98 °C; 15–30 min.

Rinse (50 °C).

Rinse (cold) until clear.

Two "soapings" are necessary in the case of deep shades.

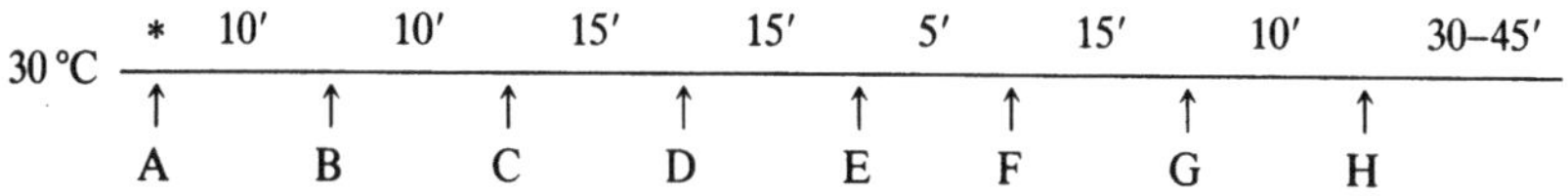

* pH of dyebath below 7

A. Procion MX (ICI) dye 2% owf
B. NaCl $2.5\,gl^{-1}$
C. NaCl $7.5\,gl^{-1}$
D. NaCl $25\,gl^{-1}$
E. Na_2CO_3 $0.125\,gl^{-1}$
F. Na_2CO_3 $0.125\,gl^{-1}$
pH 8.8–9.3
G. NaOH (flake) $0.125\,gl^{-1}$
H. NaOH (flake) $0.125\,gl^{-1}$
pH 10.5–11.0
L. R. = 20:1

Shading. Add Na_2CO_3 ($5\,gl^{-1}$) to raise pH to about 9, add the dye, and continue dyeing for 10 min. Add NaOH (flake) ($1.5\,gl^{-1}$) to raise bath to pH above 10 and continue dyeing for 20 min. Additional shading requires that half the dyebath is dropped and refilled using cold water. The dye is added over 10 min and the temperature raised to 30 °C. Dyeing is carried out for 20 min at this temperature.

Scheme 39

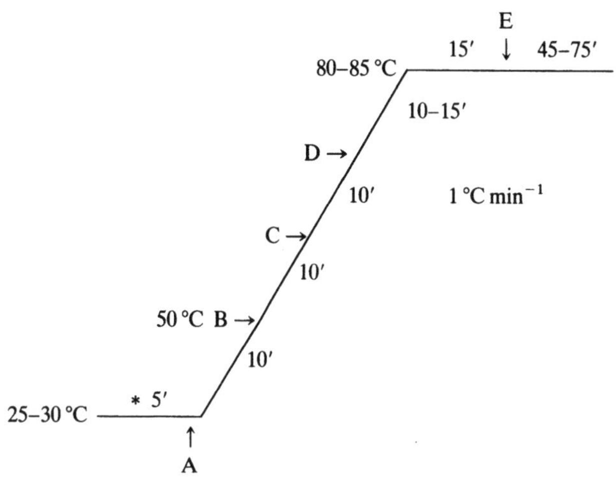

* Check pH is below 7 using CH_3COOH if necessary.

A. Procion H-E (ICI) dye 2% owf
B. NaCl $5\,gl^{-1}$
C. NaCl $20\,gl^{-1}$
D. NaCl $35\,gl^{-1}$
E. Na_2CO_3 $15\,gl^{-1}$ over 15′

Shading. The dyebath is reduced to $\frac{1}{2}$ original volume using cold water. The dye is added and dyeing continued for 10 min. The temperature is then raised to 80 °C and dyeing continued at this temperature for 15–30 min.

Scheme 40

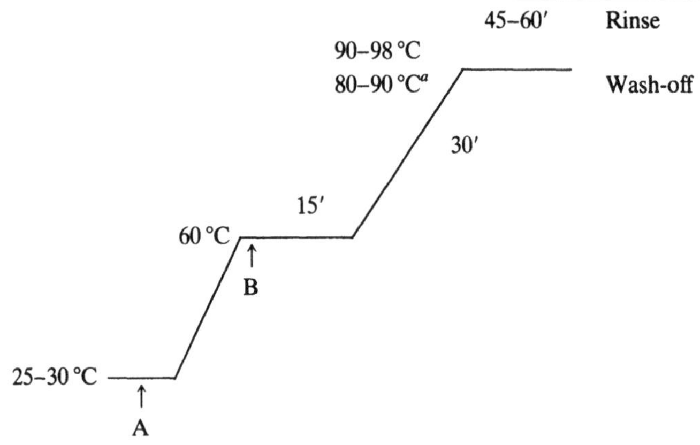

A.	Sodium sulfate (anhydrous)	$50\ g\,l^{-1}$
	Sodium carbonate	$10\ g\,l^{-1}$
	+ Revatol S (Sandoz)[b]	$1\ g\,l^{-1}$
B.	Drimarene X (Sandoz) dye	2% owf

[a] Viscose rayon.
[b] $2\ g\,l^{-1}$ used when dyeing viscose.

Scheme 41

Cotton

	20–25 °C	
	Procion MX (ICI) dye	$5\text{–}30\ g\,l^{-1}$
Pad	Matexil WA-HS (ICI) (wetting agent)	$2\ ml\,l^{-1}$
	Sodium carbonate	$5\text{–}30\ g\,l^{-1}$
↓	Urea[a]	$50\text{–}200\ g\,l^{-1}$
Batch	2–4 h *	
↓	Water 70 °C	
	Water 98 °C	
Wash off	Water 98 °C	
	Detergent (twice) 98 °C	
	Water (twice) 98 °C	
	Water (cold)	

[a] If dye solubility necessitates.

Scheme 42

As for Scheme 42 except:

* 24–48 h for fabrics that are, for example, difficult to penetrate; use:

sodium carbonate	2.5–15 $g l^{-1}$
sodium bicarbonate	2.5–15 $g l^{-1}$

instead of the sodium carbonate in Scheme 42.

Scheme 43

b. Semicontinuous Processes. Pad–Batch. High Reactivity Dyes (Scheme 42). For fabrics that are difficult to penetrate, long batching (24–48 h) permits greater dye diffusion. The following recipe (Scheme 43), using the above procedure, is recommended in such cases.

Low Reactivity Dyes (Scheme 44).

Cotton or linen

Pad	20–25 °C	
	Procion H (ICI) dye	5–30 $g l^{-1}$
	NaOH (flake)	10–15 $g l^{-1}$
	NaCl	10–30 $g l^{-1}$
	Matexil WA-HS (ICI)	2 $ml l^{-1}$
	Urea[a]	50–200 $g l^{-1}$
↓ Batch	24–48 h	
↓ Wash off		

[a] If dye solubility necessitates.

Scheme 44

c. Continuous Processes. A variety of fixation methods is used.

Pad–Dry.

Pad	20–25 °C	
	Procion MX (ICI) dye	x $g l^{-1}$
	Sodium hydrogencarbonate	10 $g l^{-1}$
	Matexil WA-HS (ICI) (wetting agent)	2 $ml l^{-1}$
	Urea[a]	0–200 $g l^{-1}$
↓ Dry	103–105 °C	
	2–5 min	
↓ Wash off		

[a] If dye solubility necessitates.

Scheme 45

Pad–Dry–Pad–Steam

	20–25 °C	
	Procion H (ICI) dye	$x\ g\,l^{-1}$
Pad	Matexil WA-HS (ICI)	$2\ ml\,l^{-1}$
↓	Matexil PA-L (ICI) (mild oxidant)	$10\ g\,l^{-1}$
Dry		
↓		
Pad	20–25 °C	
	NaOH (flake)[a]	$y\ g\,l^{-1}$
↓	NaCl	$200–300\ g\,l^{-1}$
Steam	60–75 s	
↓	100–102 °C	
Wash off		

[a] H dyes use ⅓ weight of dye (minimum $10\ g\,l^{-1}$).

Scheme 46

Pad–Dry–Bake.

	20–25 °C	
	Procion H (ICI) dye	$x\ g\,l^{-1}$
Pad	Matexil WA-HS (ICI)	$2\ ml\,l^{-1}$
	Sodium carbonate	$20\ g\,l^{-1}$
↓	Urea	$200\ g\,l^{-1}$
Dry		
↓		
Bake	135–150 °C	
↓	1–3 min	
Wash off		

Scheme 47

3. Aftertreatment

In the case of dyeings which have been inadequately washed off after dyeing, an aftertreatment with a cationic fixing agent, such as Matexil FC-ER or FC-PN (ICI), will improve the wet fastness properties of the dyeings:

Matexil FC-ER half the weight of dye applied (deep shades) 40 °C; 20 min; pH 5.0.

IV. DYEING OF SECONDARY ACETATE AND TRIACETATE FIBERS

Secondary cellulose acetate and cellulose triacetate fibers are prepared by acetylation of cellulose.[53,121,122] Secondary acetate fibers, which have a combined acetic acid content of 54–55%, were commercially introduced in 1921 and triacetate fibers (61.5–62.5% combined acetic acid content) in 1954.

The difference in degree of acetylation has a significant effect on the chemical and physical properties (Table 8) and dyeing behavior of the two fibers. Both fibers are resistant to dilute aqueous acids but are decomposed by concentrated acids.[123] The fibers are sensitive to alkali, especially at high temperatures. Secondary acetate is particularly prone to saponification (i.e., hydrolysis of the acetyl groups) which results in loss in weight and luster as well as impairment of handle. Surface saponification of triacetate (known as the "S" finish in the case of "Tricel") is, however, carried out, the advantages of which include[121] improvements in the fastness of the dyeings to gas-fume fading and rubbing (both wet and dry). Triacetate is resistant to boiling water[121] and dyeing temperatures of up to 115 °C can be used.[124] In contrast, secondary acetate is severely delustered if treated in water at temperatures above 85 °C and thus the maximum dyeing temperatures normally used is 80–85 °C. Delustered fiber can be dyed at temperatures of up to 90 °C; higher temperatures cause loss of fiber strength and quality. Since both fibers are thermoplastic, care must be observed to avoid introducing distortions (crease and rope marks) during dyeing.

The greater hydrophobicity of triacetate (Table 8) can be attributed to the polymer's greater degree of acetylation, i.e., larger number of hydrophobic acetyl groups. This higher acetyl content, however, also imparts a greater degree of crystallinity to triacetate and the fiber thereby assumes a more compact structure than the less hydrophobic secondary acetate. These two characteristics result in triacetate possessing low moisture uptake by virtue of reduced availability of less-ordered regions in the fiber where water sorption occurs together with reduced accessibility of the water molecules to these regions.

The generally superior stability of triacetate fibers to aqueous treatments can

Table 8. Some Properties of Acetate Fibers

	Sec. acetate	Triacetate
Melting point (°C)[121]	230	290–300
Moisture regain (%)[124] (60% RH; 20 °C)	6.0–6.5	3.0–4.5 2.5–3.0[a]
Water imbibition (%)[124]	25	16–17 10–11[a]
Swelling in water (%)[25]	6–30	Very slight
Tanacity (g tex^{-1})[123]		
Dry	0.09–0.11	0.11–0.13
Wet	0.05–0.07	0.06–0.08

[a] Heat-set fiber.

be attributed to its more compact structure, greater crystallinity, and hydrophobicity. The effects that these three factors have upon the dyeing behavior of triacetate are discussed later.

The superior thermal stability of triacetate permits the fiber to be heat-set, most commonly using either steam or dry heat. Setting can be carried out both before or after dyeing, depending on the form of the goods and dyeing method used. This heat treatment increases the crystallinity and orientation of the fiber and consequently presetting reduces the rate of dyeing and saturation value of the fiber (although the latter factor is of little practical value since saturation is rarely achieved under normal dyeing conditions), these effects being most marked in the case of dry heat treatment. In contrast, the fastness of dyeings on dry heat-set triacetate material is generally slightly higher than on steam-set fiber. Because of the high temperatures involved, only those dyes which are sufficiently stable to these temperatures can be employed when post heat-setting is carried out.

Both secondary acetate and triacetate fibers are mostly used in continuous filament form, staple fiber being mainly used in blends with other fibers. Both fibers are available as texturized and flat (untexturized) yarns and can be dyed as loose stock, tow, slubbing, yarn, woven, and knitted fabrics and garments. Batchwise, semicontinuous, and continuous dyeing processes are employed.

Acetate fibers are dyed using the following dye classes: disperse, azoic, and vat

A. Disperse Dyes

A disperse dye is defined[125] as "a substantially water-insoluble dye having substantivity for one or more hydrophobic fibers, for example, cellulose acetate, and usually applied from fine aqueous dispersion." In addition to their use on acetate fibers, the dyes are used on polyester, polyamide, and acrylic fibers.

When secondary cellulose acetate fibers were introduced, few of the existing water-soluble, ionic dyes would dye this relatively hydrophobic fiber. Various techniques (saponification, fiber-swelling) were explored as possible means of applying the available dyes.[126] The work by Green and Saunders of BDC resulted in the introduction of the Ionamine range of dyes for cellulose acetate in the early 1920s,[127] such as Ionamine Red KA **(74)**, which were the precursors of the disperse dyes. The "temporarily solubilized" dyes hydrolyzed in the dyebath to yield the original low solubility dye, which was then adsorbed by the cellulose acetate fiber.

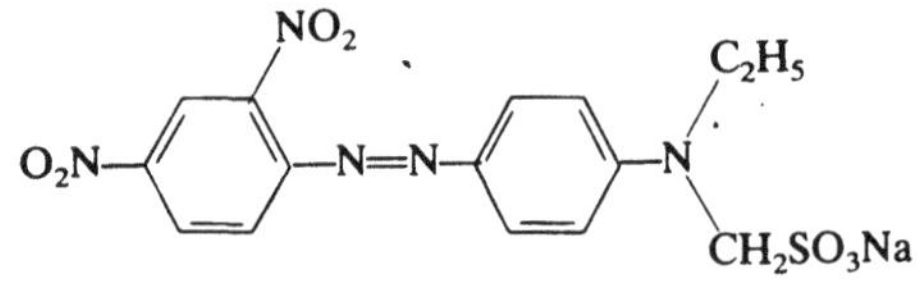

74

However, practical difficulties related to varying rates of hydrolysis of individual dyes led to the Ionamines being superseded by fine aqueous dispersions of dyes or low water solubility, or "Acetate" dyes, marketed as the SRA (Br.C)[128] and Duranol (ICI)[128] ranges in the early 1920s. The SRA dyes derived their name from the use of sulforicinoleic acid as dispersing agent. The Duranol dyes were dispersed using soap or Turkey Red Oil.[129] Ranges of dyes for secondary acetate were subsequently introduced by various manufacturers.

Characteristically, disperse dyes are small molecular size compounds which contain no ionic groups but do carry polar substituents such as —OH—, —CH_2OH. As a consequence of their low size, they possess a small but nevertheless important solubility in water.[130] The dyes are volatile due to their relative lack of cohesive energy in the solid state, and may therefore be applied via the vapor phase in heat-fixation processes such as thermofixation.

The early disperses dyes developed for dyeing cellulose diacetate were required to have adequate solubility and sufficiently small molecular size in order to diffuse at the relatively low dyeing temperatures (80–85 °C) which had to be used. With the advent of triacetate and particularly polyester fibers and the high temperatures to which these fibers are subjected, for example, during setting and ironing, such dyes sublimed readily giving rise to poor sublimation fastness. By increasing the hydrophilicity of the dyes through the introduction of polar substituents (e.g., benzoyl, amide) the vapor pressure of the dyes was reduced which improved heat fastness. However, the increased water solubility of the resulting dyes had to be offset by increasing the molecular weight which, in turn, resulted in low diffusion rates of the dyes, especially within polyester, and consequent slow rates of dyeing. Therefore, methods of accelerating dye diffusion, and thus dyeing rate, namely, high-temperature (130–140 °C) dyeing and the use of carriers were developed.

As a result of the variety of disperse dyes and their varied dyeing behavior on different fibers, many dye makers have introduced systems of classification for their products. For example, ICI[131] classify their dyes into five groups according to dyeing behavior and fastness properties on polyester. In 1964[132] the SDC proposed four tests by which the migration, buildup, temperature range, and rate of dyeing characteristics of disperse dyes on cellulose acetate under practical dyeing conditions could be determined. Tests for determining the same characteristics on triacetate were proposed in the following year.[133] These tests permit individual dyes to be rated on an A to E scale for each characteristic property, where A represents the most satisfactory and E the least satisfactory performance.

Disperse dyes are of the following chemical types, the first three of which are the most important.[134–138]

1. Azo

This very large and important class comprises mostly monoazo with a few disazo derivatives such as CI Disperse Orange 29 **(75)** and provides a full range of

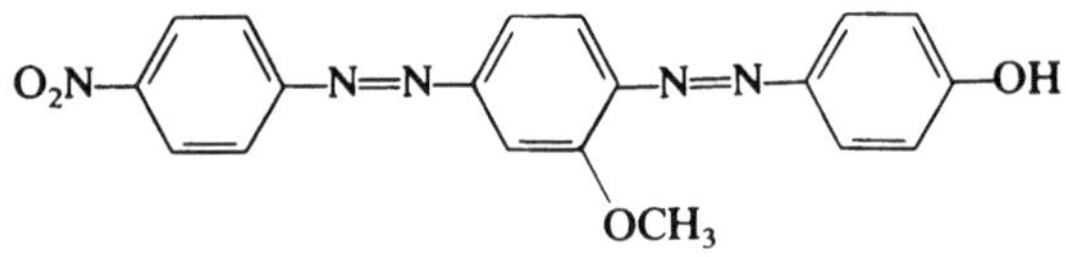

75

bright hues. The most important group consists of aminoazobenzene types. The earlier dyes of this type, such as CI Disperse Orange 3 (**76**), which gave dyeings of good light fastness on cellulose acetate fibers, generally show poor fastness to light on polyester materials.

The introduction of cyano substituents yields dyes of higher light fastness on polyester as typified by CI Disperse Blue 183 (**77**). Heterocyclic azo dyes provide a comprehensive range of characteristically bright hues.

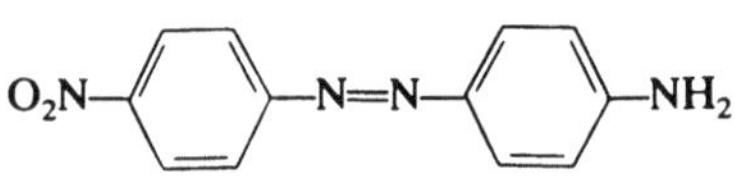

76

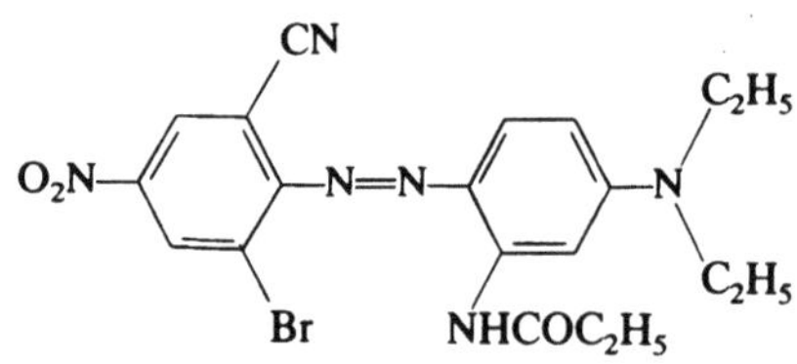

77

2. Anthraquinone

Dyes within this large and important class provide bright red to blue shades. The dyes developed for dyeing secondary acetate were relatively simple amino and/or hydroxy derivatives of AQ such as CI Disperse Blue 7 (**78**). More complex derivatives of AQ have been developed to extend the hue range and improve fastness properties, as typified by CI Disperse Blue 73 (**79**), which gives good all-round fastness on acetate, triacetate, and polyester fibers.

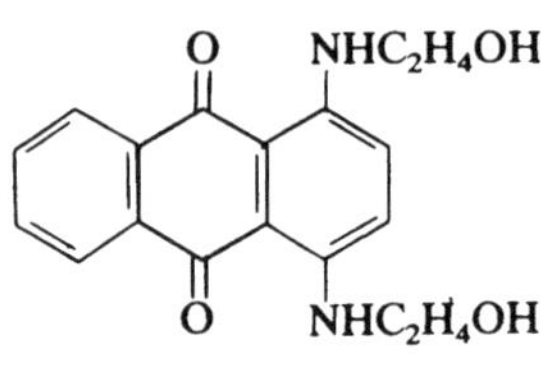

78

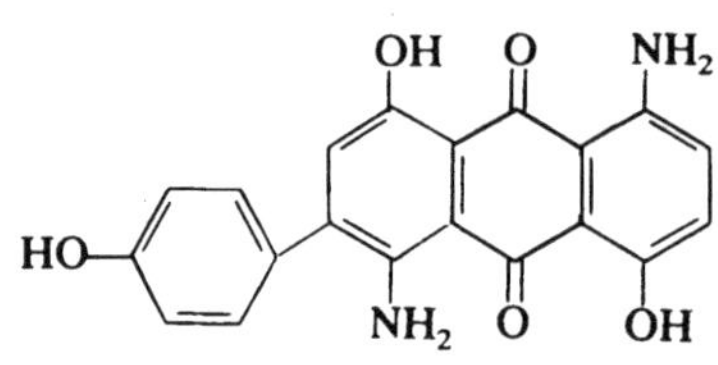

79

3. Nitrodiphenylamine

This is a relatively small group of inexpensive, mostly yellow and yellow-orange dyes. Characteristically, the dyes, as exemplified by CI Disperse Yellow 42 (**80**), exhibit very good light fastness on polyester and acetate fibers. Dyes of large molecular size and/or carrying polar groupings yield dyeings of good sublimation fastness on cellulose triacetate and polyester fibers.

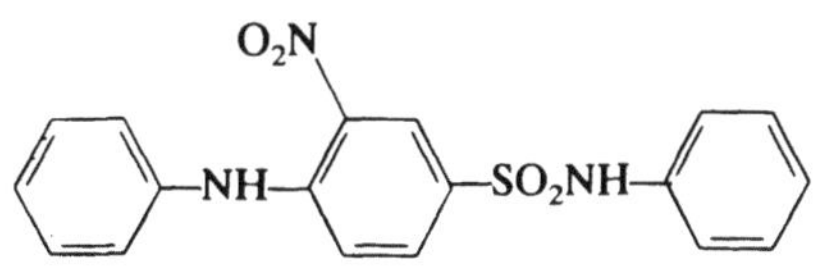

80

4. Styryl

The earlier types of these greenish-yellow dyes designed for use on secondary acetate such as CI Disperse Yellow 31 **(81)** were prone to hydrolysis during dyeing. More recent members, as typified by CI Disperse Yellow 99 **(82)**, are of improved stability to hydrolysis and are applicable to polyester fibers. Various other chemical types are available, details of which can be found elsewhere.[134–138]

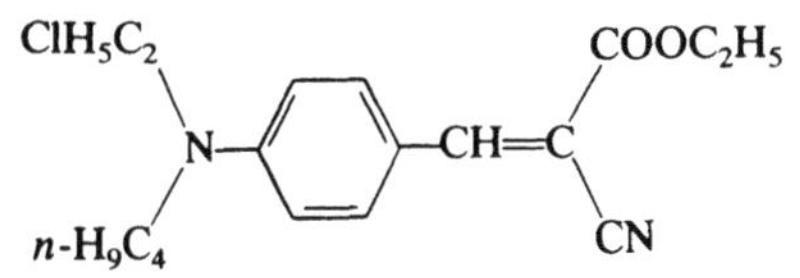

81

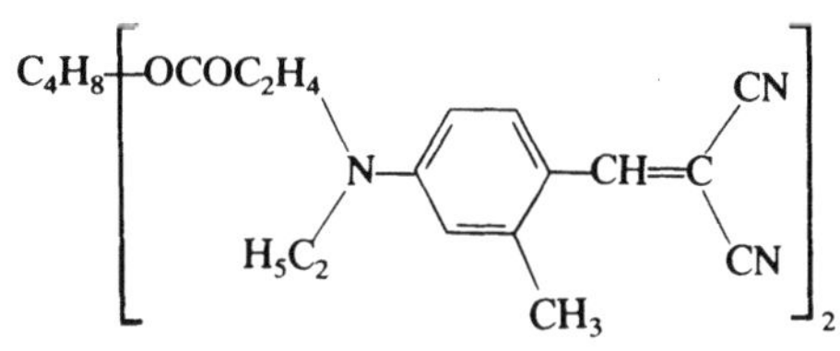

82

5. Dispersing the Dyes

Powder forms are slowly sprinkled into 10–20 times their weight of water at 30–40 °C with stirring. The ensuing dispersion is added to the dyebath through a sieve. Liquid brands are stirred into warm water and the dispersion added to the dyebath through a sieve. Alternatively, the liquid dye may be added undiluted to the dyebath through a suitable strainer.

Liquid and powder brands are usually compatible in admixture, in which case dispersion is effected using the procedure described for powder brands.

6. Dyeing Behavior

Generally, the dyes, as represented by the Serisol and Serilene (YCL), Cibacet and Terasil (Ciba-Geigy) and Palanil (BASF) ranges, exhibit good fastness to light on both secondary acetate and triacetate fibers, typically ISO BO1 5–6 or higher at 1/1 SD and 4–6 at 1/12 SD. Some dye mixtures exhibit anomalous light fastness properties on acetate fibers inasmuch as one component (activating dye) causes the abnormal fading of another component (susceptible dye). The light fastness of the dyes is generally higher on bright secondary acetate as compared with the delustered fiber, but this difference is not observed with triacetate materials.

The wash and wet fastness of dyeings on secondary acetate are generally lower than the corresponding dyeings on triacetate. For example, (1/1 SD) fastness to washing (ISO CO2) is typically 4 to 4–5 for secondary acetate and 4–5 to 5 in the case of triacetate, while alkaline perspiration fastness is typically in the region 4–5 to 5 on secondary acetate and 5 on triacetate. This generally superior fastness to wet treatments of dyeings on triacetate can be attributed to the more compact and crystalline structure of triacetate fibers and the correspondingly lower rate of dye diffusion out of the fiber during wet treatments. Furthermore, because triacetate is more hydrophobic, the fiber will swell to a much lesser extent than that of secondary acetate, resulting in lower rates of dye diffusion out of the substrate. The increase in crystallinity of triacetate which accompanies post heat-setting results in a further increase in wet and wash fastness of dyeings relative to unset material.

The fastness of the dyeings to burnt gas fumes varies from poor to very good on both fibers. Certain anthraquinone dyes which contain primary and secondary amines are particularly susceptible to fading when exposed to burnt gas fumes owing to reaction with nitrogen oxides (equations 11 and 12).

$$RNH_2 \xrightarrow{NO_2} RN_2NO_3 \text{ Diazonitrate} \tag{11}$$

$$RNHR' \xrightarrow{NO_2} RNNO \text{ } N\text{-nitrosamine} \tag{12}$$

Gas fume fading can be prevented by the use of inhibitors, e.g., diphenyl(acetamidine), which preferentially react with the nitrogen oxides. The inhibitors can be added to the dyebath, often toward the end of the dyeing process, or can be applied as an aftertreatment of the dyed material, although the inhibitors are generally less resistant to washing when applied by this latter method. Generally, the inhibitors, such as Protac 80 (YCL), are applied at temperatures of between 60–80 °C or by padding at ambient temperature followed by drying in the case of an aftertreatment. The amount of gas fume fading inhibitor required for protection of sensitive dyes is usually of the order of 1–1.5% owf.

Some dyeings are susceptible to ozone or "O" fading.[50]

The fastness to heat treatments of dyeings on triacetate also varies from poor to very good.

Shortly after their introduction, several theories were proposed to describe the mechanism of adsorption of disperse dyes from aqueous baths. Kartaschoff[139] proposed that the solid dye particles were attracted to, and formed a layer on, the secondary acetate fiber surface. The solid dye then dissolves or diffuses into the fiber by means of a solid-state diffusion process to form a *solid solution*, i.e., the dye dissolves and diffuses through the polymeric material in the same way as it would through an organic solvent.

Clavel[140] suggested that dyeing takes place from a dilute aqueous solution of dye which is replenished by dissolution of dye from the bulk dispersion. This model has gained support from many workers and is now generally accepted.

Support for the solid solution theory, which is analogous to partition of the solid dye between two immiscible solvents, has been offered by many workers[141] for a variety of fibers.

In contrast, however, Burns and Wood[142] postulated that the disperse dye was adsorbed by secondary acetate from true aqueous solution at sites within the fiber. Such an *adsorption site* mechanism for various fibers has received support[141] and suggests that the disperse dye is bound to sites within the polymer at which dispersion and hydrogen bond forces are operative.

Many workers, however, consider that there is little distinction between the solid solution and adsorption site models.[104,143,144]

Both solid solution and adsorption site models have been proposed in the case of vapor phase dyeing with disperse dyes.[143]

Thus, although the precise mode of attachment of the dye to the fiber is not entirely resolved, it is considered that the aqueous phase transfer of disperse dyes occurs from a monomolecular aqueous solution. The disperse dyebath comprises a small amount of dye in aqueous solution at the fiber surface with, initially at least, the greater proportion of dye in dispersion in the bulk of the dyebath. Dye is adsorbed at the fiber surface from the aqueous solution. As dye molecules diffuse monomolecularly from the fiber periphery to the interior, the aqueous solution is replenished by dissolution of dye from the bulk dispersion and further dye molecules are adsorbed at the fiber surface. This process continues until the fiber becomes saturated with dye. This can be represented simply as follows:

$$\text{Dye dispersion} \rightarrow \begin{matrix}\text{Aqueous dye}\\ \text{solution}\end{matrix} \rightarrow \begin{matrix}\text{Dye adsorption}\\ \text{at fiber surface}\end{matrix} \rightarrow \begin{matrix}\text{Dye diffusion}\\ \text{within the fiber}\end{matrix}$$

When applied from aqueous baths, the dye must be presented to the fiber as a fine, stable dispersion so as to obtain levelness and full color yield. The particle size of the dye molecule is of great importance in this context;[145] a particle size of the order of 0.1 to 2 μm is common.[102] The usual method of obtaining the required particle size is to grind the dye particles in the presence of a dispersing agent, typically the sodium salt of an anionic, naphthalene sulfonic acid–formaldehyde condensate.[102] The dispersing agent allows the preparation of the dye in powder form and facilitates dispersion of the dye in water as well as maintaining the dispersion during dyeing.[102,129,146] Disperse dyes are also available in liquid form as concentrated, aqueous dispersions which overcome problems of dusting and facilitate metering.

It is often necessary to add further surfactant to the dyebath in order to maintain dispersion stability and promote leveling. Such agents increase dye migration and fiber penetration especially at temperatures below 100 °C.[147] Commonly, these leveling or dispersing agents, such as Dyapol SL (YCL), are synergistic mixtures of nonionic and anionic surfactants, the latter products raising the cloud point of the nonionic component to well above the dyeing temperature used so as to avoid agglomeration of the dye.

The rate and extent of dye uptake onto triacetate fiber at temperatures below the boil are considerably less than on secondary acetate. The greater crystallinity of the former fiber results in a reduction in the availability of the less-ordered regions of the fiber where dye adsorption takes place so that the saturation value of triacetate is less than that of its counterpart. Because triacetate has a more compact structure, then the accessibility of the dye molecules to these less-ordered regions will be reduced and so result in a low rate of dye uptake. The higher hydrophobicity of triacetate fibers will also contribute to low dyeing rates because of the fiber's low water imbibition and swelling.

As mentioned earlier, the rate of uptake of disperse dyes on triacetate can be raised to acceptable levels by the use of either elevated dyeing temperatures (115 °C) or carriers.

7. High-Temperature Dyeing

The elevated dyeing temperatures of up to 115 °C

1. Increase the average kinetic energy of the dye molecules.
2. Decrease the cohesive energy between molecular chains in the fiber and thus increase the segmental mobility of the less-ordered regions within the fiber.
3. Increase the water absorption and thence swelling of the fiber.
4. Increase the aqueous solubility of the dye, as a result of which the aqueous dye solution from which dyeing takes place may become more concentrated thereby increasing the rate of transfer from the dyebath to the fiber.
5. Improve the migrating power of the dye mainly as a consequence of effects (1) and (4) above.

The overall effects of using elevated dyeing temperatures are therefore better penetration and more rapid and more level dyeing than when dyeing is carried out at the boil.

8. Carrier Dyeing

A carrier is defined[125] as "a type of accelerant, particularly used in the dyeing or printing of hydrophobic fibers with disperse dyes." The term "carrier" stems from the original suggestion that such compounds "carried" the dye molecules into the fiber. Carrier dyeing is more widely practised on polyester than triacetate fibers; certain carriers are also used in the dyeing of secondary acetate with some slow-diffusing dyes, but care must be exercised to avoid producing adverse fiber properties. A variety of compounds function as carriers for triacetate, such as butyl benzoate, diethyl phthalate, and tripropyl phosfate,[124,148] some of which are also effective for polyester fibers.

Many theories have been proposed to explain carrier action in terms of their effects on the disperse dye and the hydrophobic fiber. It is now generally considered that carrier action entails modification of fiber structure. Plasticization of the fiber has been advanced as the probable mechanism involved. The carrier is adsorbed by the hydrophobic fiber and disrupts the polymer structure, lowering T_g, the temperature at which the fiber becomes plastic. As a consequence,[149,150] in terms of the free-volume model of dye diffusion,[150,151] the dye molecules are able to diffuse more easily within the fiber. The solubility parameter concept[149] has also been involved in this context since dissolution of the carrier in the polymer (fiber) entails disruption, i.e., plasticization.

Carriers, such as Optinol TR (YCL), are usually available as self-emulsifiable liquids (see Section V) which form fine dispersions on addition to warm water and are compatible with disperse dyes. Care must be exercised in selecting the type and quantity of carrier to be used to prevent undue fiber swelling, adverse handle and "carrier marks" (see Section V). The quantity used depends on depth of shade and liquor ratio. After dyeing, carrier residues are removed from the material during rinsing, scouring, and/or reduction clearing. Further details of carriers can be found in Section V.

9. Barré Effects

Barré effects are most common in knitted fabrics of texturized yarn. Texturizing involves both mechanical and heat treatments, both of which modify the internal structure of the fiber. Thus, even small variations in the texturizing process will be highlighted in dyeing. Because dry heat-setting of triacetate prior to dyeing reduces the dyeing rate, then any variations in setting conditions will be manifest as unevenness in dyeing. The ability of a given disperse dye to cover such physical variations in the fiber will, in the main, depend on the dye's migration characteristics. Poor Barré coverage is obtained when dyeing triacetate at the boil in the presence of carrier, while high-temperature (115 °C) dyeing gives best coverage.

10. Dyeing Auxiliaries

Certain disperse dyes, notably azo types, are degraded under even mildly alkaline conditions, especially at high temperatures leading to low color yield. This is of greater significance with regard to the dyeing of triacetate owing to the higher dyeing temperatures used. The use of slightly acidic dyeing conditions is therefore preferred, typically pH 6–7 for secondary acetate and pH 5–6 in the case of triacetate. Acetic acid is commonly used to achieve the required pH since it avoids the danger of excessive dyebath acidity. Buffer systems, such as sodium dihydrogenphosphate, are also employed to control pH. When dyeing triacetate which has undergone an alkaline saponification treatment, the use of acetic acid solution (30%) (typically 0.5–3 $ml\,l^{-1}$) gives a buffering action (through the formation of sodium acetate) at pH 5–5.5.

To overcome the shade change and sometimes reduction in fastness that the presence of heavy metal ions, especially copper and iron, in the dyebath have on certain disperse dyes, sequesterants such as EDTA and Trilon B (BASF) are used.

11. Aftertreatment

In order to obtain maximum fastness of medium and heavy depth dyeings toward water and perspiration, especially in the case of slow diffusing dyes on triacetate, it may be necessary to remove surface deposited dye from the dyed material. For pale and medium depths this can be achieved by scouring:

> Detergent, e.g., Dyamol OE (YCL), 1 $g l^{-1}$,
> 40–50 °C; 30 min.

For heavy depths a reduction clear treatment can be given:

> Detergent [Dyamol OE (YCL)], 2 $g l^{-1}$
> Sodium dithionite, 2 $g l^{-1}$
> Ammonia (880), 2 $ml l^{-1}$
> 50–60 °C; 30 min.

(a) Secondary acetate (winch) and triacetate (jet)

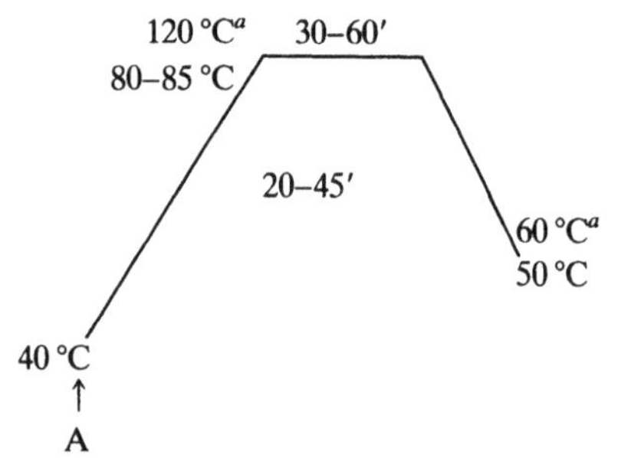

A.	Terasil or Cibacet (Ciba–Geigy) dye	x% owf
	Albatex PON conc. (Ciba–Geigy)	0.3 $g l^{-1}$
	Irgasol DAM (Ciba–Geigy)[a]	0.5 $g l^{-1}$
	Acetic acid[b]	

[a] Triacetate only.
[b] To pH 6.6–7.5 for secondary acetate or pH 5 for triacetate.

(b) Carrier dyeing of triacetate

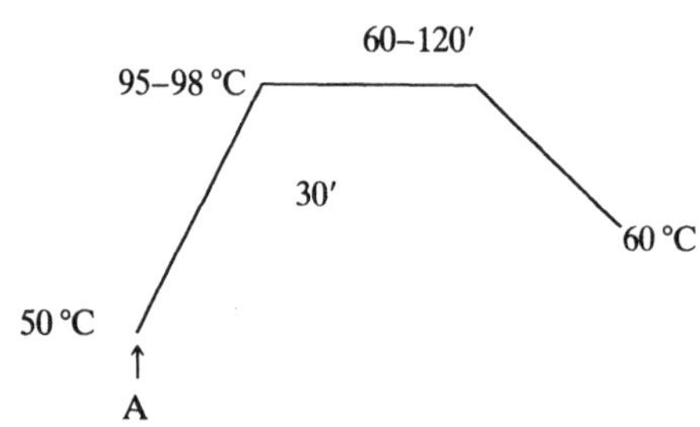

The recipe in (a) above for dyeing triacetate is used with the additional use of:
Invalon TA or HTB (Ciba–Geigy) 0–4 $g l^{-1}$.

Scheme 48

12. Dyeing Processes

Many disperse dyes are incompatible in admixture, even though their individual rates of dyeing and build-up characteristics on acetate fibers may be very similar. A possible explanation of this phenomenon is that disperse dyes may function as carriers for other disperse dyes in admixture. Thus, dye manufacturers recommend compatible dye mixtures to ensure optimum results. Furthermore, dye makers often recommend specific dyes, and mixtures, for certain application conditions, i.e., jig dyeing, jet dyeing, etc.

a. Batchwise Dyeing. The following (Scheme 48) are typical procedures.

b. Continuous Dyeing. Pad–Thermofix (Scheme 49). Further details of the thermofix process are given in Section V.

A further dyeing method for both secondary acetate and triacetate fibers which has not yet been developed commercially is the Vapacol Process.[152]

Triacetate		
	30–40 °C	
	Disperse dye	x g l^{-1}
Pad	Wetting agent	2 g l^{-1}
↓	Gas Fume Fading Inhibitor	50 g l^{-1}
	Migration Inhibitor	10–20 g l^{-1}
Dry		
↓		
Bake	190–210 °C	
↓	30–60	
Rinse		
↓	1–5 min	
Scour	60 °C	
↓	Detergent	1–2 g l^{-1}
Rinse		

Scheme 49

B. Azoic Colorants

The use of this dye class on acetate fibers is nowadays restricted to the production of deep and predominantly black shades. Blacks are most commonly obtained using diazotized and developed disperse dyes although azoic compositions are also used. Colors other than black are produced using various coupling components and bases, many popular combinations of which are available as azoic compositions.

The conventional procedure for applying azoic colorants to cellulosic fibers (see Section III.C), namely, application of the coupling component and subsequent development with the diazotized base, is however not suitable for acetate fibers. In the conventional method the two components are applied as ions from aqueous solution. However, moderately strong alkaline conditions cannot be used for applying the sodium salts of the coupling components due to saponification of the acetate fibers at the dyeing temperatures used.[153] Also, the low temperatures necessary for stability of the diazotized bases prevents their adequate diffusion into the hydrophobic fibers; higher temperatures, which increase fiber swelling, and thus diffusion, cannot be used owing to decomposition of the base.[153] In their unionized forms, both components have very low water solubility and can therefore be applied to acetate fibers from an aqueous dispersion, i.e., as disperse dyes.

Azoic colorants are applied to acetate materials using either a "reversed" conventional procedure or a so-called "concurrent" process in which the coupling component and base are applied simultaneously and diazotization and coupling are carried out *in situ.*

Owing to the predominant use of black combinations on acetate fibers, this section is confined to these dyeings. Accounts of dyeings other than black can be found elsewhere.[123,124]

The commonest coupling component or developer used is 2-hydroxy-3-naphthoic acid (CI Coupling Component 2; CI Developer 8, **61**), which is available as the water-soluble sodium salt or the free acid. In the latter case the component is usually dissolved in aqueous sodium hydroxide and reprecipitated in cold water containing a dispersing agent set at the appropriate pH for development (usually in the range pH 3.5–4.5).

Diazotization is carried out *in situ* using nitrous acid solution (sodium nitrite and either hydrochloric or acetic acids). Unlike azoic colorants on cellulosic fibers and other azoic colors on acetate fibers,[123,124] black combinations do not require a soaping treatment to stabilize the hue. The dyeings are, however, normally scoured using hot detergent solution to remove surface-deposited colorant and any decomposition products from the coupling reaction so as to achieve optimum fastness properties. A reduction clear aftertreatment is often required to secure maximum fastness properties, especially in the case of deep dyeings on triacetate.

Several dyes are used, exemplified by the Serisol Diazo (YCL) range, common examples including CI Disperse Blacks 1, 3, and 9. The dyes, which are usually available in liquid and powder forms, are dispersed by the techniques previously recounted for conventional disperse dyes and are applied by exhaustion methods that are essentially the same as those employed for conventional disperse dyes.

The following application methods (Schemes 50 and 51) are suitable for dyeing secondary and triacetate fibers, respectively. Azoic combinations comprising a mixture of the stabilized diazonium salt and coupling component or developer, such as Serilene Azoic Black B2 (YCL), may be applied (Scheme 52) to triacetate fibers at elevated temperatures (up to 120 °C).

Dyeing

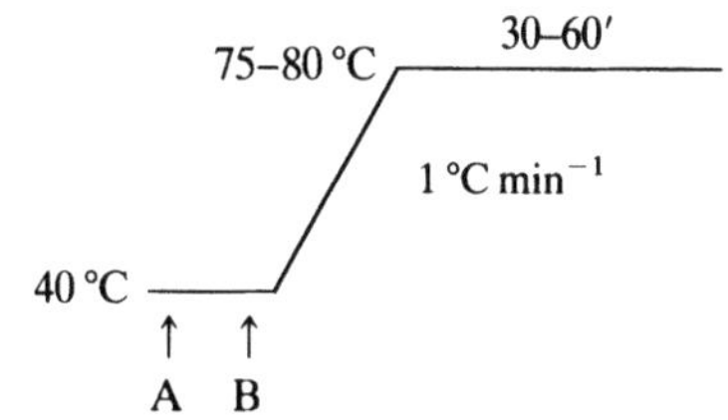

A. Dispersing agent 1 g l^{-1}
B. Dispersed dye x% owf

↓

Rinse

↓

Diazotize — Sodium nitrite 2 g l^{-1}; HCl 5 g l^{-1}; 20 °C; 30′

↓

Rinse

↓

Develop — CI Developer 8 3% owf
Prepare bath by dissolving CI Developer in hot NaOH solution with stirring. Add to cold (20 °C) liquor containing dispersing agent (1 g l^{-1}).
Adjust pH of bath to 4.5, raise temperature to 60 °C over 15 min, and continue at this temperature for 30 min.

↓

Scour — Detergent 2 g l^{-1}; 50 °C; 20–30′

Scheme 50

C. Vat Dyes

The use of this dye class is mostly confined to printing of acetate fibers for high-quality goods owing to the expensive application techniques employed;[124] their application to these fibers by dyeing is discussed by Blackburn.[124]

D. Other Dyes

This author[124] also describes the use of two other dyes, namely, Aniline Black[65] and the natural dye Logwood, applied by the afterchrome method, each of which are used only to a very small extent nowadays on secondary acetate fibers.

V. DYEING OF POLYESTER FIBERS

The first polyester fiber was marketed by ICI in 1948 under the trade name "Terylene." The poly(ethylene terephthalate) (PET) fibers, prepared typically

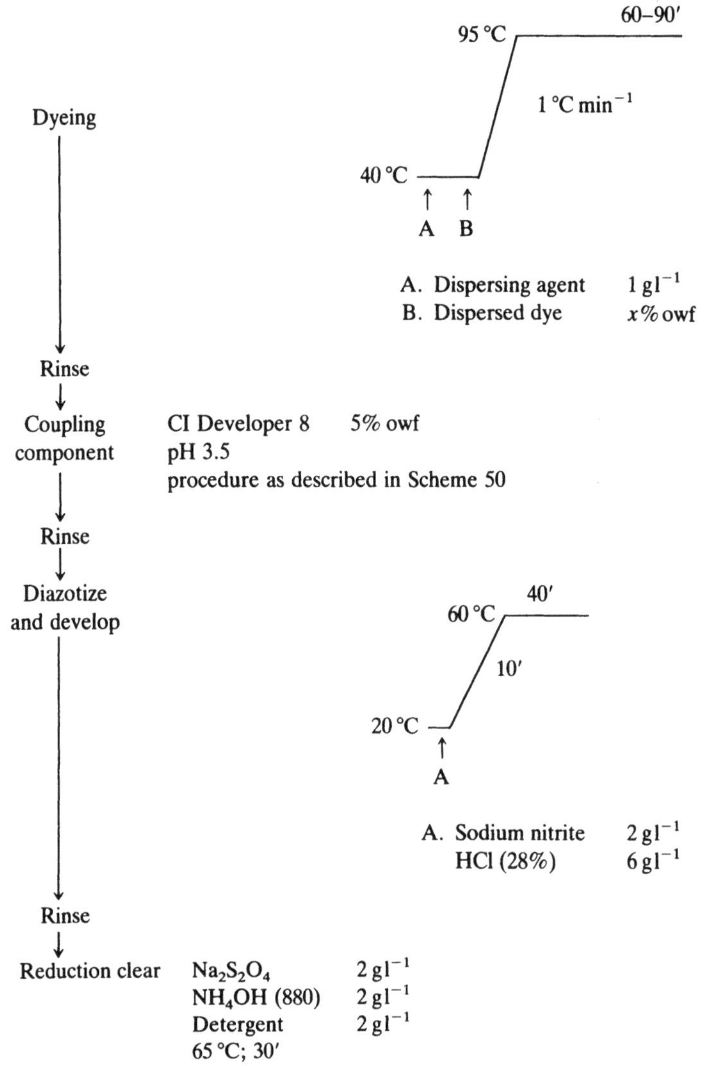

Scheme 51

from ethylene glycol and terephthalic acid.[53,121,154] are now made throughout the world and sold under various trade names. Also produced are polyester fibers derived from PET that have been modified either during fiber production by using shorter molecular chains or modified spinning conditions, or by the use of comonomers.[154,155] The incorporation of a comonomer, such as ethylene oxide or isophthalic acid, disturbs the regular crystalline structure of the fiber, reducing

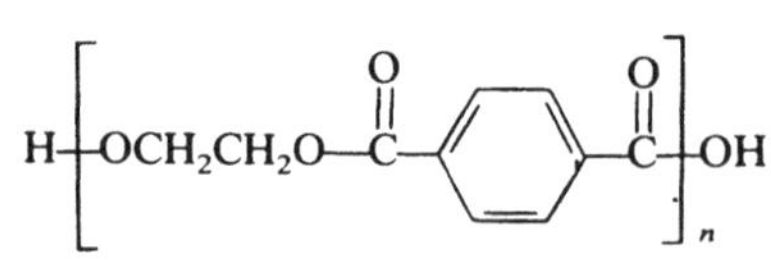

Triacetate

Dyeing	115–120 °C 60 min Serilene Azoic black B2 (paste 50) (YCL) dye	 10% owf
↓		
Rinse	Cold water	
↓		
Diazotization	Sodium nitrite Hydrochloric acid 70 °C 30 min	$2\,gl^{-1}$ $6\,ml\,l^{-1}$
↓		
Reduction clear		

Scheme 52

the degree of crystallinity and orientation. As a consequence the modified comonomer fibers have enhanced dyeability with disperse dyes, the rate of dye diffusion and the saturation value of the fiber being increased. In addition, the pilling performance of many staple fibers of the copolymers is improved. The use of anionic comonomers imparts to the PET fiber substantivity for cationic dyes. Some modified PET fibers, referred to as "carrierless" or "noncarrier dyeing" types, can be dyed using disperse dyes at the boil without carrier. Deep-dye PET fibers are often used in blends with both regular and anionic-modified PET fibers for multicolor effects.

Physical modification of the homopolyester fiber also alters the fiber structure, resulting in modified dyeability with disperse dyes and usually greater resistance to pilling.

The Kodel (Eastman) range of polyester fibers are based on poly(1,4-dimethylcyclohexane terephthalate) and possess different dyeing behavior to PET fibers, which can be attributed to their more open structure.[25] These fibers are also modified to yield basic-dyeable materials as well as fibers with reduced substantivity for disperse dyes.

Polyester fibers are produced as medium- and high-tenacity continuous filament yarns, many of which are texturized, and as staple fibers, usually of medium tenacity, which normally are crimped.

PET fibers are highly resistant to oxidizing and reducing agents. They are reasonably resistant to dilute aqueous solutions of acids but are slowly degraded by concentrated acids. The polymer is more sensitive to alkali; hot solutions of caustic alkali hydrolyze the fiber, although this is limited to surface saponification at temperatures up to the boil.[156] Although PET fibers are prone to hydrolysis, relatively little degradation takes place even at elevated dyeing temperatures (130 °C) provided that the dyebath pH is maintained within the range 4.5–6.0.[155]

The hydrophobic fibers have a very compact structure and are highly

Table 9. Some Properties of Unmodified PET Fibers

Melting point (°C)[155]	252–256
Moisture regain (%) (65% RH; 21 °C)[25]	0.4
Swelling in water (%)[25]	Virtually nil
Tenacity ($g\ tex^{-1}$)[155] (wet and dry)	
Staple fiber	2.7–3.5
Continuous filament	3.5–4.5
High tenacity	5.4–7.2

crystalline. Consequently they show low moisture regain and very little swelling in water (Table 9).

The thermoplastic fibers are heat set either before or after dyeing using hot air, steam, or hot water depending on the material in use and the setting effect required. Heat setting is more often carried out prior to dyeing which is advantageous with respect to the heat fastness of the dyes employed. Presetting, or rather the uniformity of the process, has a significant effect on the dyeing behavior of the fiber as is discussed later.

PET fibers are dyed in the form of loose stock, tow, slubbing, yarn, and piece. Batchwise, semicontinuous and continuous dyeing processes are used. The fibers are dyed using the following classes of dye: disperse, azoic, vat, and basic (anionic-modified polyester).

A. Disperse dyes

This is by far the most widely used class of dye for dyeing PET fibers.

1. Dispersing the Dyes

The dyes are dispersed using the methods described in IV.A.6.

2. Dyeing Behavior

Generally the dyes, as represented by, for example, the Palanil (BASF), Samaron (H), Serilene (YCL), Foron (Sandoz), and Dispersol (ICI) ranges, offer a wide range of hues and exhibit good build-up and very good fastness to light [typically 5–6 to 6–7 1/12 SD; 6 to 6–7 1/1 SD (ISO BO1)], washing (4 to 4–5 ISO CO4), and alkaline perspiration (4–5) on PET. However, some dye combinations give rise to anomalous light fading.

The mechanism of dyeing PET fibers with disperse dyes is identical to that described previously for the dyeing of cellulose triacetate.

The rate of dye uptake onto unmodified PET fiber is very slow at temperatures below the commercial boil (98 °C) owing to the low rate of dye diffusion within the compact and highly crystalline fiber. Indeed, unmodified PET fiber adsorbs disperse dyes considerably less rapidly than do other fibers (e.g., triacetate, polyamides) on which this dye class is used. The dyeing rate is raised to an acceptable level by the use of high temperatures, carriers, and vapor phase transfer (thermofixation).

The dyeing behavior of disperse dyes on polyester fibers varies markedly between dyes. This is of particular importance with regard to compatibility of dyes in admixture and the use of "optimized" and "rapid-dyeing" application methods.

The immersion (exhaustion) dyeing characteristics of disperse dyes on polyester fibers have been classified by the S.D.C.[157] in accordance with four tests for determining the migration, build-up, critical temperature, and diffusion characteristics. The information obtained from these tests can be used to select the appropriate dyes and dyeing conditions for a given dye–polyester fiber system. Many dye manufacturers, however, have adopted their own classification systems for their dye ranges.

Disperse dyes have high substantivity for PET fibers. Dye is rapidly adsorbed at the surface of the fiber and quickly comes to equilibrium with dye in the saturated, aqueous solution located at the fiber surface. As with all dye–fibre systems, the rate at which this initial adsorption occurs depends on the substantivity of the dye for the fiber under the particular dyeing conditions in operation (e.g., leveling agent concentration, temperature). Further dye adsorption and penetration cannot proceed until dye has diffused from the surface to the interior of the fiber. This exhaustion stage is determined by the diffusional behavior of the dye within the fiber and, as in all commercial dyeing processes, is the slowest step. The exhaustion stage is then followed by a leveling phase during which the adsorbed dye becomes evenly distributed throughout the material.

Optimization of the dyeing process therefore involves controlling the exhaustion stage so that the time required for leveling is as small as possible. Many dye makers have developed such optimized dyeing systems for their dye ranges. Such systems typically provide details, under various dyeing conditions (machine, carrier, fiber characteristics), of the optimum starting temperature, heating rate, and final temperature for the exhaustion stage together with the time required at the maximum dyeing temperature to achieve adequate penetration and leveling.

3. Barré Effects

In PET fibers, these arise from physical variations introduced during primary spinning, drawing, texturizing, and heat setting prior to dyeing. Each of these processes influences the crystallinity of the fiber. An increase in crystallinity reduces not only the accessibility of the dye to those regions in the fiber where dye adsorption occurs, and thus the rate of dye uptake, but also the availability of these particular regions and thereby the saturation value of the fiber. The latter property is of little practical significance, with the exception of very highly drawn materials, since dyeing is rarely carried to equilibrium. Hence, variation in the conditions of texturizing, drawing, or heat setting prior to dyeing will be reflected as variations in crystallinity and therefore as unevenness in dyeing. Generally,

Barré effects arising from variation during drawing are less prevalent than are those introduced during texturizing and heat setting.

In general, continuous dyeing processes (thermofixation) cover such physical variations better than exhaust dyeing methods. In the latter case, poor coverage of Barré material is obtained when using dyes which possess low migration power or when dyeing is carried out at the boil using a carrier. Good coverage of fiber variations is achieved by dyeing at elevated temperatures (up to 135 °C), using prolonged dyeing times and by the use of appropriate carriers under high-temperature conditions (e.g., 110–125 °C).

BASF divide their Palanil disperse dyes for polyester fibers into three groups according to their ability to cover variations in texturized PET material:

Group 1 dyes—cover fiber variations very well.
Group 2 dyes—cover variations in the material satisfactorily.
Group 3 dyes—can be used under appropriate conditions for dyeing texturized PET fiber.

4. Oligomers

PET fibers contain a small proportion, typically 0.5–1.5%, of low molecular weight compounds or oligomers. The principal oligomer in unmodified PET is the cyclic trimer of ethylene terephthalate, although other sparingly water-soluble compounds are also present in smaller quantities. These substances migrate from the material during dyeing and steam setting and to a lesser extent during dry-heat setting. Little migration occurs below 110 °C but at the temperatures normally used in high-temperature dyeing and steam setting, migration is more pronounced.

The oligomers may initially be dispersed in the dyebath at 125–130 °C but, as the bath cools, become deposited as crystals on the surface of the material and the dyeing vessel. Although these crystals are not dyed by disperse dyes, they can cause nucleation and growth of crystals of disperse dyes under certain conditions. Furthermore, the deposition of crystals of oligomer on the fiber surface, although of little significance for many forms of dyed PET material, can cause problems with certain types of material. For example, oligomer precipitation can impair the spinning characteristics of fibers and yarns and also reduce liquor flow in package dyeing. The liberation of oligomers during dyeing can be minimized by using the lowest practicable dyeing temperature (i.e., below 125 °C) and the shortest dyeing times; the use of a carrier is advantageous. Discharge of the dyebath without prior cooling, by means of special drains, reduces deposition on both fiber and dyeing equipment. Dyeing machines should be regularly cleared of oligomer deposits. The addition of proprietary dispersing agents, such as Uniperol OE (BASF), to the dyebath maintains the oligomers in fine dispersion, thereby reducing their deposition on fiber and vessel during high-temperature dyeing.

Reduction clearing followed by rinsing and, if necessary, by neutralization using aqueous acetic acid solution, effectively removes oligomer deposits from the fiber surface. More severe conditions, such as boiling sodium hydroxide solution (5% w/w), can be used to remove deposits from dyeing vessels.

5. Dyeing Auxiliaries

As discussed earlier in Section IV, commercial disperse dyes normally contain a dispersing agent. This is usually present in greater concentration in powder forms than in liquid forms. Depending on the particular dyeing method used and the depth of shade applied, further dispersing agent may be added to the dyebath to maintain stability of the dye dispersion throughout dyeing and promote leveling (migration) of the dye. The quantity of this additional dispersing agent required depends on the applied depth of shade and the application method employed. Typically, a concentration of 0.75–1.0 $g l^{-1}$ is usually satisfactory but, with low liquor ratio (e.g., 5:1) dyeing and in the case of deep shades, these concentrations can be doubled. Typical dispersing agents are Dyapol NS (YCL) and Uniperol W (BASF). Certain agents, such as Dyapol SL and Dylac L (YCL), are mixtures of nonionic and anionic compounds which, in addition to enhancing the migration power of the dye, inhibit precipitation of unexhausted dye during cooling of the dyebath. Such dispersing agents are commonly used in conjunction with a more conventional agent such as Dyapol NS (YCL). Generally, the dispersing agents, which are commonly supplied in liquid form, are readily soluble in warm water. Some agents have a tendency to foam and in certain dyeing processes the use of an antifoam may be required.

Disperse dyes are usually applied to PET fibers in the pH range 4–5.5. It is important that the desired pH is maintained throughout the dyeing process to prevent fiber hydrolysis and dye degradation. The appropriate dyebath pH is most commonly achieved using acetic acid (30%), although a buffer system such as ammonium sulfate/formic acid can be used if required.

Since some disperse dyes are sensitive to heavy metal ions, the addition of a suitable sequesterant, e.g., EDTA or Trilon B (BASF), can be made to the dyebath.

Other additives, such as antisoiling and antistatic agents, may also be sometimes used. The dye manufacturers' literature should be consulted to establish the recommended concentrations of each additive required for specific dyes and dyeing methods.

In all cases, additions, suitably diluted in water, should be made to the dyebath at the appropriate temperature prior to adding the dye dispersion.

6. Aftertreatment

Following dyeing, the dyed material is aftertreated to remove unfixed dye, carrier residues, etc. so as to develop optimum fastness properties.

a. Alkaline Scour. This may be adequate in the case of pale depth dyeings: detergent (nonionic), 1 $g l^{-1}$; sodium carbonate, 2 $g l^{-1}$; 70–80 °C; 15–20 min. The material is then rinsed and, if necessary, neutralized by rinsing with dilute acetic acid solution (30%).

b. Reduction Clear. This treatment is normally given batchwise to medium and heavy depth dyeings: sodium hydroxide (flake), $2\,g l^{-1}$; sodium dithionite, $2\,g l^{-1}$; detergent (nonionic), $1\,g l^{-1}$; 50–70 °C; 20–30 min. The material is rinsed and, if necessary, neutralized using dilute acetic acid solution (30%). The above quantities can be approximately doubled when lower liquor ratios are used. In the case of modified-PET fibers which are more sensitive to alkali, ammonia can be substituted for sodium hydroxide.

The reduction-clear treatment may be followed by an alkaline scour in the case of heavy deposits; the above method is applicable. Reduction clearing may only partially remove carrier residues from the dyed substrate. A heat treatment (e.g., 140–170 °C; 30–60 s) effectively removes all traces of carrier. With dyed loose fiber, a reduction clear is always carried out regardless of depth of shade to remove surface deposits of oligomers which otherwise will impair subsequent processing performance.

7. Batchwise Dyeing

a. High-Temperature Dyeing. As previously mentioned, elevated dyeing temperatures (125–135 °C) promote leveling and also improve coverage of fiber irregularities. The technique (Scheme 53) offers several advantages over carrier dyeing at 98 °C including better penetration of the material, shorter dyeing times, and, in some instances, superior fastness properties. Certain substrates, however, especially texturized PET, are liable to lose bulk and elasticity when dyed at elevated temperatures and in such cases recourse is made to the use of lower dyeing temperatures (e.g., 110–125 °C) in the presence of carrier.

b. Carrier Dyeing. As discussed earlier in Section IV, many theories have been postulated to explain carrier action. While many carriers such as Dilatin TCR (Sandoz) and Optinol B (YCL) are suitable for use at 98 °C, other carriers,

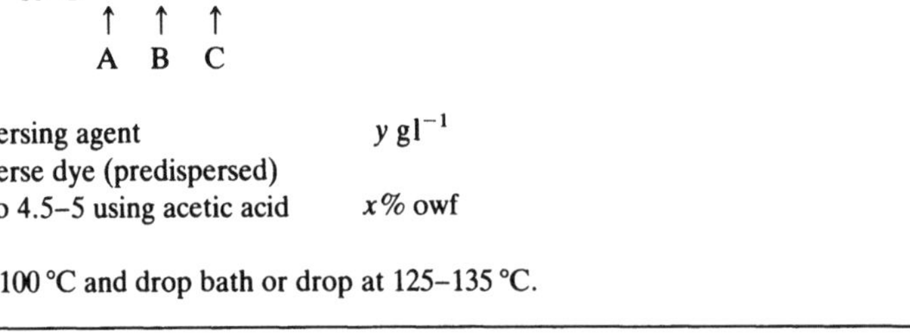

A. Dispersing agent $y\,g l^{-1}$
B. Disperse dye (predispersed)
C. pH to 4.5–5 using acetic acid $x\%$ owf

[a] Cool to 90–100 °C and drop bath or drop at 125–135 °C.

Scheme 53

such as Optinol ENV (YCL), are less efficient at this temperature but are suitable for dyeing temperatures in excess of 98 °C.

Most carriers contain emulsifying agents to facilitate dispersion of the usually sparingly water-soluble compounds. Such agents are commonly anionic and thus, although normally compatible with anionic and nonionic auxiliaries, they are incompatible with cationic products which can result in precipitation of the carrier.

Carriers that have a swelling action on PET fibers promote shrinkage of the material under hot aqueous conditions. This consolidation of the material can impede liquor flow and cause unlevelness and poor penetration; this is of major significance in, for example, package dyeing. Small amounts of carrier can also be used at high dyeing temperatures (i.e., in excess of 98 °C) to promote levelness and coverage of substrate irregularities. However, depending on the type of carrier used, the shrinking effect increases with increasing temperature so that care must be exercised in selecting the most suitable carrier and dyeing temperature. In the case of texturized PET, dyeing at a temperature of 110–125 °C in the presence of carrier is often preferred so as to minimize loss of bulk and elasticity incurred during high-temperature dyeing. Furthermore, such dyeing conditions reduce oligomer liberation.

Carriers which are volatile in steam, such as Dilatin TCR (Sandoz) and Optinol PW (YCL), can, on certain dyeing machines such as winches, give carrier spots or carrier stains on the material (i.e., regions that are more deeply colored than the rest of the material) caused by condensed carrier vapor failing to emulsify on its return to the dyebath. Carrier marking is also obtained by using inadequately emulsified carrier, or when the emulsion breaks down during dyeing, or as a result of mistakes in pretreating the goods with the carrier emulsion.

Residual carrier in the dyed material can give rise to malodor and impair the light fastness of dyeings. This can be minimized by thorough rinsing and reduction clearing (see later); a hot-air treatment (e.g., 140–180 °C for 30–120 s) removes virtually all carrier residue from the substrate.

The amount of carrier used depends on the particular carrier itself, the dye and quantity applied, liquor ratio, dyeing temperature, and material used. Generally, the quantity of carrier required decreases with increasing dyeing temperature and increasing liquor ratio. For example, BASF[155] suggest the following general concentration range for their Palanil Carriers A and AN for use at a liquor ratio of between 10 and 20 : 1, depending on depth of shade:

Dyeing temperature (°C)	Quantity of carrier ($g l^{-1}$)
98	4–6
103–108	2–4
110–120	1–2

YCL[158] suggest that the amount of Optinol B required for winch dyeing is in the range 2–4 $g l^{-1}$, according to depth of shade, and for shorter liquor ratios

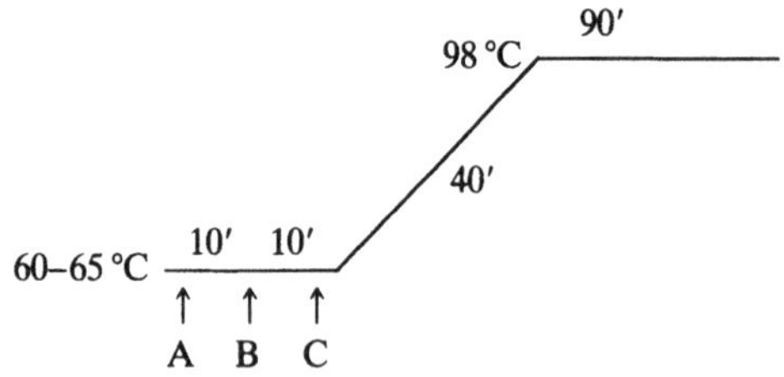

A. Ammonium sulfate — $2\ g l^{-1}$
pH to 5–5.5 using formic acid
Sandozol KB (Sandoz) — $1–2\ ml l^{-1}$
(Dispersing agent)
B. Dilatin NAN (Sandoz) — 6–18% owf
(Carrier)
C. Foron E or SE (Sandoz) dye — x% owf

Scheme 54

(e.g., jig dyeing) double these quantities can be used. Scheme 54 shows a typical carrier dyeing process.

Emulsification of Carrier. The carrier liquid is poured with stirring into warm or hot water. Some carriers may require boiling for a short period to effect complete emulsification. Any residual, nonemulsified carrier must not be added to the dyebath in order to avoid carrier marking of the dyed material.

c. Rapid Dyeing Processes. Dye manufacturers have developed such processes. The Palegal (BASF) method utilizes a proprietary dyeing auxiliary Palegal P (BASF) for use with the Palanil range of disperse dyes. The Serilene V and VX (YCL) and Foron RD (Sandoz) dye ranges are particularly suitable for

A. Permalose TM — 2% owf
(optional)
pH to 4.5–5.5 using acetic acid
B. Dispersol (ICI) dye — x% owf

Scheme 55

application by a rapid-dyeing technique. A typical example is shown in Scheme 55.

8. Semicontinuous Dyeing

The Pad–Roll process[155] which typically entails batching the padded material for several hours at 100 °C is used relatively little.

9. Continuous Dyeing

Thermofixation. The basis of this process, first introduced under the name "Thermosol" by du Pont in 1949,[159] is the use of high temperatures (190–225 °C) to obtain rapid fixation of the dyes. The rate of dyeing of PET fibers increases with increasing temperature so that a dyeing which requires several hours at 100 °C can be achieved in a matter of seconds at 200 °C.[65] This pad–thermofix or pad–bake process makes use of the rapid vapor-phase transfer (or diffusion) of disperse dyes. In essence the process comprises:

Pad → Dry → Bake → Wash off

The PET fiber (usually woven or knitted cloth, although continuous filament tow can also be dyed using this process) is impregnated with a dispersion of the dye together with a wetting agent, thickener, and migration inhibitor. After drying, dye fixation is carried out, typically at 210 °C for 1–2 min. Much of this time is involved in raising the impregnated material to the required temperature so that the actual time that the substrate spends at this elevated temperature is about 30–60 s. The amount of dye that is fixed during this stage depends on the dye itself and the time and temperature of treatment.

Precise control of the latter two elements must be exercised since, if fixation is prolonged unduly, then a slow and progressive loss of color yield occurs due to excessive sublimation of the dye from the cloth. Also, for short periods of fixation, the color yield increases with increasing temperature to a maximum and then decreases due to excessive sublimation. The temperature at which this maximum color yield occurs decreases as the time of fixation is increased.[65]

The particular conditions required for fixation depend on several factors, in particular the equipment available, PET material and the dyes used, and the heat fastness requirements of the dyeing. Although the common temperature range used is 200–210 °C, lower temperatures can be used for dyes of low heat fastness; dyes of high fastness to heat may require temperatures of 215–220 °C for optimum color yield.

In the ICI classification of their Dispersol dyes,[131] transfer rate decreases and heat fastness increases in the order A to D. BASF[155] utilize the contact-heat number (CHN) as a measure of the thermal stability and transfer rate of their Palanil range of disperse dyes, a dye with a high CHN having very high heat fastness but low transfer rate on PET fiber.

In the case of texturized PET knitgoods, while thermofixation gives excellent coverage of fiber variations, the temperature of fixation is reduced to 160–180 °C,

Pad	20–30 °C	
	Disperse dye	x% owf
	Migration inhibitor	y g l^{-1}
	wetting agent	z g l^{-1}
	50–90% pick up	
↓ Dry	110–120 °C	
↓ Bake	30–60 s	
	180–220 °C	
↓ Cool		
↓ Aftertreat		

Scheme 56

according to fiber type, in order to reduce stiffness and loss of bulk of the goods. At these fixation temperatures, superheated steam often gives more rapid fixation than does dry heat. If superheated steam is used for fixation, the pad liquor must have adequate acidity (i.e., pH 4–5) to prevent damage to the dyes and the substrate by the alkaline vapor.

Following fixation, the dyed material is cooled and then scoured or, in the case of medium and heavy depths, reduction cleared to remove unfixed dye. Scheme 56 shows a typical thermosol process for dyeing polyester fabric.

B. Azoic Colorants

Because disperse dyes yield dyeings of high fastness in a variety of hues,[50] the use of azoic combinations on polyester fibers is restricted to the production of black shades. This, in turn, refers to the use of diazotized and developed disperse dyes (see Section IV.B).

C. Vat Dyes

The use of this dye class on polyester fiber is mostly restricted to blends with cotton.[156]

D. Dyeing of Modified Polyester Fibers

Differential dyeing polyester fibers comprise fibers which have differing substantivity for disperse dyes and, in some cases, fibers that are dyeable with both basic and disperse dyes. A major use of differential dyeing polyester fiber is in the carpet trade.

"Noncarrier dyeing" or "carrierless" modified fibers can be dyed without recourse to carriers or at temperatures below 100 °C, while "deep-dyeing" fibers may require a small amount of carrier to achieve level dyeing at these temperatures. Basic-dyeable polyester fibers may, depending on fiber type, be

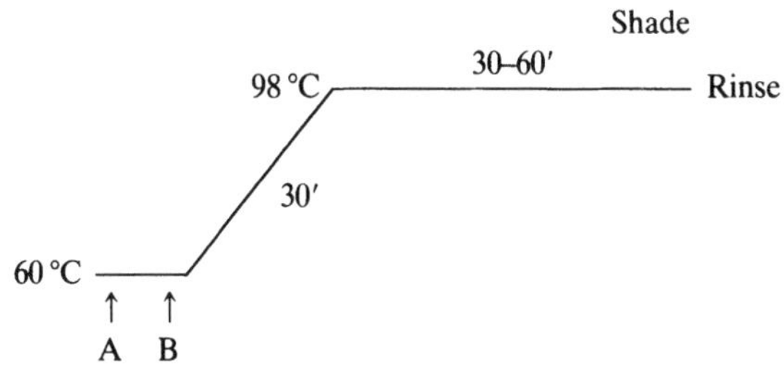

A. Matexil DA-GD (ICI) (dispersing agent)	$0.5\ g l^{-1}$
Acetic acid (80%)	$0.2\ ml l^{-1}$
Silcolapse 5001 (ICI) (antifoam)	$0.1\ g l^{-1}$
Matexil CA-DPL (ICI) (carrier)	$2\ g l^{-1}$
B. Dispersol (ICI) dye (dispersed)	x% owf

Shading. The dyebath is cooled to 80 °C, the dye is added, and dyeing continued at this temperature for 5 min. The bath is raised to 98 °C over 20 min and dyeing continued at this temperature for 20 min.

Aftertreatment. Heavy shades are soaped to remove surface dye, e.g., Synperonic BD (ICI), $1\ g l^{-1}$; 20′; 70 °C.

Scheme 57

dyed under both atmospheric (98–100 °C) and pressurized conditions (120 °C) with the aid of a carrier. Polyester fibers for carpets are available as staple and are dyed as loose stock, yarn, or in piece forms.

1. Dyeing of Deep-Dyeing Polyester Fibers

The following ICI[166] procedure (Scheme 57) yields dyeings of good color yield at the commercial boil.

2. Dyeing of Noncarrier Dyeing Polyester Fibers

The procedure given in Scheme 58 is suggested by ICI to yield deep dyeings on the winch.

3. Dyeing of Anionic-Modified PET Fibers

Such PET fibers modified by the incorporation of typically 5-sulfoisophthalic acid can be dyed by using both disperse and basic (cationic) dyes. Anionic-modified PET fibers are very sensitive to hydrolysis and, consequently, the dyeing pH must be strictly controlled and the dyeing temperature is limited to 120 °C or lower.

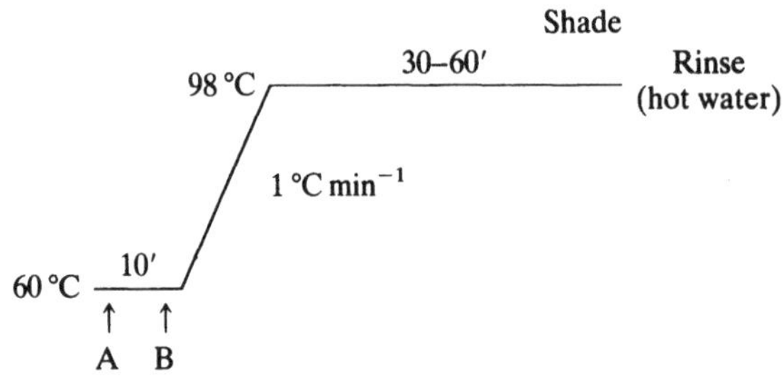

A. Matexil DA-GD (ICI) (dispersing agent) — 0.5 g l^{-1}
   Acetic acid to give pH 5–5.5
B. Dispersol (ICI) dye (dispersed) — x% owf
   Adjust pH to 5–5.5 if necessary

Shading. Cool to 80 °C, add the dye and continue dyeing at this temperature for 10 min. The bath is raised to 98 °C at 1 °C min^{-1} and dyeing continued at this temperature for 30 min.

Aftertreatment. Either "soap" as described in Scheme 57 or reduction clear as follows:

Matexil DN-VD (ICI)	2 g l^{-1}
NaOH (38 Be)	6 g l^{-1}
$Na_2S_2O_4$	2 g l^{-1}
30′; 70 °C	
"Soap" as in Scheme 57	

Scheme 58

Dyeing with Disperse Dyes. The fibers possess slightly higher substantivity for disperse dyes than unmodified PET fiber. The carriers used are those for unmodified homopolymer but the quantity of carrier required is approximately 50% of that used for dyeing the unmodified fiber; at high temperatures (110–120 °C) the carrier may be omitted. Glauber's salt is added to the dyebath to protect the basic dyeable fiber from hydrolysis. The following BASF[155] procedure (Scheme 59) can be used for dyeing yarn to medium depths.

Dyeing with Basic Dyes. Although, generally, the rate of diffusion of basic dyes in anionic-modified PET is less than that in unmodified polyacrylonitrile fibers even in the presence of carrier,[25] good build-up of shade is obtained. The carrier, if required, should contain a nonionic emulsifying agent to avoid precipitation of the cationic dye. The light fastness of basic dyes on basic-dyeable polyester fibers is markedly reduced by the presence of carrier residues in the dyed material and therefore a heat aftertreatment is usually carried out to remove carrier (see Aftertreatment). The wash and wet fastness properties of the dyeings are generally very good, for example, 4–5 (ISO CO3), 4–5 (alkaline perspiration).

The basic dyes are dissolved as detailed in Section VII. A typical dyeing procedure[155] is, essentially, the same as that described above for the application

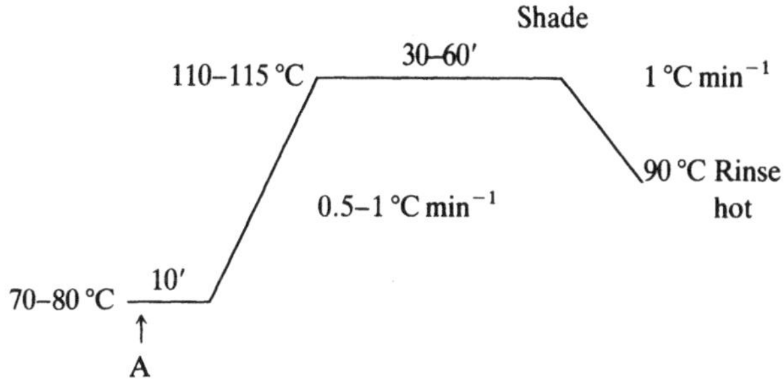

A. Uniperol W (BASF) (dispersing agent)	0.5 g l^{-1}
Acetic acid (30%) to give pH 5	
Palanil carrier B (BASF)	0–1% owf
Palanil (BASF) dye	x% owf
Sodium sulfate	3–6 g l^{-1}

Shading. Cold bath to 70–80 °C to compensate for the higher dye substantivity and thus avoid unlevelness. Add the dye and continue dyeing for 20–30 min at 100–105 °C.

Aftertreatment. Reduction clear
Rinse
Acidify (acetic acid)
Bake 140–160 °C
1–2 min to remove carrier

Scheme 59

of disperse dyes except that dyeing is carried out at pH 4–5 (acetic acid) and shading additions are made at 70 °C.

Many dye manufacturers recommend the use of selected disperse and basic dyes for dyeing these fibers.

4. Dyeing of Differential-Dyeing PET Fibers

Tone-in-tone color effects on carpet piece tufted from both unmodified and deep-dyeing PET fibers can be produced on the winch using the procedure described in Subsection a above.

Biscolor effects on polyester fiber comprising disperse dyeing and basic-dyeable fibers can be produced using the procedure in Scheme 60.

VI. DYEING OF POLYAMIDE FIBERS

The term nylon refers generically to any fiber-, film-, or bristle-forming polyamide. A variety of polyamides is presently manufactured and marketed under several different trade names. The two most important textile repre-

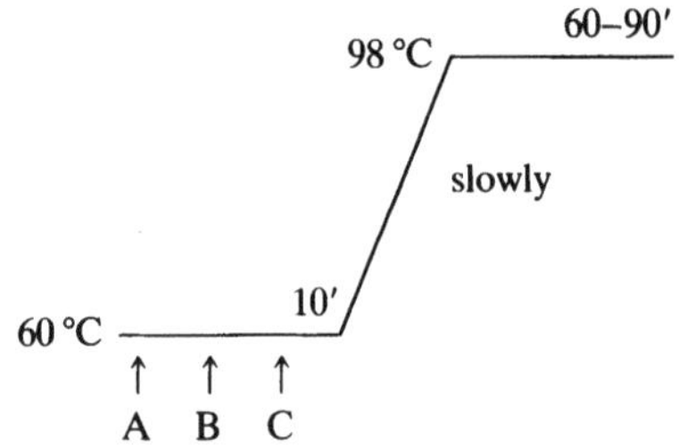

A. Uniperol W (BASF) (dispersing agent)	0.5–1.0 $g l^{-1}$
Trilon B (BASF)	0.25–0.5 $g l^{-1}$
Acetic acid (30%) to give pH 4–5	
Sodium sulfate (anhydrous)	3 $g l^{-1}$
Carrier	2 $g l^{-1}$
B. Palanil (BASF) dye	x% owf
C. Basacryl (BASF) dye	y% owf

Scheme 60

sentatives are nylon 6.6 and nylon 6; other types such as nylon 6.10 and nylon 11 enjoy relatively limited, mostly speciality usage.

Nylon 6.6

$$H\{HN(CH_2)_6NHCO(CH_2)_4CO\}_nOH \quad n = 60\text{–}80$$

This, the first commercially-available wholly-synthetic fiber was introduced by du Pont in 1939 and is made by the polymerization of adipic acid and hexamethylene diamine.[53,121,154,161]

Nylon 6

$$H\{NH(CH_2)_6CO\}_nOH \quad n = 130\text{–}160$$

This was commercially introduced in 1939 by IG Farbenindustrie under the name Perlon L and is prepared by the polymerization of ε-caprolactam.[53,121,154,161]

Although generally very similar, nylon 6.6 has a more compact and crystalline structure than nylon 6. This is manifest in differences in physical properties (Table 10) and dyeing behavior of the two fibers. Characteristically, polyamides have high tenacity (both wet and dry) and relatively high resistance to aqueous alkali although at high temperatures hydrolysis of the amide linkage occurs.[162] Nylons are more sensitive to acids, which cause degradation of the polymer. The hydrophobic polyamide fibers, which are produced as continuous filament and staple fiber, swell little when immersed in water (Table 10).

The thermoplastic fibers undergo a heat- or steam-setting treatment prior to dyeing which has a significant effect on dyeing behavior as is discussed later.

Typically, polyamide fibers contain four different types of polar group. A typical analysis of nylon 6.6 is shown in Table 11. The difference in acidic (—COOH) and basic (—NH_2) group content stems from the use of low molecular

Table 10. Some Properties of Nylon 6.6 and Nylon 6 Fibers

	Nylon 6.6	Nylon 6
Melting point (°C)[122]	263	215
Moisture regain (%)[25] (65% RH; 21 °C)	4–4.5	3.5–5
Swelling in water (%)[25]	2	2

weight carboxylic acids as "chain-stoppers" during polymerization. The number of the different polar groups will vary between fibers from different sources.

The presence of terminal amino groups gives nylon fibers substantivity toward the following types of dye: nonmetallized acid, premetallized, mordant, direct, and reactive. The hydrophobic fiber is also dyeable with disperse dyes. In addition, basic dyes are used for dyeing anionic-modified (i.e., basic-dyeable) polyamides.

Nylon fibers are dyed by using predominantly batchwise methods, although continuous methods are also employed, as loose fiber, yarn woven, and knitted fabric and garment forms.

Polyamides with modified dyeing properties find use predominantly in yarns to produce tone-in-tone and multicolor effects in tufted carpeting and in woven and knitted fabrics. The fibers are produced by chemical modification of the regular nylon 6 or 6.6. Deep- (and ultradeep-) dyeing nylon has considerably more amino end-groups than its regular dyeing progenitor and thus displays higher substantivity for anionic dyes with increased rate of dye uptake and saturation levels. The fiber is produced, typically, by protecting the free amino groups during polymer and yarn manufacture by the addition of *p*-toluene sulfonic acid or phenyl phosfinic acid derivatives in the polymerization stage. Basic-dyeable nylon fibers contain a significantly low proportion of terminal amino groups and have a reduced substantivity for anionic dyes, but are readily dyed by using basic (cationic) dyes. They are prepared by the introduction of sulfonic or carboxylic acid groups into the fiber during polymerization by the use of, for example, 5-sulfoisophthalic acid.[154] Although exhibiting different substantivity for anionic and cationic dyes, both deep- (and ultradeep-) dyeing and basic dyeable polyamides can be dyed to equal capacity using disperse dyes.

Table 11. Typical Analysis of Nylon 6.6[65]

Polar group		Quantity (g-equiv kg^{-1})
Amino end groups	(—NH_2)	0.036
Carboxylic end groups	(—COOH)	0.09
Acylamino end groups	(—$NHCOCH_3$)	0.063
Chain amide groups	(—NHCO—)	0.85

Many factors influence the selection of dyes for nylon fibers, including not only the characteristics of the dye type itself but also the nature of the substrate.

Barré Effects. The chemical and physical properties of nylon fiber are greatly influenced by the processes involved in its manufacture. Chemical (e.g., amino end-group content) variations arising during yarn production give rise to affinity variations with anionic dyes. The drawing process significantly increases the orientation of the fiber and thus the accessibility of the amino end-groups, although there is little evidence that the overall crystallinity of the fiber is altered as a result of this treatment.[53] Variations in the drawing stage will therefore affect the rate of dye uptake of anionic dyes.

Heat setting under tension increases the crystallinity of the fiber with an accompanying reduction in the rate and equilibrium uptake of various dye classes. When setting is carried out using steam, the fiber is swollen and the opposite effect is observed.[163] Thus, any variation in setting conditions leads to unlevel dyeing.

A. Nonmetallized Acid Dyes

This is the most important and versatile dye class used for dyeing polyamide fibers, being suitable for virtually all forms of nylon goods.

1. Dissolving the Dyes

The powder is pasted with cold water to which is added boiling water with stirring; the solution may be boiled. The ensuing dye solution is added to the dyebath through a sieve.

2. Dyeing Behavior

In general, the dyes, as represented by the Nylosan (Sandoz) and Nylomine (ICI) ranges, are applied in the pH range 4–7 and display moderate to very good light fastness, typically (ISO BO1) 4–5 to 6–7 1/1 SD, 4 to 6–7 1/25 SD, and poor to moderate fastness to washing and wet treatments, for example, (1/1 SD) 3–4 to 5 (ISO CO3), 3-4 to 5 (perspiration) (nylon 6.6).

At its isoelectric point, the nylon fiber is considered to be in the zwitterion form. On immersion in an acidic dye solution, protons are adsorbed onto the dissociated carboxylic acid groups imparting an overall positive potential to the substrate.

$$
\begin{array}{c}
H_3N^+\text{—Nylon—}COO^- \\
\text{HX}\ \big\downarrow\ DSO_3^-Na^+ \\
\underset{X^-}{H_3N^+}\text{—Nylon—}COOH \\
\big\downarrow \\
\underset{DSO_3^-}{H_3N^+}\text{—Nylon—}COOH
\end{array}
$$

The rapidly-diffusing acid anion (X^-) is readily adsorbed onto the protonated amino end-groups but is gradually displaced by the larger, slower-diffusing, higher affinity dye anion (DSO_3^-).

In the case of relatively simple, small molecular size dye anions, adsorption at the protonated amino groups will predominate. This electrostatic interaction is favored by the use of low pH values and, since this particular dye–fiber bonding is relatively weak, such dyes exhibit very good migration and leveling properties. The affinity of such dyes is low and, as a consequence, dyeings possess relatively poor fastness to washing and to wet treatments.

The more complex, larger size dye anions possess greater substantivity for the substrate at higher pH values arising from the operation of dispersion forces and will therefore tend to exhibit lower migrating power and level dyeing characteristics. The low diffusional properties of large size dye anions coupled with their high affinity results in the dyeings possessing higher fastness to washing and wet treatments. Generally, the more hydrophobic the dye anion the higher the wet fastness properties of the dyeing.

Because of the more compact and crystalline structure of nylon 6.6, this fiber will generally be dyed less rapidly than nylon 6. Furthermore, the fastness of a given dye to washing and to wet treatments will be slightly higher on nylon 6.6 than on nylon 6.

The mechanism of adsorption of this dye class by polyamide fibers therefore closely resembles that of the dyes by wool (Section II.A). The saturation value of wool, however (i.e., the point at which all the available protonated amino end-groups are occupied by adsorbed dye anions), occurs at depths of shade in the region of 20% owf, which is far outside practical application depths. In contrast, the saturation value of polyamide fibers can in some cases be reached at depths of shade as low as 2–3% owf. The limited number of available dye sites in nylon is of practical significance. Nonmetallized acid dyes vary markedly in molecular size and degree of sulfonation. Since build-up is related to saturation value, then high affinity, large molecular size dyes will generally have build-up characteristics superior to their lower affinity counterparts whose adsorption is mostly confined to the limited number of protonated terminal amino groups.

The dyes are sensitive to both chemical and physical variations in the substrate and vary in their ability to cover such fiber irregularities. Physical variations can be greatly reduced or even eliminated by:

- Using low affinity (high migrating power) dyes.
- Using elevated dyeing temperatures (110–115 °C for nylon 6 and 110–120 °C for nylon 6.6) which facilitate fiber swelling and thus dye penetration and, in addition, promote dye migration. Since polyamide fibers are prone to oxidative degradation during high-temperature dyeing, particularly under neutral and alkaline conditions, an antioxidant, such as Lanalbin B (Sandoz), is commonly employed.
- Pretreating the fiber with specific auxiliaries, typically anionic compounds of moderate affinity that are adsorbed onto the more accessible protonated amino end groups. When dyeing is subsequently commenced, these

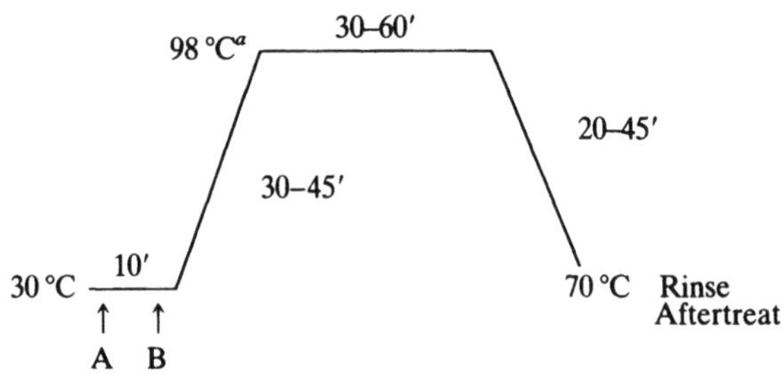

A. Lyogen PAM (Sandoz)	1% owf
Acetic acid (80%)[b]	0.3 $ml\,l^{-1}$
Sodium acetate[b]	2 $g\,l^{-1}$
B. Nylosan N (Sandoz) dye	*x*% owf

[a] Nylon 6, 105 °C; Nylon 66, 120 °C.
[b] To pH 5.5.

Scheme 61

anionic blocking agents must be displaced by the higher affinity dye anions before dye uptake at the more accessible dye sites can occur. Thus, variations in site accessibility are evened out. Such agents are usually applied as a pretreatment prior to dyeing (Scheme 61).

Chemical variations in the substrate are virtually impossible to cover completely, but their effects can be minimized by adopting the above precautions.

As a class, nonmetallized dyes are incompatible when applied in admixture because of the variation in dyeing behavior. In general, dyes of higher affinity are adsorbed at the expense of those with lower affinity. Owing to the variation in dyeing behavior (e.g., build-up, coverage of barriness, pH dependence), many dye makers categorize their ranges. For example, the Nylomine (ICI) dye range is divided into four groups A to D, the latter group comprising metal-complex dyes, and the Nylosan (Sandoz) range includes E, N, F, and Nylosan (no suffix) dyes, further divided into six groups.

Controlled rate of dye exhaustion is necessary even in the absence of fiber variations to obtain level dyeing. This can be achieved by the use of the following.

a. Leveling Agents. In addition to the use of anionic "blocking" agents, auxiliaries are also employed to promote dye migration. For example, Matexil LC-CWL (ICI) is a weakly cationic ethylene oxide condensate that retards dye uptake by forming a complex with the anionic dye.

b. Control of Temperature Rise. The dyebath is set and maintained at the appropriate pH required for the particular dye–fiber combination, as detailed in the dye manufacturer's literature, and the rate of temperature rise of the dyebath is then carefully controlled. The recommended optimum application pH usually

Nylon 6.6

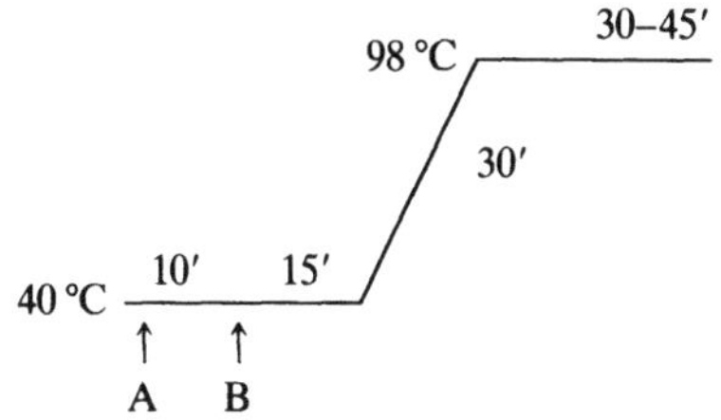

A. Matexil LA-NS (ICI) 1% owf
   pH 5–5.5
B. Nylomine A, B or C (ICI) dye[a] *x*% owf
   Matexil LC-CWL (ICI)[a] 1% owf

Shading. The bath is cooled to 60 °C, the dye is added, and dyeing continued for 10 min at this temperature. The bath is then raised to 98 °C and dyeing continued at this temperature for at least a further 15 min.

[a] Dissolved together.

Scheme 62

depends on the amount of dye applied, the lower pH values being used for heavy depth dyeings. This pH is maintained within close limits throughout the dyeing process using, for example, acetic acid, ammonium sulfate, or a suitable buffer system such as acetic acid/sodium acetate or a mixture of phosphates. Scheme 62 shows a typical application process.

c. Controlled Lowering of Dyebath pH. The principle of this technique is that the dyebath is set usually under alkaline conditions (to reduce substantivity of the dye for the fiber). A constant temperature is maintained throughout most of the dyeing process and the dyebath pH is then reduced to increase substantivity and thus dye exhaustion. It is important that the added acid is evenly distributed throughout the material, otherwise unlevel dyeing will result. Several processes are available, an example being the Sandacid V/VS system developed by Sandoz that is suitable for all their Nylosan dyes.

Sandacid V and VS are esters which, when added to a boiling dyebath at pH 8–10, gradually hydrolyze over 45–60 min to liberate acid which reduces the dyebath pH by up to three units. This process, shown in Scheme 63, is recommended for dyeing carpet fiber where good penetration and levelness are required.

d. Rapid Dyeing Methods. In the case of package dyeing of yarn and where the substrate and machine are suitable, a rapid dyeing method can be used (Scheme 64).

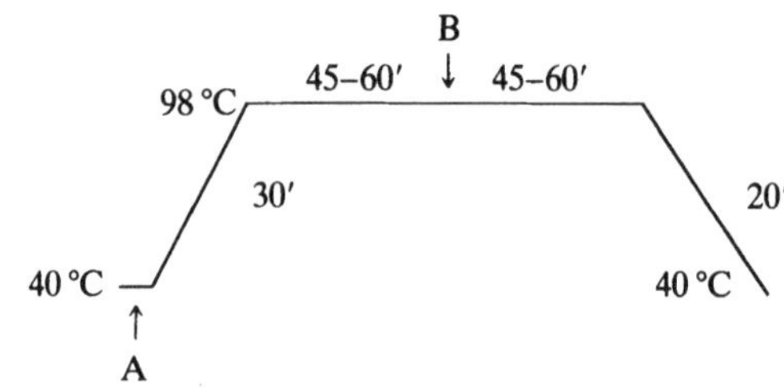

A. Lyogen PAM (Sandoz)	0.5–2% owf
Ammonia[a]	0.2–1.0 ml l^{-1}
Nylosan Groups I–VI (Sandoz) dye	x% owf
B. Sandacid V (Sandoz) liquid	0.2–1.0 ml l^{-1}
to final pH of 5.5–5.8	

[a] To pH 8–10 depending on fiber substantivity for 70–80% bath exhaustion.

Scheme 63

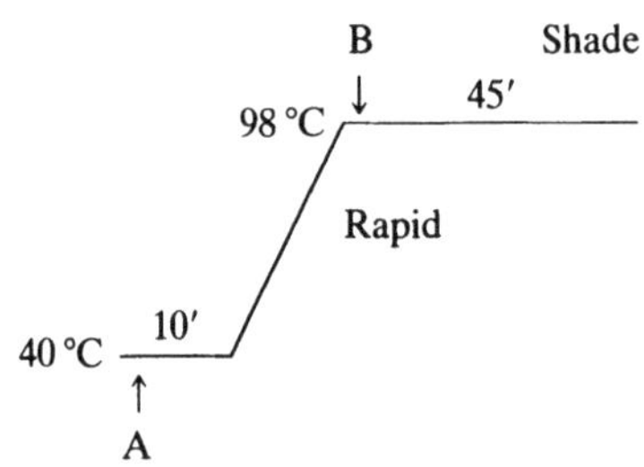

A. Matexil LA-NS (ICI)	1% owf
Ammonium acetate to pH 6–6.5	
B. Nylomine C (ICI) dye[a]	x% owf
Matexil LC-CWL (ICI)[a]	1% owf

Shading. The dye[a] and Matexil LC-CWL[a] are added and dyeing is continued for 30 min at 98 °C.

[a] Dissolved together.

Scheme 64

3. Aftertreatment

It is common to aftertreat the dyed material with either a natural or a synthetic tanning agent to improve the wet and wash fastness properties of the dyeings.[164]

a. Natural Tanning Agent. A typical treatment is shown in Scheme 65. It is considered that this treatment, known as the "full back-tan," results in the formation of a large molecular size, sparingly water-soluble potassium antimonyl tartrate complex located at the periphery of the dyed fiber, which restricts

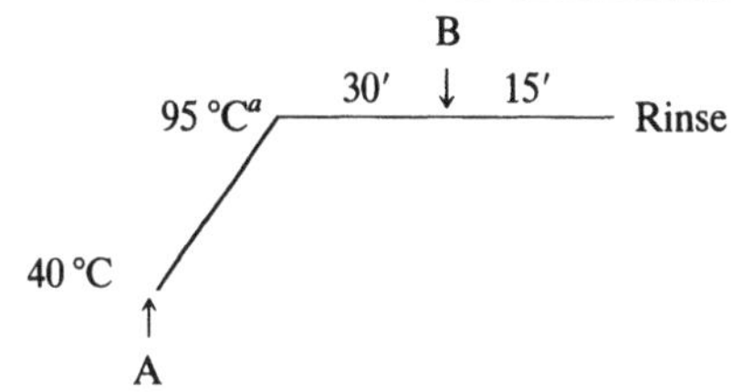

A. Formic acid (85%) 4% owf
Tannic acid 2% owf
B. Potassium antimony tartrate 1% owf
(predissolved)

[a] Nylon 6, 60 °C; Nylon 6.6, 95 °C.

Scheme 65

diffusion of anionic dye molecules out of the substrate during subsequent washing and wet treatments.

In the case of aftertreatment being carried out at low liquor ratio (e.g., on beam or jig machines) the tartar emetic treatment should be carried out in a fresh bath to avoid precipitation of the potassium antimonyl tartrate complex which is very difficult to remove from the substrate.

Although the full back-tan results quite often in considerable improvement in wet and wash fastness, the aftertreatment suffers several disadvantages: tartar emetic is poisonous; the process is time-consuming and can be two-stage; the yellow/fawn impurities in commercial tannic acid can cause dulling of shade; tannic acid discolors due to oxidation under alkaline washing conditions.

Thus, use of the full back-tan is usually restricted to heavy, deep shades.

b. Synthetic Tanning Agents (Syntans). Typically, syntans, as exemplified by Cibatex PA (Ciba-Geigy), Nylofixan P (Sandoz), and Matexil FA-SNX (ICI), are high molecular weight formaldehyde polycondensates of sulfonated dihydroxydiphenylsulfone. It is considered that, unlike the full back-tan aftertreatment, the syntan does not form a "skin" as such around the outer surface of the dyed fiber, but instead is located within the periphery of the dyed fiber and thereby presents a "barrier" to movement of dye molecules out of the material during washing and wet treatments.

Although not as effective as the full back-tan, an aftertreatment with a syntan nevertheless results in a marked improvement in wash and wet fastness of nonmetallized dyes on nylon. Aftertreatment can be carried out also in the exhausted dyebath. For example, Ciba-Geigy recommend that the exhausted dyebath be cooled to 80 °C and the pH adjusted to 5.5. A dilute solution of Cibatex PA (1–3% owf) is added and the goods are treated for 15 min at 70–80 °C. The syntanned material is then rinsed and dried.

Aftertreatment can also be carried out in a fresh bath. For example, ICI suggest that the dyed material is entered into a fresh bath set with 4% formic acid

(85%) and 0.5–2% Matexil FA-SNX, and the temperature then raised to 95 °C in the case of nylon 6.6 and to 70 °C for nylon 6. Treatment is continued at the appropriate temperature for 30 min, after which the syntanned material is rinsed and dried.

The single-stage treatment with syntan is simpler and less time-consuming than the full back-tan and does not affect shade. The aftertreatment can, however, result in a slight decrease in light fastness of the dyeing.

B. Disperse Dyes

Although they cover chemical variations present in the fiber but are more sensitive to physical variations, disperse dyes nevertheless cover fiber irregularities better than any other dye class. The dyes in general exhibit very good migration and leveling properties and moderate to very good buildup. Although the light fastness of dyeings is usually adequate (typically (Nylon 6 and 6.6) 4–5 1/1 SD; 4 1/25 SD (ISO BO1), fastness to washing and wet treatments are, with the exception of 2:1 metal-complex disperse dyes, generally very poor, for example, ISO CO3, 4 nylon 6.6; 3–4 nylon 6 and, consequently, the use of disperse dyes on polyamides is restricted to pale shades and applications where high wet and wash fastness is not required.

Perhaps the most common use of disperse dyes is in nylon hosiery dyeing on drum and paddle machines; woven and knitted fabric is also dyed batchwise on jet, jig, and winch equipment.

The dyes are dispersed as previously discussed in Section IV.A.6. The mechanism of dyeing polyamide fibers with disperse dyes is identical to that discussed in Section IV.A.5.

Typical processing conditions for dyeing fabric are shown in Scheme 66.

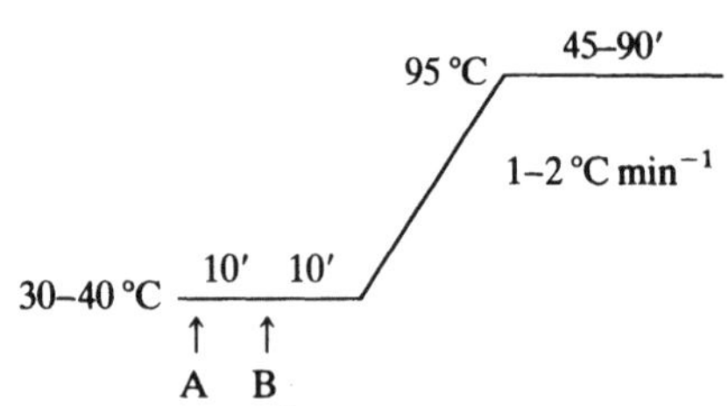

A.	Nonionic dispersing agent	$1\ g l^{-1}$
	Sodium hexaphosphate	$1\ g l^{-1}$
	pH adjusted to 6–7	
	Antifoam agent	$0.5\ g l^{-1}$
B.	Disperse dye (predispersed)	x% owf

Scheme 66

C. Mordant Dyes

These dyes cover physical variations well but are sensitive to chemical irregularities. They yield economical shades on polyamides of very good fastness to washing (e.g., 4–5 ISO CO3) and light [e.g., 6–7 1/1 SD, 6 1/12 SD (ISO BO1)].

The dyes are dissolved as described in Section II.B.6.

Chromium(VI), as chromate or dichromate ions, is adsorbed onto the terminal amino groups in the substrate.[165] Since these groups are present in relatively low concentration, the amount of chromium(VI) adsorbed is correspondingly low. Furthermore, reduction of chromium(VI) to chromium(III), which is necessary for complex formation with the mordant dye (see Section II.B), occurs very slowly. Thus, both pre- and metachrome dyeing methods give low color yields and consequently mordant dyes are applied to polyamide fibers using the afterchrome dyeing method. Because of the difficulty in achieving complete conversion of Cr(VI) into Cr(III) during the chroming stage, a reducing agent, usually sodium thiosulfate, is commonly added as a final treatment. Addition of reducing agent is, however, unnecessary if high temperatures (up to 130 °C) are employed.[166]

For environmental reasons, however, and also because of the difficulties involved in color matching afterchromed dyeings, the use of this dye class on nylon fibers is mostly restricted to a relatively few specific end uses. A typical procedure for the dyeing of half-hose is provided in Scheme 67.

A.	Mordant dye	x% owf
	Acetic acid (30%)	3–5% owf
	Sodium sulfate	10% owf
B.	Formic acid (85%)[a]	1–2% owf
C.	Sodium dichromate	0.25–2.5% owf
	Sodium thiosulfate	0.5% owf

[a] If necessary to exhaust the dye.

Scheme 67

D. Premetallized Acid Dyes

The dyes are dissolved as described in Section III.C.1.

1. 1:1 Metal-Complex Dyes

The low pH (approximately 2) conditions normally employed for the application of these dyes, such as the Palatin Fast (BASF) range, to wool cannot be used for their application to nylon owing to the fiber's sensitivity to acid. Selected 1:1 premetallized dyes can, however, be applied in the region pH 4–6 in the presence of a weakly cationic leveling agent and yield dyeings of good fastness to washing, for example, 4–5 to 5 ISO CO3 (1/1 SD), and to light, typically 5–6 to 6 1/1 SD (ISO BO1). Typical application conditions are given in Scheme 68.

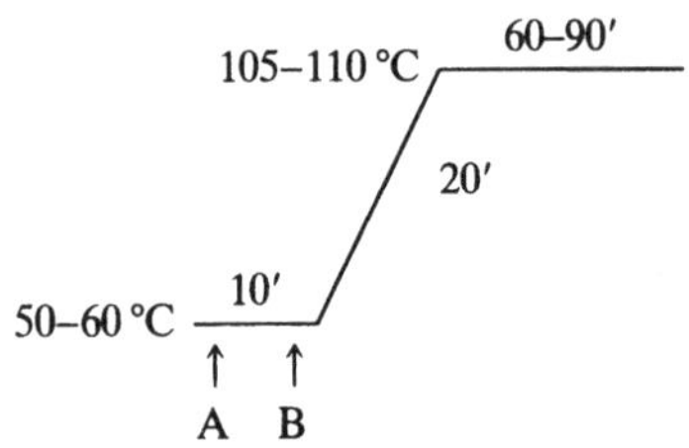

A.	Uniperol PA (BASF) (levelling agent)	1–2% owf
	Ammonium sulfate	2 g l^{-1}
	Formic acid (85%)[a]	1–2% owf
B.	Palatin Fast (BASF) dye	x% owf

[a] If necessary.

Scheme 68

2. 2:1 Metal-Complex Dyes

These dyes, as exemplified by the weakly polar Avilon (Ciba-Geigy), monosulfonated Lanasyn S (Sandoz), and disulfonated Acidol M (BASF) ranges, although providing relatively dull hues, are widely used on polyamide fibers. The dyes generally exhibit very good buildup properties, but their ability to cover fiber irregularities varies according to dye type. They are applied to various forms of material, from loose fiber to piece goods, and yield dyeings of typically good fastness to washing [e.g., 5 (ISO CO3)], wet treatments (4–5 to 5 alkaline perspiration), and light (ISO BO1) 6–7 to 7 or higher 1/1 SD, 6 to 7 1/3 SD.

The application conditions vary according to the type of 2:1 premetallized dye used as shown in Schemes 69 to 71. An aftertreatment with the full back-tan or a syntan may be given when high wet fastness of the dyeings is demanded.

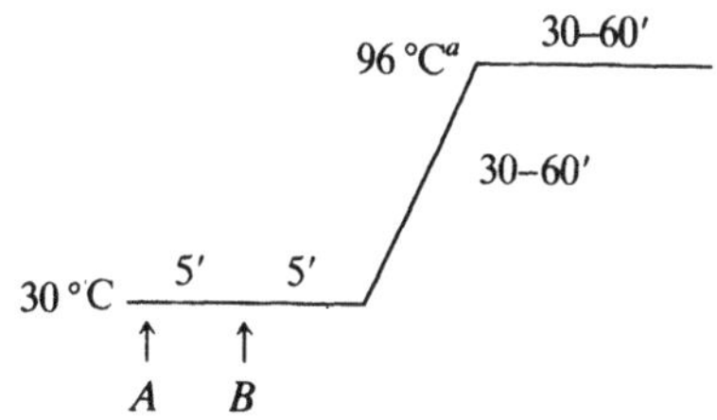

A. Ammonium sulfate 1–3%
Albegal SW (Ciba–Geigy) 1–2.5% owf
B. Avilon (Ciba–Geigy) dye x% owf

[a] Dyeing temperatures:
96 °C Nylon 6 texturized
120 °C Nylon 6 nontexturized

108 °C Nylon 6.6 texturized
130 °C Nylon 6.6 nontexturized

Scheme 69

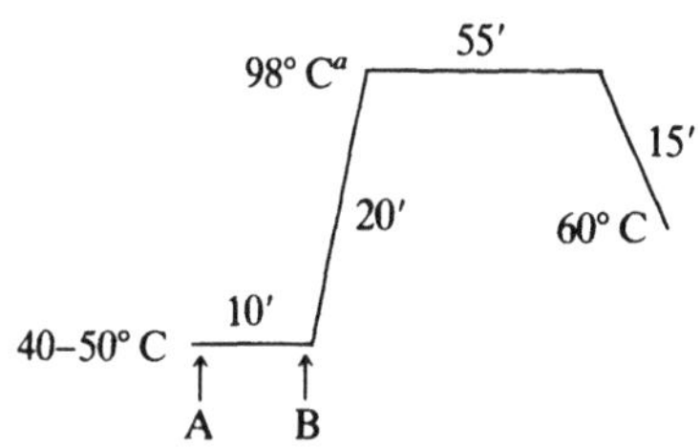

A Lyogen PAM (Sandoz) 0.5–2% owf
Ammonium sulfate 1–4% owf

B Lanasyn S (Sandoz) dye x% owf

Shading cool the bath to 60–70° C prior to adding the dye.

[a] Dyeing temperatures of up to 130° C can be used according to the type and form of substrate.

Scheme 70

E. Direct Dyes

Direct dyes are sensitive to both physical and chemical variations in the fiber and possess poor compatibility in admixture. Selected members, usually applied individually, are however used for the production of economic, heavy shades, and yield dyeings of good fastness to washing and wet treatments.[161]

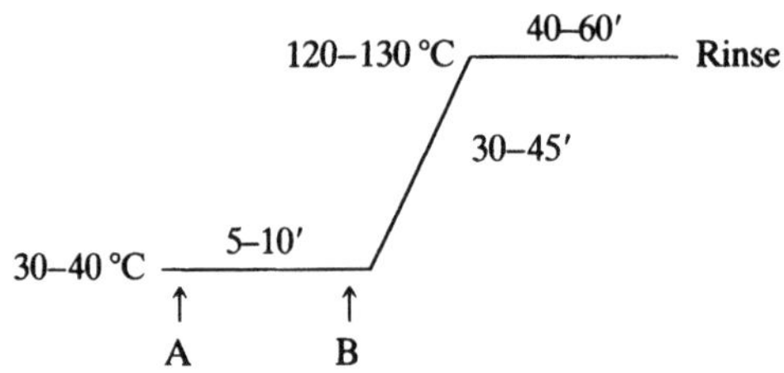

A. Uniperol SE (BASF) $0.5\ g l^{-1}$
Acetic acid (30%) or acetic acid/sodium acetate to pH 4
B. Acidol M (BASF) dye x% owf

Shading. The bath is cooled to about the boil prior to addition of the dye.

Scheme 71

F. Reactive Dyes

Reactive dyes enjoy relatively little use on polyamide fibers owing to their sensitivity to chemical variations in the fiber and their poor migration properties on these fibers. Ginns and Silkstone[161] discuss the use of this dye class.

G. Dyeing of Modified Nylons

Tone-in-tone effects on regular and deep-dyeing nylons can be produced using selected anionic dyes that highlight the contrast between the different fibers.[167] Strict control of the pH and temperature rise of the dyebath as well as duration of dyeing must be exercised to secure optimum contrast effects. When required, the wet and washing fastness of the dyeings can be improved by an appropriate aftertreatment. Scheme 72 shows a typical dyeing process.

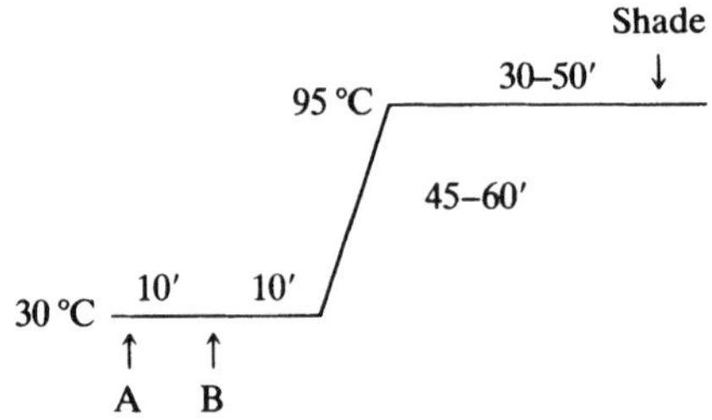

A. Sodium hydrogen phosfate[a] 6% owf
Disodium hydrogen phosfate[a] 1% owf
B. Nylomine B (ICI) dye x% owf

Shading. The bath is cooled to 70 °C and the dye added. After 10 min at this temperature the bath is raised to 95 °C and dyeing continued at this temperature for 30 min.

[a] To pH 6.

Scheme 72

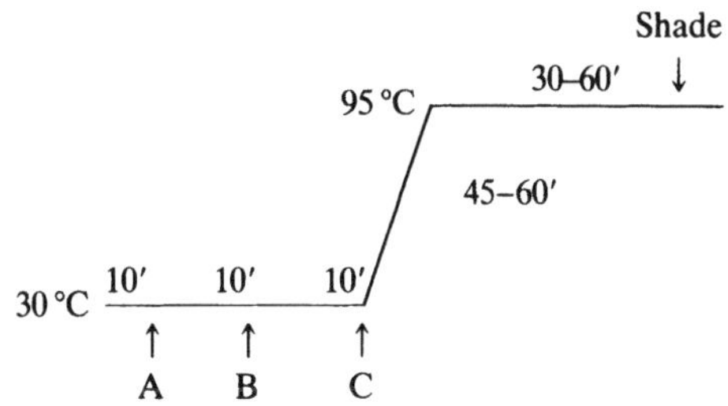

A.	Sodium hydrogen phosfate[a]	6% owf
	Disodium hydrogen phosfate[a]	1% owf
	Matexil DN-VL 200 (ICI)	1 $g l^{-1}$
B.	Synacril (ICI) dye	x% owf
C.	Nylomine (ICI) dye	x% owf

Shading as for Scheme 72.

[a] To pH 6–6.5.

Scheme 73

Multicolor effects[167] can be obtained using combinations of differential-dyeing and basic-dyeable nylon fibers and selected anionic and basic dyes (Scheme 73). An aftertreatment with syntan or full back-tan can be given to improve wash and wet fastness of the dyeings.

VII. DYEING OF ACRYLIC FIBERS

The first commercial acrylic fiber was introduced by du Pont in 1948 under the trade name "Orlon."[154] The fiber is now manufactured throughout the world under various trade names.

Production of the fiber, a regular polymer of acrylonitrile prepared using redox catalysts (potassium persulfate and sodium bisulfite), presented considerable difficulties owing to the polymer's high melting point and poor solubility in the limited number of solvents available. Furthermore, because of the compact structure and high glass transition temperature (T_g) of the homopolymer (in the region of 104 °C[168]), the fiber had to be dyed at high temperatures.

Consequently, copolymerization was used to improve dyeability and to increase the solubility in solvents. Acrylic fibers contain 5–10% of one or more comonomers. Acrylic comonomers such as methyl acrylate, vinyl acetate, and methyl methacrylate improve solvent solubility and reduce the structural regularity of the polymer, thereby increasing the thermoplasticity, i.e., the T_g value of the polymer is reduced to 75–85 °C. This reduced compactness of the fiber structure thereby enhances dye diffusion and therefore dyeability. The fiber contains chain terminal acid groups resulting from the redox polymerization initiators. Additional comonomers containing carboxylic, sulfonic, or phosphoric

Table 12. Some Properties of Acrylic Fiber

Moisture regain (%) (65% RH; 21 °C)	1–2
Swelling in water (%)[25]	Very slight

acid groups, such as sodium vinyl sulfonate, enhance the substantivity of the fiber for basic (cationic) dyes. Basic comonomers containing amino or pyrimidino groups, such as 2-vinylpyridine, confer substantivity for anionic dyes.

Thus, differential dyeing acrylic fibers are available comprising, typically, acid-dyeable and basic-dyeable types both of which can be dyed using disperse dyes, but only either anionic or cationic dyes, respectively.

Various chemical modifications of acrylic fibers have been developed, including graft polymerization. Physical modification of the fiber, brought about by developments in the polymerization, dope preparation, and spinning stages, has resulted in, for example, high bulk fibers, porous (absorbent) fibers, and hollow fibers.[169]

Modacrylic fibers contain more than 15% of comonomers to confer special properties, for example, reduced flammability by the incorporation of 40–60% vinyl chloride. Such fibers enjoy much less textile usage than acrylic fibers and generally exhibit similar, but often reduced, dyeability.

Acrylic fibers are both dry and wet spun, the latter system being preferred as it gives fibers with a more accessible microstructure[168] and consequently better dyeability. The fibers, produced as continuous filament and as staple fiber, are usually set either using dry heat or steam to improve tenacity and impart bulk. The majority of acrylic fiber is used in staple form and is dyed as loose stock, tops or tow, woven and knitted fabrics and garments. Modacrylic fiber is commonly dyed as loose stock, tops or tow, and yarn.

Although acrylic fibers have a low degree of crystallinity, they are highly orientated[25] and consequently swell to a small extent in water at low temperature (Table 12). Of greater significance with regard to water absorption, fiber swelling, and dyeing is the influence of T_g. At temperatures below T_g, little water absorption and dyeing occurs owing to the compact, glass-like structure of the fiber. At temperatures above T_g, water and dye molecules can more readily penetrate the fiber as a result of the increased segmental mobility of the composite polymer chains.[149]

Acrylic fibers are dyed using the following classes of dye: basic and disperse. Batchwise and continuous dyeing methods are employed.

A. Basic Dyes

This is by far the most important class of dye used on acrylic fibers. Basic dyes dissociate in water to yield colored cations and are characterized by their brilliance and very high tinctorial strength. This dye class, which includes some of the earliest synthetic dyes, was originally used for dyeing wool, silk, and mordanted cotton, but the generally poor light fastness of the dyeings suppressed their use until the introduction of acrylic fibers on which the dyes exhibited higher light fastness and very good fastness to wet treatments. Currently, in addition to their primary use for dyeing acrylic fibers, basic dyes are used for dyeing acid-modified polyamide (Section VI) and polyester (Section V) fibers.

The chemical constitutions of cationic dyes have recently been reviewed comprehensively by Raue.[170] Further details can be found elsewhere.[59,171,172] Basic dyes belong to a variety of chemical classes and can be divided into two groups according to whether the positive charge on the cation is delocalized over the chromophoric system or is localized on an ammonium group that is external to the chromophoric system.

1. Positive Charge Delocalized over the Dye Cation

These dyes, which include some of the oldest synthetic dyes, are tinctorially the strongest but are the least fast to light and include, for example, triarylmethane dyes such as CI Basic Violet 14 **(83)**. Methine dyes such as CI Basic Yellow 28 **(84)** are also represented, as are Azines. Nitro dyes provide a small number of examples for use on acrylic fibers.

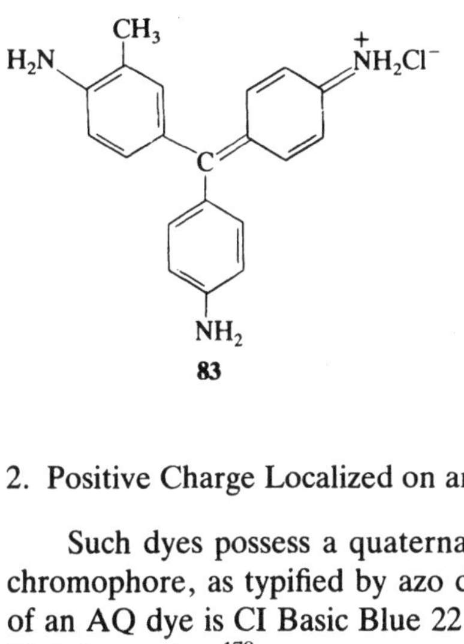

83

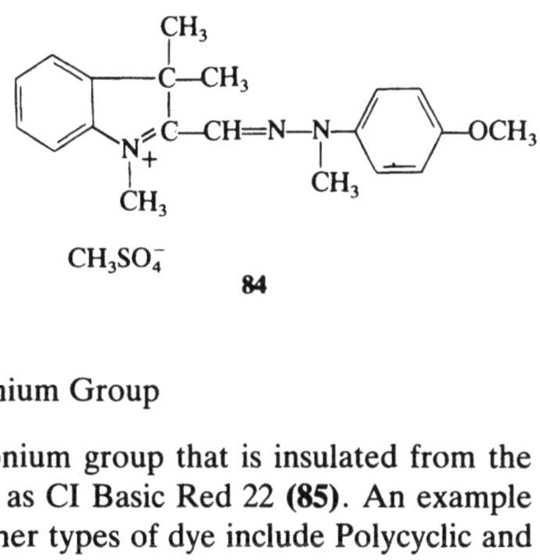

84

2. Positive Charge Localized on an Ammonium Group

Such dyes possess a quaternary ammonium group that is insulated from the chromophore, as typified by azo dyes such as CI Basic Red 22 **(85)**. An example of an AQ dye is CI Basic Blue 22 **(86)**. Other types of dye include Polycyclic and Phthalocyanine.[170]

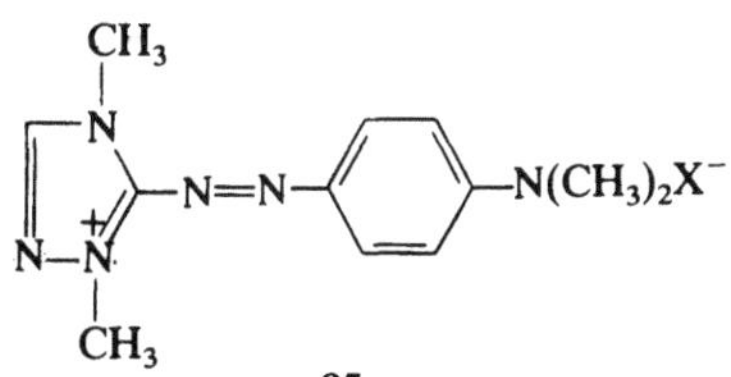

85

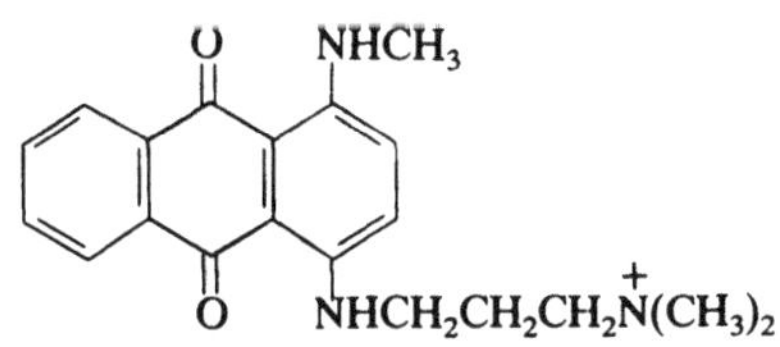

86

3. Dissolving the Dyes

To facilitate rapid dissolution and to avoid problems presented by dusting with powders, many basic dyes are marketed in liquid form as highly concentrated solutions of the dye in aqueous solutions of acids or organic solvents in the presence of additives to promote stability of the solution.

Powder forms are pasted with a portion (e.g., half) of the acetic acid used for dyeing, to which is then added (e.g., 10–40 times the amount of) hot or boiling water. The ensuing dye solution is sieved when added to the dyebath.

Liquid brands can be added directly to the dyebath or may be diluted by pouring into 2–3 times their volume of water.

4. Dyeing Behavior

While many of the earlier basic dyes are still of interest for the economic production of deep shades on acrylic fibers, their use has been superseded by that of new dyes, typified by the Basacryl (BASF), Yoracryl (YCL), Maxilon (Ciba-Geigy), and Astrazon (Bayer) ranges, which have been specifically developed for acrylic dyeing, with attention being paid to improved light fastness and dyeing behavior, achieved by means of either modification of established chromogens or new chromophoric systems. A relatively recent introduction has been that of the more hydrophilic migrating cationic dyes for acrylic fibers, as exemplified by the Maxilon M and BM (Ciba-Geigy), Basacryl MX (BASF), and Remacryl E (Hoechst) ranges.

Generally, basic dyes provide a wide range of hues and exhibit excellent fastness to light [e.g., 7 to 7–8 1/1 SD; 6 to 6–7 1/12 SD (ISO BO1)], washing [typically, (ISO CO3) 5 1/1 SD], and wet treatments (alkaline perspiration, 5) on acrylic fibers (for example, Orlon 42).

Most acrylic fibers contain chain terminal sulfate and sulfonic acid groups derived from the redox initiators used in polymerization. Additional weakly acidic (e.g., carboxyl) or strongly acidic (e.g., sulfonate) groups may also be present as a result of copolymerization. The number and nature of the acidic groups varies markedly between different commercial fibers, as shown in Table 13. Since the fiber contains acidic groups, it is considered to act as an ion-exchange medium,[65] i.e., the cationic dye molecules displace the cations associated with the fiber.

$$F \begin{cases} SO_4^{2-} \\ SO_3^- \\ COO^- \end{cases} \begin{matrix} K^+ \\ Na^+ \\ H^+ \end{matrix} \rightleftharpoons D^+X^- \qquad F = \text{fibre},\ D = \text{dye}$$

Basic dyes have very high substantivity for acrylic fibers and therefore binding forces other than electrostatic may be considered to operate, for example, dispersion forces and, due to the presence of cyano groups in the substrate, dipole forces.

The high substantivity of the dyes results in rapid dye adsorption onto sites at

Table 13. Acidic Group Content of Some Acrylic Fibers[168]

Fiber	Group content m-equiv kg^{-1}	
	Strongly acid	Weakly acid
Orlon 42 (du Pont)	46	17
Dralon (Bayer)	48	53
Acrilan 16 (Monsanto)	31	21
Acribel (Fabelta)	28	30

the fiber surface. When these surface sites are saturated (totally occupied) the dyeing rate is determined by the rate of diffusion of dye from the surface to the interior of the substrate. Diffusion of basic dyes in acrylic fibers is highly dependent on temperature. At temperatures below T_g, little or no diffusion occurs owing to the compact structure of the fiber. At or about T_g, however, the rate of dye diffusion increases dramatically due to the onset of increased segmental mobility of the component polymer chains. Further increase in temperature results in a marked increase in segmental mobility with a concomitant increase in diffusion rate and hence dyeing rate. Indeed, at 100 °C the rate of dyeing can change by approximately 30% for a 1°C change in dyeing temperature.[65]

The generally high wet and wash fastness properties of basic dyes on acrylic can be attriubted to both the high affinity of the dyes and to the influence of the T_g of the fiber. Although dyeing is carried out at temperatures above T_g to facilitate dye diffusion and adsorption, domestic washing is carried out at temperatures below T_g to avoid fiber deformation. At such temperatures, dye diffusion out of the fiber will be very low giving rise to high fastness to aqueous agencies.

The migrating power of the majority of basic dyes on acrylic fibers is low owing to their high affinity. The high thermoplasticity of the fiber at temperatures above T_g precludes the use of prolonged dyeing at high temperature to promote dye migration and thus leveling. Consequently, leveling is achieved by regulating the rate of dye uptake; various approaches are employed.

The adsorption of basic dyes depends on both the quantity and nature of the acidic groups in the fiber. The degree of dissociation of weakly acid groups is hlghly pH-dependent while the strongly acidic groups are fully dissociated over a wide range of pH values. Thus, at any given pH the extent of dye uptake is determined by the number of accessible, ionized acid groups present in the substrate. In practice, the dyebath is usually buffered in the pH region 4–5.5 to suppress ionization of the weakly acidic groups and thereby effect some measure of control of dyeing rate, and hence leveling. The rate of dye uptake cannot, however, adequately be controlled by using pH alone due, in the majority of fibers, to the presence of strongly acidic groups and to the very high affinity of the dyes. Furthermore, the marked variation between fiber types (Table 13) does not make pH control a practical proposition.

In practice, optimum results (penetration, leveling, etc.) are secured by means of control of the pH and temperature of dyeing and the use of dyebath additives.

a. Dyebath Additives.

Electrolytes. The addition of neutral electrolytes, such as sodium sulfate, to the dyebath reduces the rate of dye uptake by lowering the initially high negative surface potential of the fiber. The fast diffusing metal ions (Na^+) also compete with the dye cations for negative sites in the fiber, but the retarding effect given by electrolyte alone is, however, often inadequate because of the low affinity of the small metal ions. Consequently, recourse is usually made to more efficient retarding agents, Glauber's salt sometimes being used in conjunction with such retarding agents.

Retarding Agents. Anionic Retarders. These operate in the dyebath by forming a dye-retarder complex which has little or no affinity for the fiber.

$$D^+ + R^{2-} \rightleftharpoons \underset{\text{Complex}}{[DR]^-} \qquad D = \text{dye},\ R = \text{anionic retarder}$$

The anionic retarder usually contains one or more sulfonic acid group and so the complex retains some aqueous solubility. In practice, a surplus of retarder is usually used in conjunction with a nonionic dispersing agent to prevent precipitation of the complex.

At low temperatures, the complex becomes distributed evenly on the fiber surface (i.e., migrates) but as the temperature is raised the complex breaks down, releasing dye molecules for adsorption. Since only "free" dye will be adsorbed, the degree of retardation obtained depends on the concentration of the anionic agent and on the position of the above equilibrium. For high affinity dyes, this will be in favor of complex formation and so the compatibility values of the dyes are altered.[173]

Cationic Retarders. These colorless, cationic compounds which have affinity for the fiber and which compete with dye cations for anionic sites in the substrate are generally more widely used than are anionic retarders. Several types are available which differ in their affinity for the fiber and mode of action.[169,173,174] For optimum results, the quantity of retarder used must be carefully calculated according to the manufacturer's recommendations.

Polymeric cationic retarders containing up to several hundred cationic groups are adsorbed at the fiber surface, forming a nondiffusing electric potential barrier to dye uptake. Although dye migration is little affected, both the rate and extent of dye uptake are reduced. Such compounds modify the dyeing behavior and compatability value, especially of low affinity dyes.[173]

b. Dyeing Temperature. Owing to the marked temperature dependence of dye diffusion and therefore dyeing rate above T_g, careful control of dyeing temperature is essential in achieving level dyeing. As a result of the effect of T_g upon dyeing rate, if a uniform heating rate is employed, then virtually all of the

Table 14. The *A* or *S* values for Some Acrylic Fibers[176]

Fiber	*A* or *S* value
Acrilan 16 (Monsanto)	1.4
Dralon (Bayer)	2.1
Orlon 42 (du Pont)	2.1

dyebath exhaustion occurs within a small temperature region, i.e., 85–105 °C. Various application methods have been developed that involve temperature control.

c. Dye–Fiber Characteristics. To determine whether a given mixture of basic dyes will give the desired hue and depth of shade, the saturation characteristics of both fiber and dyes must be known. The S.D.C. has devised a method for determining the saturation characteristics of dye–fiber combinations.[175]

Fiber Saturation Value (*A or S*). This value, which is a constant specific to the fiber under consideration, represents the number of accessible anionic sites per unit weight of the fiber. The *A* or *S* value varies between 1 and 3 for most commercial fibers (Table 14).

Dye Saturation Factor (f). This dye constant represents the saturation characteristics of a given commercial basic dye and is related to the purity and molecular weight of the dye.

Saturation Concentration (C). This is the quantity of commercial dye (expressed as a percent on weight of fiber) which saturates the given fiber, i.e., $C = A/f$.

These three constants are used to ensure that the particular fiber is not oversaturated with dye. Should oversaturation occur, then although dye uptake in excess of the fiber saturation value (A) will usually take place,[65] dyebath exhaustion will be slow and the excess of adsorbed dye will be only loosely bound to the fiber, giving rise to reduced fastness of the dyeing to rubbing, washing, and wet treatments.

In the case of dye combinations, the concentration of each dye (owf) is multiplied by the respective dye saturation factor (f), and the total of the products used to determine whether oversaturation will arise. For example, the saturation factors (f) of three commercial dyes are as follows: dye A, $f = 0.3$; dye B, $f = 0.4$; dye C, $f = 0.6$.

The following recipe is suitable for Dralon (Bayer), which has a saturation value (A) of 2.1: dye A, 2.5% owf; dye B, 1.5% owf; dye C, 1.0% owf.

The saturation concentration (C) of this combination is:

$$(2.5 \times 0.3) + (1.5 \times 0.4) + (1.0 \times 0.6) = 1.95$$

The value of C is therefore less than that of A and oversaturation will not occur. This particular recipe, however, is not suitable for Acrilan 16, since C exceeds the fiber saturation value (A) of this particular fiber which is 1.4.

d. Compatibility of Dyes. The adsorption of an individual cationic dye is influenced by the presence of other dyes in the bath.[173] Because of this and the limited number of dye sites in the fiber, compatibility in admixture is of vital importance if the desired hue is to be obtained. Furthermore, because many basic dyes show virtually no migration due to their high affinity and since uneven initial dye strike cannot be remedied by prolonged dyeing times, it is essential that dyeing is as level as possible from the start of the operation.

The S.D.C. has proposed a test[177] that enables commercial dyes to be characterized according to their compatibility. The ensuing compatibility value of the dye, denoted as CV or K, ranges from 1 to 5. Dyes of the same K value are compatible in admixture: satisfactory dyeing can in most cases be obtained if the dyes used possess K values that differ by no more than one half unit. In general, the lower the K value the more likely is the dye to exhaust rapidly at the fiber surface, while the higher the K value the greater the tendency of the dye to diffuse rapidly into the fiber. When a recipe approaches the saturation value of the fiber (A or S), dyes of low K value can "block" the adsorption of dyes of higher K value. Since the compatibility value is determined mainly by affinity, the effect of retarding agents and electrolytes is more pronounced with dyes of high K value.

5. Aftertreatment

A rinse with warm or cold water is usually sufficient but, when required, surplus dye can be removed by a treatment with hot detergent solution: detergent, 1 $g l^{-1}$; 60 °C; 30 min.

6. Cooling of the Dyebath

Owing to the high thermoplasticity of acrylic fibers at temperatures above T_g (i.e., at the dyeing temperatures used), it is essential that the exhausted dyebath, containing the dyed material, is cooled very slowly and uniformly (e.g., at 1–2 °C per min) to about 60 °C so as to avoid the formation of creases and the loss of bulk and handle of the goods. Preferably, cooling is done indirectly; cooling water should not contact the material.

7. Exhaustion Dyeing

As a result of the poor leveling characteristics of most basic dyes on acrylic, several dyeing procedures have been developed by dye manufacturers. Dye selection depends on the type and form of fiber as well as the desired hue, depth of shade, and fastness requirements of the dyeing. Many dye makers recommend suitable, compatible dye combinations for various uses. Some typical application methods are given below.

Ciba-Geigy "ABC" Method.[178,179] A standard dyeing cycle is used to which the Maxilon (Ciba-Geigy) dyes are made to conform by the use of a cationic retarder, Tinegal AC or CAC. The three parameters used are:

A, the quantity of Tinegal AC or CAC required for the desired rate of exhaustion:

$$A = C - B \qquad \text{for Tinegal AC}$$

$$A = C - B/2 \qquad \text{for Tinegal CAC}$$

B, the total amount of dye applied expressed as a percent which is an equivalent strength to the retarder. The percent of each dye used is multiplied by the appropriate "*b*" factor of the dye (supplied in table form) to convert it into an amount relative to the retarder:

$$B\% = \text{sum of } \%\text{dye} \times b \text{ factor for each dye}$$

C, the maximum amount of Tinegal AC or CAC (or equivalent quantity of dye) expressed as a percent adsorbed by the particular fiber over the standard dyeing cycle. The *C* values for various acrylic fibers are provided in tabular form.[178,179]

Dyes of CV values of 2–3.5 are recommended for use with this dyeing method. The starting temperature for dyeing depends on the type of acrylic fiber to be dyed and is available from tables provided by the dye maker.

A typical procedure is shown in Scheme 74.

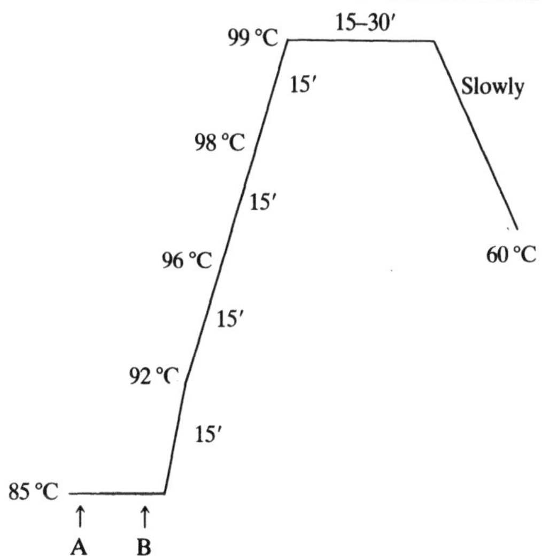

A.	Sodium sulfate	10% owf
	Acetic acid (80%) to pH 4–4.5	
B.	Maxilon (Ciba–Geigy) dye	*x*% owf
	Tinegal AC or CAC (Ciba–Geigy)	*y*% owf

Scheme 74

Pad	Basic dye	$x\ g l^{-1}$
	Thickener	$5\text{–}10\ g l^{-1}$
	Glacial acetic acid	$5\ g l^{-1}$
	Wetting agent	$10\text{–}15\ g l^{-1}$
↓		
Steam	100 °C; 30 min or 110 °C; 5 min	
↓		
Wash off		

Scheme 75

8. Continuous Dyeing

Dye fixation is most usually carried out using saturated steam.[173] A typical procedure is shown in Scheme 75.

9. Migrating Cationic Dyes[181]

Migrating cationic dyes have quite recently been introduced and, as their name suggests, possess greater migrating power on acrylic fiber than conventional basic dyes. Their level dyeing characteristics enable faster rates of temperature rise to be employed than for conventional basic dyes.

The poor migrating power of conventional dyes results from their very high affinity for the fiber and is primarily determined by the size, hydrophobicity, and basicity of the dye molecules. Migrating cationic dyes are generally small size, hydrophilic molecules which, characteristically, have low affinity for acrylic and therefore exhibit a high rate of diffusion. The dyes are adsorbed rapidly by the fiber but can migrate to yield level dyeings.

For example, the Maxilon M (Ciba-Geigy) dyes, with few exceptions, yield dyeings of very good fastness to light, washing, and wet treatments. Owing to their small molecular size and relatively low affinity, the fastness of dyeings to wet treatments is low at temperatures above T_g of the fiber, due to ready diffusion of dye out of the substrate. In practice this is of relatively little consequence, since domestic laundering is often carried out at temperatures below T_g so as to avoid fiber deformation.

The dyes are applied (Scheme 76) using electrolyte to promote migration in conjunction with a cationic retarding agent, which has a similar K value and migration value[173] to that of the dyes used. Conventional cationic retarding agents are not suitable. The admixture of migrating and conventional (nonmigrating) dyes is also not recommended.

B. Disperse Dyes

Disperse dyes are little used on acrylic and modacrylic fibers. They yield pale to medium depth dyeings of moderate to good fastness and are applied at

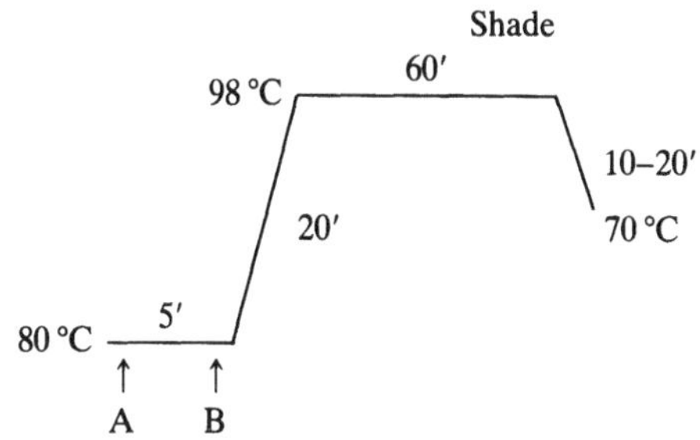

A. Acetic acid (80%)[a] 2% owf
Sodium acetate if required[a] 3% owf
Sodium sulfate (anhydrous) 10% owf
B. Maxilon M (Ciba–Geigy) dye x% owf
Tinegal MR (Ciba–Geigy) 0.5–0% owf
(i.e., 0.5 − x% dye)

Shading. Add Tinegal MR 0.5 − y% owf and Maxilon M dye y% owf and continue dyeing at the boil for 30 min.

[a] pH 4–4.5.

Scheme 76

temperatures of up to the boil, using similar conditions to those used for dyeing polyester fibers (Section V.A).

C. Dyeing Modacrylic Fibers

As with their "regular" acrylic counterparts, these fibers are predominantly dyed by using basic dyes. Although the application of cationic dyes to modacrylic fibers is generally similar to that for acrylic fibers, care must be exercised in selecting suitable dyes and dyeing methods.

The fibers have more pronounced thermoplasticity than "regular" acrylic fibers and are prone to delustering during dyeing at temperatures above T_g. Furthermore, some modacrylic fibers have a small number of anionic dye sites, resulting in low build-up and compatibility of dyes in admixture. Although the fastness of dyeings on these fibers to washing and wet treatments is generally of the same order as that on regular acrylic fiber, the light fastness of dyeings is usually lower.

Consequently, dye manufacturers recommend selected dye combinations for optimum results on these fibers. Also, the dyeing conditions used are often specific to particular fiber types. For certain modacrylic fibers, carriers are employed to promote dyeing at the low temperatures permissible.

References

1. F. Brunello, *Art of Dyeing in the History of Mankind,* Eng. Edn., Vicenza (1973).
2. I. I. Ziderman, *Rev. Prog. Color. Relat. Top. 16,* 46 (1986).

3. G. W. Taylor, *Rev. Prog. Color. Relat. Top. 16,* 53 (1986).
4. R. Meldola, *J. Soc. Dyers Colour. 26,* 103 (1910).
5. S. M. Edelstein and H. C. Borghetty, transl. First edn., MIT Press, Cambridge (Massachusetts) and London (1969).
6. J. Hubner, *J. Soc. Dyers Colour. 29,* 344 (1913); *30,* 211 (1914).
7. S. D. Forester, *J. Soc. Dyers Colour. 91,* 217 (1975).
8. M. Tordoff, *The Servant of Colour,* Society of Dyers and Colourists, Bradford (1984).
9. F. M. Rowe, *J. Soc. Dyers Colour. 54,* 551 (1938).
10. M. R. Fox, *Dyemakers of Great Britain 1856–1976,* ICI (1987).
11. J. Shore, in: *The Dyeing of Synthetic Polymer and Acetate Fibers* (D. M. Nunn, ed.), pp. 411–489, Dyers' Co. Publ. Trust, Bradford (1979).
12. D. Cheetham, *Dyeing Fibre Blends,* Van Nostrand, London (1966).
13. L. W. C. Miles, ed., *Textile Printing,* Dyers' Co. Publ. Trust, Bradford (1981).
14. A. Howarth, *Rev. Prog. Color. Relat. Top. 1,* 53 (1969).
15. L. W. C. Miles, *Rev. Prog. Color. Relat. Top. 4,* 44 (1973).
16. K. Consterdine, *Rev. Prog. Color. Relat. Top. 7,* 34 (1976).
17. W. A. Ratcliffe, *Rev. Prog. Color. Relat. Top. 7,* 43 (1976).
18. B. Glover, *Rev. Prog. Color. Relat. top. 8,* 36 (1981).
19. K. Dunkerley, *Rev. Prog. Color. Relat. Top. 11,* 74 (1981).
20. F. R. Alsberg, *Rev. Prog. Color. Relat. Top. 12,* 66 (1982).
21. W. Clarke and L. W. C. Miles, *Rev. Prog. Color. Relat. Top. 13,* 27 (1983).
22. J. S. Schofield, *Rev. Prog. Color. Relat. Top. 14,* 69 (1984).
23. W. Schwindt and G. Faulhaber, *Rev. Prog. Color. Relat. Top. 14,* 166 (1984).
24. *Colour Index,* Second Edition, Vol. 1. p. xii (1956).
25. B. C. Burdett, in: *The Theory of Coloration of Textiles* (C. L. Bird and W. S. Boston, eds.), pp. 111–162, Dyers' Co. Publ. Trust, Bradford, (1975).
26. I. Holme, *Rev. Prog. Color. Relat. Top. 1,* 31 (1970).
27. M. Fox, *Rev. Prog. Color. Relat. Top. 4,* 18 (1973).
28. I. Holme, *Rev. Prog. Color. Relat. Top. 7,* 1 (1976).
29. K. Silkstone, *Rev. Prog. Color. Relat. Top. 12,* 22 (1982).
30. P. Ackroyd, *Rev. Prog. Color. Relat. Top. 5,* 86 (1974).
31. G. Clarke, *A Practical Introduction to Fibre and Tow Coloration,* Society of Dyers and Colourists, Bradford (1982)
32. J. Park, *A Practical Introduction to Yarn Dyeing,* Society of Dyers and Colourists, Bradford (1981).
33. R. Fleming and J. F. Gaunt, *Rev. Prog. Color. Relat. Top. 8,* 47 (1977).
34. I. Bearpark, F. W. Marriott, and J. Park, *A Practical Introduction to the Dyeing and Finishing of Wool Fabrics,* Society of Dyers and Colourists, Bradford (1986).
35. D. Haigh, *Rev. Prog. Color. Relat. Top. 2,* 27 (1970).
36. M. Patterson, *Rev. Prog. Color. Relat. Top. 4,* 80 (1972).
37. G. A. Richardson, *Rev. Prog. Color. Relat. Top. 9,* 19 (1978).
38. J. D. Ratcliffe, *Rev. Prog. Color. Relat. Top. 9,* 58 (1978).
39. H. W. Partridge, *Rev. Prog. Color. Relat. Top. 6,* 56 (1975).
40. C. Duckworth, ed., *Engineering in Textile Coloration,* Dyers' Co. Publ. Trust, Bradford (1983).
41. C. Duckworth, *Rev. Prog. Color. Relat. Top. 1,* 3 (1970).
42. J. Rayment, *Rev. Prog. Color. Relat. Top. 2,* 26 (1971).
43. G. Seigrist, *Rev. Prog. Color. Relat. Top. 8,* 24 (1977).
44. D. H. Wyles, *Rev. Prog. Color. Relat. Top. 14,* 139 (1984).
45. J. Park, *Rev. Prog. Color. Relat. Top. 15,* 25 (1985).
46. J. Shore, *Rev. Prog. Color. Relat. Top. 10,* 33 (1980).
47. H. Ellis, *Rev. Prog. Color. Relat. Top. 14,* 145 (1984).
48. C. B. Stevens, in: *The Dyeing of Synthetic-Polymer and Acetate Fibres* (D. M. Nunn, ed.), pp. 53–75. Dyers' Co. Publ. Trust, Bradford (1975).
49. M. Langton, *Rev. Prog. Color. Relat. Top. 14,* 176 (1984).
50. ISO 105 *Textiles—Tests for color fastness,* ISO, Geneva, 1982.
51. J. R. Macphee and T. Shaw, *Rev. Prog. Color. Relat. Top. 14,* 58 (1984).

52. C. L. Bird, *The Theory and Practice of Wool Dyeing,* 4th edn., Society of Dyers and Colourists, Bradford (1972).
53. R. H. Peters, *Textile Chemistry,* Vol. 1, Elsevier, London (1963).
54. R. S. Asquith, ed., *Chemistry of Natural Fibres,* Plenum, London (1977).
55. J. A. Maclaren and B. Milligan, *Wool Science,* Science Press, Marrickville (1981).
56. P. Alexander and R. F. Hudson, *Wool its Chemistry and Physics,* 2nd edn. (C. Earland, ed.), Chapman and Hall, London (1963).
57. *Progress in Coloration,* IWS, Ilkley (1986).
58. C. L. Bird, in: *Chemistry of Natural Fibres* (R. S. Asquith, ed.), Plenum, London (1977).
59. E. N. Abrahart, *Dyes and Their Intermediates,* 2nd edn., Arnold, London (1977).
60. P. Rhys and H. Zollinger, *Fundamentals of the Chemistry and Application of Dyes,* Wiley, London (1972).
61. P. A. Duffield and D. M. Lewis, *Rev. Prog. Color. Relat. Top. 15,* 38 (1985).
62. *Wool A Sandoz Manual,* Sandoz, Basle (1979).
63. K. Venkataraman, *The Chemistry of Synthetic Dyes,* Vol. II, pp. 818–833, Academic Press, New York (1952).
64. C. H. Giles, *J. Soc. Dyers Colour. 60,* 303 (1944).
65. R. H. Peters, *Textile Chemistry,* Vol. III, Elsevier, Amsterdam (1975).
66. F. R. Hartley, *J. Soc. Dyers Colour. 86,* 209 (1970).
67. F. R. Hartley, *J. Soc. Dyers Colour. 85,* 66 (1969).
68. E. Race, F. M. Rowe, and J. B. Speakman, *J. Soc. Dyers Colour. 62,* 372 (1946).
69. G. Meier, *S.D.C. Scottish Region Symposium* (1978).
70. G. Schetty, *J. Soc. Dyers Colour. 71,* 705 (1955).
71. F. Beffa and G. Back, *Rev. Prog. Color. Relat. Top. 14,* 33 (1984).
72. I. D. Rattee and D. Lemin, *J. Soc. Dyers Colour. 65,* 217 (1949).
73. F. R. Hartley, *Aust. J. Chem. 23,* 287 (1970).
74. C. H. Giles, T. H. MacEwan, and N. McIver, *Text. Res. J. 44,* 580 (1974).
75. E. Rexroth, *Arbeitstagung (Aachen),* 89 (1972).
76. J. F. Graham and R. R. D. Holt, *Proc. 6th Int. Wool Text. Res. Conf. V,* 501 (1980).
77. D. M. Lewis, *J. Soc. Dyers Colour. 98,* 165 (1982).
78. P. Rhys and H. Zollinger, in: *The Theory of Coloration of Textiles* (C. L. Bird and W. S. Boston, eds.), pp. 326–358, Dyers' Co. Publ. Trust, Bradford (1975).
79. F. W. Beech, *Fiber Reactive Dyes,* Logos Press, London (1970).
80. M. Fox and H. H. Sumner, in: *Dyeing of Cellulosic Fibers* (C. Preston, ed.), pp. 142–195, Dyers' Co. Publ. Trust, Bradford (1986).
81. I. D. Rattee, in: *The Chemistry of Synthetic Dyes* (K. Venkataraman, ed.), Vol. VIII, Academic Press, London (1978).
82. K. Venkataraman, ed., *The Chemistry of Synthetic Dyes,* Vol. VI, Academic Press, London (1972).
83. D. Evans, *Text. J. Aus. 46,* 20 (1971).
84. K. R. F. Cockett and D. M. Lewis, *J. Soc. Dyers Colour. 92,* 141 (1976).
85. D. M. Lewis and I. Seltzer, *J. Soc. Dyers Colour. 86,* 298 (1970).
86. D. M. Lewis and I. Seltzer, *J. Soc. Dyers Colour. 84,* 501 (1968).
87. D. M. Lewis and I. Seltzer, *Textilveredlung 6,* 87 (1971).
88. D. M. Lewis, *Wool Sci. Rev. 50,* 22 (1974).
89. F. A. Barkhuysen and N. J. Rensburg, *Proc. 7th Int. Wool Text. Res. Conf., Japan, V, 201* (1985).
90. G. A. Smith, *Proc. 7th Int. Wool Text. Res. Conf., Japan, V, 211* (1985).
91. H. Flensberg, W. Mosimann, and H. Salathe, *Colourage (Annu. Ed.),* 43 (1986).
92. *Sandolan MF Dyes,* Sandoz Tehnical Information, 9220.00.83.
93. T. P. Nevell, in: *The Dyeing of Cellulosic Fibers* (C. Preston, ed.), pp. 1–54. Dyers' Co. Publ. Trust, Bradford (1986).
94. K. Dickinson, in: *The Dyeing of Cellulosic Fibers* (C. Preston, ed.), pp. 55–105. Dyers' Co. Publ. Trust, Bradford (1986).
95. I. Holme, in: *The Dyeing of Cellulosic Fibres,* (C. Preston, ed.), pp. 106–41, Dyers' Co. Publ. Trust, Bradford (1986).

96. ICI Product Data Sheet, CAFT 4(R).
97. K. Venkataraman, *The Chemistry of Synthetic Dyes,* Vol. II, pp. 861–1058, Academic Press, London (1952).
98. *Ciba Review,* No. 85, 3066 (1951).
99. H. A. Lubs, ed., *The Chemistry of Synthetic Dyes and Pigments,* pp. 431–576, Reinhold, New York (1956).
100. U. Baumgarte, *Rev. Prog. Color. Relat. Top. 5,* 17 (1974).
101. U. Baumgarte, in: *The Dyeing of Cellulosic Fibers* (C. Preston, ed.), pp. 224–255, Dyers' Co. Publ. Trust, Bradford (1986).
102. S. Heimann, *Rev. Prog. Color. Relat. Top. 11,* 1 (1981).
103. *Cellulosic Fibers,* BASF Manual, B 375 (1979).
104. T. Vickerstaff, *The Physical Chemistry of Dyeing,* 2nd edn., Oliver and Boyd, London (1954).
105. H. H. Sumner, T. Vickerstaff, and E. Waters, *J. Soc. Dyers Colour. 69,* 181 (1953).
106. H. Wegmann, *J. Soc. Dyers Colour. 76,* 282 (1960).
107. H. Wegmann, *Am. Dyest. Rep. 51,* 276 (1962).
108. D. G. Orton, in: *The Chemistry of Synthetic Dyes* (K. Venkataraman, ed.), Vol. VII, pp. 1–34, Academic Press, London (1974).
109. W. E. Wood, *Rev. Prog. Color. Relat. Top. 7,* 80 (1976).
110. K. Venkataraman, *The Chemistry of Synthetic Dyes,* Vol. II, pp. 1059–1100, Academic Press, New York (1952).
111. H. A. Lubs, *The Chemistry of Synthetic Dyes and Pigments,* Reinhold, New York, pp. 302–334 (1956).
112. *Colour Index,* Third Edition.
113. C. D. Weston and W. S. Griffith, *Text. Chem. Color. 1,* 462 (1969).
114. C. D. Weston, in: *The Chemistry of Synthetic Dyes* (K. Venkataraman, ed.), Vol. VII, pp. 35–68, Academic Press, London (1974).
115. C. Senior and D. A. Clarke, in: *The Dyeing of Cellulosic Fibers* (C. Preston, ed.), pp. 256–289, Dyers' Co. Publ. Trust, Bradford (1986).
116. H. Herzog, in: *The Dyeing of Cellulosic Fibers* (C. Preston ed.), pp. 290–319, Dyers' Co. Publ. Trust, Bradford (1986).
117. S. R. Cockett and K. A. Hilton, *The Dyeing of Cellulosic Fibers and Related Processes,* pp. 165–178, Leonard Hill, London (1961).
118. A. Johnson, in: *The Theory and Practice of Textile Coloration* (C. L. Bird and W. S. Boston, eds.), pp. 359–421, Dyers' Co. Publ. Trust, Bradford (1975).
119. I. D. Ratee, *Rev. Prog. Color. Relat. Top. 14,* 50 (1984).
120. C. V. Stead, *Dyes and Pigments 3,* 161 (1982).
121. R. W. Moncrieff, *Man-made Fibers,* 6th edn., Newnes-Butterworth, London (1975).
122. A. J. Hall, *The Standard Handbook of Textiles,* 8th edn., Newnes-Butterworth, London (1975).
123. H. Leube, *Dyeing and Finishing of Acetate and Triacetate and their blends with other Fibers,* BASF, S 388e (1970).
124. D. Blackburn, in: *The Dyeing of Synthetic-Polymer and Acetate Fibers* (D. M. Nunn, ed.), pp. 71–130, Dyers' Co. Publ. Trust, Bradford (1979).
125. J. S. Ward, *J. Soc. Dyers Colour. 89,* 411 (1973).
126. J. F. Briggs, *J. Soc. Dyers Colour. 37,* 287 (1921).
127. W. M. Gardner, *J. Soc. Dyers Colour. 38,* 171 (1922); A. G. Green and K. H. Saunders, *J. Soc. Dyers Colour. 39,* 10 (1923); *J. Soc. Dyers Colour. 40,* 138 (1924).
128. G. H. Ellis, *J. Soc. Dyers Colour. 40,* 285 (1924); *42,* 184 (1926); *57,* 353 (1941).
129. R. K. Fourness, *J. Soc. Dyers Colour. 69,* 513 (1956).
130. C. L. Bird, *J. Soc. Dyers Colour. 70,* 68 (1954).
131. *ICI Technical Information,* D1389 (1973).
132. C. L. Bird, *J. Soc. Dyers Colour. 80,* 237 (1964).
133. C. L. Bird, *J. Soc. Dyers Colour. 81,* 209 (1965).
134. J. F. Dawson, *Rev. Prog. Color. Relat. Top. 4,* 18 (1973).
135. J. F. Dawson, *Rev. Prog. Color. Relat. Top. 9,* 5 (1978).
136. J. F. Dawson, *Rev. Prog. Color. Relat. Top. 14,* 90 (1984).

137. C. V. Stead, in: *The Chemistry of Synthetic Dyes,* Vol. III (K. Venkataraman, ed.), pp. 385–462, Academic Press, New York (1970).
138. *The Chemistry of Synthetic Dyes and Pigments* (H. A. Lubs, ed.), Reinhold, New York (1956).
139. V. Kartaschoff, *Helv. Chim. Acta 8,* 938 (1925).
140. R. Clavel, *Rev. Gen. Matiere Color. 28,* 145, 167 (1923).
141. S. R. Sivaraja Iyer, in: *The Chemistry of Synthetic Dyes,* Vol. VII (K. Venkataraman, ed.), pp. 240–260, Academic Press, New York (1974).
142. H. H. Burns and j. K. Wood, *J. Soc. Dyers Colour. 25,* 12 (1929).
143. C. H. Giles, in: *The Theory of Coloration of Textiles* (C. L. Bird and W. S. Boston, eds.), pp. 41–110, Dyers' Co. Publ. Trust, Bradford (1975).
144. L. Peters, in: *The Theory of Coloration of Textiles* (C. L. Bird and W. S. Boston, eds.), pp. 163–236, Dyers' Co. Publ. Trust, Bradford (1975).
145. H. Braun, *Rev. Prog. Color. Relat. Top. 13,* 62 (1983).
146. R. K. Fourness, *J. Soc. Dyers Colour. 90,* 15 (1974).
147. A. Murray and K. K. Mortimer, *J. Soc. Dyers Colour. 87,* 173 (1971).
148. A. Murray and K. Mortimer, *Rev. Prog. Color. Relat. Top. 4,* 67 (1973).
149. W. Ingamells, in: *The Theory of Coloration of Textiles* (C. L. Bird and W. S. Boston, eds.), pp. 285–325, Dyers' Co. Publ. Trust, Bradford (1975).
150. R. H. Peters and W. Ingamells, *J. Soc. Dyers Colour. 89,* 397 (1973).
151. F. Jones, in: *The Theory of Coloration of Textiles* (C. L. Bird and W. S. Boston, eds.), pp. 237–284, Dyers' Co. Publ. Trust, Bradford (1975).
152. D. A. Garrett, *J. Soc. Dyers Colour. 73,* 365 (1957).
153. I. M. S. Walls, *J. Soc. Dyers Colour. 70,* 429 (1954).
154. J. S. Ward, *Rev. Prog. Color. Relat. Top. 14,* 98 (1984).
155. *Dyeing and Finishing of Polyester Fibers,* BASF Manual, B 363e (1975).
156. R. Broadhurst, in: *The Dyeing of Synthetic-polymer and Acetate Fibers* (D. M. Nunn. ed.), pp. 131–242. Dyers' Co. Publ. Trust, Bradford (1979).
157. A. N. Derbyshire, *J. Soc. Dyers Colour. 93,* 228 (1977).
158. *Serisol and Serilene Dyes on Polyester,* YCL Pattern Card.
159. Du Pont, *Dyes Chem. Tech. Bull. 5,* 82 (1949).
160. *ICI Technical Information D*1521 (1977).
161. P. Ginns and K. Silkstone, in: *The Dyeing of Synthetic-polymer and Acetate Fibers* (D. M. Nunn, ed.), Dyers' Co. Publ. Trust, Bradford (1979).
162. R. Hill, in: *Fibers from Synthetic Polymers* (R. Hill, ed.), Elsevier, Amsterdam (1953).
163. H. W. Peters and T. R. White, *J. Soc. Dyers Colour. 77,* 601 (1961).
164. C. C. Cook, *Rev. Prog. Color. Relat. Top. 12,* 73 (1982).
165. H. R. Hadfield and D. N. Sharing, *J. Soc. Dyers Colour. 64,* 381 (1948).
166. H. R. Hadfield and h. Seaman, *J. Soc. Dyers Colour. 74,* 392 (1958).
167. *Dyeing and Finishing of Polyamide Fibers and their Blends with Other Fibers,* BASF Manual B 358e, BASF (1972).
168. I. Holme, *Chimia 34,* No. 3, 110 (1980).
169. I. Holme, *Rev. Prog. Color. Relat. Top. 13,* 10 (1983).
170. R. Raue, *Rev. Prog. Color. Relat. Top. 14,* 187 (1984).
171. U. Mayer and E. Siepmann, *Rev. Prog. Color. Relat. Top. 5,* 65 (1974).
172. C. B. Stevens, in: *The Dyeing of Synthetic-polymer and Acetate Fibers* (D. M. Nunn, ed.), Dyers' Co. Publ. Trust, Bradford (1979).
173. W. Beckmann, in: *The Dyeing of Synthetic-polymer and Acetate Fibers* (D. M. Nunn, ed.), Dyers' Co. Publ. Trust, Bradford (1979)
174. J. Park, *Dyer 167,* No. 1, 16 (1982).
175. D. G. Evans, *J. Soc. Dyers Colour. 89,* 292 (1973).
176. *Synacryl Dyes for Acrylic Fibers,* ICI Pattern Card 22810.
177. D. G. Evans, *J. Soc. Dyers Colour. 88,* 220 (1972).
178. *Maxilon Dyes on Acrylic Fibers.* Ciba-Geigy Pattern card 3100 UK.
179. *Tinegal CAC,* Ciba-Geigy Technical Information 4657 UK.
180. *ICI Technical Information D1586.*
181. W. Biedermann, *Rev. Prog. Color. Relat. Top. 10,* 1 (1979).

8

Nontextile Applications of Dyes

P. F. GORDON

I. INTRODUCTION

The previous chapters in this book have described the dyestuffs industry with reference to textiles. It is fair to say that, as far as textile dyes are concerned, most of the work carried out in dyestuffs research in this century has focused upon the development of chromogens already invented in the last century, with a few notable exceptions, e.g., phthalocyanine, triphendioxazines, and benzodifuranones. Much of this research effort has been directed at modifying the chromogens so that they could be applied to the natural as well as synthetic fibers that started appearing earlier in this century. A great deal of manpower has also been devoted to improving the efficiency of the dyeing process and the quality of the dyes themselves.

However, the interest in dyestuffs is confined not merely to the textile industry, but also extends to the photographic industry, which is another major user of dyes, and to a variety of other smaller outlets such as for foodstuffs, drugs, histological stains, etc. Various aspects of these latter uses have already been highlighted and will therefore be ignored in this chapter.[1] Instead, the use of dyes in the "new technology" industries will be considered by taking a few illustrative examples of their application in the electronic and electrophotographic industries.

Rapid development in R&D in electronics has resulted in a multibillion dollar industry with high growth and high returns to the envy of much of the established and slower-growing older industries. In fact there can be few people in the developed world who remain unaffected by the ever-increasing range of products containing electronic components. Chemistry has had an important role to play in facilitating many of these rapid developments, such as in liquid crystals, integrated circuits, printed circuit boards, and semiconductors . Chemistry has

P. F. GORDON • Fine Chemicals Research Centre, ICI Organics Division, Hexagon House, Blackley, Manchester M9 3DA, England.

also played a vital part in another high-growth industry, electrophotography, especially in the manufacture of the photoconductive drums and developing materials. In the last few years there has also been a perceptible increase in the role played by an important branch of chemistry, i.e., dyestuffs, in these new technology fields and this chapter will select a few examples to illustrate this growing importance. It is not intended that the following sections be comprehensive, but it is hoped that the reader will gain a flavor of the type of research and the nature of the dyestuffs being used.

II. DYES FOR DISPLAYS

Over the last few decades there has been a significant push to develop new display technologies to interface with the ever-increasing number of electronic devices. The role of the display is simply to render the output from the device in a form which is intelligible to the human reader and its use is manifold as can be gauged from the number of instruments that require one, such as computers, calculators, watches, public signboards, petrol pumps, and many more. The present demand for displays is so high that the displays industry is now a multibillion dollar earner supporting high growth and occupying a strategic position in the electronics industry.

Although the cathode ray tube (CRT) still dominates the market, some of the new display technologies have made a noticeable impact. One of the most notable successes is the liquid crystal display,[2] which now dominates in areas where a compact display with low energy consumption is essential, e.g., watches, clocks, and calculators. Considerable research effort is, even now, being targeted at providing a liquid crystal equivalent of the CRT in televisions. The liquid crystal materials and the cell configuration, in which they are used, have changed constantly to improve properties such as contrast, viewing angle, multiplexing, and of course cost effectiveness. It is during this development process that liquid crystal dyes have very recently begun to make an impact, since they allow certain properties to be improved. Their most important contribution is probably their ability to be used in displays with fewer polarizers than previously, thereby resulting in a cheaper display device and opening the way to larger area displays.

Figure 1 describes schematically how a liquid crystal cell containing a dye works. In simple terms, the liquid crystal cell is filled with a mixture of the liquid crystal material (called the host) and a compatible dye (the guest). The liquid crystal material itself is usually composed of a specially designed mixture of several single liquid crystalline materials to allow the cell to be switched quickly with low fields over a wide temperature range.[2] An important class of liquid crystals is the thin rodlike cyanobiphenyls (**1**, R = alkyl), which were among the first practical liquid crystalline materials. They possess an ordered liquid state in which the molecules are highly aligned (for a liquid), at a temperature just above their melting point, and will remain in this ordered state over a temperature range which depends on the nature of R until they eventually pass into the more familiar isotropic liquid state at a higher temperature. The molecules in this

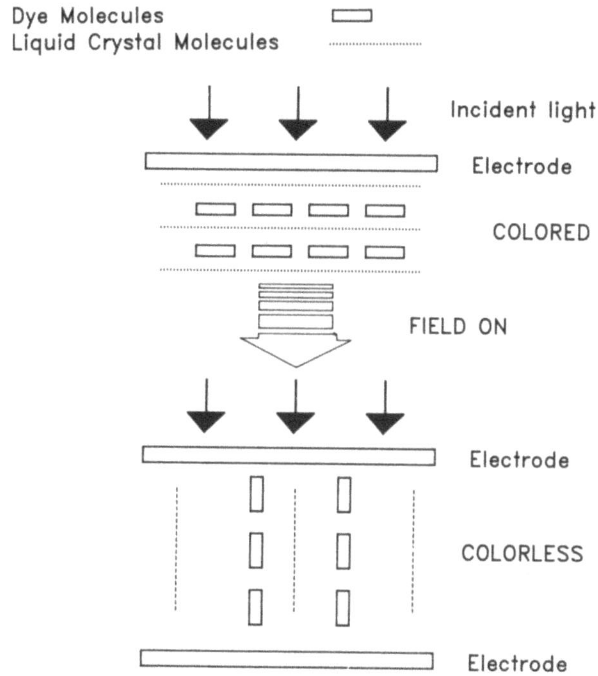

Figure 1. Schematic representation of a dyed liquid crystal display.

highly ordered or "liquid crystalline" state exhibit several important properties, including their ability to align with specially prepared surfaces and also electric fields—a consequence of their dielectric anisotropy.[2] Hence, a simple cell would be composed of two glass plates separated by a gap, to contain the liquid crystalline material. The insides of the glass plates are treated so that the liquid crystalline molecules align in one direction (homeotropically in Figure 1) and the plates also contain electrodes so that, when a field is applied across them, the molecules are "switched" into an alternative (homogeneous) alignment (see Figure 1).

This cell configuration forms the basis for a device in which dyes can be used. The device works when the dye molecules align themselves readily with the liquid crystal molecules as shown in Figure 1(a) and (b). If it is assumed that the transition dipole moment of the dye lies along its long axis, then it can be seen that the switch from (a) to (b) in Figure 1 would involve a change from a colored display to a colorless one. This is so because the light entering the cell has its

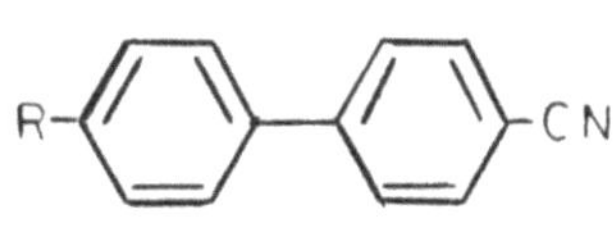

1

oscillating electric vector perpendicular to the direction of the traveling wave and parallel to the dye molecules. Hence, the light would be absorbed (in Figure 1a) since its electric vector would lie in the same plane as the transition dipole moment of the dye molecule. Upon switching (Figure 1b) the dye molecules are forced to align parallel to the applied electric field and the transition dipole moment of the dye would now be perpendicular to the electric vector associated with the light wave. In this orientation (Figure 1b) the dye would no longer absorb and the light would therefore pass through the cell which would then appear colorless.

In point of fact, commercially feasible guest–host displays work on the above principles, although cell design is considerably more complicated than shown. In general, dyes which are to be used in such displays must possess several properties among which are:

1. Nonionic character—to avoid turbulence, etc.
2. Solublity in, and compatibility with, the liquid crystal host.
3. High purity.
4. A high-order parameter.

Several of the above properties are quite obvious, especially the requirement for a high-order parameter. The order parameter is a measure of how well the dye aligns itself with the liquid crystal. If no alignment is present a value of 0 is recorded, and for perfect alignment a value of 1. Dyes with order parameters as close to 1 as possible are highly desirable, since this gives a truly high-contrast colored to colorless transition upon switching—anything less than 1 results in a display which always shows some color in the off state and hence poorer contrast upon switching.

Bearing in mind the nature of the liquid crystal material which is a thin, long rodlike molecule, it is not surprising that azo compounds were the first dyes to be tried in dyed liquid crystal mixtures, since they too can be designed to have a rodlike shape. Indeed, several azo dyes have been developed successfully for this application and dyes **2–4**[3a–d] are fairly typical.

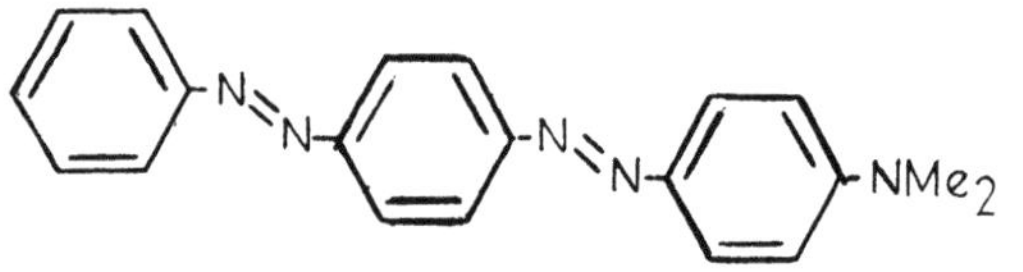

2

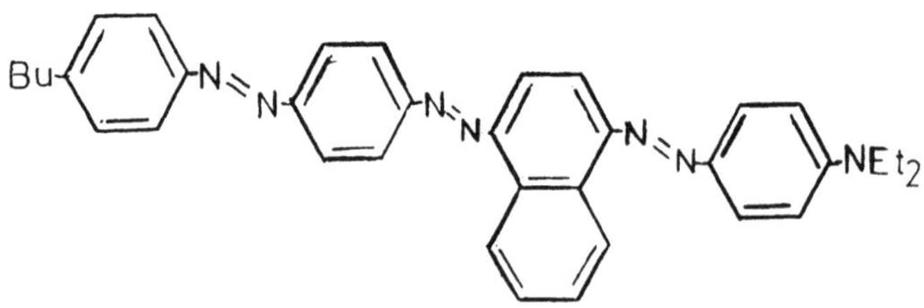

3

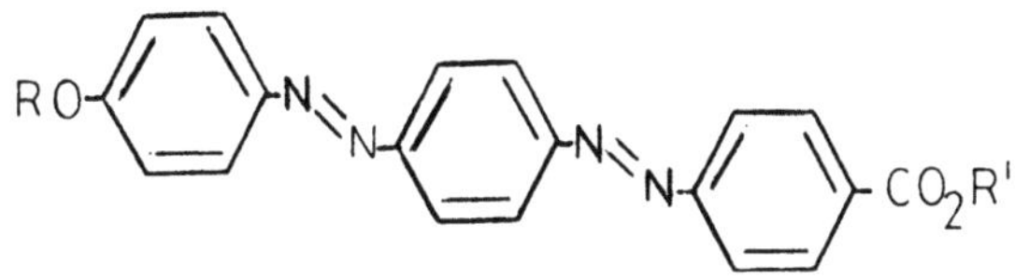

4

A quite surprising and more recent development is the discovery that suitably modified anthraquinones can also be used in dyed liquid crystal displays even though they can hardly be described as thin rodlike molecules. Several anthraquinone dyes have been developed for use in liquid crystal displays, such as the yellow (**5**), red (**6**), blues (**7**, n = 1–12; **8**), and green (**9**, n = 1–12) dyes.[4] A combination of all four results in a black dye and the trend in the textile industry of developing a range of different colored dyes for color mixing is therefore parallel for liquid crystals. A second parallel can also be seen in their photostability, where it has long been known that anthraquinone dyes possess high lightfastness on fibers; not surprising, then, is the finding that they also exhibit high photostability in liquid crystal media. No doubt the knowledge gained from the textile industry has been applied to the benefit of the electronics industry. However, these dyes do have several features which differ from textile

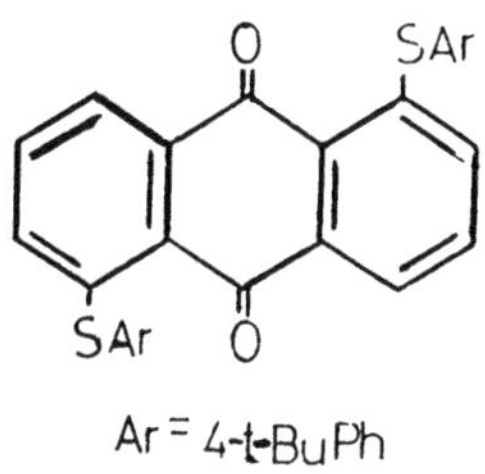

5

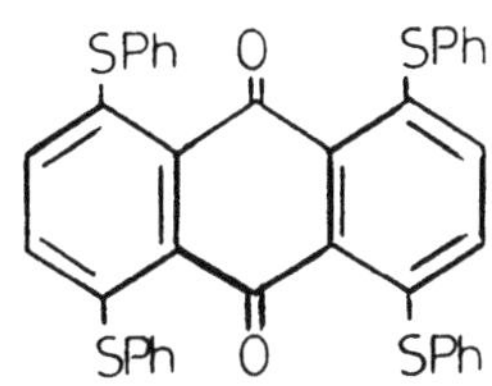

6

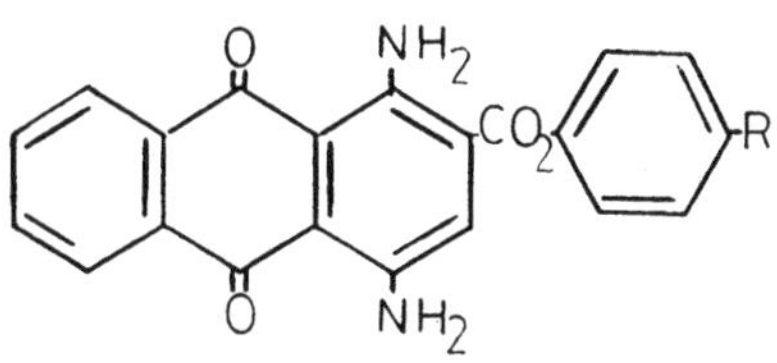

7 R = $(CH_2)_nMe$

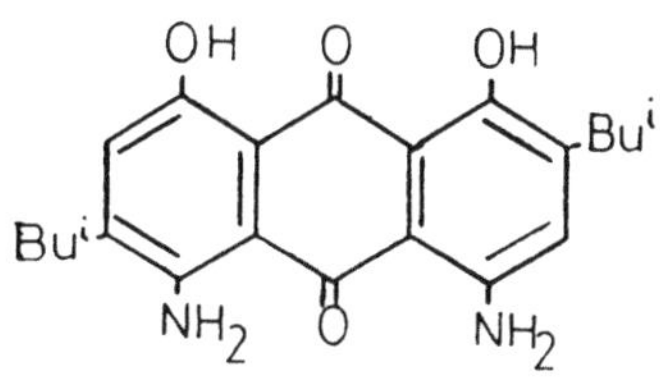

8

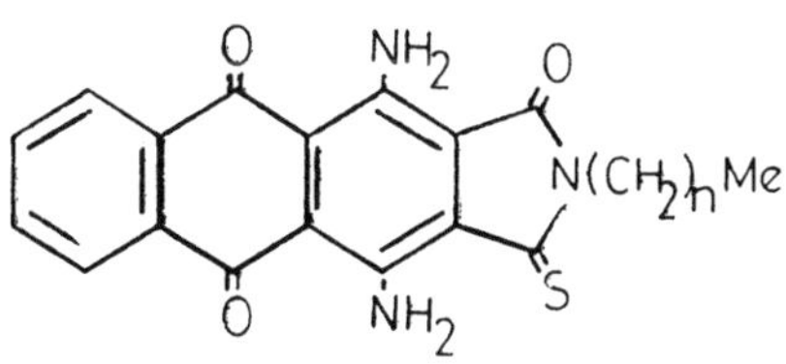

9

dyes, such as the long alkyl chains present in dyes **7**, **8**, and **9**. The *n*-alkyl chains confer higher-order parameters on the dyes and enhanced solubility (in the LC mixture). Surprisingly, the sulfur derivatives **5** and **6** do not need the long alkyl chains to possess high-order parameters. Despite the superior photostability of the anthraquinones over the azo dyes, they have one major important disadvantage in that they are inherently weaker and must therefore be used at higher concentration in the liquid crystal medium. It therefore seems likely that mixtures of both types of dye will find application in practical devices.

Although azo and anthraquinone dyes dominate in guest–host liquid crystal displays, fluorescent dyes such as perylenes.(**10**, $n = 1$–16 and **11**, $n = 1$–16) are fairly recent newcomers to the scene of dyed liquid crystal displays.[5] These dyes, because they are fluorescent, give bright colored displays that have a very attractive appearance and it will be interesting to follow their progress over the next few years.

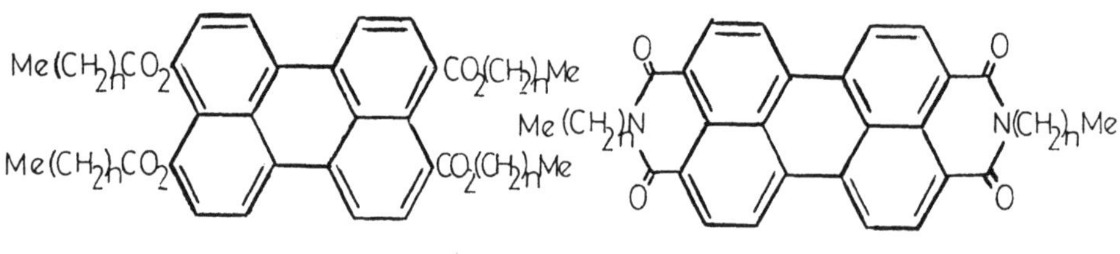

10 **11**

It thus seems that the dyes developed for liquid crystal displays, so far, are simply slightly modified textile dyes, a point underlined by the fact that many of the patentees in the area are acknowledged dyestuff companies.

A somewhat different picture is painted by electrochomic dyes which have no counterparts in textiles at all; yet they represent an application of dyes in the electronics industry and underline the fact that there is much scope for innovation in dye research. In simple terms, an electrochromic dye must possess at least two different "colored" states interconvertible by electrochemical means.

Although electrochromic displays have not made a significant impact, as yet, the viologen system based upon the bispyridinium (**12**) offers some promise and will therefore be used to briefly illustrate the principles involved.[6] The process for the color change is demonstrated below and involves the electrochemical reduction of the colorless bispyridinium salt (**12**) to the highly-colored radical cation (**13**). The reaction is completely reversible since the cation (**13**) can be

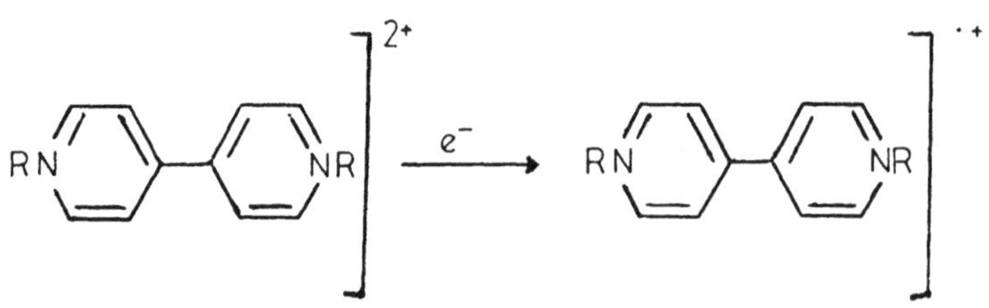

12 **13**

electrochemically oxidized back to the colorless form **(12)**. Apart from this completely reversible color change, the viologens offer an added advantage in that the color of the radical cation can be varied merely by altering the nature of the substituent at the quaternary centers; for example, the presence of alkyl groups results in a purple coloration while *p*-cyanophenyl substituted radical cations are green. A major advantage of the electrochromic displays is their ability to exhibit a full gray scale simply by controlling the extent of the electrochemical reaction.

The electrochromic cell is a fairly complex device in which a silicon wafer is processed to produce an active array and integrated drives necessary to control the formation of the image.[6] The silicon wafer containing the necessary circuitry is then protected from the electrolyte by a number of photolithographic and electroplating steps. In the final steps the electrolyte containing the electrochromic material is added and the cell is sealed to give a very compact unit. The cell is then ready for interfacing to a suitable output, e.g., a computer.

How electrochromic displays fare in the next few years depends upon a demonstration of their performance, dependability, versatility, and, importantly, their cost, since they suffer savage competition from liquid crystal displays and the well-established cathode ray tube.

III. LASER DYES

The vast majority in the developed world will have heard of lasers and have some concept of their use, if not the mechanism by which they work. The word laser is an acronym for Light Amplification by Stimulated Emission of Radiation, and lasers have been on the scene since about 1960.[7] They have gradually penetrated all levels of our life from industry, where they are used in welding and hole boring, to medical therapy, where they are used in, for example, cancer treatments, and even to our leisure life. An example of the latter is the use of lasers in optical disks (see later), and more recently in compact disks for audio reproduction. It seems likely therefore that in the near future a significant proportion of us will actually own a laser.

There are several lasers currently available emitting light from the ultraviolet to the infrared as shown in Table 1. However, close inspection of the table

Table 1. Emission of Common Lasers

Laser	Peak emission (nm)
Helium–cadmium	325
Argon	488, 514
Helium–neon	632
Semiconductor	780–850 (approx)
Yttrium–aluminum–garnet (YAG)	1060
Carbon dioxide	10,600

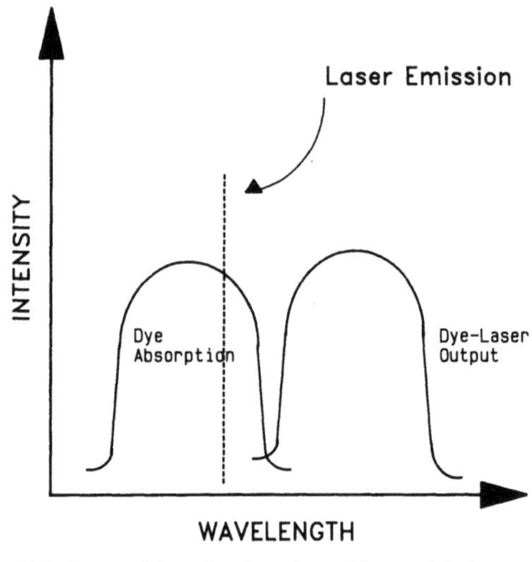

Figure 2. Relative positions for dye absorption and lasing wavelength.

reveals that these lasers are extremely restricted in their emission spectrum (± 1 nm) and so the greater proportion of the UV-visible and infrared spectrum is not served by a laser at all. The sensible course is to have just a few tunable lasers, individually covering a wide spectrum, rather than several thousand each with a narrow bandwidth. This is where laser dyes play an important role, since they allow the present fixed lasers to be tuned over a much greater range. The mechanism by which such laser dyes work is fairly well understood and has provided a rationale for choosing appropriate dyes. For a dye to lase, its individual molecules must be capable of being excited, or pumped, into one of its higher states. Once a good population inversion has been attained, i.e., many dye molecules are in the excited state, the dye must release the stored energy from its excited state to the ground state by a *radiative* process. This is a very important condition, since the efficiency of the laser dye is critically dependent upon an efficient radiative decay; if there are competing nonradiative decay processes, e.g., internal conversion, then the dye is of very limited use. In practical terms this means that good laser dyes will generally be efficient fluorescers, though the corollary is not necessarily true. A second practical consideration is that the dye must absorb at the output wavelength of an appropriate laser, since the laser is used to "pump" the dye into its excited state. Figure 2 illustrates a typical relationship between dye absorption, laser output, and dye laser emission. It can be seen from the figure that the output from the laser dye (dotted line) is at longer wavelengths than both its absorption (full line) and the pump lasers output, and the range of laser frequencies emitted by the dye laser is considerably wider than that of the pump laser's. Because the dye laser emits over a significant range, the essentially monochromatic emission of the pump laser has been effectively converted into a tunable source; by using a set of filters, any wavelength of laser light inside the envelope of the dye lasers emission can be chosen.

Ideally, one dye would be used to generate a very wide range of laser wavelengths with equally high efficiency using just one laser source. Unfortunately this has proved impossible and just as in textiles, where several dyes are used to build up a shade gamut, several laser dyes are required to provide a "shade gamut" of laser wavelengths. Examination of suppliers' catalogs reveals several tens of suitable dyes covering the range <400 nm to >1000 nm, the eventual choice of dye being dictated by the required laser wavelength, nature of the pump laser, cost, stability of the dye, efficiency, etc. The dyes are usually used at low concentration (10^{-4} *M*) in solvents, e.g., alcohols, DMF, to avoid overheating and must be very pure to avoid premature quenching of the excited state. Consequently, they command a high price, orders of magnitude higher than textile dyes, although the quantities required are a minute proportion of the millions of tons of textile dyes sold each year.

Considering that the application of laser dyes is a very recent one, it might seem surprising that many of the most effective laser dyes belong to dye classes which were among the first to be developed in the last century. The following paragraphs list a few of the more important. The blue region of the spectrum is dominated by dyes of type **14**; the carbostyril **(15)** and the coumarin **(16)** are just two representatives of the class. The carbostyril absorbs at 360 nm and shows its maximum lasing efficiency at 425 nm (+65 nm shift) while **16** absorbs at longer wavelength (366 nm) and shows maximum lasing wavelength at the even longer wavelength of 446 nm (80 nm shift).

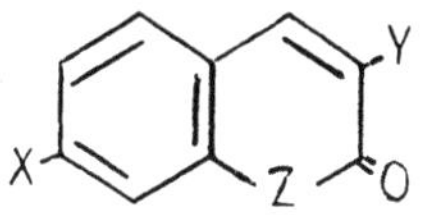

14

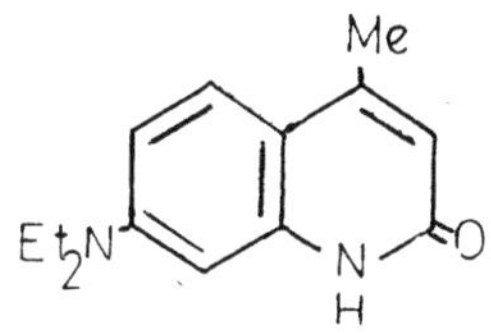

15

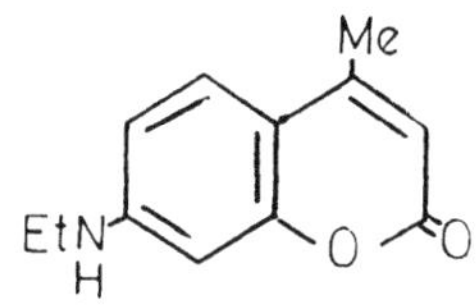

16

In the green region of the spectrum, coumarin-type dyes can still be used as illustrated by the coumarin **(17)** which absorbs at 436 nm and lases at 523 nm (87 nm shift). However, the most useful class of dyes is based on the generalized structure **18**; dyes of this type will cover not just the green region of the spectrum but also the red; fluorescein **(19)**, rhodamine 6G **(20)**, cresyl violet **(21)**, and nile blue **(22)** absorb respectively at 501, 530, 601, and 628 nm and show lasing maxima at 560, 590, 650, and 710 nm, respectively.

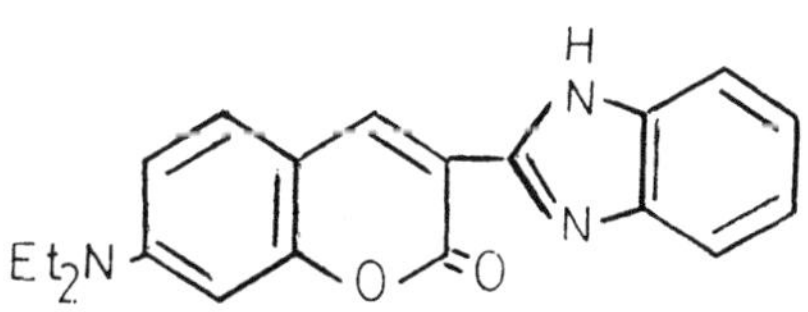

17

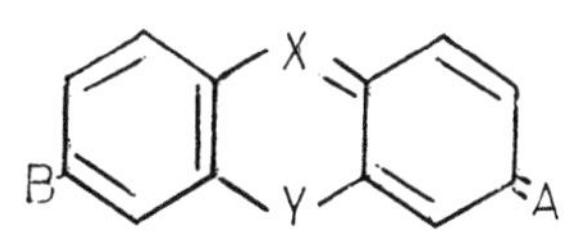

18

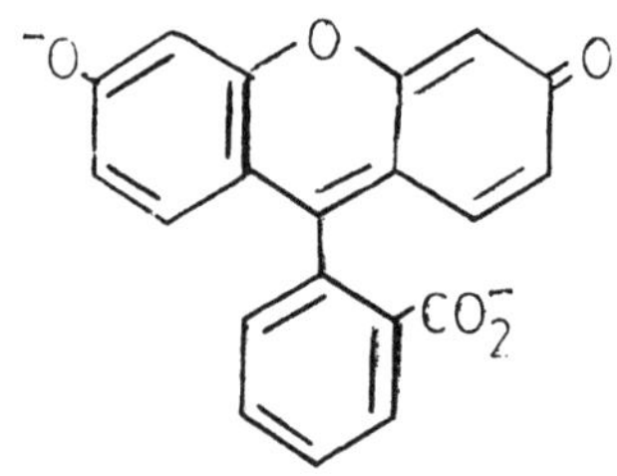

19

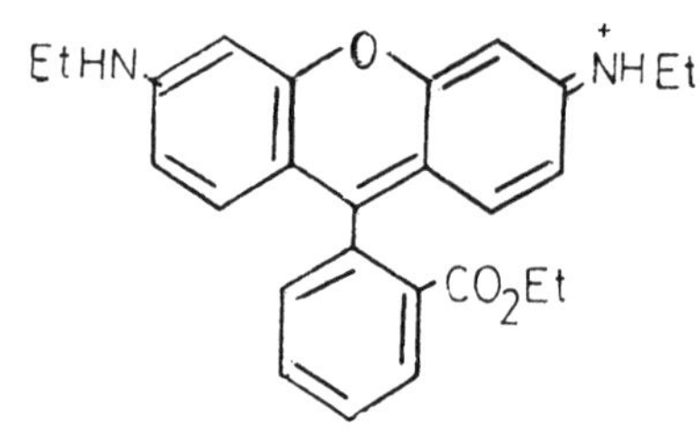

20

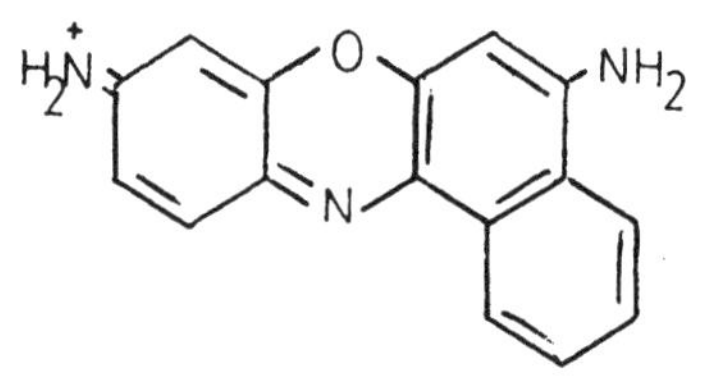

21

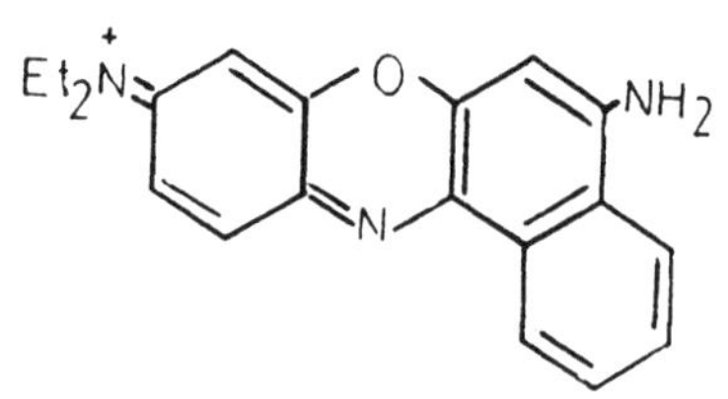

22

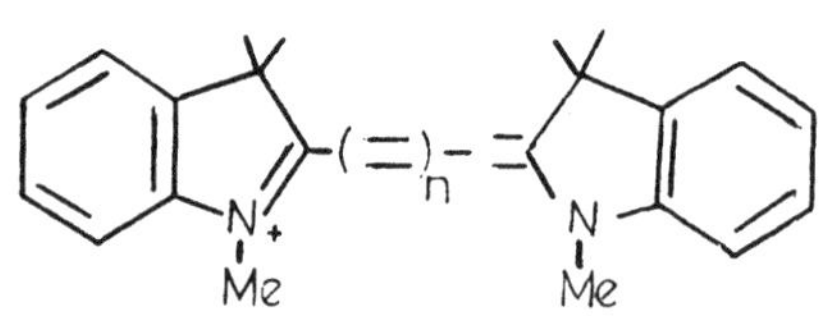

23 n=2
24 n=3

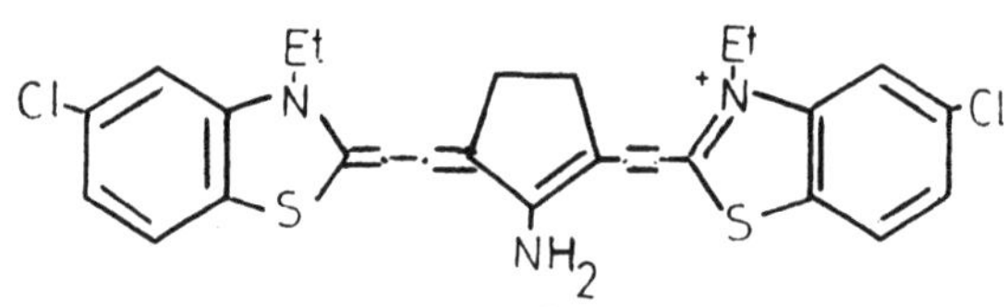

25

The near infrared is an important area of the laser spectrum in total contrast to textile dyes, where it is pretty well neglected, except perhaps for camouflage applications. Cyanines find wide application as laser dyes in the near infrared and cyanines **23**, **24**, and **25** illustrate how various structural changes can be made to predictably alter the lasing wavelength. For instance, by incorporating extra double bonds the lasing wavelength is shifted from 740 nm in **23** to 850 nm in **24**, and by incorporating further heterocycles in the rings as well as substituents at the conjugated chain, as in **25**, a further large shift can be obtained to 980 nm. It is even possible to obtain dyes which lase at wavelengths over one micron by such manipulation of the structure.

Most dyes used in laser applications are not used for textiles because of their poor light stability. However, it is worth noting that they are not used purely for their ability to absorb electromagnetic radiation as textile dyes are, but must also be capable of emitting the radiation so absorbed at a longer and broader wavelength range. Thus we see an important difference between laser dyes and textile dyes; laser dyes are not used for their ability to confer color upon a substrate but for an application which depends critically upon their special electronic properties. This feature arises in all of the following sections.

Laser dyes have been readily accepted as an important part of laser technology; however, they remain in a fairly small and specialized market sector. Nevertheless, there is still a call for better laser dyes which have greater stability and wider wavelengths over which they will lase efficiently.

IV. DYES FOR OPTICAL DATA STORAGE

Different civilizations have been actively involved in storing data for several thousands of years. Initially, the human memory sufficed and information was passed down from generation to generation by word of mouth. Eventually, data storage media were used in the form of stone inscription, papyrus leaves, all the way through to paper. For several hundred years, paper and ink have been the favored data storage media but, with the advent of the computer, magnetic data storage media have made a major impact, offering advantages such as very high storage density and rapid access to, and retrieval of, information. More recently, optical data storage has appeared on the scene offering unbelievably high levels of data storage density.[8] Photographic-type processes can still offer a high level of storage density, but it is really the optical disk linked with laser optics that has opened up the ultrahigh densities possible in optical disks. A schematic representation of a simple optical disk is shown in Figure 3. The substrate is typically made with a plastic such as polycarbonate, upon which is deposited a very thin recording layer. Further layers are then coated upon the recording layer to protect it from dust, fingerprints, and extraneous contaminants. Information is stored in the recording layer by altering specific areas of it such that they interact with light differently from the unaltered parts. Typically this is done by ensuring that there is a large change in reflectivity between these two areas of the disk. The compact audio disk (CD), which is currently exciting much attention, uses just this technology. In practice, the recording layer is a metal film through which holes have been punched so that a laser scanning the surface of the disk is either reflected from the metal or passes through the holes; a detector then picks up signals for eventual processing and thence information retrieval.

The laser is, of course, the secret for the extremely high storage density since it can be focused to spot sizes of about one micron. The compact disk uses a semiconductor laser emitting at approximately 800 nm and this type of laser has gained widespread acceptance in the optical data storage field.

While the above approach of using a prestamped metallized disk is perfectly acceptable for the compact disk format, it is not so for optical disks. As already mentioned, the CD contains all the information already encoded just like a record; however, for many data storage uses it is preferable to supply a blank disk, which can then be recorded on by the user. It is far too costly and

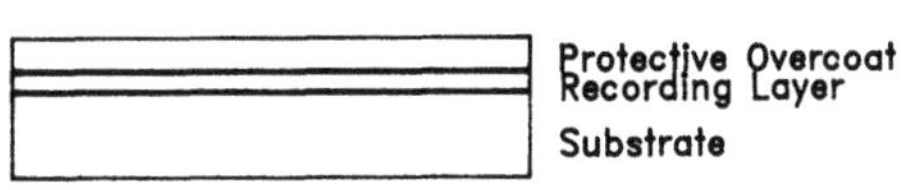

Figure 3. Configuration of a simple optical disk.

inconvenient to make masters and pressings, as happens for CDs, to produce just one or two copies of the required data. For this reason dyes have made an impact, since they allow the user to encode his information directly via the laser. The method of writing and reading is, in principle, very simple. A dye is coated onto the disk substrate to provide the recording layer. In the writing mode the laser is used near to full power and "blasts" holes in the very thin layer of the dye at appropriate parts of the disk, as dictated by the digitally encoded information. When the information is to be retrieved, the power of the laser is turned down to avoid damaging the recording layer and then played over the surface of the disk to retrieve the encoded signal with the aid of a detector.

The above outline of the writing/reading process immediately suggests some properties the dye must possess. For example, since the mechanism of hole formation relies upon the dye absorbing the laser radiation, heating up and migrating from the hot region during the hole formation stage, then the dye must absorb at the wavelength of the laser output. In fact a good dye must fulfil several criteria, including:

1. Absorb at laser output.
2. Possess good reflectivity.
3. Possess good stability.

It is the first point that has really sparked new research into dyes because, as previously mentioned, the solid-state lasers used emit in the near infrared at approximately 800 nm—a region of the spectrum where conventional textile dyes are not needed and hence do not absorb strongly. However, many of the dyes that have been proposed for this application are based upon conventional dye structures, suitably modified. In fact, the tricks used to push the known chromogens into the near infrared are essentially the same ones used by generations of dyestuff chemists for obtaining differently colored dyes and so near-infrared dyes can really be seen as a mere continuation of the visible spectrum.

The types of dyes currently receiving most attention can be divided into four classes:

1. Cyanines and related dyes.
2. Triarylmethanes and related dyes.
3. Quinone-based dyes.
4. Metal-complex dyes.

The following discussion outlines a few typical structures for each category. Cyanine dyes which absorb in the near infrared have been known for a long time, because of their use in photography as sensitizers. A major advantage of cyanine dyes is their predictable color–structure relationship such that dyes can be effectively "tuned" to absorb at the wavelength of the emitting laser by varying the conjugated chain, the nature of the end groups, and substituents in the chain. For example, the simple cyanine (**26**, $n = 0$) absorbs at 224 nm while cyanine (**26**, $n = 5$), containing more conjugated double bonds, absorbs at 735 nm. In fact, as a rule of thumb, every additional double bond results in a 100 nm bathochromic

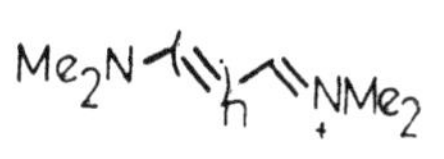

26

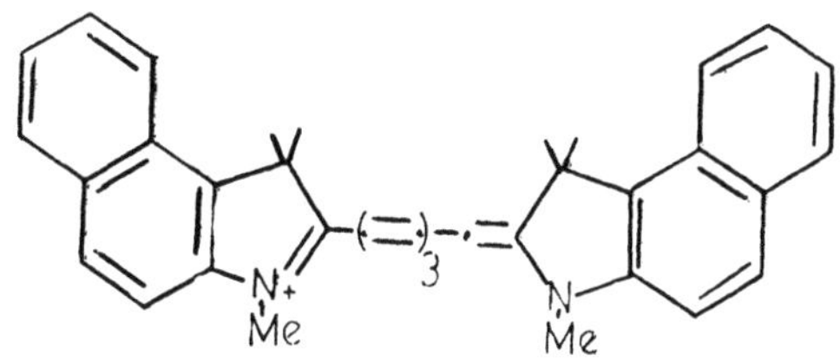

27

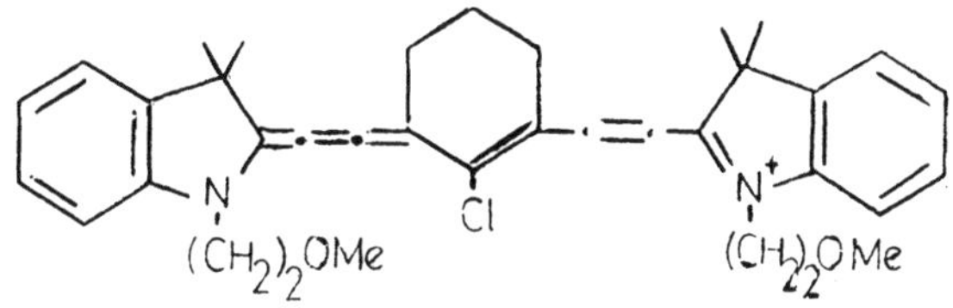

28

shift.[9] For instability reasons, dyes of type **26** are unlikely to be used in optical data storage. However, more stable dyes are attainable as exemplified by cyanine dye **(24)**, which absorbs at 755 nm. A closely related dye **(27)** absorbs at even longer wavelength (780 nm) by virtue of the extra conjugation resulting from the two benzene rings annelated to the indoline rings, and is the subject of several patents.[10] Almost the same wavelength can be obtained by cyanine **(28)** simply by constraining the central double bond in a ring.

Several variations upon conventional cyanine dyes have received a considerable amount of attention from companies interested in optical data storage, as adjudged by the number of recent patents and publications. Thiopyrylium dyes, such as **29**, absorb in the near infrared (**29**, λ_{max} = 755 nm) with very few conjugated double bonds in the chain connecting the two heterocyclic rings,[11] and if a squarylium nucleus is substituted at the center of the conjugated chain then interesting near-infrared dyes of potential application in video disks are obtained, e.g., **30**, λ_{max} = 800 nm[12] and **31**, λ_{max} = 745 nm.[13] Indeed, intense activity has centered around squarylium-substituted dyes and their analogs such as **32**, which

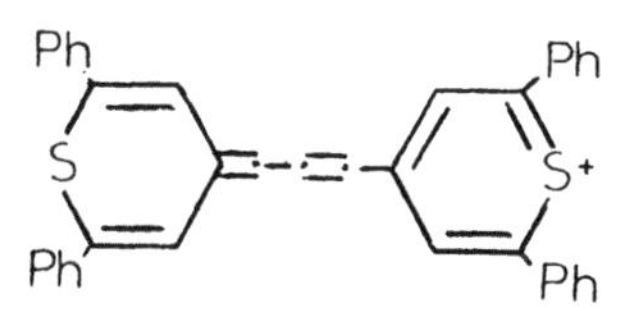

29

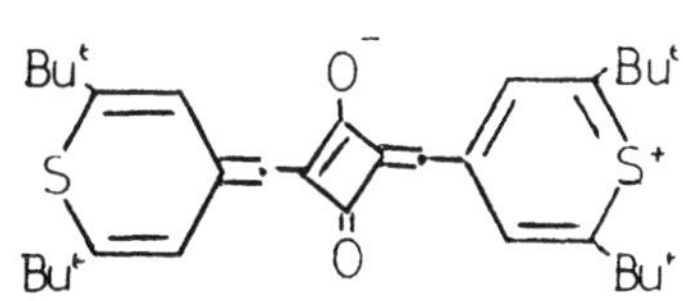

30

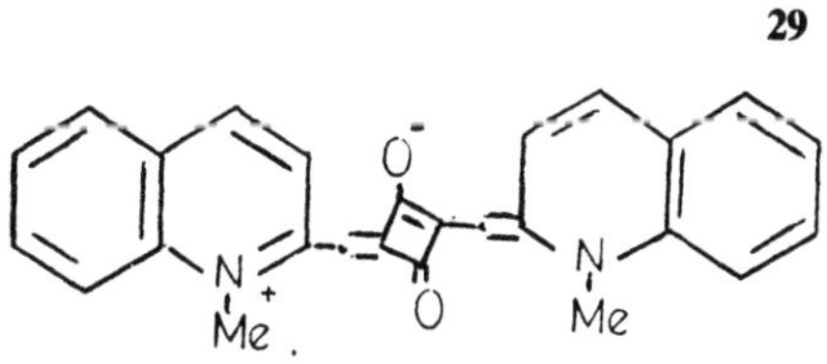

31

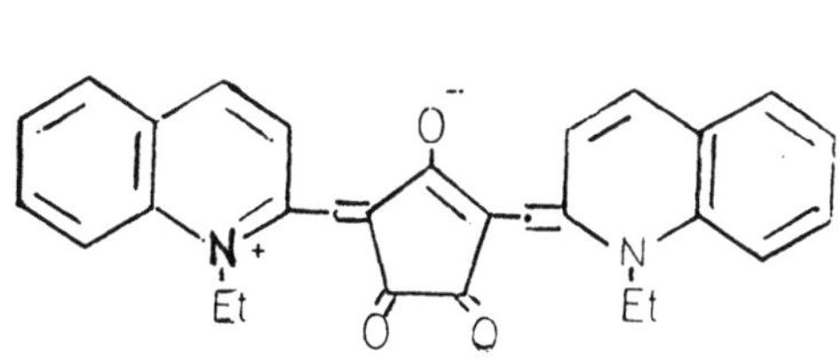

32

also absorb at wavelengths corresponding to the emission of semiconductor lasers.

Somewhat related to the cyanines are the triarylmethane dyes **(33)**, which were among the very first synthetic dyes. Indeed, a few are still used today on textiles in the violet to green shade region although they are still relatively unimportant, as gauged from the previous chapters. However, the absorption maximum of these dyes can be shifted into the near infrared by enforcing planarity in the rings. The size of the shift that can be so accomplished may be seen by comparing crystal violet (**34**, $n = 0$, $\lambda_{max} = 589$ nm) with the infrared dye (**35**, $\lambda_{max} = 850$ nm).[14] Alternatively, and by analogy with the cyanines, increasing conjugation also results in a bathochromic shift as evident in the dye (**34**, $n = 1$) which absorbs at 770 nm, 200 nm more bathochromic than **34**, $n = 0$. Additionally, varying the central atom, i.e., replacing carbon by nitrogen, also results in a very large bathochromic shift so that the aza compound **(36)** absorbs at 920 nm.[15]

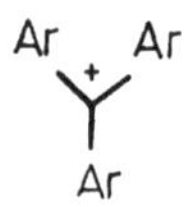

33

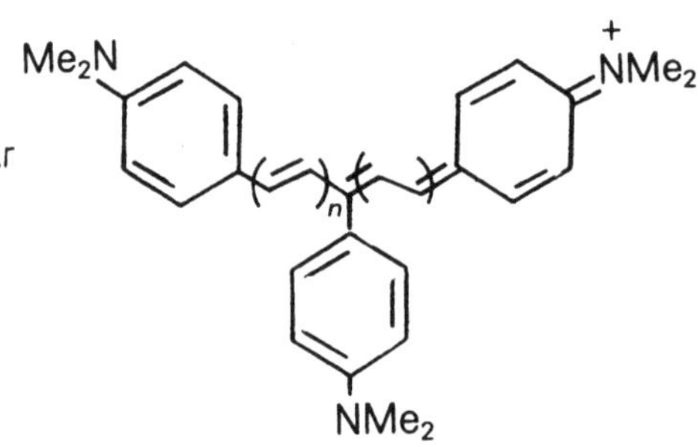

34

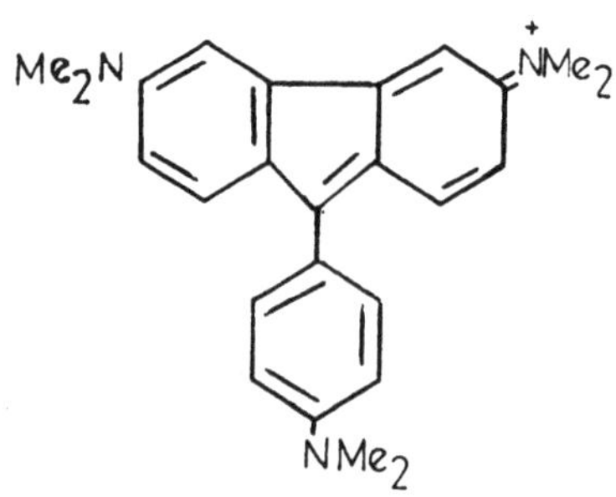

35

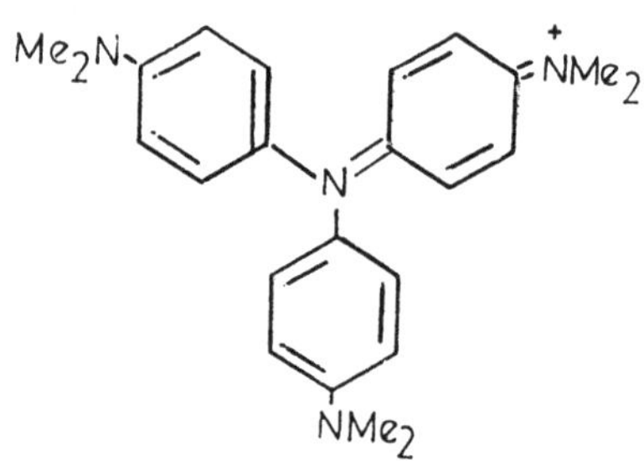

36

The search for suitable infrared absorbers for optical disks has not been restricted to cyanines and triarylmethanes since research on more conventional-looking textile dye types has been carried out, for example, quinones, azo compounds, and phthalocyanines. Several patents pertaining to the use of substituted anthraquinones in optical disks have recently appeared and rely upon heavily (donor) substituted anthraquinones to gain the desired shift into the near infrared. Thus, anthraquinone **(37)** absorbs at 750 nm and the heterocyclic derivative **(38)** absorbs at even longer wavelength, 770 nm.[16,17] Very similar approaches in the structurally related benzoquinones have borne fruit, in that several substituted benzoquinones have been described which absorb in the

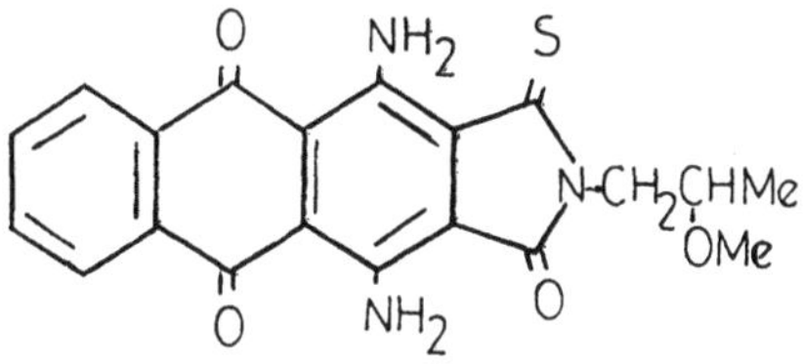

37

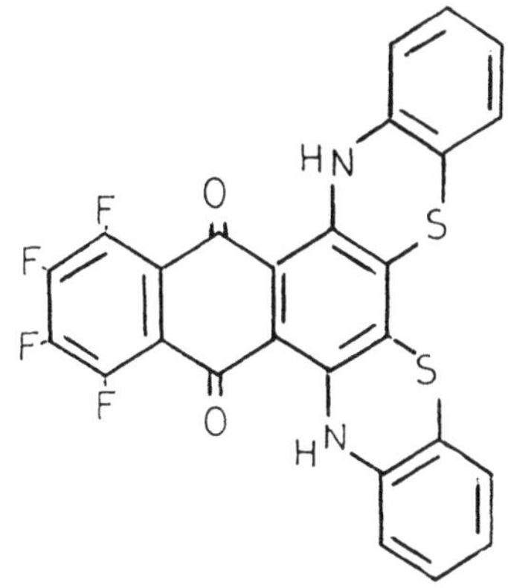

38

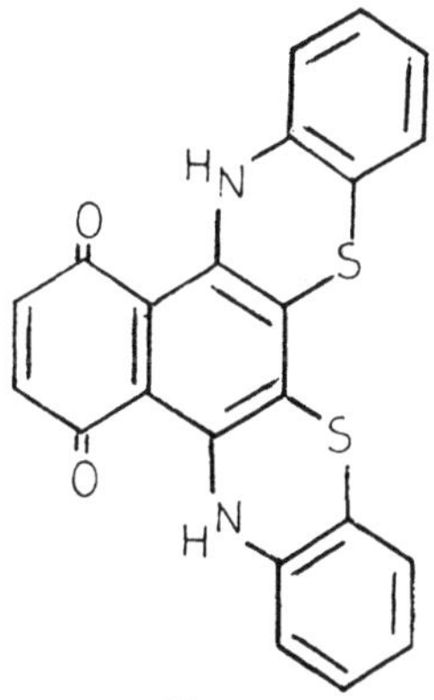

39

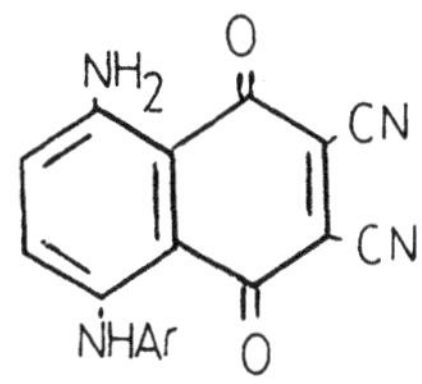

40

near-infrared region and should therefore be suitable for optical data storage purposes. The heterocyclic derivative **(39)** absorbs at 735 nm and the dicyano compound **(40)** absorbs at 770 nm, very close to the output of the semiconductor lasers.[18a,b]

The phthalocyanines are renowned for their high strength and light stability and, not surprisingly, research effort has been directed at modifying them to absorb in the near infrared. Copper phthalocyanine is a blue pigment absorbing at 678 nm and the corresponding dihydrogen compound absorbs at 698 nm. Simply by annelating additional benzene rings to the phthalocyanine results in a large bathochromic shift into the near infrared as found in naphthalocyanine (λ_{max} = 780 nm).[19] The choice of the central metal atom also affects the electronic absorption, and metals such as vanadium[20] and lead[21] both seem suitable for optical data storage. Substitution of electron donors at the periphery of the phthalocyanine ring system is another useful ploy for inducing bathochromic shifts.[22] Alternatively, near-infrared absorbing phthalocyanines can be obtained by altering their crystal morphology; that different crystalline forms of phthalocyanines absorb at different wavelengths is a well-known fact. Indeed α and β forms of phthalocyanines have been manipulated in the dyestuffs industry for some time now. This fact has been capitalized upon and various metastable crystalline forms of phthalocyanines have been produced which will absorb at 780 and 830 nm; examples are the X and T forms.[23a,b]

While phthalocyanines have been known and used in textiles for over half a century, nickel complexes **(41)** are a fairly recent innovation and do not find application as textile dyes. These complexes have excited much interest because

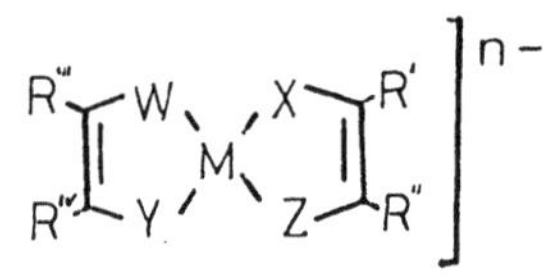

41

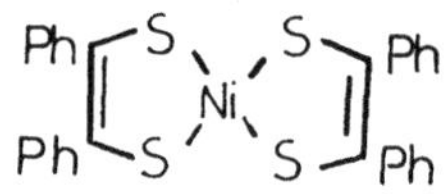

42

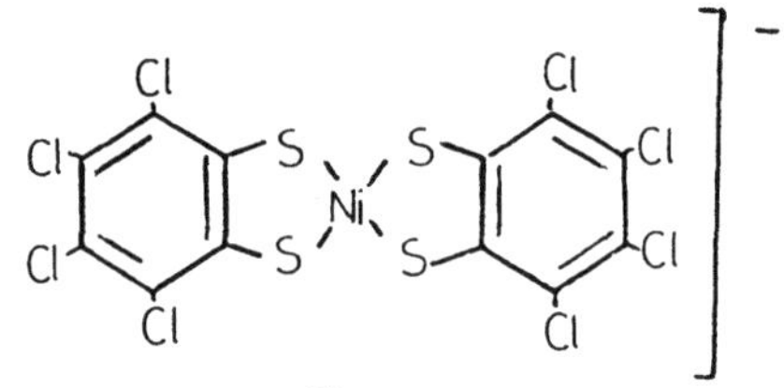

43

of their strong electronic absorption at such long wavelengths.[24] By altering the nature of the heteroatoms X, Y and substituents R, R^1, dyes absorbing anywhere between 700 and 1100 nm can be prepared. The neutral complex **42**[25] and the charged complex **43**[26] both absorb in the right region for use in optical data storage and reports of their use are present in the patent literature.

From the above it can be seen that there is a large number of dyes which now absorb in the infrared and can, theoretically, be used in optical data storage. However, the successful dye will not only have to absorb at the appropriate place but will have to satisfy a whole range of properties, some of which have already been mentioned, and so it is almost certain that only a fraction of the above dyes will be suitable. Dye-based systems themselves also face competition from other types of materials[27] and optical data storage itself faces a stiff challenge from the ever-improving magnetic data storage media. Nevertheless, this area illustrates the rejuvenation in dye research and the potential for dyestuffs to play an important role in high-technology industries in the future. It can also be seen that once again dyes need no longer be considered just as colored molecules but truly as effect chemicals capable of many different roles. Dyes in optical data storage are not used at all for purposes of coloration but rather because they absorb laser radiation, generate heat in doing so, and melt or deform to form the pits used to record the information. This feature was noted in the last section and will be further exemplified in the remaining two sections.

V. ORGANIC PHOTOCONDUCTORS

Although photography has been deliberately omitted in this chapter, electrophotography is certainly worthy of inclusion among the category of new uses of dyes (in high-technology industries). The electrophotographic process is a relatively new innovation that emerged only a few decades ago. However, its progress has been astonishing from very humble origins to the present multi-billion dollar industry; few offices, schools, administration buildings, libraries etc., will be found without a photocopier, and with the possibility of using

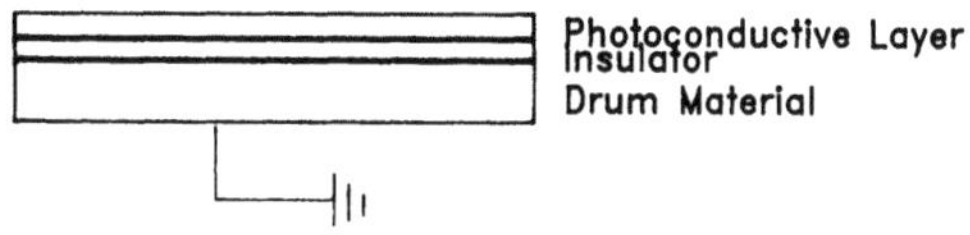

Figure 4. Photoconductive drum.

photocopiers in the form of laser printers attached directly to computers the potential market is immense.

The most familiar electrophotographic process is the transfer xerographic process pioneered by Carlson.[28] Basically, the process involves (a) formation of an electrostatic latent image; (b) development; (c) transfer of developed image to a substrate. For the sake of convenience the entire process can be split into five separate steps:

1. The xerographic plate is charged by ions, e.g., by corona discharge.
2. The plate is exposed to the subject to form an electrostatic latent image.
3. The latent image is developed.
4. The developed image is transferred to a substrate, e.g., paper or plastic.
5. The image is fixed, for instance, by heat.

The xerographic plate consists, in its simplest form, of a photoconductive layer and a base electrode separated by an insulator (Figure 4). In the first stage the xerographic plate, which is frequently in the form of a drum or continuous belt, is charged uniformly. In the second stage the original to be photocopied is illuminated by light and the reflected image focused onto the plate. In places where the light intensity is high, the charge is dissipated, and where no light falls on the plate, the charge remains. Thus, an electrostatic image of the original is formed whereby light sections of the original are represented by low or no charge density and dark sections of the original by high charge density on the xerographic plate.

For quite a long time selenium has been the photoconductor of choice; however, in the last few years the photoconductor market has been increasingly penetrated by organic photoconductors (OPCs). OPCs offer a number of advantages over selenium—they are relatively inexpensive, nontoxic, can be tuned to provide sensitivity in any spectral region, including the near infrared, and may be fabricated into various designs. They are therefore ideally suited for personal and smaller copying machines and this is the market where they are currently making greatest impact.

The majority of OPCs are combined in a dual-layer assembly with the charge generation layer (CGL) nearer to the electrode and a transparent charge transfer layer coated on top. When light falls upon the xerographic plate during the latent image formation stage, it passes through the transparent charge transfer layer and is incident upon the charge generating layers whereupon an electron or a hole (depending on the charge at the surface) is injected back into the change transfer layer (CTL). The hole/electron is then transported by the CTL to the surface, canceling out the charge previously deposited during corona discharge (see Figure 5). In areas where light does not fall, no such mechanism operates, leaving the

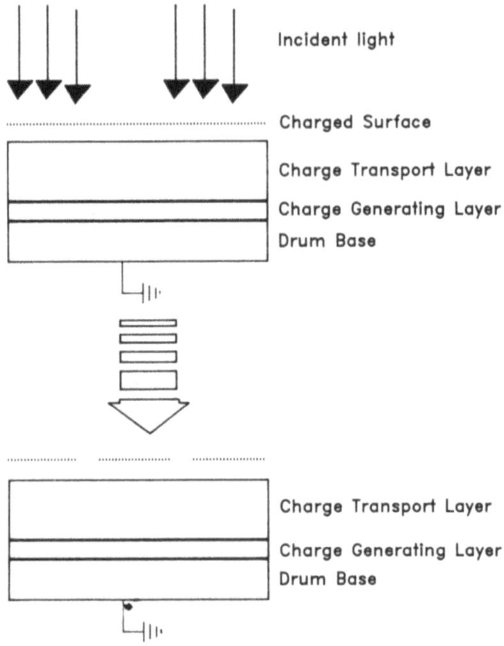

Figure 5. Latent image formation.

charge at the surface. Of course, if the charge at the surface does leak to ground by some mechanism when light does not fall upon the OPC, i.e., there is dark decay, then the efficacy of the OPC is seriously impaired.

Of the OPCs in use, or proposed, the majority are modified dyestuffs. This is not surprising since the nature of the light source (visible or near infrared) used in the exposure step requires the OPC to be sensitive to light in these regions. Once again, then, we see that dyestuffs are not being used simply as colorants. Of course, not every dyestuff is a successful organic photoconductor with the full range of desirable properties. However, several different classes of dyestuffs have been found to be very useful, such as azos, perylenes, phthalocyanines, vat types, and cyanine dyes. For example, thiopyrylium dyes, such as **44**, have been claimed as useful organic photoconductors.[29] Azo dyestuffs, especially in pigmentary form, have also been very heavily covered by patenting and several companies use them as photoconductors. Most of the azo compounds are symmetrical dis- or trisazo compounds, such as **45–48**.[30] Other dye types which seem to be useful are the perylenes such as **49**, and vat dyes exemplified by dibromoanthanthrone

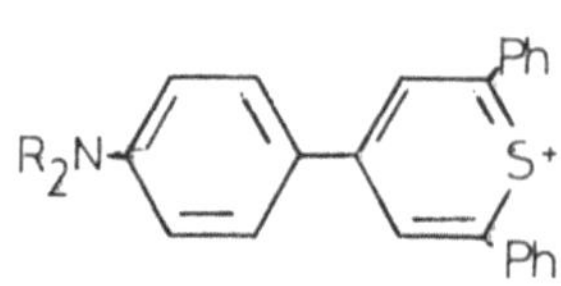

44

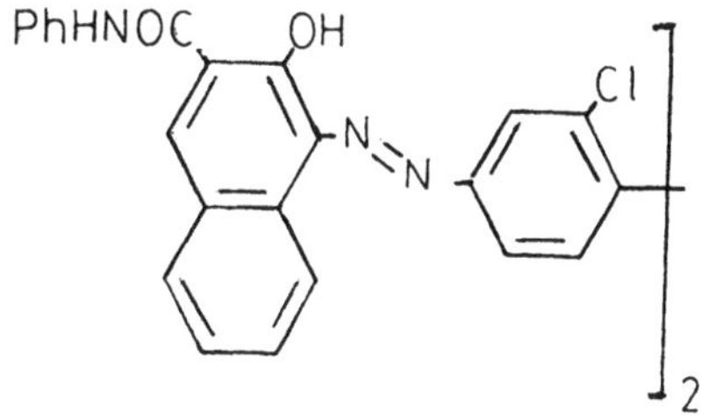

45

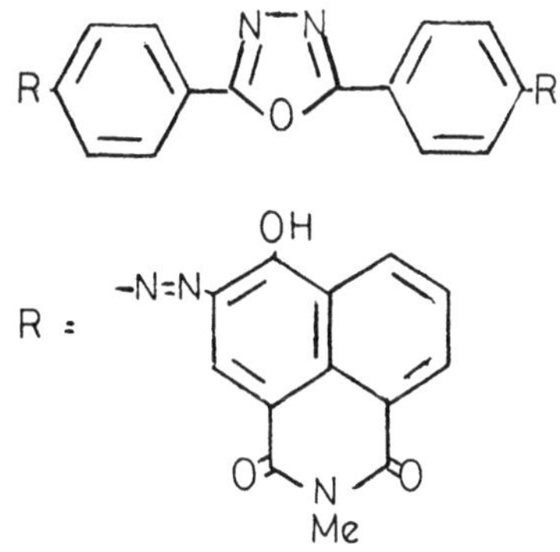

46

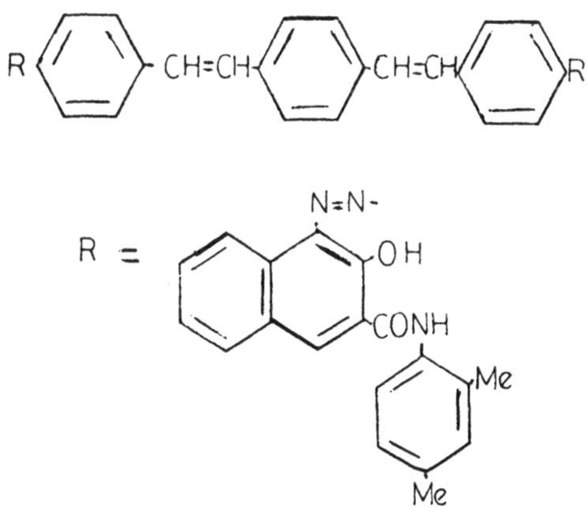

47

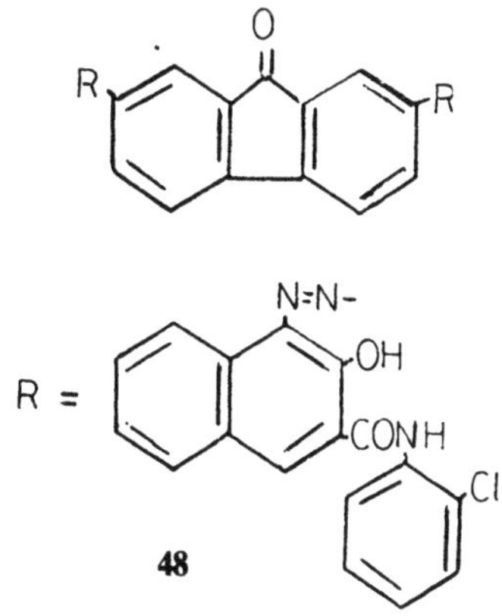

48

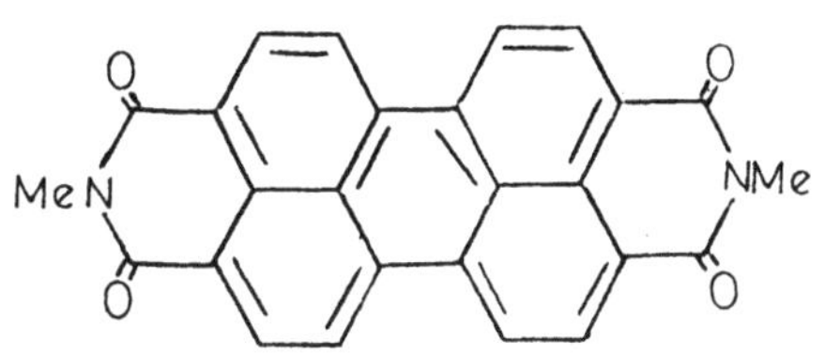

49

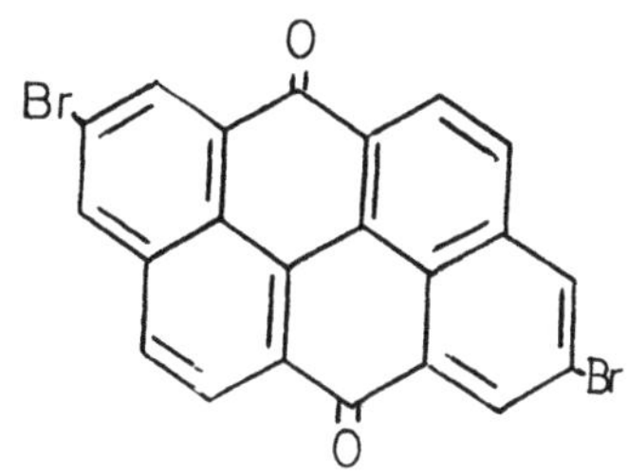

50

(50).[31,32] All of these aforementioned organic photoconductors are usually matched to a charge transport layer containing electron-rich compounds such as pyrazolines **(51)** or hydrazones **(52)**.

A much more recent trend in electrophotography is the development of near-infrared organic photoconductors, which can be matched with the semiconductor laser diodes and then used in laser printers. As discussed in the previous section, these lasers emit at approximately 800 nm and several materials are therefore capable of acting as organic photoconductors. Among these are polyazo compounds such as **53**.[33] Interestingly, and in common with optical data

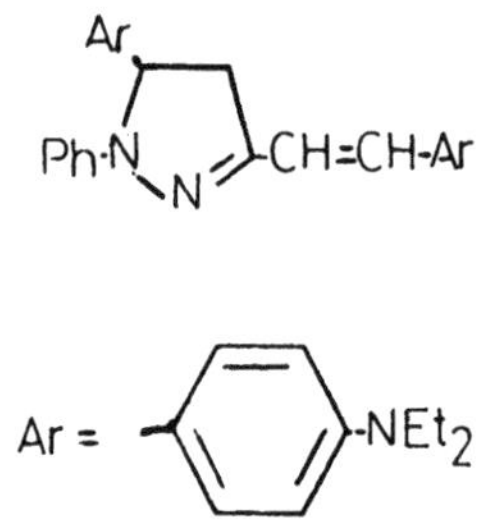

51

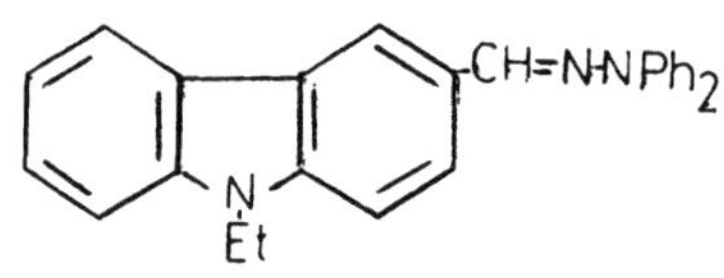

52

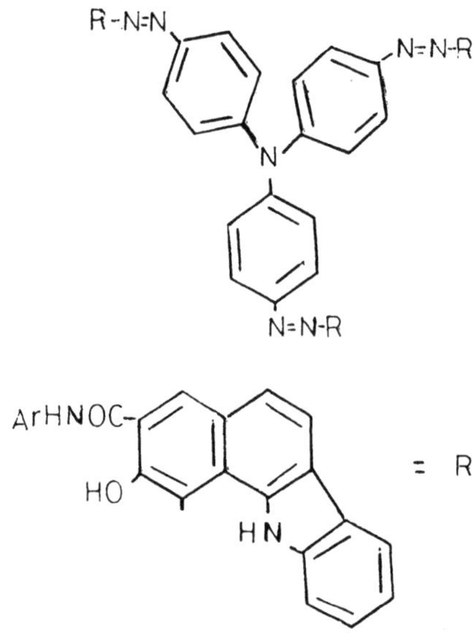

53

storage applications, phthalocyanines have been modified to act as OPCs. For instance, different crystal forms of dihydrogen phthalocyanines act as very successful photoconductors and at least one type of machine uses this material.[34] Chloroaluminum phthalocyanine is also capable of being used as an OPC material, as are various other metal-substituted phthalocyanines.[35]

From the above discussions and brief description of OPC materials, it is quite apparent that dyes are playing a new and exciting role in xerography as organic photoconductors. However, dyestuffs are not just limited to the photoconductor application but also function in another step in the xerographic process—the development stage. At the development step, the invisible electrostatic image is changed into a visible image on the photoconductor drum by a developer. The latter consists usually of a carrier (~95%) and a toner (~5%). The carrier consists of iron milled to 50–500 μ size particles. The toner accounts for 5% of the developer composition and imparts the color to the final copy. The toner particles are 10–15 μ size particles and are coated onto the larger particles of the iron carrier. Thus, when the charged xerographic plate approaches, an opposite

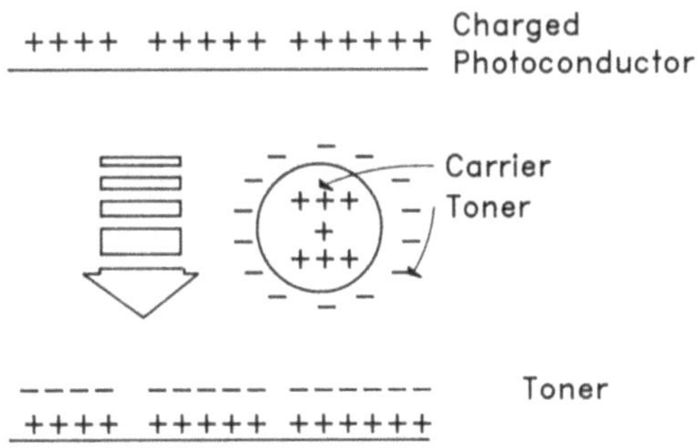

Figure 6. Development of latent image.

charge is induced in the toner which then sticks to the electrostatic image on the plate when the developer is brushed on. Figure 6 shows schematically this process for a positively charged selenium drum.

The toner consists of three components: a thermoplastic heat fusible resin (90%), a pigment (8%), and a charge control agent (2%). Amazingly, the dyes are not used as the colorant; carbon black is the usual choice here, but rather as the charge control-agent (CCA). The CCA seems to act as a sort of buffer in controlling the charge in the toner particles allowing an even, contrasty, and dense final copy. The precise mechanism of how they act seems to be unknown; however, for positively charging xerographic plates such as selenium metal, complexed azo dyes seem to be ideal. Typically, the dyes are chromium complexes of azo compounds as exemplified by compound **54**. On the other hand, for negatively charging photoconductors, nigrosines[36] appear to be useful.

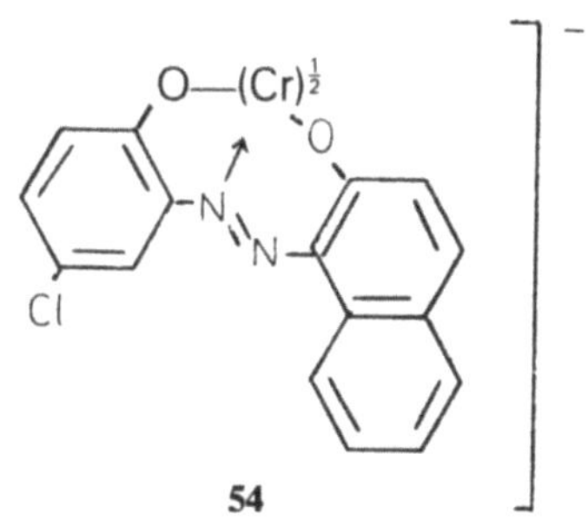

54

From the discussion above it can be seen that dyes can play a very important role in two vital steps of the xerographic process—as photoconductors and in toner formulation. The market seems to be expanding at a significant growth rate and it would therefore appear that dyestuffs have a bright future in this particular outlet, and especially if color photocopiers make an impact in the next few years, since this will require the formulation of colored toners instead of the present carbon black based toners.

VI. NONLINEAR OPTICS

In the first of the previous four sections we have seen dyestuffs used entirely for their ability to act as colorants. The three sections that followed used dyestuff molecules for other applications, where their color is irrelevant but their ability to absorb light is important. In these cases, once the light has been absorbed, it is reemitted at longer wavelength, it effects a local heating leading to pit formation, or it facilitates a change in the conductivity of the medium. Nonlinear optics is another application of dyestuffs in which their color is largely irrelevant, however, for nonlinear optics the dyestuff is not even needed to absorb the radiation.

Nonlinear optics are basically concerned with the interaction of light waves with dielectric media so that certain properties of the light, such as phase or frequency, are modified. The oscillating electric vector associated with the incident light induces an oscillating polarization in the dielectric medium which can then reirradiate at its oscillation frequency. It can therefore be seen that the nature of the reirradiated light depends in some way upon the interaction between the incident light and the medium. In a so-called linear optical medium the polarization (P) is related "linearly" to the polarizing radiation (E) as shown in equation (1), giving rise to observable effects such as refractive index.

$$P = \varepsilon X E \tag{1}$$

In a nonlinear optical medium, as the name suggests, the polarization is no longer related in a linear sense to the polarizing field as equation (2) illustrates.

$$P = \varepsilon X(1)E + \varepsilon X(2)E^2 + \varepsilon X(3)E^3 + \cdots \tag{2}$$

The difference between equations (1) and (2) is readily apparent, since extra

Table 2. Possible Applications for Nonlinear Effects

Effect	Application
Second Order	
Electrooptic (pockels)	Modulators, Q-switches, integrated optics switches and couplers
Second harmonic generation	New laser sources
Frequency mixing	Parametric amplifiers for optical repeaters, parametric oscillators, for tunable laser sources and for IR up-conversion
Third Order	
Quadratic electrooptic	Phase retarders
Degenerate four-wave mixing	Optical conjugation, real time holography, optical bistability and optical logic
Third harmonic generation	UV laser sources

terms have been introduced into the second equation and give rise to additional and useful effects (see Table 2).

Just why dyestuffs should be useful for these rather exotic applications can be seen by considering equations (2) and (3) with respect to the second term.

$$p = \alpha E + \beta E^2 + \cdots \tag{3}$$

Equation (2) shows a relationship between the macroscopic polarization and the incident electric field; a rather similar relationship holds at the molecular level as shown in equation (3). The hyperpolarizability term, β, can be calculated using molecular orbital programs[37] and it appears that molecular features which give rise to a high β are the presence of: electron donors, electron acceptors, and a conjugated pathway—the very features possessed by many colored molecules. Unfortunately, designing efficient materials is not quite as simple as making compounds with high β since, for X^2 to be large, a symmetry requirement must be satisfied. The symmetry rule requires that the molecules are assembled in a noncentrosymmetric arrangement with all the molecules pointing in approximately the same direction, otherwise cancellation occurs to give a zero X^2, whether or not the individual molecules possess a high β. Hence, an ideal material would have the following attributes:

1. High β.
2. Noncentric structure.

The first of these requirements is already satisfied by many dyes since many contain donors, acceptors, and a conjugated pathway, e.g., azo dyes, nitro dyes, and anthraquinones, etc. The remaining problem is to organize the molecules into some sort of noncentrosymmetric system. A very attractive option is to crystallize the molecules since crystals can be cut, polished, and incorporated into devices. The major problem is that only a minority of compounds have noncentrosymmetric crystal structure. Nonetheless, several colored compounds *have* been found which possess a noncentric structure and a high nonlinear value, higher than the inorganic materials used currently. In particular, nitroanilines, reminiscent of the nitro dyes, have shown themselves to be useful nonlinear optic materials. Unfortunately, 4-nitroaniline itself is of no use because it crystallizes to give a centric structure; however, simply by incorporating a methyl group at the 2-position (*ortho* to the amine) a very useful material is obtained. Other tricks have been used to assure noncentric structures, such as incorporation of chiral groups (see **55** and **56**) or hydrogen bonding effects (see **57** and **58**).

Nitroanilines are not the only colored compounds which may be used as nonlinear optical materials. Several other well-recognized dye classes are represented, such as stilbenes, azo compounds, and hemicyanine dyes, examples of which have been reported to give high second-order effects. Furthermore, their molecular efficiency would appear to be even higher than the nitroanilines described above. The stilbazolium salt **(59)** crystallizes in a noncentrosymmetric space group and shows an effect far greater than the molecules described previously. The stilbene **(60)**, azo molecule **(61)**, and the hemicyanine **(62)** all

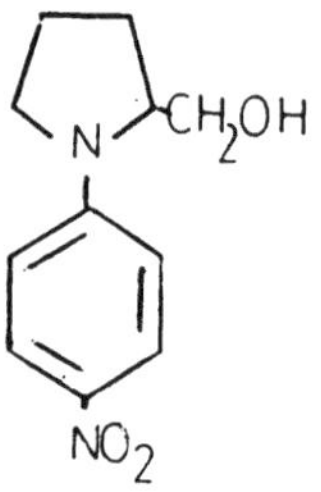

55

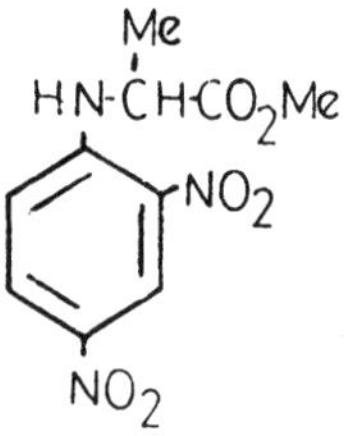

56

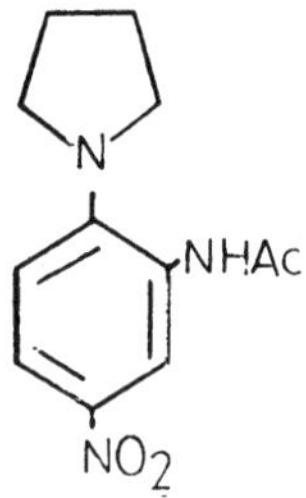

57

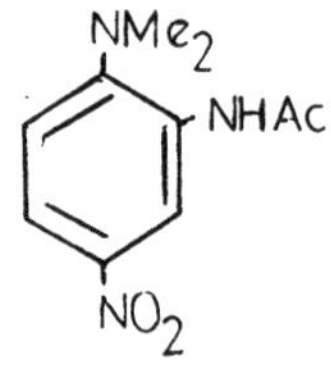

58

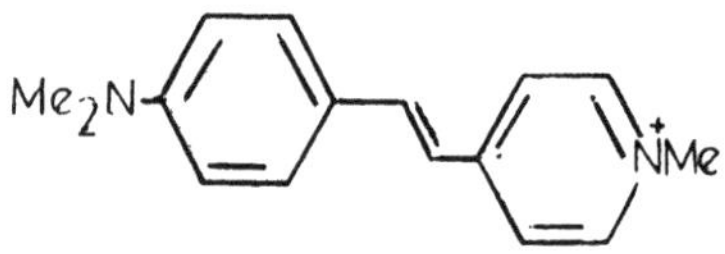

59

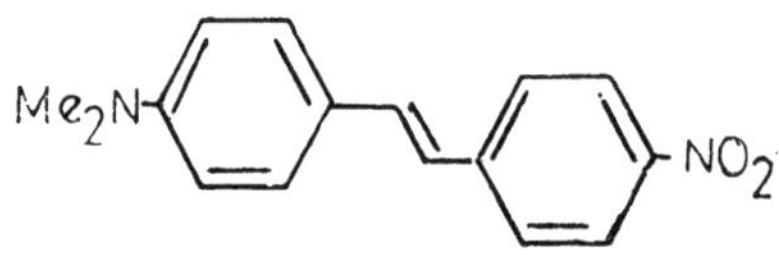

60

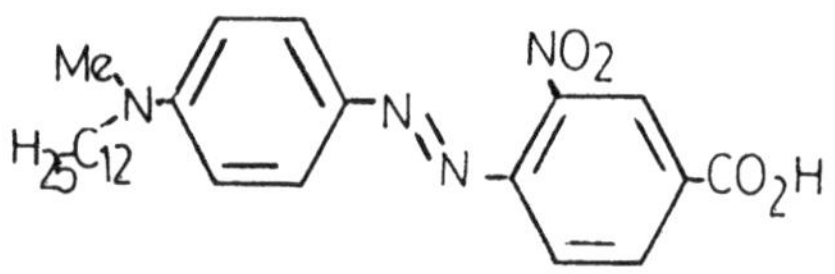

61

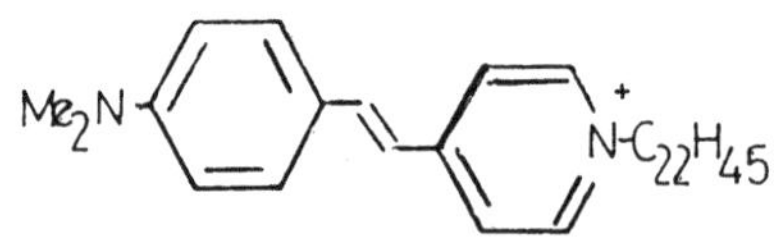

62

show useful effects, but none has been used as a simple crystal since alternative methods to attain a noncentrosymmetric structure have been utilized. For instance, the stilbene **(60)** has been added to a liquid crystal mixture and then aligned with an electric field to gain a suitable noncentrosymmetric arrangement.[38] Alternatively, by incorporating hydrophobic and hydrophilic groups the azo compound **(61)**[39] and hemicyanine **(62)**[40] have been successfully modified to permit their assembly into multilayers, one molecular layer at a time being deposited by the Langmuir–Blodgett technique,[41] to give highly efficient, nonlinear optical systems.

Thus, it can be seen that several dye classes are already represented in the list of useful nonlinear optical materials and an interesting correlation is already emerging in that the more highly colored the molecule, the more efficient it appears to be as a nonlinear material. It would therefore seem that dyestuffs have a lot to offer this area of science, with the best still to come. It is also highly likely that the knowledge appropriate to designing colored molecules gained over the last century should prove most useful in designing new organic nonlinear materials.

VII. CONCLUSION

As the previous five sections indicate, dyestuffs research is far from dead and is, in fact, blossoming, partly through its role in several of the new technology industries. Although the topics in this chapter are covered at a fairly superficial level and this list of new uses for dyes is not comprehensive, it is apparent that many of the new applications require the dyestuffs chemist to use his experience in designing dyes which absorb at specific wavelengths. However, on top of this is the requirement that the dyes fulfil some extra criteria whether it be lasing, pit formation for optical disks, or photoconductivity. Nevertheless, the dyestuffs chemist is well able to meet these requirement since many of these phenomena make use of properties known to and studied by him/her. Even in nonlinear optics where the absorption of light is not required, the dyestuffs chemist seems to be able to make a significant contribution since the very factors used to push dyes more bathochromically, e.g., increasing conjugation and electron donor/acceptor strengths, appear to result in materials with higher molecular nonlinearities.

It is likely that new applications for colored molecules will arise in the future, whether it be for their unique electronic properties or genuinely for their color response. Additionally, the molecules designed for optical data storage, laser dyes, and organic photoconductors may have a spin-off value in other areas. Dyestuffs research would therefore seem to have a bright future and nontextile uses of dyes should increase even further. Perhaps in another century from now nontextile uses of dyes will dominate and textile applications will be seen as a minor, though no doubt important, application for dyes!

REFERENCES

1. P. F. Gordon and P. Gregory, in: *Developments in the Chemistry and Technology of Organic Dyes* (J. Griffiths, ed.), pp. 66–110, Blackwell, Oxford (1984).
2. D. B. Dupre, in: *Encyclopedia of Chemical Technology* (Kirk–Othmer, eds), pp. 395–427, Wiley, New York (1981).
3. a. G. W. Gray, *Molecular Structure and Properties of Liquid Crystals,* Academic Press, London (1979). b. G. W. Gray, *Chimia 34,* 47 (1980) and references cited therein. c. 3M, British Patent 2090274 (1982). d. Bayer A. G., European Patent 55838B (1984).
4. a. ICI, U.S. Patent 4391754 (1983). b. ICI, British Patent 2106125 (1983). c. Nippon Kayaku, Japanese Patent, 59026293A (1984). d. Hoffman–La-Roche, European Patent 110205A (1984).
5. BASF, German Patent DE 3148206 (1983).
6. D. J. Barclay and D. H. Martin, in: *Technology of Chemicals and Materials for Electronics* (E. R. Howells, ed.), pp. 226–277, Ellis Horwood, Chichester (1984) and references cited therein.
7. F. P. Schafer, *Dye Lasers,* Springer-Verlag, Berlin (1977).
8. *Techniques and Materials for Optical Data Storage,* Society of Chemical Industry, London (October 1985).
9. P. F. Gordon and P. Gregory, *Organic Chemistry in Colour,* Springer-Verlag, Berlin (1983).
10. TDK, Japanese Patent 59024690A (1984).
11. J. Fabian and H. Hartmann, *Light Absorption of Organic Colourants,* Springer-Verlag, Berlin (1980).
12. D. J. Grauesteijn and J. van der Veen, *Society of Photo-optical and Instrumentation Engineers, 420,* 327 (1983).
13. H. E. Sprenger and W. Ziegenbein, *Angew. Chem., Int. Ed. Engl. 6,* 552 (1967).
14. Canon, Japanese Patent Appl. 58-173696 (1983).
15. J. Griffiths, *Colour and Constitution of Organic Molecules,* Academic Press, London (1976).
16. Nippon Kayaku, Japanese Patent Appl. 60-95891 (1983).

17. M. Matsuoka, S. H. Kim and T. Kitao, *J. Chem. Soc., Chem. Commun.*, 1195 (1985).
18 a. T. Kitao and M. Matsuoka, *J. Soc. Dyers Colour. 101,* 140 (1985). b. NEC, European Patent 97929A (1984).
19. a. Xerox, U.S. Patent 4492750-A (1985). b. Yamamoto Kagaku, European Patent, 134518A (1985). c. Nippon Kayaku, Japanese Patent J60043605A (1985). d. S. A. Mikhalenko and E. A. Luk'yanets, *Zh. Obshch. Khim. 39,* 2554 (1969).
20. P. Kivits, R. de Bont, and J. van der Veen, *Appl. Phys. A26,* 101 (1981).
21. Ricoh, Japanese Patent 57082093 (1982).
22. Imperial Chemical Industries, European Patent 186404-A (1986).
23. a. Hitachi, European Patent 0093331-A (1983). b. Xerox, U.S. Patent 4098795 (1978).
24. J. A. McCleverty, *Prog. Inorg. Chem. 10,* 49 (1969).
25. Polaroid, U.S. Patent 3663089 (1972).
26. Mitsui Toatsu, Japanese Pat. Appl. 85 023996 (1985).
27. C. M. J. van Uijen, *Proc. SPIE—Int. Soc. Opt. Eng. 529,* 2 (1985).
28. J. H. Dessauer and H. E. Clark, *Xerography and Related Processes,* Focal Press, London (1965).
29. Eastman Kodak, U.S. Patent 3141770 (1971).
30. a. IBM, U.S. Patent 3898084 (1975). b. Mitsubishi, Japanese Patent 11751 (1980). c. Ricoh, U.S. Patents 4274598 (1981), 4314015 (1982), and 4507471 (1985).
31. Kalle, U.S. Patent 3871882 (1975).
32. Xerox, U.S. Patent 3977935 (1976).
33. Ricoh, U.S. Patent 3838084 (1975).
34. a. Xerox, U.S. Patent 4098795 (1978). b. Hitachi, U.S. Patent 4507374 (1985).
35. N.T.T., U.S. Patent 4426434 (1984).
36. H. A. Lubs, *The Chemistry of Synthetic Dyes and Pigments,* Reinhold, New York (1955).
37. a. V. J. Docherty, D. Pugh, and J. O. Morley, *J. Chem. Soc., Faraday Trans. 2 81,* 1179 (1985). b. S. J. Lalamar and A. F. Garito, *Phys. Rev. A 20,* 1179 (1979).
38. D. J. Williams, *Angew. Chem., Int. Ed. Engl. 23,* 690 (1984).
39. R. A. Hann, P. F. Gordon, S. K. Gupta, S. Allen, I. Ledoux, P. Vidakovic, J. Zyss, P. Robin, and J. C. Dubois, *Europhys. Lett.*, in press.
40. I. R. Girling, N. A. Cade, P. V. Kolinsky, J. O. Earls, G. H. Cross, and I. R. Peterson, *Thin Solid Films 132,* 101 (1985).
41. G. G. Roberts, *Adv. in Phys. 34* (4), (1985).

Index of Dyes

This is an index of Colour Index dyes discussed in the book.

Subject Index

MIX
Papier aus verantwortungsvollen Quellen
Paper from responsible sources
FSC® C105338

If you have any concerns about our products,
you can contact us on
ProductSafety@springernature.com

In case Publisher is established outside the EU,
the EU authorized representative is:
Springer Nature Customer Service Center GmbH
Europaplatz 3, 69115 Heidelberg, Germany

Printed by Libri Plureos GmbH
in Hamburg, Germany